Fractional Integrals and Derivatives: "True" versus "False"

Fractional Integrals and Derivatives: "True" versus "False"

Editor

Yuri Luchko

MDPI • Basel • Beijing • Wuhan • Barcelona • Belgrade • Manchester • Tokyo • Cluj • Tianjin

Editor
Yuri Luchko
Beuth Technical University of
Applied Sciences Berlin
Germany

Editorial Office
MDPI
St. Alban-Anlage 66
4052 Basel, Switzerland

This is a reprint of articles from the Special Issue published online in the open access journal *Mathematics* (ISSN 2227-7390) (available at: https://www.mdpi.com/journal/mathematics/special_issues/Fractional_Integrals_Derivatives).

For citation purposes, cite each article independently as indicated on the article page online and as indicated below:

LastName, A.A.; LastName, B.B.; LastName, C.C. Article Title. *Journal Name* **Year**, *Volume Number*, Page Range.

ISBN 978-3-0365-0495-7 (Hbk)
ISBN 978-3-0365-0494-0 (PDF)

© 2021 by the authors. Articles in this book are Open Access and distributed under the Creative Commons Attribution (CC BY) license, which allows users to download, copy and build upon published articles, as long as the author and publisher are properly credited, which ensures maximum dissemination and a wider impact of our publications.

The book as a whole is distributed by MDPI under the terms and conditions of the Creative Commons license CC BY-NC-ND.

Contents

About the Editor vii

Preface to "Fractional Integrals and Derivatives: "True" versus "False"" ix

Yuri Luchko and Masahiro Yamamoto
The General Fractional Derivative and Related Fractional Differential Equations
Reprinted from: *Mathematics* **2020**, *8*, 2115, doi:10.3390/math8122115 1

Virginia Kiryakova
Unified Approach to Fractional Calculus Images of Special Functions—A Survey
Reprinted from: *Mathematics* **2020**, *8*, 2260, doi:10.3390/math8122260 21

Min Cai and Changpin Li
Numerical Approaches to Fractional Integrals and Derivatives: A Review
Reprinted from: *Mathematics* **2020**, *8*, 43, doi:10.3390/math8010043 57

Rudolf Hilfer and Yuri Luchko
Desiderata for Fractional Derivatives and Integrals
Reprinted from: *Mathematics* **2019**, *7*, 149, doi:10.3390/math7020149 111

R. Hilfer and T. Kleiner
Maximal Domains for Fractional Derivatives and Integrals
Reprinted from: *Mathematics* **2020**, *8*, 1107, doi:10.3390/math8071107 117

Manuel Ortigueira and José Machado
Fractional Derivatives: The Perspective of System Theory
Reprinted from: *Mathematics* **2019**, *7*, 150, doi:10.3390/math7020150 121

Vasily E. Tarasov and Svetlana S. Tarasova
Fractional Derivatives and Integrals: What Are They Needed For?
Reprinted from: *Mathematics* **2020**, *8*, 164, doi:10.3390/math8020164 135

Kai Diethelm, Roberto Garrappa and Martin Stynes
Good (and Not So Good) Practices in Computational Methods for Fractional Calculus
Reprinted from: *Mathematics* **2020**, *8*, 324, doi:10.3390/math8030324 157

Dumitru Baleanu and Arran Fernandez
On Fractional Operators and Their Classifications
Reprinted from: *Mathematics* **2019**, *7*, 830, doi:10.3390/math7090830 179

Jocelyn Sabatier, Christophe Farges and Vincent Tartaglione
Some Alternative Solutions to Fractional Models for Modelling Power Law Type Long Memory Behaviours
Reprinted from: *Mathematics* **2020**, *8*, 196, doi:10.3390/math8020196 189

Daniel Cao Labora
Fractional Integral Equations Tell Us How to Impose Initial Values in Fractional Differential Equations
Reprinted from: *Mathematics* **2020**, *8*, 1093, doi:10.3390/math8071093 205

Eyaya Fekadie Anley and Zhoushun Zheng
Finite Difference Method for Two-Sided Two Dimensional Space Fractional Convection-Diffusion Problem with Source Term
Reprinted from: *Mathematics* **2020**, *8*, 1878, doi:10.3390/math8111878 223

Christopher N. Angstmann, Byron A. Jacobs, Bruce I. Henry, and Zhuang Xu
Intrinsic Discontinuities in Solutions of Evolution Equations Involving Fractional Caputo–Fabrizio and Atangana–Baleanu Operators
Reprinted from: *Mathematics* **2020**, *8*, 2023, doi:10.3390/math8112023 251

About the Editor

Yuri Luchko is a Full Professor at the Faculty of Mathematics - Physics - Chemistry of the Technical University of Applied Sciences Berlin. He studied Mathematics at the Belarussian State University in Minsk and received his PhD degree from this University. In 1994, he was awarded a DAAD research grant at the Free University of Berlin. The main field of his research is Applied Mathematics with a special focus on Fractional Calculus and its applications. Yuri Luchko published more than 100 papers in the international peer-reviewed scientific journals and about 20 books and books chapters as author or editor. He is associate editor of the international journal "Fractional Calculus and Applied Analysis" and editor of a dozen of other reputable mathematical journals.

Preface to "Fractional Integrals and Derivatives: "True" versus "False""

Even if the Fractional Calculus (FC) is nearly as old as the conventional calculus, for long time it was addressed and used just sporadically and only by few scientists. Within the last few decades, the situation changed dramatically and nowadays we observe an exponential growth of FC publications, conferences, and scientists involved into this topic. One of the explanations for this phenomenon is in active attempts to introduce a new kind of mathematical models containing fractional order operators into physics, chemistry, engineering, biology, medicine, and other sciences. This speeds up the development of the mathematical theory of FC, including fractional ordinary and partial differential equations, fractional calculus of variations, inverse problems for the fractional differential equations, fractional stochastic models, etc. Unfortunately, most of these new models and results are just formal "fractionalisations" of the known conventional theories, often without any justification and motivation.

Additionally, a recent trend in FC is in introducing "new fractional derivatives and integrals" and considering classical equations and models with these fractional order operators in place of the conventional integrals and derivatives. This development led to an uncontrolled flood of FC publications both in mathematical and physical journals. Some of these publications contain trivial, well-known, and sometimes even wrong results that threaten the image of FC in the scientific community. Thus, the FC researches have to think about and to answer questions like "What are the fractional integrals and derivatives?", "What are their decisive mathematical properties?", "What fractional operators make sense in applications and why?", etc. These and similar questions were mostly unanswered until now and the main aim of this Special Issue is a contribution to resolving of some of these questions.

The Special Issue opens with three surveys. The review article [1], by Yuri Luchko and Masahiro Yamamoto, presents an in deep discussion of the general fractional derivatives and integrals as well as the fractional ordinary and partial differential equations with the general fractional derivatives. For the fractional partial differential equations with the general fractional derivatives, both direct and inverse problems are discussed. The survey paper [2], by Virginia Kiryakova, addresses another important topic in FC: a unified approach to evaluation of images of the special functions under action of different FC operators. The general scheme proposed in the paper is based on a few classical results combined with ideas and developments from more than 30 years of author's research. The review article [3], by Min Cai and Changpin Li, focuses on numerical approximations to fractional integrals and derivatives. Almost all relevant results known up to now are presented, discussed, and compared each to other.

The research part of the Special Issue contains a series of important contributions that provide some partial answers to the questions already mentioned above and others as e.g., "What are the FC operators needed for?". This series starts with the article [4], by Rudolf Hilfer and Yuri Luchko, which proposes desiderata for calling an operator a fractional derivative or a fractional integral. The desiderata are based on a small number of time honored and well established criteria. However, they are not axioms and do not define fractional derivatives or integrals uniquely. The short communication [5], by Rudolf Hilfer and Tillmann Kleiner, announces existence of fractional calculi on precisely specified domains of distributions that satisfy the desiderata proposed in [4]. The contribution [6], by Manuel Duarte Ortigueira and José Tenreiro Machado, aims to provide the answers to the same questions as the ones considered in [4]. However, by doing so, the authors act from the perspective of the classical system theory. The article [7], by Vasily E. Tarasov and Svetlana S. Tarasova, addresses usefulness of the FC operators from the viewpoint of applied mathematics. The key idea of the authors is an attempt to introduce a correspondence between the kernel properties of the FC operators and the types of the phenomena that they can adequately describe.

The next three contributions of the Special Issue are devoted to some important practical aspects of FC. The article [8], by Kai Diethelm, Roberto Garrappa, and Martin Stynes, addresses the numerical treatment of the fractional differential equations. In particular, the authors provide a description of some common pitfalls in the use of numerical methods in fractional calculus, explain their nature, and list some good practices. In the article [9], by Dumitru Baleanu and Arran Fernandez, a classification idea for the FC integrals and derivatives into distinct classes of operators is discussed. The contribution [10], by Jocelyn Sabatier, Christophe Farges, and Vincent Tartaglione, considers modeling of the processes exhibiting a power law long memory behavior and discusses the alternatives to the models in form of the fractional differential equations.

The final part of the Special Issue includes a series of three original contributions. The article [11], by Daniel Cao Labora, addresses the role of the initial values in the fractional differential equations, their form, and the amount of necessary initial values and the orders of differentiability where these conditions need to be imposed. In the contribution [12], by Eyaya Fekadie Anley and Zhoushun Zheng, a numerical difference approximation for solving a two-dimensional space-fractional convection-diffusion equation with a source term is suggested. Finally, the article [13], by Christopher Nicholas Angstmann, Byron Alexander Jacobs, Bruce Ian Henry, and Zhuang Xu, raises concerns about using some of the "new fractional derivatives" with the non-singular kernels in modelling because the solutions to the fractional differential equations with these derivatives have an intrinsic discontinuity at the origin.

Even if the articles collected in this Special Issue provide a certain contribution to resolving of some current problems related to FC and its applications, the discussions regarding "true" and "false" fractional integrals and derivatives are not yet completed. On the contrary, they are nowadays more urgent than ever before. Thus, let us continue to debate about the questions like "What are the fractional integrals and derivatives?", "What are their decisive mathematical properties?", "What fractional operators make sense in applications and why?", etc. There are still many open problems both in the foundations of FC and in its usability for applications.

References

1. The General Fractional Derivative and Related Fractional Differential Equations by Yuri Luchko and Masahiro Yamamoto, Mathematics 2020, 8(12), 2115; https://doi.org/10.3390/math8122115.
2. Unified Approach to Fractional Calculus Images of Special Functions—A Survey by Virginia Kiryakova, Mathematics 2020, 8(12), 2260; https://doi.org/10.3390/math8122260.
3. Numerical Approaches to Fractional Integrals and Derivatives: A Review by Min Cai and Changpin Li, Mathematics 2020, 8(1), 43; https://doi.org/10.3390/math8010043.
4. Desiderata for Fractional Derivatives and Integrals by Rudolf Hilfer and Yuri Luchko, Mathematics 2019, 7(2), 149; https://doi.org/10.3390/math7020149.
5. Maximal Domains for Fractional Derivatives and Integrals by R. Hilfer and T. Kleiner, Mathematics 2020, 8(7), 1107; https://doi.org/10.3390/math8071107.
6. Fractional Derivatives: The Perspective of System Theory by Manuel Duarte Ortigueira and José Tenreiro Machado, Mathematics 2019, 7(2), 150; https://doi.org/10.3390/math7020150.

7. Fractional Derivatives and Integrals: What Are They Needed For? by Vasily E. Tarasov and Svetlana S. Tarasova, Mathematics 2020, 8(2), 164; https://doi.org/10.3390/math8020164.

8. Good (and Not So Good) Practices in Computational Methods for Fractional Calculus by Kai Diethelm, Roberto Garrappa and Martin Stynes, Mathematics 2020, 8(3), 324; https://doi.org/10.3390/math8030324.

9. On Fractional Operators and Their Classifications by Dumitru Baleanu and Arran Fernandez, Mathematics 2019, 7(9), 830; https://doi.org/10.3390/math7090830.

10. Some Alternative Solutions to Fractional Models for Modelling Power Law Type Long Memory Behaviours by Jocelyn Sabatier, Christophe Farges and Vincent Tartaglione, Mathematics 2020, 8(2), 196; https://doi.org/10.3390/math8020196.

11. Fractional Integral Equations Tell Us How to Impose Initial Values in Fractional Differential Equations by Daniel Cao Labora, Mathematics 2020, 8(7), 1093; https://doi.org/10.3390/math8071093.

12. Finite Difference Method for Two-Sided Two Dimensional Space Fractional Convection-Diffusion Problem with Source Term by Eyaya Fekadie Anley and Zhoushun Zheng, Mathematics 2020, 8(11), 1878; https://doi.org/10.3390/math8111878.

13. Intrinsic Discontinuities in Solutions of Evolution Equations Involving Fractional Caputo–Fabrizio and Atangana–Baleanu Operators by Christopher Nicholas Angstmann, Byron Alexander Jacobs, Bruce Ian Henry and Zhuang Xu, Mathematics 2020, 8(11), 2023; https://doi.org/10.3390/math8112023.

Yuri Luchko
Editor

Review

The General Fractional Derivative and Related Fractional Differential Equations

Yuri Luchko [1,*] and Masahiro Yamamoto [2]

1. Department of Mathematics, Physics, and Chemistry, Beuth Technical University of Applied Sciences Berlin, Luxemburger Str. 10, 13353 Berlin, Germany
2. Graduate School of Mathematical Sciences, The University of Tokyo, 3-8-1 Komaba, Meguro-ku, Tokyo 153-8914, Japan; myama@next.odn.ne.jp
* Correspondence: luchko@beuth-hochschule.de

Received: 29 September 2020; Accepted: 23 November 2020; Published: 26 November 2020

Abstract: In this survey paper, we start with a discussion of the general fractional derivative (GFD) introduced by A. Kochubei in his recent publications. In particular, a connection of this derivative to the corresponding fractional integral and the Sonine relation for their kernels are presented. Then we consider some fractional ordinary differential equations (ODEs) with the GFD including the relaxation equation and the growth equation. The main part of the paper is devoted to the fractional partial differential equations (PDEs) with the GFD. We discuss both the Cauchy problems and the initial-boundary-value problems for the time-fractional diffusion equations with the GFD. In the final part of the paper, some results regarding the inverse problems for the differential equations with the GFD are presented.

Keywords: general fractional derivative; general fractional integral; Sonine condition; fractional relaxation equation; fractional diffusion equation; Cauchy problem; initial-boundary-value problem; inverse problem

MSC: 26A33; 35A05; 35B30; 35B50; 35C05; 35E05; 35L05; 35R30; 45K05; 60E99

---

1. Introduction

In functional analysis, the integral operators with the weakly singular kernels have been an important topic for research for many years. They are defined in the form

$$(Tf)(t) = \int_\Omega K(t,\tau) f(\tau) d\tau, \qquad (1)$$

where Ω is an open subset of $\mathbb{R}^n$ and the kernel $K = K(t,\tau)$ is a real or complex valued continuous function on $\overline{\Omega} \times \overline{\Omega} \setminus D$, $D = \{(t,t) : t \in \overline{\Omega}\}$ being the diagonal of $\overline{\Omega} \times \overline{\Omega}$ that satisfies the condition

$$|K(t,\tau)| \le \frac{c}{|t-\tau|^\alpha} \text{ with } \alpha < n. \qquad (2)$$

In case Ω is a bounded domain, the operator (1) is called the Schur integral operator. It is compact from $C(\overline{\Omega})$ to $C(\overline{\Omega})$ [1]. If, additionally, $\alpha < \frac{n}{q}$, $\frac{1}{p} + \frac{1}{q} = 1$, $1 \le p < +\infty$, then (1) is a Hilbert-Schmidt operator that is compact from $L^p(\Omega)$ to $C(\overline{\Omega})$.

In Fractional Calculus (FC), the operators of type (1) with special weakly singular kernels are studied on both bounded and unbounded domains. For instance, the classical left-hand sided Riemann-Liouville fractional integral of order $\alpha \in \mathbb{R}_+$ of a function f on a finite or infinite interval (a, b) is defined as follows:

$$(I_{a+}^\alpha f)(t) = \frac{1}{\Gamma(\alpha)} \int_a^t (t-\tau)^{\alpha-1} f(\tau)\, d\tau, \quad t \in (a,b). \tag{3}$$

In the case $\alpha = 0$, this integral is interpreted as the identity operator

$$\left(I_{a+}^0 f\right)(t) = f(t) \tag{4}$$

because of the relation ([2])

$$\lim_{\alpha \to 0+} (I_{a+}^\alpha f)(t) = f(t), \tag{5}$$

that is valid in particular for $f \in L^1(a,b)$ in every Lebesgue point of f, i.e., almost everywhere on (a,b).

Evidently, the Riemann-Liouville fractional integral is a generalization of the well-known formula for the n-fold definite integral

$$(I_{a+}^n f)(t) = \int_a^t d\tau \int_a^t d\tau \cdots \int_a^t f(\tau)\, d\tau = \frac{1}{(n-1)!} \int_a^t (t-\tau)^{n-1} f(\tau)\, d\tau, \; n \in \mathbb{N}.$$

In a certain sense, this generalization is unique. In the case of a finite interval (without any restriction of generality, we fix the interval $(0,1)$), the following result was proved in [3]:

Let E be the space $L^p(0,1)$, $1 \le p < +\infty$, or $C[0,1]$. Then there exists precisely one family I_α, $\alpha > 0$ of operators on E satisfying the following conditions:

(CM1) $(I_1 f)(t) = \int_0^t f(\tau)\, d\tau$, $f \in E$ (interpolation condition),
(CM2) $(I_\alpha I_\beta f)(t) = (I_{\alpha+\beta} f)(t)$, $\alpha, \beta > 0$, $f \in E$ (index law),
(CM3) $\alpha \to I_\alpha$ is a continuous map of $\mathbb{R}_+$ into the space $\mathcal{L}(E)$ of the linear bounded operators from E to E for some Hausdorff topology on $\mathcal{L}(E)$, weaker than the norm topology (continuity),
(CM4) $f \in E$ and $f(t) \ge 0$ (a.e. for $E = L^p(0,1)$) $\Rightarrow$ $(I_\alpha f)(t) \ge 0$ (a.e. for $E = L^p(0,1)$) for all $\alpha > 0$ (non-negativity).

That family is given by the Riemann-Liouville Formula (3) with $a = 0$ and $b = 1$. From the present viewpoint ([4]), the conditions (CM1)–(CM4) are very natural for any definition of the fractional integrals defined on a finite interval. As proved in [3], they are also sufficient for uniqueness of the family of the Riemann-Liouville fractional integrals. Thus, in this sense, the Riemann-Liouville fractional integrals are the only "right" one-parameter fractional integrals defined on a finite one-dimensional interval.

The problem regarding the "right" fractional derivatives is more delicate and has no unique solution. Presently, the main approach for introducing the fractional derivatives is to define them as the left-inverse operators to the fractional integrals ([4–6]). However, even for the Riemann-Liouville fractional integral, there exist infinitely many different families of operators that fulfill this property ([6]). In particular, for $0 < \alpha \le 1$, the Riemann-Liouville fractional derivative

$$(D_{RL}^\alpha f)(t) = \frac{d}{dt}(I_{0+}^{1-\alpha} f)(t), \tag{6}$$

the Caputo fractional derivative

$$(D_C^\alpha f)(t) = (I_{0+}^{1-\alpha} \frac{df}{d\tau})(t), \tag{7}$$

and the Hilfer fractional derivative

$$(D_H^\alpha f)(t) = (I_{0+}^{\beta(1-\alpha)} \frac{d}{d\tau} I_{0+}^{(1-\alpha)(1-\beta)} f)(t), \; 0 \le \beta \le 1 \tag{8}$$

are the left-inverse operators to the Riemann-Liouville fractional integral on the suitable nontrivial spaces of functions including the space of the absolutely continuous functions on $[0,1]$, i.e., the Fundamental Theorem of FC holds true ([6]):

$$(D_X^\alpha I_{0+}^\alpha f)(t) = f(t), \ t \in [0, 1], \ X \in \{RL, C, H\}. \tag{9}$$

Moreover, in [6], infinitely many other families of the fractional derivatives in the sense of Formula (9) called the nth level fractional derivatives were introduced. Let the parameters $\gamma_1, \gamma_2, \ldots, \gamma_n \in \mathbb{R}$ satisfy the conditions

$$0 \leq \gamma_k \text{ and } \alpha + s_k \leq k \text{ with } s_k := \sum_{i=1}^{k} \gamma_i, \ k = 1, 2, \ldots, n. \tag{10}$$

The nth level fractional derivative of order α, $0 < \alpha \leq 1$ and type $\gamma = (\gamma_1, \gamma_2, \ldots, \gamma_n)$ is defined as follows:

$$(D_{nL}^{\alpha,(\gamma)} f)(t) = \left(\prod_{k=1}^{n} (I_{0+}^{\gamma_k} \frac{d}{d\tau}) \right) (I_{0+}^{n-\alpha-s_n} f)(t). \tag{11}$$

This derivative satisfies the Fundamental Theorem of FC, i.e., the relation

$$(D_{nL}^{\alpha,(\gamma)} I_{0+}^\alpha f)(t) = f(t), \ t \in [0, 1] \tag{12}$$

holds true on a nontrivial space of functions (see [6] for details).

To keep an overview of these and many other fractional derivatives, it is very natural to consider some general integro-differential operators of convolution type and to clarify the question under what conditions can they be interpreted as a kind of the fractional derivatives. In particular, one expects that for these derivatives and the appropriate defined fractional integrals the Fundamental Theorem of FC holds true. Moreover, for the sake of possible applications, one wants to keep some fundamental properties of solutions to the differential equations with these derivatives. In particular, the property of complete monotonicity of solutions to the appropriate relaxation equation or the positivity of the fundamental solution to the Cauchy problem for the fractional diffusion equation with the time-derivatives of this type.

In the theory of the abstract Volterra integral equations in the Banach spaces, the evolution equations including the integro-differential operators of convolution type

$$(D_k u)(t) = \frac{d}{dt} \int_0^t k(t - \tau) u(\tau) \, d\tau, \ t \in [0, T], \ 0 < T \leq +\infty \tag{13}$$

have been a subject for research for more than a half century. In particular, in [7] (see also the references therein), the abstract Volterra integral equations including the operators (13) with the completely positive kernels $k \in L^1(0, T)$ have been studied. This class of the kernels can be characterized as follows: A function $k \in L^1(0, T)$ is completely positive on $[0, T]$ if and only if there exist $a \geq 0$ and $l \in L^1(0, T)$, non-negative and non-increasing, satisfying the relation

$$a\, k(t) + \int_0^t k(t - \tau) l(\tau) \, d\tau = 1, \ t \in (0, T]. \tag{14}$$

In particular, the completely monotone kernels are completely positive. The notion of completely positive kernels originated from the "positivity preserving property" that is valid for the corresponding Volterra integral equations in the case of the Banach space $X = \mathbb{R}$ with the usual norm.

In [8,9], the properties of the appropriate defined weak solutions to the linear and quasi-linear evolutionary partial integro-differential equations of second order with the time-operators of type (13) in the form

$$(D_k u)(t) = \frac{d}{dt} \int_0^t k(t - \tau)(u(\tau) - u_0) \, d\tau, \ t \in \mathbb{R}_+ \tag{15}$$

were addressed in the case of the kernels $k \in L_{loc}^1(\mathbb{R}_+)$ that satisfy the following conditions:

(Z1) The kernel k is non-negative and non-increasing on $\mathbb{R}_+$,
(Z2) There exists a kernel $l \in L_{loc}(\mathbb{R}_+)$ such that $\int_0^t k(t-\tau)l(\tau)\,d\tau = 1$, $t \in \mathbb{R}_+$.

An important example of a kernel k that satisfies the conditions (Z1)–(Z2) is the following generalization of the Riemann-Liouville kernel ([8]):

$$k(t) = h_{1-\alpha}(t)\exp(-\mu t), \mu \geq 0,\ 0 < \alpha < 1, \tag{16}$$

where

$$h_\beta(t) = \frac{t^{\beta-1}}{\Gamma(\beta)},\ t > 0,\ \beta > 0. \tag{17}$$

For this kernel k, the kernel l from the property (Z2) takes the form

$$l(t) = h_\alpha(t)\exp(-\mu t) + \mu \int_0^t h_\alpha(\tau)\exp(-\mu\tau)\,d\tau,\ t > 0. \tag{18}$$

However, in [8,9] and in earlier publications, the operators of type (13) were not interpreted as a kind of the generalized fractional derivatives. In particular, no construction of the corresponding fractional integral was presented and no conditions that ensure the physically relevant properties of solutions to the time-fractional differential equations including these derivatives were suggested. Both tasks along with a series of other useful properties were addresses in [10–13], where a very nice theory of the general fractional derivative of type (13) was developed and applied for studying properties of the ordinary and partial differential equations with this derivative. In this survey, we present some selected results obtained in these and other related publications.

The rest of the paper is organized as follows. In Section 2, we introduce the GFD and the related fractional integral and discuss some of their basic properties. Section 3 is devoted to the Cauchy problems for the fractional ODEs with the GFD. In particular, the fractional relaxation equation and properties of its solution and the fractional growth equation and long time asymptotic of its solution are considered [10,11,13]. Moreover, existence and uniqueness of solutions to the Cauchy problem for the nonlinear fractional ODEs with the GFD and their continuous dependence on the problem data are also addressed following [14]. In Section 4, we present some results regarding the fractional PDEs with the GFD. We start with the Cauchy problem for the linear fractional diffusion equation and address its well-posedness with the focus on an interpretation of its fundamental solution as a probability density function [10,11]. Then we proceed with a treatment of the initial-boundary-value problems for the time-fractional diffusion equation including the GFD. Based on a suitable estimate for the GFD of a function at its maximum point, a weak maximum principle for the general time-fractional diffusion equation with the GFD is deduced. Then, following [15], the maximum principle is employed to show uniqueness of the strong and the weak solutions to the initial-boundary-value problems for the general time-fractional diffusion equations. Existence of a weak solution in the sense of Vladimirov [16] is also discussed. Finally, some important results from the recent publications [17–19] regarding inverse problems for the fractional differential and integral equations with the GFD are shortly presented.

2. General Fractional Derivative and Integral

Following [10], in this paper we consider the GFD of the Riemann-Liouville type in the form (compare to (13))

$$(\mathbb{D}_k^{RL} f)(t) = \frac{d}{dt}\int_0^t k(t-\tau)f(\tau)\,d\tau \tag{19}$$

and of the Caputo type in the form

$$(\mathbb{D}_k^C f)(t) = \int_0^t k(t-\tau)f'(\tau)\,d\tau, \tag{20}$$

where k is a non-negative locally integrable function. For an absolutely continuous function f satisfying $f' \in L^1_{loc}(\mathbb{R}_+)$, the relation (compare to (15))

$$(\mathbb{D}^C_k f)(t) = (\mathbb{D}^{RL}_k (f - f(0)))(t) = (\mathbb{D}^{RL}_k f)(t) - k(t)f(0) = \frac{d}{dt}\int_0^t k(t-\tau)f(\tau)\,d\tau - k(t)f(0) \quad (21)$$

between the Caputo and Riemann-Liouville types of GFDs holds true. In [10], the Caputo type GFD was introduced in form (21) that is well defined for a lager class of functions (in particular, for absolutely continuous functions) compared to the definition (20) that requires the inclusion $f' \in L^1_{loc}(\mathbb{R}_+)$. In what follows, we mainly address the GFD of the Caputo type in the sense of the right-hand side of Formula (21).

The Riemann-Liouville and Caputo fractional derivatives defined by (6) and (7), respectively, are particular cases of the GFDs (19) and (20) with the kernel

$$k(t) = h_{1-\alpha}(t), \ 0 < \alpha < 1, \quad (22)$$

the power function h_β being defined by (17). Other important particular cases of (19) and (20) are the multi-term fractional derivatives and the fractional derivatives of the distributed order. They are generated by (19) and (20) with the kernels

$$k(t) = \sum_{k=1}^n a_k h_{1-\alpha_k}(t), \ 0 < \alpha_1 < \cdots < \alpha_n < 1, \ a_k \in \mathbb{R}, \ k = 1, \ldots, n \quad (23)$$

and

$$k(t) = \int_0^1 h_{1-\alpha}(t)\,d\rho(\alpha), \quad (24)$$

where ρ is a Borel measure on $[0, 1]$.

Even if the operators (19) and (20) have been employed in the theory of the abstract Volterra integral equations for many years, the main advantage of the Kochubei's approach was to establish a connection of these operators to FC and to introduce a special class of the kernels that ensures both existence of the corresponding fractional integrals and physically relevant properties of the fractional differential equations with these time-fractional derivatives. Moreover, the results presented in [10] and in the subsequent publications were derived using a completely different technique, namely the theory of the complete Bernstein functions ([20]).

The kernels of the GFDs (19) and (21) considered in [10] satisfy the following conditions:

(K1) The Laplace transform $\tilde{k}$ of k,

$$\tilde{k}(p) = (\mathcal{L}k)(p) = \int_0^\infty k(t)\,e^{-pt}\,dt$$

exists for all $p > 0$,
(K2) $\tilde{k}(p)$ is a Stieltjes function,
(K3) $\tilde{k}(p) \to 0$ and $p\tilde{k}(p) \to \infty$ as $p \to \infty$,
(K4) $\tilde{k}(p) \to \infty$ and $p\tilde{k}(p) \to 0$ as $p \to 0$.

In what follows, we denote the set of the kernels that satisfy the conditions (K1)–(K4) by $\mathcal{K}$. As we see, the condition of type (Z2) (Sonine condition) does not belong to the set of the conditions (K1)–(K4). However, it is one of the consequences of these conditions and especially of the strong condition (K2). Roughly speaking, a function defined on $\mathbb{R}_+$ is a Stieltjes function if it can be represented as a restriction of the Laplace transform of a completely monotone function to the real positive semi-axis. Any completely monotone function is non-negative and thus any Stieltjes function is completely monotone as the Laplace transform of a non-negative function. For the strict definition and properties of the Stieltjes functions see e.g., [20,21]. The kernel functions (22) and (23) as well as the function (24) under some suitable conditions on the measure ρ belong to the class $\mathcal{K}$. In [10], another example of a function $k \in \mathcal{K}$ was introduced in terms of its Laplace transform $\tilde{k}(p) = p^{-1}\log(1+p^\beta)$, $0 < \beta < 1$.

As shown in [10], for each $k \in \mathcal{K}$, there exists a completely monotone function κ such that the Sonine condition holds true:

$$(k * \kappa)(t) = \int_0^t k(t-\tau)\kappa(\tau)\,d\tau = 1. \tag{25}$$

Henceforth by

$$(\mathbb{I}_k f)(t) = \int_0^t \kappa(t-\tau) f(\tau)\,d\tau \tag{26}$$

we denote a general fractional integral (GFI) with the kernel κ associated with the kernel k of the GFD by means of the relation (25).

The notion of the GFI is justified by the following Fundamental Theorem of FC:

Theorem 1 ([10]). *If f is a locally bounded measurable function on $\mathbb{R}_+$, then*

$$(\mathbb{D}_k^C \mathbb{I}_k f)(t) = f(t). \tag{27}$$

If f is absolutely continuous on $[0, +\infty)$, then

$$(\mathbb{I}_k \mathbb{D}_k^C f)(t) = f(t) - f(0). \tag{28}$$

The Formula (27) and the relation (21) between the GFDs of the Caputo and the Riemann-Liouville types lead to the identity

$$(\mathbb{D}_k^{RL} \mathbb{I}_k f)(t) = f(t), \tag{29}$$

i.e., the Riemann-Liouville GFD is also a left inverse operator to the GFI defined by (26).

In the case of the Riemann-Liouville and the Caputo fractional derivatives that are particular cases of the GFDs (19) and (20), respectively, with the power function kernel k defined by (22), the kernel κ in the GFI (25) is also the power function $\kappa(t) = h_\alpha(t)$ and thus in this case the GFI (26) is nothing else as the conventional Riemann-Liouville fractional integral.

As shown in [22], the functions that satisfy the Sonine condition (25) cannot be continuous at the point $t = 0$ and thus the "new fractional derivatives" with the continuous kernels introduced recently in the FC literature do not belong to the class of the GFDs that are discussed in this paper.

In the next sections, we consider other physically relevant properties of the GFD including complete monotonicity of the solutions to the fractional relaxation equation with this derivative, positivity of the fundamental solution to the Cauchy problem for the fractional diffusion equation with the time-derivative in form of the GFD, and a maximum principle for the initial-boundary-value problems for the fractional diffusion equation with this derivative.

3. Fractional ODEs with the GFD

3.1. Fractional Relaxation Equation

In this subsection, we consider the fractional relaxation equation

$$(\mathbb{D}_k^C u)(t) = -\lambda u(t),\ \lambda > 0,\ t > 0 \tag{30}$$

subject to the initial condition

$$u(0) = 1. \tag{31}$$

As discussed in [23] (see also references therein), in the framework of the linear viscoelasticity models, the solutions to the relaxation equations are expected to be completely monotone. Only in this case the relaxation processes can be interpreted as superpositions of (infinitely many) elementary, i.e., exponential, relaxation processes. For the fractional relaxation equation with the GFD of the Caputo type, the following result holds true:

Theorem 2 ([10]). *Let the kernel k of the Caputo type GFD belong to the class $\mathcal{K}$.*
Then the Cauchy problem (30), (31) has a unique solution $u_\lambda = u_\lambda(t)$, continuous on $[0, +\infty)$, infinitely differentiable and completely monotone on $\mathbb{R}_+$.

We remind the readers that a function $u : \mathbb{R}_+ \to \mathbb{R}$ is called completely monotone if the conditions

$$(-1)^n u^{(n)}(t) \geq 0, \quad t > 0, \quad n = 0, 1, 2 \ldots \tag{32}$$

hold true.

In the case of the Cauchy problem (30), (31) with the Caputo fractional derivative (7), the solution can be expressed in terms of the Mittag-Leffler function

$$u_\lambda(t) = E_{\alpha,1}(-\lambda t^\alpha),$$

where $E_{\alpha,\beta}$ stands for the two-parameters Mittag-Leffler function that is defined by the following convergent series:

$$E_{\alpha,\beta}(z) = \sum_{k=0}^{\infty} \frac{z^k}{\Gamma(\alpha k + \beta)}, \quad \alpha > 0, \, \beta, z \in \mathbb{C}. \tag{33}$$

It is worth mentioning that the fractional relaxation equations with the Riemann-Liouville fractional derivative (6) and the Hilfer derivative (8) have the solutions

$$u_\lambda(t) = t^{\alpha-1} E_{\alpha,\alpha}(-\lambda t^\alpha),$$

and

$$u_\lambda(t) = t^{\alpha+\gamma_1-1} E_{\alpha,\alpha+\gamma_1}(-\lambda t^\alpha),$$

respectively, provided that we set suitable initial conditions. These solutions are continuous, infinitely differentiable and completely monotone on $\mathbb{R}_+$, but have an integrable singularity at the point $t = 0$. As to the relaxation equation with the nth level fractional derivative (11), its solution is given by the following theorem:

Theorem 3 ([24]). *The fractional relaxation equation*

$$(D_{nL}^{\alpha,(\gamma)} u)(t) = -\lambda\, u(t), \quad \lambda > 0, \, t > 0 \tag{34}$$

subject to the initial conditions

$$\left(\prod_{i=k+1}^{n} \left(I_{0+}^{\gamma_i} \frac{d}{dt} \right) I_{0+}^{n-\alpha-s_n} u \right)(0) = u_k, \quad k = 1, \ldots, n \tag{35}$$

has a unique solution, continuous and infinitely differentiable on $\mathbb{R}_+$, given by the formula

$$u(t) = \sum_{k=1}^{n} u_k\, t^{\alpha+s_k-k}\, E_{\alpha,\alpha+s_k-k+1}(-\lambda t^\alpha), \quad s_k = \sum_{i=1}^{k} \gamma_i. \tag{36}$$

If the initial conditions are non-negative ($u_k \geq 0$, $k = 1, \ldots, n$ in (35)) and the inequalities

$$k - 1 \leq s_k = \sum_{i=1}^{k} \gamma_i, \ k = 1, \ldots, n \tag{37}$$

hold true, the solution (36) is completely monotone on $\mathbb{R}_+$.

In [10], an important probabilistic interpretation of Theorem 2 has been provided. Let $D(t)$ be a subordinator of the Lévy process with the Laplace exponent $\Psi = \Psi(s)$ ([25]):

$$\mathbf{E}\left[e^{-sD(t)}\right] = e^{-t\Psi(s)},$$

where Ψ is a Bernstein function ([20]) with the representation

$$\Psi(s) = bs + \int_0^{+\infty} (1 - e^{-s\tau})\Phi(d\tau),$$

where $b \geq 0$ is the drift coefficient and Φ is the Lévy measure, such that either $b > 0$, or $\Phi(\mathbb{R}_+) = +\infty$, or both.

Because the process D is strictly increasing, it possesses an inverse function

$$E(t) = \inf\{r > 0 : D(r) > t\}.$$

Now we consider a Poisson process $N(t)$ with the intensity λ. It is known that $N(E(t))$ is a renewal process with the waiting times J_n and

$$\mathbf{P}[J_n > t] = \mathbf{E}\left[e^{-\lambda E(t)}\right]. \tag{38}$$

As shown in [10], if the restriction $\Psi = \Psi(p)$, $p > 0$ of the Laplace exponent $\Psi = \Psi(s)$ to the real semi-axes $\mathbb{R}_+$ is a complete Bernstein function that satisfies the conditions (K3) and (K4), then the right-hand side of Formula (38) can be interpreted as the solution to the Cauchy problem (30), (31). According to Theorem 2, it is continuous on $[0, +\infty)$ and completely monotone.

3.2. Fractional Growth Equation

In this subsection, some of the results derived in [13] are shortly addressed. We consider the fractional growth equation with the GFD of Caputo type

$$(\mathbb{D}_k^C u)(t) = \lambda u(t), \ \lambda > 0, \ t > 0 \tag{39}$$

subject to the initial condition

$$u(0) = 1. \tag{40}$$

In the case of the conventional growth equation, the solution is $u_\lambda(t) = \exp(\lambda t)$. The problem (39), (40) with the Caputo fractional derivative (7) is solved by the function $u_\lambda(t) = E_{\alpha,1}(\lambda t^\alpha)$ that is known to be of exponential growth as $t \to +\infty$ ([26]). It turns out that the solution to the problem (39), (40) with the GFD with a kernel $k \in \mathcal{K}$ (k fulfills the conditions (K1)–(K4)) is also of exponential growth as $t \to +\infty$.

For formulation of the corresponding result, the notation

$$\Phi(p) = p\tilde{k}(p)$$

is introduced, $\tilde{k}$ being the Laplace transform of k. Because $\tilde{k}$ is a Stieltjes function, Φ is a Bernstein function and Φ' is completely monotone ([20]). The made assumptions ensure that Φ is strictly monotone and thus for every $\lambda > 0$ there exists a unique $p_0 = p_0(\lambda)$ such that $\Phi(p_0) = \lambda$. Then we have the following result regarding the asymptotic of the unique solution u_λ of the Cauchy problem (39), (40) as $t \to +\infty$.

Theorem 4 ([13]). Let $k \in \mathcal{K}$ and the condition

$$\int_1^{+\infty} \frac{dp}{p\,\Phi(p)} < +\infty \tag{41}$$

hold true. Then the solution u_λ of the Cauchy problem (39), (40) has the following exponential asymptotic behavior:

$$u_\lambda(t) = \frac{\lambda}{\Phi'(p_0(\lambda))p_0(\lambda)} e^{p_0(\lambda)t} + o\left(e^{p_0(\lambda)t}\right), \quad t \to +\infty. \tag{42}$$

In the case of the Cauchy problem (39), (40) with the Caputo derivative (7), the function Φ takes the form $\Phi(p) = p^\alpha$ and $p_0(\lambda) = \lambda^{1/\alpha}$. The condition (41) is evidently fulfilled and Formula (42) takes the form

$$u_\lambda(t) = \frac{1}{\alpha} e^{\lambda^{\frac{1}{\alpha}} t} + o\left(e^{\lambda^{\frac{1}{\alpha}} t}\right),$$

that corresponds to the main term of the asymptotic of the Mittag-Leffler function $u_\lambda(t) = E_{\alpha,1}(\lambda t^\alpha)$ as $t \to +\infty$ (see e.g., [26]).

Another important particular case is the Cauchy problem (39), (40) with the distributed order derivative (the GFD with the kernel (24)). It turns out that under some standard assumptions Theorem 4 is applicable also in this case (see [13] for details).

3.3. The Cauchy Problem for a Nonlinear Fractional ODE

Following [14], in this subsection, we address a nonlinear fractional differential equation with the GFD of the Caputo type

$$(\mathbb{D}_k^C u)(t) = f(t, u(t)), \quad t > 0 \tag{43}$$

subject to the initial condition

$$u(0) = u_0 \in \mathbb{R}. \tag{44}$$

In what follows, we again assume that the kernel k of $\mathbb{D}_k^C$ defined by (21) belongs to the class $\mathcal{K}$, i.e., that the conditions (K1)–(K4) are fulfilled.

For derivation of results regarding existence and uniqueness of the solution to the Cauchy problem (43), (44), we first transform it into an integral equation by applying the GFI (26). Let $L > 0$ and $f : [0, L] \times \mathbb{R} \to \mathbb{R}$ be a continuous function. Then the Cauchy problem (43), (44) is equivalent to the integral equation

$$u(t) = u_0 + \int_0^t \kappa(t - \tau) f(\tau, u(\tau))\, d\tau, \quad t > 0, \tag{45}$$

where the kernel function κ is determined by the relation (25).

The sufficient conditions for existence of the local and global solutions to the integral Equation (45) and thus to the Cauchy problem (43), (44) are provided in the following two theorems.

Theorem 5 ([14]). Let $L, Q > 0$, f be a continuous function on the closed domain $G = \{(t, \tau) : 0 \leq t \leq L, |\tau - u_0| \leq Q\}$, and $l \in (0, L]$ satisfy the inequality

$$\max_{(t,\tau) \in G} |f(t,\tau)| \int_0^l \kappa(\tau)\, d\tau \leq Q.$$

Then the Cauchy problem (43), (44) has a solution absolutely continuous on the interval $[0, l]$.

Theorem 6 ([14]). *Let $L > 0$ and $f : [0, L] \times \mathbb{R} \to \mathbb{R}$ be a continuous function that satisfies the inequality*

$$|f(t, \tau)| \leq b_0 + b_1 |\tau|^p, \ 0 \leq t \leq L, \ \tau \in \mathbb{R}$$

with $b_0, b_1 > 0$ and $0 < p \leq 1$.
Then the Cauchy problem (43), (44) has a solution absolutely continuous on the interval $[0, L]$.

As a corollary from Theorem 6, we get the following result: Let $f : [0, +\infty) \times \mathbb{R} \to \mathbb{R}$ be a continuous function that satisfies the inequality

$$|f(t, \tau)| \leq b_0 + b_1 |\tau|^p, \ 0 \leq t < +\infty, \ \tau \in \mathbb{R}$$

with $b_0, b_1 > 0$ and $0 < p \leq 1$. Then the Cauchy problem (43), (44) has a global solution absolutely continuous on $[0, +\infty)$.

As to uniqueness of the solution, it was proved under some stronger conditions compared to the ones required for its existence.

Theorem 7 ([14]). *Let $L > 0$ and $f : [0, L] \times \mathbb{R} \to \mathbb{R}$ be a continuous function that satisfies the Lipschitz condition with respect to its second variable*

$$|f(t, s) - f(t, \tau)| \leq b|s - \tau|, \ 0 \leq t \leq L, \ s, \tau \in \mathbb{R}, \ b > 0. \tag{46}$$

Then the Cauchy problem (43), (44) has a unique absolutely continuous solution on the interval $[0, L]$.

As before, the result formulated in Theorem 7 can be extended to the case of the infinite interval $[0, +\infty)$: Let $f : [0, +\infty) \times \mathbb{R} \to \mathbb{R}$ be a continuous function that satisfies the Lipschitz condition

$$|f(t, s) - f(t, \tau)| \leq b|s - \tau|, \ 0 \leq t < +\infty, \ s, \tau \in \mathbb{R}, \ b > 0.$$

Then the Cauchy problem (43), (44) has a unique absolutely continuous solution on the interval $[0, +\infty)$.

To address a continuous dependence of solutions to the Cauchy problem (43), (44) on the problem data, in [14], a Gronwall-type inequality was derived. It is important by itself and we formulate it below.

Lemma 1 ([14]). *Let $l, v_0, b > 0$ and $v \in C[0, l]$. If the inequality*

$$v(t) \leq v_0 + b \int_0^t \kappa(t - \tau) v(\tau) \, d\tau \tag{47}$$

holds true for $t \in [0, l]$, then

$$v(t) \leq u(t), \ t \in [0, l], \tag{48}$$

where $u = u(t)$ is the solution to the fractional relaxation equation

$$(\mathbb{D}_k^C u)(t) = -b \, u(t) \tag{49}$$

subject to the initial condition

$$u(0) = v_0. \tag{50}$$

Based on Lemma 1, in [14], the continuous dependence of the solution to the Cauchy problem (43), (44) on the problem data was proved.

Theorem 8 ([14]). *Let the conditions of Theorem 7 be fulfilled, u be the solution to the Cauchy problem (43), (44), and u_k be the solution to the Cauchy problem (43), (44) with the initial condition $u(0) = u_{0k}$. Then*

$$\|u - u_k\|_{C[0,L]} \to 0 \text{ as } u_{0k} \to u_0.$$

Theorem 9 ([14]). *Let the conditions of Theorem 7 be fulfilled for the functions f and g, u be the solution to the Cauchy problem (43), (44), and v be the solution to the Cauchy problem (43), (44) with the right-hand side g and with the same initial condition u_0. Then*

$$\|u - v\|_{C[0,L]} \to 0 \text{ as } \max_{(t,\tau)\in[0,L]\times[-\Omega,\Omega]} |f(t,\tau) - g(t,\tau)| \to 0,$$

where Ω is the upper bound of the supremum norm of the solution u.

4. Time-Fractional PDEs with the GFD

In this section, we present some important results concerning the direct and inverse problems for the time-fractional PDEs with the GFD of the Caputo type.

4.1. Cauchy Problem for the Time-Fractional Diffusion Equation

Following [10], in this subsection we address the properties of solutions to the Cauchy problem for the general time-fractional diffusion equation in the form

$$(\mathbb{D}_k^C u(x,\cdot))(t) = \Delta u(x,t) \ t > 0, \ x \in \mathbb{R}^n, \ u(x,0) = u_0(x), \tag{51}$$

where u_0 is a bounded globally Hölder continuous function on $\mathbb{R}^n$.

In [10], solutions to the Cauchy problem (51) were understood in the following sense: Applying formally the Laplace transform in the variable t to the equation in (51), we arrive at the equation

$$p\tilde{k}(p)\tilde{u}(x,p) - \tilde{k}(p)u_0(x) = \Delta \tilde{u}(x,p), \ p > 0, \ x \in \mathbb{R}^n \tag{52}$$

for the Laplace transform $\tilde{u}(x,p)$ of a solution to the Cauchy problem (51). A bounded function $u = u(x,t)$ is called an LT-solution of (51), if u is continuous in t on $[0, +\infty)$ uniformly with respect to $x \in \mathbb{R}^n$, satisfies the initial condition $u(x,0) = u_0(x)$, while its Laplace transform $\tilde{u}(x,p)$ is twice continuously differentiable in x, for each $p > 0$, and satisfies Equation (52).

Theorem 10 ([10]). *Let the kernel k of the GFD of the Caputo type belong to the class $\mathcal{K}$. Then there exist a non-negative function $Z = Z(x,t)$, $t > 0$, $x \in \mathbb{R}^n$, $x \neq 0$, locally integrable in t and infinitely differentiable in $x \neq 0$ that satisfies the relation*

$$\int_{\mathbb{R}^n} Z(x,t)\,dx = 1, \ t > 0, \tag{53}$$

and for any bounded globally Hölder continuous u_0, the function

$$u(x,t) = \int_{\mathbb{R}^n} Z(x-\zeta,t)\,u_0(\zeta)\,d\zeta \tag{54}$$

is an LT-solution to the Cauchy problem (51).

The function $Z = Z(x,t)$ is what is usually called the fundamental solution to the Cauchy problem (51), i.e., the one that formally corresponds to the initial condition $u(x,0) = u_0(x) = \delta(x)$, δ being the Dirac delta function. As stated in Theorem 10, the fundamental solution to the Cauchy problem (51) can be interpreted as a probability density function for each $t > 0$ and thus the time-fractional diffusion Equation (51) with the GFD could be potentially useful for modeling of the anomalous diffusion processes.

The explicit form of the fundamental solution is as follows ([10]):

$$Z(x,t) = \int_0^{+\infty} (4\pi s)^{-n/2} e^{-\frac{|x|^2}{4s}} G(s,t)\, ds, \ x \neq 0,$$

where

$$G(s,t) = \int_0^t k(t-\tau)\mu_s(d\tau)$$

with the probability measure $\mu_s(d\tau)$ that satisfies the relation

$$e^{-s\tilde{k}(p)} = \int_0^{+\infty} e^{-p\tau} \mu_s(d\tau).$$

Under the conditions of Theorem 10, the measure $\mu_s(d\tau)$ always exists because the function $p \to e^{-s\tilde{k}(p)}$ is completely monotone on $\mathbb{R}_+$ for each $s \geq 0$.

For the validity of Theorem 10, the condition $k \in \mathcal{K}$ is essential. However, it is worth mentioning that for the well-posedness of the Cauchy problem for the equations with the operators of type $\mathbb{D}_k^C$ much weaker conditions than (K1)–(K4) are sufficient (see e.g., [27]). Here we formulate the uniqueness result for the Cauchy problem (51) in the class of the LT-solutions proved in [10].

Theorem 11 ([10]). *Let the kernel k of the GFD of the Caputo type be non-negative, locally integrable, nonzero on a set of positive measure, and its Laplace transform $\tilde{k} = \tilde{k}(p)$ exist for all $p > 0$.*

If $u = u(x,t)$ is a polynomially bounded LT-solution to the Cauchy problem (51) with $u_0(x) \equiv 0$, then $u(x,t) \equiv 0$.

In the definition of the LT-solutions, their boundedness was required. However, this definition makes sense also for the polynomially bounded solutions, i.e., such solutions $u = u(x,t)$ that satisfy the inequality $|u(x,t)| \leq P_u(|x|)$, where P_u are some polynomials independent on t.

4.2. Initial-Boundary-Value Problems for the Time-Fractional Diffusion Equation

In this subsection, we present some results regarding the initial-boundary-value problems for the time-fractional diffusion equation with the GFD of the Caputo type in the form

$$(\mathbb{D}_k^C u(x,\cdot))(t) = D_2(u) + D_1(u) - q(x)u(x,t) + F(x,t), \ (x,t) \in \Omega \times (0,T], \tag{55}$$

subject to the initial condition

$$u(x,t)\big|_{t=0} = u_0(x), \ x \in \bar{\Omega} \tag{56}$$

and the boundary condition

$$u(x,t)\big|_{(x,t)\in\partial\Omega\times(0,T]} = v(x,t), \ (x,t) \in \partial\Omega \times (0,T]. \tag{57}$$

In Equations (55)–(57), Ω is a bounded open domain in $\mathbb{R}^n$ with a smooth boundary $\partial\Omega$, $q \in C(\bar{\Omega})$, $q(x) \geq 0$ for $x \in \bar{\Omega}$, and

$$D_1(u) = \sum_{i=1}^n b_i(x) \frac{\partial u}{\partial x_i}, \ D_2(u) = \sum_{i,j=1}^n a_{ij}(x) \frac{\partial^2 u}{\partial x_i \partial x_j}. \tag{58}$$

Moreover we assume that D_2 is a uniformly elliptic differential operator.

A function $u \in S(\Omega, T)$ is called a strong solution to the initial-boundary-value problem (55)–(57) if it satisfies both Equation (55) and the initial and boundary conditions (56) and (57), respectively.

By $S(\Omega, T)$, we denoted the space of functions $u = u(x,t)$, $(x,t) \in \bar{\Omega} \times [0, T]$ that satisfy the inclusions $u \in C(\bar{\Omega} \times [0, T])$, $u(\cdot, t) \in C^2(\Omega)$ for any $t > 0$, and $\partial_t u(x, \cdot) \in C(0, T] \cap L^1(0, T)$ for any $x \in \Omega$.

First we discuss a maximum principle for the time-fractional diffusion Equation (55). For its validity, we assume that the following conditions for the kernel k of $\mathbb{D}_k^C$ are satisfied:

(LY1) $k \in C^1(\mathbb{R}_+) \cap L^1_{loc}(\mathbb{R}_+)$,
(LY2) $k(\tau) > 0$ and $k'(\tau) < 0$ for $\tau > 0$,
(LY3) $k(\tau) = o(\tau^{-1})$, $\tau \to 0$.

Let us note that the Kochubei's conditions (K1)–(K4) are not needed for validity of the maximum principle for the general diffusion Equation (55). However, if the condition (K3) holds true, then it follows from the Feller-Karamata Tauberian theorem for the Laplace transform ([21]) that the condition (LY3) is also satisfied.

The maximum principle for the general diffusion Equation (55) is based on an appropriate estimate of the GFD of a function f at its maximum point. It is given in the following theorem.

Theorem 12 ([28]). *Let the conditions (LY1)–(LY3) be fulfilled, a function $f \in C[0, T]$ attain its maximum over the interval $[0, T]$ at the point t_0, $t_0 \in (0, T]$, and $f' \in C(0, T] \cap L^1(0, T)$.*
Then the inequality

$$(\mathbb{D}_k^C f)(t_0) \geq k(t_0)(f(t_0) - f(0)) \geq 0 \tag{59}$$

holds true.

In the case of the Caputo fractional derivative, the inequality (59) takes the known form

$$(D_C^\alpha f)(t_0) \geq \frac{t_0^{-\alpha}}{\Gamma(1-\alpha)}(f(t_0) - f(0)) \geq 0. \tag{60}$$

In what follows, we use the notation

$$\mathbb{P}_k(u) := (\mathbb{D}_k^C u)(t) - D_2(u) - D_1(u) + q(x)u(x,t). \tag{61}$$

Theorem 13 ([28]). *Let the conditions (LY1)–(LY3) be fulfilled and a function $u \in S(\Omega, T)$ satisfy the inequality*

$$\mathbb{P}_k(u) \leq 0, \ (x,t) \in \Omega \times (0, T]. \tag{62}$$

Then the following maximum principle holds true:

$$\max_{(x,t) \in \bar{\Omega} \times [0,T]} u(x,t) \leq \max\{\max_{x \in \bar{\Omega}} u(x,0), \max_{(x,t) \in \partial\Omega \times [0,T]} u(x,t), 0\}. \tag{63}$$

The maximum principle formulated in Theorem 13 can be applied, among other things, for derivation of some a priori estimates for the strong solutions of the initial-boundary-value problem (55)–(57).

Theorem 14 ([28]). *Let the conditions (K1)–(K4) and (LY1)–(LY3) be fulfilled and u be a strong solution to the initial-boundary-value problem (55)–(57).*
Then

$$\|u\|_{C(\bar{\Omega} \times [0,T])} \leq \max\{M_0, M_1\} + M f(T), \tag{64}$$

where

$$M_0 = \|u_0\|_{C(\bar{\Omega})}, \ M_1 = \|v\|_{C(\partial\Omega \times [0,T])}, \ M = \|F\|_{C(\Omega \times [0,T])}, \tag{65}$$

and

$$f(t) = \int_0^t \kappa(\tau) d\tau, \tag{66}$$

where the function κ is the kernel of the GFI defined by (25).

The uniqueness of the strong solution to the initial-boundary-value problem (55)–(57) and its continuous dependence on problem data easily follow from the solution norm estimate (64).

Theorem 15 ([28]). *The initial-boundary-value problem (55)–(57) possesses at most one strong solution.*

This solution—if it exists—continuously depends on the problem data in the sense that if u and $\tilde{u}$ are strong solutions to the problems with the sources functions F and $\tilde{F}$ and the initial and boundary conditions u_0 and $\tilde{u}_0$ and v and $\tilde{v}$, respectively, and

$$\|F - \tilde{F}\|_{C(\bar{\Omega} \times [0,T])} \leq \epsilon,$$

$$\|u_0 - \tilde{u}_0\|_{C(\bar{\Omega})} \leq \epsilon_0, \quad \|v - \tilde{v}\|_{C(\partial\Omega \times [0,T])} \leq \epsilon_1,$$

then the norm estimate

$$\|u - \tilde{u}\|_{C(\bar{\Omega} \times [0,T])} \leq \max\{\epsilon_0, \epsilon_1\} + \epsilon f(T) \tag{67}$$

holds true, where the function f is defined by (66).

In the rest of this subsection, we address uniqueness and existence of a weak solution to the initial-boundary-value problem (55)–(57) defined in the sense of Vladimirov [16]: We call $u \in C(\bar{\Omega} \times [0, T])$ a weak solution to the initial-boundary-value problem (55)–(57) in the sense of Vladimirov, if there exist $F_k \in C(\bar{\Omega} \times [0, T])$, $u_{0k} \in C(\bar{\Omega})$ and $v_k \in C(\partial\Omega \times [0, T])$, $k = 1, 2, \ldots$ satisfying (V1) and (V2) below such that

$$\|u_k - u\|_{C(\bar{\Omega} \times [0,T])} \to 0 \text{ as } k \to +\infty. \tag{68}$$

(V1)   There exist the functions F, u_0, and v, such that

$$\|F_k - F\|_{C(\bar{\Omega} \times [0,T])} \to 0 \text{ as } k \to +\infty, \tag{69}$$

$$\|u_{0k} - u_0\|_{C(\bar{\Omega})} \to 0 \text{ as } k \to +\infty, \tag{70}$$

$$\|v_k - v\|_{C(\partial\Omega \times [0,T])} \to 0 \text{ as } k \to +\infty. \tag{71}$$

(V2)   For each $k = 1, 2, \ldots$ there exists a strong solution $u_k = u_k(x, t)$ to the general time-fractional diffusion equation

$$(\mathbb{D}_k^C u_k(x, \cdot))(t) = D_2(u_k) + D_1(u_k) - q(x)u_k(x, t) + F_k(x, t), \ (x, t) \in \Omega \times (0, T]. \tag{72}$$

subject to the initial condition

$$u_k|_{t=0} = u_{0k}(x), \ x \in \bar{\Omega} \tag{73}$$

and the boundary condition

$$u_k|_{\partial\Omega \times (0,T]} = v_k(x, t), \ (x, t) \in \partial\Omega \times (0, T]. \tag{74}$$

In [28], the correctness of the definition of a weak solution was shown. A weak solution to the problem (55)–(57) in the sense of Vladimirov is a continuous function, not a distribution. However, the weak solutions are not required to be smooth.

Any strong solution to the problem (55)–(57) is evidently also its weak solution. If the problem (55)–(57) possesses a weak solution, then the functions F, u_0 and v from the problem formulation have to belong to the spaces $C(\bar{\Omega} \times [0, T])$, $C(\bar{\Omega})$ and $C(\partial\Omega \times [0, T])$, respectively, as the limits of the sequences of the continuous functions in the uniform norm.

The estimate (64) for the strong solutions holds true also for the weak solutions. To show this, we just let $k \to +\infty$ in the inequality

$$\|u_k\|_{C(\bar{\Omega}_T)} \leq \max\{M_{0k}, M_{1k}\} + M_k f(T), \ k = 1, 2 \ldots \tag{75}$$

with

$$M_{0k} := \|u_{0k}\|_{C(\bar{\Omega})}, \ M_{1k} := \|v_k\|_{C(\partial\Omega \times [0,T])}, \ M_k := \|F_k\|_{C(\bar{\Omega} \times [0,T])}.$$

The estimate (64) for the weak solutions is employed to prove the following uniqueness result.

Theorem 16 ([28]). *The initial-boundary-value problem (55)–(57) possesses at most one weak solution. The weak solution—if it exists—continuously depends on the data given in the problem in the sense of the estimate (67).*

In the rest of this subsection, we address the question of existence of a weak solution to the initial-boundary-value problem (55)–(57) in the case of the homogeneous Equation (55) without the first order spatial differential operator D_1 subject to the initial condition (56) and the homogeneous boundary condition (57), i.e., we consider the initial-boundary-value problem

$$(\mathbb{D}_k^C u(x,\cdot))(t) = D_2(u) - q(x)u(x,t), \quad (x,t) \in \Omega \times (0,T], \tag{76}$$

$$u(x,t)|_{t=0} = u_0(x), \ x \in \bar{\Omega}, \tag{77}$$

$$u(x,t)|_{(x,t) \in \partial\Omega \times (0,T]} = 0, \ (x,t) \in \partial\Omega \times (0,T] \tag{78}$$

under the same conditions on the coefficients of the operator D_2 and the function q that we assumed at the beginning of the subsection. Moreover, we also assume that the kernel k of $\mathbb{D}_k^C$ is from $\mathcal{K}$, i.e., the conditions (K1)–(K4) are satisfied.

First, a formal solution to the initial-boundary-value problem (76)–(78) is constructed in form of the Fourier series

$$u(x,t) = \sum_{k=1}^{\infty} (u_0, X_k) U_k(t) X_k(x), \tag{79}$$

where X_k, $k = 1, 2, \ldots$ are the eigenfunctions corresponding to the eigenvalues λ_k of the eigenvalue problem

$$L(X(x)) = \lambda X(x), \ x \in \Omega, \tag{80}$$

$$X(x)|_{x \in \partial\Omega} = 0 \tag{81}$$

for the operator L, $L(x) = -D_2(X) + q(x)X(x)$. Because of the conditions posed on the operator D_2 and the function q, the differential operator L is positive definite and self-adjoint. Thus the eigenvalue problem (80)–(81) has a countable number of the positive eigenvalues $0 < \lambda_1 \leq \lambda_2 \leq \ldots$ with the finite multiplicities and—if the boundary $\partial\Omega$ of Ω is smooth—any function $f \in \mathcal{M}_L$ can be represented through its Fourier series in the form

$$f(x) = \sum_{k=1}^{\infty} (f, X_k) X_k(x), \tag{82}$$

where $X_k \in \mathcal{M}_L$ are the eigenfunctions corresponding to the eigenvalues λ_k:

$$L(X_k) = \lambda_k X_k, \ k = 1, 2, \ldots. \tag{83}$$

By $\mathcal{M}_L$, the space of the functions f that satisfy the boundary condition (81) and the inclusions $f \in C^1(\bar{\Omega}) \cap C^2(\Omega)$, $L(f) \in L^2(\Omega)$ is denoted.

As to the functions $U_k = U_k(t)$, they are solutions to the fractional relaxation equations

$$(\mathbb{D}_k^C U_k)(t) = -\lambda_k U_k(t),\ t > 0,\ k = 1, 2, \ldots \tag{84}$$

subject to the initial conditions

$$U_k(0) = 1,\ k = 1, 2, \ldots. \tag{85}$$

According to Theorem 2 from Section 3, for any $\lambda = \lambda_k > 0$, $k = 1, 2, \ldots$ this initial-value problem has a unique solution $U_k = U_k(t)$ that belongs to the class $C^\infty(\mathbb{R}_+)$ and is a completely monotone function. In particular, any U_k is non-negative and non-increasing and thus the inequalities

$$0 \leq U_k(t) \leq U_k(0) = 1 \tag{86}$$

hold true. Let us mention that in the case of the single-term time-fractional diffusion equation with the Caputo fractional derivative ($k(\tau) = \frac{\tau^{-\alpha}}{\Gamma(1-\alpha)}$, $0 < \alpha < 1$), the solution to the initial-value problem (84), (85) with $\lambda = \lambda_k$, $k = 1, 2, \ldots$ has the form ([29])

$$U_k(t) = E_{\alpha,1}(-\lambda_k t^\alpha). \tag{87}$$

Under some standard assumptions, the formal solution (79) is a weak solution to the initial-boundary-value problem (76)–(78) in the sense of Vladimirov.

Theorem 17 ([15]). *Let the function u_0 in the initial condition (77) be from the space $\mathcal{M}_L$. Then the formal solution (79) of the problem (76)–(78) is its weak solution in the sense of Vladimirov.*

For a survey of other results regarding the maximum principles for the time-fractional PDEs of different types see the recent publication [30].

4.3. Inverse Problems Involving GFD

The starting point of this subsection is a reconstruction problem for a function based on its values and the values of its GFD in a neighborhood of the final time ([19]):

IP1. Let $0 < t_0 < T < +\infty$. Given ϕ, $g : (t_0, T) \to \mathbb{R}$, find a function $u : (0, T) \to \mathbb{R}$ such that

$$u(t)|_{(t_0,T)} = \phi(t),\ \text{and}\ (\mathbb{D}_k^C u)(t)|_{(t_0,T)} = g(t). \tag{88}$$

The inverse problems of type IP1 are potentially useful for applications. For instance, in the framework of the Scott-Blair model of viscoelasticity, the stress is proportional to a time-fractional derivative of the strain ([23]). In this context, the IP1 means a reconstruction of the strain history based on the measurements of strain and stress starting from a certain time t_0.

In [19], a uniqueness result for the IP1 was proved under the following conditions on the kernel k of $\mathbb{D}_k^C$:

(KJ1) $\exists \mu \in \mathbb{R} : \int_0^{+\infty} e^{-\mu t} |k(t)|\, dt < +\infty$,
(KJ2) k is real analytic on $\mathbb{R}_+$,
(KJ3) the Laplace transform $\tilde{k}$ of k cannot be meromorphically extended to the whole complex plane $\mathbb{C}$.

Theorem 18 ([19]). *Let the kernel k of $\mathbb{D}_k^C$ fulfill the conditions (KJ1)–(KJ3). Then the following uniqueness results for the IP1 hold true:*

(i) *If $u \in L^1(0, T)$, $k * u \in W^{1,1}(0, T)$, and $u(t)|_{(t_0,T)} = (\mathbb{D}_k^C u)(t)|_{(t_0,T)} = 0$, then $u = 0$,*
(ii) *If $u \in W^{1,1}(0, T)$ and $u(t)|_{(t_0,T)} = (\mathbb{D}_k^C u)(t)|_{(t_0,T)} = 0$, then $u = 0$.*

The results formulated in Theorem 18 were employed in [19] for studying uniqueness of solution to the following source reconstruction problem for the fractional PDEs with the GFD of Caputo type:

IP2. Let $0 < t_0 < T < +\infty$ and $\Omega \subseteq \mathbb{R}^n$. Given ϕ, $\Phi : \Omega \times (t_0, T) \to \mathbb{R}$, find the functions $u, F : \Omega \times (0, T) \to \mathbb{R}$ such that they fulfill the equation

$$(\mathbb{D}_k^C B(u))(t) + D^l(u) - A(u) = F(x,t), \quad x \in \Omega, \ t \in (0, T) \tag{89}$$

and the relations

$$u(x,t)|_{\Omega \times (t_0, T)} = \phi, \text{ and } F|_{\Omega \times (t_0, T)} = \Phi \tag{90}$$

hold true.

In the formulation of IP2, $D^l = \sum_{j=1}^l q_j \frac{\partial^j}{\partial t^j}$ is a differential operator of order l with respect to the time variable t and with $q_j \in \mathbb{R}$ and A and B are some operators that act with respect to the spatial variable x. Moreover, we assume that $\mathcal{D}(A) \subseteq C(\Omega) \to C(\Omega)$, $\mathcal{D}(B) \subseteq C(\Omega) \to C(\Omega)$ and the operator B is invertible.

In particular, the time-fractional PDE (89) includes the time-fractional diffusion Equations (51) and (55) that were considered in the previous subsections of this section.

As shown in [19], IP2 can be reduced to IP1. Indeed, let the pair of functions (u, F) solve the IP2. The Equation (89) restricted to $\Omega \times (t_0, T)$ has the form $(\mathbb{D}_k^C B(u))(t) + D^l(\phi) - A(\phi) = \Phi(x,t)$ and thus the function Bu is a solution to the following inverse problem of IP1 type:

$$Bu|_{\Omega \times (t_0, T)} = B\phi, \text{ and } \mathbb{D}_k^C Bu|_{\Omega \times (t_0, T)} = g, \tag{91}$$

where

$$g(x,t) = \Phi(x,t) + A(\phi) - D^l(\phi), \ x \in \Omega, \ t \in (t_0, T).$$

The solution (u, F) of IP2 can be explicitly expressed in terms of the solution Bu of the IP1 formulated above as follows: $u = B^{-1} Bu$, $F = \mathbb{D}_k^C Bu + D^l(u) - A(u)$. Accordingly, a uniqueness result for the IP2 immediately follows from Theorem 18 (see [19] for details).

It is worth mentioning that the inverse problems IP1 and IP2 are severely ill-posed ([19]) and thus appropriate regularization methods are needed for their numerical treatment.

Next, we consider the following evolutionary integral equation:

$$u(x,t) = \int_0^t \kappa(t-\tau) \Delta u(x,\tau) \, d\tau + f(x,t), \ x \in \mathbb{R}^n, \ t \geq 0. \tag{92}$$

Please note that the Cauchy problem (compare to (51))

$$(\mathbb{D}_k^C u(x,\cdot))(t) = \Delta u(x,t) + F(x,t) \ t > 0, \ x \in \mathbb{R}^n, \ u(x,0) = u_0(x) \tag{93}$$

can be reduced to an evolutionary integral equation of type (92) by applying the GFI (26) to both sides of this equation and by using Formula (28) from Theorem 1:

$$u(x,t) = \int_0^t \kappa(t-\tau) \Delta u(x,\tau) \, d\tau + (\mathbb{I}_k^C F(x,\cdot))(t) + u_0(x), \ x \in \mathbb{R}^n, \ t \geq 0, \tag{94}$$

where the kernel κ is connected with the kernel k of $\mathbb{D}_k^C$ by means of the relation (25).

In [17], an important inverse problem of kernel identification in the boundary value problems for an equation associated with the evolutionary integral Equation (92) was addressed. Let $\Omega \subset \mathbb{R}^n$ be a bounded domain with sufficiently smooth boundary $\partial \Omega$. The direct boundary value problem formulated in [17] is as follows:

$$\begin{cases} u(x,t) = \int_0^t \kappa(t-\tau)\Delta u(x,\tau)\,d\tau + f(x,t), & x \in \Omega,\ t \in [0,T], \\ \mathcal{B}u(x,t) = 0, & x \in \partial\Omega,\ t \in [0,T], \end{cases} \quad (95)$$

where $\mathcal{B}$ is a boundary operator of the Dirichlet, Neumann, or Robin type, respectively:

$$\mathcal{B}v(x) = v(x),\ \mathcal{B}v(x) = n(x)\cdot\nabla v(x),\ \mathcal{B}v(x) = n(x)\cdot\nabla v(x) + \theta v(x),\ \theta \geq 0,$$

$n(x)$ being the unit outer normal of $\partial\Omega$ at the point $x \in \Omega$.

The inverse problem addressed in [17] is formulated via the so called observation functional Φ that maps the functions defined on $\overline{\Omega}$ onto $\mathbb{R}$. Usually, the functional Φ is defined in one of the following ways:

$$\Phi[v] = v(x_0),\ \Phi[v] = n(x_0)\cdot\nabla v(x_0),\ \Phi[v] = \int_\Omega \mu(x)v(x)\,dx,$$

where $x_0 \in \overline{\Omega}$ and $\mu : \Omega \to \mathbb{R}$ are given. In the case $x_0 \in \partial\Omega$, the observation functional has to be different from the boundary operator, i.e., $\Phi[v] \neq \mathcal{B}v(x_0)$.

The inverse problem considered in [17] is as follows:

IP3. Given $h : (0,T) \to \mathbb{R}$ find a kernel κ such that the solution u of the boundary value problem (95) satisfies the condition

$$\Phi[u(t,\cdot)] = h(t),\ t \in (0,T). \quad (96)$$

In [17], existence, uniqueness, and stability of solutions to the IP3 were studied for a certain class of kernels (see [17] for details).

Finally, we mention that in [18] two other inverse problems for a time-fractional PDE with the GFD of Caputo type were addressed. Let $\Omega \subset \mathbb{R}^n$ be a bounded domain with the boundary $\partial\Omega$. The direct initial-boundary-value problem is formulated as follows:

$$\begin{cases} \frac{d}{dt}(k*(U-\Phi))(x,t) = L_x U(x,t) + H(x,t),\ x \in \Omega,\ t \in (0,T), \\ U(x,0) = \Phi(x),\ x \in \Omega, \\ \mathcal{B}(U-b)(x,t) = 0,\ x \in \partial\Omega,\ t \in (0,T), \end{cases} \quad (97)$$

where Φ and b are given functions, the operator L_x is a linear second order differential operator with respect to the variable x in the form

$$L_x U(x,t) = \sum_{i,j=1}^n a_{ij}(x)\frac{\partial^2 U}{\partial x_i \partial x_j} + \sum_{j=1}^n a_j(x)\frac{\partial U}{\partial x_j} + r(x)U(x,t),$$

and $\mathcal{B}$ is a boundary operator of the Dirichlet or Neumann type, respectively:

$$\mathcal{B}v(x) = v(x)\ \text{or}\ \mathcal{B}v(x) = w(x)\cdot\nabla v(x)$$

with $w \cdot n > 0$, n being the unite outer normal of $\partial\Omega$ at the point $x \in \Omega$.

The inverse problems considered in [18] are formulated in terms of the given observation function Φ at the final time T in the form

$$U(x,T) = \Phi(x),\ x \in \Omega. \quad (98)$$

In [18], the following inverse problems were addressed:

IP4. (inverse source problem). Let

$$H(x,t) = g(x,t)f(x) + h_0(x,t), \quad (99)$$

where the components gf and h_0 may correspond to different sources or sinks. The factor f is unknown and has to be reconstructed by means of the data given by the observation function Φ from (98). The inverse problem consists in determination of a pair of functions (f, U) that satisfies (97), (98), and (99).

Another inverse problem considered in [18] is determination of the coefficient r in the operator $L_x U$:

IP5. Determine a pair of functions (r, U) that satisfies (97) and (98).

In [18], existence, uniqueness, and stability of solutions to IP4 and to IP5 were shown under some additional conditions posed on the problem data (see [18] for details).

For the surveys of the recent results concerning the inverse problems for the fractional PDEs including different kinds of the conventional fractional derivatives we refer the readers to [31–33].

Finally we mention that the theory of the fractional PDEs with the GFDs is still far away from being completed. In particular, the regularity of their solutions is not yet investigated in detail. Another interesting problem for further research would be to address the abstract fractional evolution equation in the form

$$(\mathbb{D}_k^C u)(t) = Au(t) \qquad (100)$$

subject to the initial condition $u(0) = x$. In (100), A stands for a linear closed unbounded operator densely defined in a Banach space X and the initial condition x belongs to the space X. This problem will be considered elsewhere.

Author Contributions: Y.L. and M.Y. have contributed equally to this work. All authors have read and agreed to the published version of the manuscript.

Funding: M.Y. was supported by the Grant-in-Aid for Scientific Research (S) 15H05740 of the Japan Society for the Promotion of Science and by the National Natural Science Foundation of China (No. 11771270, 91730303). This work was prepared under the support of the "RUDN University Program 5-100".

Conflicts of Interest: The authors declare no conflict of interest.

References

1. Alt, H.W. *Lineare Funktionalanalysis*; Springer: Berlin, Germany, 2006.
2. Samko, S.G.; Kilbas, A.A.; Marichev, O.I. *Fractional Integrals and Derivatives: Theory and Applications*; Gordon and Breach: New York, NY, USA, 1993.
3. Cartwright, D.I.; McMullen, J.R. A note on the fractional calculus. *Proc. Edinb. Math. Soc.* **1978**, *21*, 79–80. [CrossRef]
4. Hilfer, R.; Luchko, Y. Desiderata for Fractional Derivatives and Integrals. *Mathematics* **2019**, *7*, 149. [CrossRef]
5. Diethelm, K.; Garrappa, R.; Giusti, A.; Stynes, M. Why fractional derivatives with nonsingular kernels should not be used. *Fract. Calc. Appl. Anal.* **2020**, *23*, 610–634. [CrossRef]
6. Luchko, Y. Fractional derivatives and the fundamental theorem of fractional calculus. *Fract. Calc. Appl. Anal.* **2020**, *23*, 939–966. [CrossRef]
7. Clément, P. On abstract Volterra equations in Banach spaces with completely positive kernels. In *Lecture Notes in Math*; Kappel, F., Schappacher, W., Eds.; Springer: Berlin, Germany, 1984; Volume 1076, pp. 32–40.
8. Zacher, R. Boundedness of weak solutions to evolutionary partial integro-differential equations with discontinuous coefficients. *J. Math. Anal. Appl.* **2008**, *348*, 137–149. [CrossRef]
9. Zacher, R. Weak solutions of abstract evolutionary integro-differential equations in Hilbert spaces. *Funkcial. Ekvac.* **2009**, *52*, 1–18. [CrossRef]
10. Kochubei, A.N. General fractional calculus, evolution equations, and renewal processes. *Integr. Equa. Oper. Theory* **2011**, *71*, 583–600. [CrossRef]
11. Kochubei, A.N. General fractional calculus. In *Handbook of Fractional Calculus with Applications*; Volume 1: Basic Theory; Kochubei, A., Luchko, Y., Eds.; Walter de Gruyter: Berlin, Germany; Boston, MA, USA, 2019; pp. 111–126.
12. Kochubei, A.N. Equations with general fractional time derivatives. Cauchy problem. In *Handbook of Fractional Calculus with Applications*; Volume 2: Fractional Differential Equations; Kochubei, A., Luchko, Y., Eds.; Walter de Gruyter: Berlin, Germany; Boston, MA, USA, 2019; pp. 223–234.

13. Kochubei, A.N.; Kondratiev, Y. Growth Equation of the General Fractional Calculus. *Mathematics* **2019**, *7*, 615. [CrossRef]
14. Sin, C.-S. Well-posedness of general Caputo-type fractional differential equations. *Fract. Calc. Appl. Anal.* **2018**, *21*, 819–832. [CrossRef]
15. Luchko, Y.; Yamamoto, M. General time-fractional diffusion equation: Some uniqueness and existence results for the initial-boundary-value problems. *Fract. Calc. Appl. Anal.* **2016**, *19*, 675–695. [CrossRef]
16. Vladimirov, V.S. *Equations of Mathematical Physics*; Nauka: Moscow, Russia, 1971.
17. Janno, J.; Kasemets, K. Identification of a kernel in an evolutionary integral equation occurring in subdiffusion. *J. Inverse Ill-Posed Probl.* **2017**, *25*, 777–798. [CrossRef]
18. Kinash, N.; Janno, J. Inverse problems for a generalized subdiffusion equation with final overdetermination. *Math. Model. Anal.* **2019**, *24*, 236–262.
19. Kinash, N.; Janno, J. An Inverse Problem for a Generalized Fractional Derivative with an Application in Reconstruction of Time- and Space-Dependent Sources in Fractional Diffusion and Wave Equations. *Mathematics* **2019**, *7*, 1138. [CrossRef]
20. Schilling, R.L.; Song, R.; Vondracek, Z. *Bernstein Functions. Theory and Application*; Walter de Gruyter: Berlin, Germany, 2010.
21. Feller, W. *An Introduction to Probability Theory and Its Applications*; Wiley: New York, NY, USA, 1966; Volume 2.
22. Hanyga, A. A comment on a controversial issue: A generalized fractional derivative cannot have a regular kernel. *Fract. Calc. Anal. Appl.* **2020**, *23*, 211–223. [CrossRef]
23. Mainardi, F. *Fractional Calculus and Waves in Linear Viscoelasticity*; Imperial College Press: London, UK, 2010.
24. Luchko, Y. On Complete Monotonicity of Solution to the Fractional Relaxation Equation with the nth Level Fractional Derivative. *Mathematics* **2020**, *8*, 1561. [CrossRef]
25. Bertoin, J. *Lévy Processes*; Cambridge University Press: Cambridge, UK, 1996.
26. Gorenflo, R.; Mainardi, F.; Rogosin, S. Mittag-Leffler function: Properties and applications. In *Handbook of Fractional Calculus with Applications*; Volume 1: Basic Theory; Kochubei, A., Luchko, Y., Eds.; Walter de Gruyter: Berlin, Germany; Boston, MA, USA, 2019; pp. 269–298.
27. Gripenberg, G. Volterra integro-differential equations with accretive nonlinearity. *J. Differ. Equ.* **1985**, *60*, 57–79. [CrossRef]
28. Luchko, Y.; Yamamoto, M. On the maximum principle for a time-fractional diffusion equation. *Fract. Calc. Appl. Anal.* **2017**, *20*, 1131–1145. [CrossRef]
29. Luchko, Y.; Gorenflo, R. An operational method for solving fractional differential equations with the Caputo derivatives. *Acta Math. Vietnam.* **1999**, *24*, 207–233.
30. Luchko, Y.; Yamamoto, M. Maximum principle for the time-fractional PDEs. In *Handbook of Fractional Calculus with Applications*; Volume 2: Fractional Differential Equations; Kochubei, A., Luchko, Y., Eds.; Walter de Gruyter: Berlin, Germany; Boston, MA, USA, 2019; pp. 299–326.
31. Liu, Y.; Li, Z.; Yamamoto, M. Inverse problems of determining sources of the fractional partial differential equations. In *Handbook of Fractional Calculus with Applications*; Volume 2: Fractional Differential Equations; Kochubei, A., Luchko, Y., Eds.; Walter de Gruyter: Berlin, Germany; Boston, MA, USA, 2019; pp. 411–430.
32. Li, Z.; Liu, Y.; Yamamoto, M. Inverse problems of determining parameters of the fractional partial differential equations. In *Handbook of Fractional Calculus with Applications*; Volume 2: Fractional Differential Equations; Kochubei, A., Luchko, Y., Eds.; Walter de Gruyter: Berlin, Germany; Boston, MA, USA, 2019; pp. 431–442.
33. Li, Z.; Yamamoto, M. Inverse problems of determining coefficients of the fractional partial differential equations. In *Handbook of Fractional Calculus with Applications*; Volume 2: Fractional Differential Equations; Kochubei, A., Luchko, Y., Eds.; Walter de Gruyter: Berlin, Germany; Boston, MA, USA, 2019; pp. 443–464.

Publisher's Note: MDPI stays neutral with regard to jurisdictional claims in published maps and institutional affiliations.

© 2020 by the authors. Licensee MDPI, Basel, Switzerland. This article is an open access article distributed under the terms and conditions of the Creative Commons Attribution (CC BY) license (http://creativecommons.org/licenses/by/4.0/).

Review

Unified Approach to Fractional Calculus Images of Special Functions—A Survey

Virginia Kiryakova

Institute of Mathematics and Informatics, Bulgarian Academy of Sciences, 1113 Sofia, Bulgaria; virginia@diogenes.bg

Received: 17 November 2020; Accepted: 14 December 2020; Published: 21 December 2020

Abstract: Evaluation of images of special functions under operators of fractional calculus has become a hot topic with hundreds of recently published papers. These are growing daily and we are able to comment here only on a few of them, including also some of the latest of 2019–2020, just for the purpose of illustrating our unified approach. Many authors are producing a flood of results for various operators of fractional order integration and differentiation and their generalizations of different special (and elementary) functions. This effect is natural because there are great varieties of special functions, respectively, of operators of (classical and generalized) fractional calculus, and thus, their combinations amount to a large number. As examples, we mentioned only two such operators from thousands of results found by a Google search. Most of the mentioned works use the same formal and standard procedures. Furthermore, in such results, often the originals and the images are special functions of different kinds, or the images are not recognized as known special functions, and thus are not easy to use. In this survey we present a unified approach to fulfill the mentioned task at once in a general setting and in a well visible form: for the operators of generalized fractional calculus (including also the classical operators of fractional calculus); and for all generalized hypergeometric functions such as $_p\Psi_q$ and $_pF_q$, Fox H- and Meijer G-functions, thus incorporating wide classes of special functions. In this way, a great part of the results in the mentioned publications are well predicted and appear as very special cases of ours. The proposed general scheme is based on a few basic classical results (from the Bateman Project and works by Askey, Lavoie–Osler–Tremblay, etc.) combined with ideas and developments from more than 30 years of author's research, and reflected in the cited recent works. The main idea is as follows: From one side, the operators considered by other authors are cases of generalized fractional calculus and so, are shown to be (m-times) compositions of weighted Riemann–Lioville, i.e., Erdélyi–Kober operators. On the other side, from each generalized hypergeometric function $_p\Psi_q$ or $_pF_q$ ($p \leq q$ or $p = q+1$) we can reach, from the final number of applications of such operators, one of the simplest cases where the classical results are known, for example: to $_0F_{q-p}$ (hyper-Bessel functions, in particular trigonometric functions of order $(q-p)$), $_0F_0$ (exponential function), or $_1F_0$ (beta-distribution of form $(1-z)^\alpha z^\beta$). The final result, written explicitly, is that any GFC operator (of multiplicity $m \geq 1$) transforms a generalized hypergeometric function into the same kind of special function with indices p and q increased by m.

Keywords: fractional calculus operators; special functions; generalized hypergeometric functions; integral transforms of special functions

MSC: 26A33; 33C60; 33E12; 44A20

1. Introduction

Special functions (SF) have always been unavoidable tools for mathematicians, physicists, astronomers, applied scientists and engineers while looking to express and study (theoretically, in tables or by numerical algorithms) the solutions of treated mathematical models. On the other side, recently there has been an increased interest in fractional calculus (FC) and its applications, as evidence for which we refer the readers to the data in the survey by Machado–Kiryakova [1]. Fractional calculus is nowadays a favorite, and even a sort of fashionable research area, although the boom of publications and attempts to "fractalize" any kinds of integer order models can bring some threats to the prestige of this discipline, especially in cases of weak or wrong results and not adequate innovations. Let us mention also the phenomenon of hundreds of papers of the last few years (only a few of them can be cited here) dealing with "evaluation of FC images of SF", most of which use the same standard techniques with changing only the particular special function (SF) and the particular case of the FC operator. Furthermore, it often happens that in such results the originals and the images are special functions of different kinds, or the images are not recognized as known special functions, and thus are not easy to use. In recent papers, such as [2–5], we share our criticism on this practice and show that all such results can be derived at once by following a general approach, based on ideas from older author's works on generalized fractional calculus (GFC), since [6].

Here we try to collect the ideas, results and examples from our recent works on the subject. The survey starts with Preliminaries (Section 2) providing a short background on the considered SF and FC operators; followed by Section 3 with results for images of the generalized hypergeometric functions $_pF_q$ and $_p\Psi_q$ and their simpler cases under the operators the classical FC operators (Riemann–Liouville and Erdélyi–Kober integrals and derivatives of fractional order). Then, in Section 4 we present our *unified approach for evaluation of GFC operators of arbitrary generalized hypergeometric functions* ($_pF_q$ and $_p\Psi_q$), resulting in the main Theorems 3 and 4. This allows to handle very wide classes of operators of generalized (m-tuple, $m \geq 1$) fractional integration and differentiation and of considered special functions. In Sections 5–7 we consider specifications of these results for the Erdélyi–Kober, Saigo and Marichev–Saigo–Maeda (M-S-M) operators, that appear as cases of our GFC, resp. for $m = 1$, $m = 2$, $m = 3$, give their images for the Wright generalized hypergeometric functions, and many illustrative examples for particular results by other authors. Section 8 considers more general cases of GFC operators with arbitrary multiplicity $m \geq 1$, as the multiple Gel'fond–Leontiev operators related to the multi-index Mittag–Leffler functions, and the hyper-Bessel operators related to the hyper-Bessel functions of Delerue. In Section 9 we comment on works of other authors on introducing some "new" special functions and show that these are again Wright generalized hypergeometric functions $_p\Psi_q$. Therefore, the various FC images they propose come as simple corollaries of our general results. To show the effectiveness of the proposed unified approach, in this survey we collected some 21 examples for FC images of SF, and referred to a long list of other authors' works on the subject. Section 10 summarizes some conclusions.

2. Preliminaries

Here we provide a short and only necessary background on the considered classes of special functions (SF) and of operators of classical FC and of generalized fractional calculus (GFC), so as to explain the general ideas. All details on defining the single-valued branches of the considered functions, functional spaces, and necessary conditions on appearing parameters, can be found in our previous works, as cited, and for example in ([6], Section 5.5.i). Basically, we consider functions in the complex plane of the form $\left\{ f(z) = z^\mu \widetilde{f}(z),\ \mu \geq 0,\ \widetilde{f}(z) \text{ analytic and single valued in } \Omega \right\}$, where Ω is a starlike domain with respect to $z = 0$, usually a disk $\Delta_R : |z| < R$. Most of the considered special functions are entire functions, or analytic ones in disks in $\mathbb{C}$.

The results we consider are for the classes of so-called generalized hypergeometric functions (g.h.f) with Mellin–Barnes type integral representations, namely the Fox H-function, Meijer G-function and their most widely used cases of Wright g.h.f. $_p\Psi_q$ and g.h.f. $_pF_q$. Even if our aim is to incorporate as large as possible classes of special functions, let us mention that other transcendental functions as the elliptic integrals, Lambert W-, Mathiew-, Zeta-, etc. functions are outside of our studies. Also, we emphasize on results for LHS integrals, although for the RHS ones similar techniques and results are applied; and consider Riemann–Liouville type fractional derivatives. For the Caputo-type differentiation operators, similar but different results will be exposed in a separate work.

2.1. Special Functions of Fractional Calculus

Under "classical" Special Functions (SF) we mean these "mathematical functions" and orthogonal polynomials of which the origin goes back to 18th and 19th centuries and are named after great mathematicians like Euler, Gauss, Riemann, Bessel, Kummer, Legendre, Laguerre. These "Special Functions of Mathematical Physics" appeared with the needs of applied sciences and serve as solutions of integer order (most commonly 2nd order) differential equations from models in mathematical physics. In the last two centuries it was observed that modeling of many phenomena of the physical and social world can be reflected much more adequately by means of differential equations of arbitrary fractional or higher integer orders, and the so-called special functions of fractional calculus (SF of FC) as providing tools for their explicit solutions became unavoidable tools in the hands of theoretical and applied scientists recognizing the power of fractional calculus (FC).

Recently, many handbooks and surveys appeared as dedicated not only to classical SF but also to the SF of FC, to mention some of them: Prudnikov–Brychkov–Marichev [7], Marichev [8], Srivastava–Gupta–Goyal [9], Kilbas–Srivastava–Trujillo [10], Podlubny [11], Kiryakova [6], Yakubovich–Luchko [12], Mathai–Haubold [13], Gorenflo–Kilbas–Mainardi–Rogosin [14]. Such a list cannot be full here, and for more sources see also the survey paper Machado–Kiryakova [1]. In the papers on the topic and in this survey, we limit ourselves to the Fox H-functions of one complex variable, as enough of a general level to expose the proposed approach.

Definition 1 (Ch. Fox 1960). *see books such as [6,7,9,10], and earlier and latest ones) The Fox H-function is a generalized hypergeometric function, defined by means of the Mellin–Barnes type contour integral*

$$H_{p,q}^{m,n}\left[z \left| \begin{array}{c} (a_i, A_i)_1^p \\ (b_j, B_j)_1^q \end{array} \right. \right] = \frac{1}{2\pi i} \int_{\mathcal{L}} \mathcal{H}_{p,q}^{m,n}(s) z^{-s} ds, \text{ with } \mathcal{H}_{p,q}^{m,n}(s) = \frac{\prod_{j=1}^{m} \Gamma(b_j + B_j s) \prod_{i=1}^{n} \Gamma(1 - a_i - A_i s)}{\prod_{j=m+1}^{q} \Gamma(1 - b_j - B_j s) \prod_{i=n+1}^{p} \Gamma(a_i + A_i s)}, \quad (1)$$

$z \neq 0$, where $\mathcal{L}$ is a suitable contour (of three possible types in $\mathbb{C}$: $\mathcal{L}_{-\infty}, \mathcal{L}_{\infty}, (\gamma - i\infty, \gamma + i\infty)$), the orders (m, n, p, q) are non negative integers so that $0 \leq m \leq q, 0 \leq n \leq p$, the parameters $A_i > 0, B_j > 0$ are positive, and $a_i, b_j, i = 1, \ldots, p; j = 1, \ldots, q$ can be arbitrary complex such that $A_i(b_j + l) \neq B_j(a_i - l' - 1), l, l' = 0, 1, 2, \ldots; i = 1, \ldots, n; j = 1, \ldots, m$. Note that the integrand $\mathcal{H}_{p,q}^{m,n}(s)$ with $s \mapsto -s$ is the Mellin transform of the H-function (1).

The details on the properties of the Fox H-function can be found in many contemporary handbooks on SF such as [7,9,10], where its behavior is described in term of the denotations:

$$\rho = \prod_{i=1}^{p} A_i^{-A_i} \prod_{j=1}^{q} B_j^{B_j} \; ; \; \Delta = \sum_{k=1}^{q} B_j - \sum_{i=1}^{p} A_i; \qquad (2)$$
$$\mu = \sum_{j=1}^{q} b_j - \sum_{i=1}^{p} a_i + \frac{p-q}{2} \; ; \; a^* = \sum_{i=1}^{n} A_i - \sum_{i=n+1}^{p} A_i + \sum_{j=1}^{m} B_j - \sum_{j=m+1}^{q} B_j.$$

Note that the H-function is an analytic function of z in circle domains $|z| < \rho$ or outside them (or in sectors of them, or in the whole $\mathbb{C}$), depending on the above parameters and the contours.

If all $A_i = B_j = 1$, $i = 1, ..., p$; $j = 1, ..., q$, the H-function $H_{p,q}^{m,n}\left[z \left| \begin{array}{c} (a_i, 1)_1^p \\ (b_j, 1)_1^q \end{array}\right.\right]$ reduces to the *Meijer's G-function* (C.S. Meijer (1936), see details in ([15], Vol.1) and all above-mentioned books)

$$G_{p,q}^{m,n}\left[z \left| \begin{array}{c} (a_i)_1^p \\ (b_j)_1^q \end{array}\right.\right] = \frac{1}{2\pi i} \int_{\mathcal{L}} \mathcal{G}_{p,q}^{m,n}(s) z^{-s} ds = \frac{1}{2\pi i} \int_{\mathcal{L}} \frac{\prod_{j=1}^{m} \Gamma(b_j + s) \prod_{i=1}^{n} \Gamma(1 - a_i - s)}{\prod_{j=m+1}^{q} \Gamma(1 - b_j - s) \prod_{i=n+1}^{p} \Gamma(a_i + s)} z^{-s} ds, \quad z \neq 0. \quad (3)$$

Although simpler than (1), the G-function is yet enough general as it incorporates the Classical SF (known also as Named SF) and many elementary functions. See lists of examples, for example, in ([15], Vol.1), ([6], Appendix C).

Now, we attract the readers' attention to the most typical examples of SF of FC, which are Fox H-functions but *not* reducible to Meijer G-functions in the general case (of *irrational* A_j, B_k). These originate from works of Sir Edward Maitland (E.-M.) Wright in a series of his works (1935–1940).

Definition 2 (see, e.g., ([6,7,14], App.E)). *The Wright generalized hypergeometric function $_p\Psi_q(z)$, called also Fox–Wright function (F-W g.h.f.) is defined as:*

$$_p\Psi_q \left[\begin{array}{c} (a_1, A_1), \ldots, (a_p, A_p) \\ (b_1, B_1), \ldots, (b_q, B_q) \end{array} \bigg| z \right] = \sum_{k=0}^{\infty} \frac{\Gamma(a_1 + kA_1) \ldots \Gamma(a_p + kA_p)}{\Gamma(b_1 + kB_1) \ldots \Gamma(b_q + kB_q)} \frac{z^k}{k!} \quad (4)$$

$$= H_{p,q+1}^{1,p}\left[-z \left| \begin{array}{c} (1 - a_1, A_1), \ldots, (1 - a_p, A_p) \\ (0, 1), (1 - b_1, B_1), \ldots, (1 - b_q, B_q) \end{array}\right.\right]. \quad (5)$$

In terms of parameters (2), the $_p\Psi_q$-function is an entire function of z if $\Delta > -1$, while for $\Delta = -1$, it is an absolutely convergent series in the disk $\{|z| < \rho\}$, and also for $|z| = \rho$ if $\text{Re}(\mu) > 1/2$, see, for example, [16].

If all $A_1 = \cdots = A_p = 1$, $B_1 = \cdots = B_q = 1$, the Wright g.h.f. reduces to the generalized hypergeometric $_pF_q$-function, which is a case of the G-function (3), see details in ([15], Vol.1):

$$_p\Psi_q \left[\begin{array}{c} (a_1, 1), \ldots, (a_p, 1) \\ (b_1, 1), \ldots, (b_q, 1) \end{array} \bigg| z \right] = c \, _pF_q(a_1, \ldots, a_p; b_1, \ldots, b_q; z) = \sum_{k=0}^{\infty} \frac{(a_1)_k \ldots (a_p)_k}{(b_1)_k \ldots (b_q)_k} \frac{z^k}{k!}$$

$$= G_{p,q+1}^{1,p}\left[-z \left| \begin{array}{c} 1 - a_1, \ldots, 1 - a_p \\ 0, 1 - b_1, \ldots, 1 - b_q \end{array}\right.\right]; \quad \text{where} \quad c = \left[\prod_{i=1}^{p} \Gamma(a_i) / \prod_{j=1}^{q} \Gamma(b_j)\right], \, (a)_k := \Gamma(a+k)/\Gamma(a). \quad (6)$$

The Mittag-Leffler (M-L) function, introduced by G. Mittag-Leffler (1902–1905), with extended 2-parameters' definition by R.P. Agarwal (1953), was presented yet in Bateman Project's [15], Vol.3 (1954), in a chapter for "Miscellaneous Functions". However, it was ignored for a long time in the books on special functions because the applied scientists suffered from a lack of tables for its Laplace transforms. Although appearing from studies not related to fractional calculus, nowadays the M-L function has become the most popular and most exploited SF of FC, honored to be the "Queen"-function of FC. See details, for example, in [14], also in [6,17,18].

Definition 3. *The Mittag-Leffler (M-L) functions E_α and $E_{\alpha,\beta}$, are entire functions of order $\rho = 1/\alpha$ and type 1, defined by the power series*

$$E_\alpha(z) = \sum_{k=0}^{\infty} \frac{z^k}{\Gamma(\alpha k + 1)}, \quad E_{\alpha,\beta}(z) = \sum_{k=0}^{\infty} \frac{z^k}{\Gamma(\alpha k + \beta)}, \quad \alpha > 0, \beta > 0. \quad (7)$$

As "fractional index" ($\alpha > 0$) analogs of the exponential and trigonometric functions that satisfy ODEs of 1st and 2nd order ($\alpha = 1, 2$), the M-L functions serve as solutions of fractional order differential equations. A M-L type function with three indices, known as the Prabhakar function (T.R. Prabhakar, 1971) is also often studied and used, for details see [14,17–19], and other contemporary books and surveys on M-L type functions:

$$E^{\gamma}_{\alpha,\beta}(z) = \sum_{k=0}^{\infty} \frac{(\gamma)_k}{\Gamma(\alpha k + \beta)} \frac{z^k}{k!}, \quad \alpha, \beta, \gamma \in \mathbb{C}, \operatorname{Re} \alpha > 0, \tag{8}$$

where $(\gamma)_0 = 1, (\gamma)_k = \Gamma(\gamma + k)/\Gamma(\gamma)$ denotes the Pochhammer symbol. For $\gamma = 1$ we get the M-L function $E_{\alpha,\beta}$, and if additionally $\beta = 1$, it is E_α.

These M-L type functions are simple cases of the Wright g.h.f. and of the H-function, namely:

$$E_{\alpha,\beta}(z) = {}_1\Psi_1 \left[\begin{array}{c} (1,1) \\ (\beta, \alpha) \end{array} \bigg| z \right] = H^{1,1}_{1,2} \left[-z \bigg| \begin{array}{c} (0,1) \\ (0,1), (1-\beta, \alpha) \end{array} \right],$$

$$E^{\gamma}_{\alpha,\beta}(z) = \frac{1}{\Gamma(\gamma)} {}_1\Psi_1 \left[\begin{array}{c} (\gamma,1) \\ (\beta, \alpha) \end{array} \bigg| z \right] = H^{1,1}_{1,2} \left[-z \bigg| \begin{array}{c} (1-\gamma,1) \\ (0,1), (1-\beta, \alpha) \end{array} \right].$$

A vector index extension of (7) appeared in the works by Luchko et al. (for example, Yakubovich-Luchko [12]) on operational calculus' methods for some fractional order PDE. Under the name multi-index (multiple) M-L function, it was introduced by Kiryakova [20] using a different approach, via the Gelfond–Leontiev generalized integration and differentiation operators (see Section 8). Further, these functions are studied in detail by Kiryakova [21,22], by Kilbas–Koroleva–Rogosin [23], Paneva–Konovska [19], and many other followers.

Definition 4 (Kiryakova [21,22]). *Let $m > 1$ be an integer, $(\alpha_1 > 0, \alpha_2 > 0, ..., \alpha_m > 0)$ and $(\beta_1, \beta_2, ..., \beta_m)$ be arbitrary real parameters. By means of these "multi-indices", the multi-index Mittag-Leffler function (multi-M-L f.) is the entire function defined as:*

$$E_{(\alpha_i),(\beta_i)}(z) := E^{(m)}_{(\alpha_i),(\beta_i)}(z) = \sum_{k=0}^{\infty} \frac{z^k}{\Gamma(\alpha_1 k + \beta_1)\ldots\Gamma(\alpha_m k + \beta_m)} \tag{9}$$

$$= {}_1\Psi_m \left[\begin{array}{c} (1,1) \\ (\beta_i, \alpha_i)_1^m \end{array} \bigg| z \right] = H^{1,1}_{1,m+1} \left[-z \bigg| \begin{array}{c} (0,1) \\ (0,1), (1-\beta_i, \alpha_i)_1^m \end{array} \right]. \tag{10}$$

Under weakened restrictions on the α's not obligatory to be all nonnegative, the study was extended by Kilbas et al; see Kilbas–Koroleva–Rogosin [23].

The basic properties and results for the functions (9) and long lists of their examples, all of them having wide applications in solutions of integer- and fractional-order models, are provided in our previous papers like Kiryakova [21,22,24]. Let us shortly mention particular cases like: for $m = 1$, we have the classical M-L function $E_{\alpha,\beta}$ with all its particulars (error-, incomplete gamma-, Rabotnov, etc., functions); and for $m > 1$ (many of them treated in the examples in next sections) (see ([15], Vol.2, Section 7.5.4, Section 7.5.5), [25]):

$$
\begin{aligned}
&\text{the \textit{Wright function / Bessel-Maitland function}} \quad \phi(\kappa, \nu+1; z) := J_\nu^\kappa(-z) = E^{(2)}_{(\kappa,1),(\nu+1,1)}(z) \\
&= {}_1\Psi_1\left[\begin{array}{c}(1,1)\\(\nu+1,\kappa),(1,1)\end{array}\bigg|\, z\right] = {}_0\Psi_1\left[\begin{array}{c}-\,-\\(\nu+1,\kappa)\end{array}\bigg|\, z\right]; \\
&\text{the \textit{Mainardi function}} \quad M(z; \beta) = \phi(-\beta, 1-\beta; -z) = E^{(2)}_{(-\beta,1),(1-\beta,1)}(-z); \\
&\text{and its examples, as } M(z; 1/2) = 1/\sqrt{\pi}\exp(-z^2/4), \text{ and \textit{Airy f.}} \ M(z; 1/3) = 3^{2/3} Ai(z/3^{1/3}); \\
&\text{\textit{Pathak's gen. Wright-Bessel-Lommel f.}} \ J^\mu_{(\nu,\lambda)}(z) = \ldots = (z/2)^{\nu+2\lambda} E^{(2)}_{(1/\mu,1),(\nu+\lambda+1,\lambda+1)}(-z^2/4); \\
&\text{of which, for } \mu = 1, \text{ the \textit{Lommel f.}} \text{ appears as } s_{2\lambda+\nu-1,\nu}(z) = \text{const} \cdot J^1_{\nu,\lambda}(z), \\
&\text{and, thus its particular case, the \textit{Struve f.}} \ H_\nu(z); \\
&\text{\textit{Dzrbashjan's function} with } 2 \times 2 \text{ parameters} \quad \Phi_{1/\alpha_1,1/\alpha_1}(z; \beta_1, \beta_2) := E^{(2)}_{(\alpha_1,\alpha_2),(\beta_1,\beta_2)}(z); \\
&\text{if all } \alpha_i = 1, i=1,\ldots,m\text{: \textit{hyper-Bessel f. of Delerue}} \ J^{(m)}_{\nu_1,\ldots,\nu_m}(z), \text{ as multi-index ext. of the Bessel f.; etc.}
\end{aligned} \quad (11)
$$

Recently, in Kilbas–Koroleva–Rogosin [23] the definition (9) has been extended for arbitrary values of the α's parameters. Paneva-Konovska introduced and studied generalizations of the Prabhakar function (8) by means of three sets of parameters $(\alpha_1 > 0, \alpha_2 > 0, \ldots, \alpha_m > 0)$, $(\beta_1, \beta_2, \ldots, \beta_m)$, $(\gamma_1, \gamma_2, \ldots, \gamma_m)$, called *3m-parametric M-L functions*, see, for example, [19,26] and references therein. Multivariate and matrix extensions of the M-L and multi-index M-L functions are also explored.

In another survey paper, Kiryakova [27], we are exposing many other details on the theory of the SF of FC, in the sense of Wright generalized hypergeometric functions ${}_p\Psi_q$ and multi-index Mittag-Leffler functions, and provide an extensive list of their particular cases, studied in theoretical and applicable aspects by various authors.

Remark 1. *The techniques of the Mellin transform*

$$\mathfrak{M}\{f(z); s\} := F^*(s) = \int_0^\infty f(z)\, z^{s-1} dt$$

is one of the main tools to evaluate integrals and various integral transforms of special functions, including their images under operators of FC. After some classical publications of previous centuries, the main contribution to this approach is due to Marichev [8]. He proposed a natural but wide ranged scheme, based on the contour integral representations of Mellin–Barnes type for the H- and G-functions, like (1) *and* (3). *Note that the integrands* $\mathcal{H}^{m,n}_{p,q}$ *and* $\mathcal{G}^{m,n}_{p,q}$ *are their Mellin transforms (of variable $s \mapsto -s$) are fractions of products of 2×2 groups of Gamma-functions, and each special function being a special case of the generalized hypergeometric functions, has a particular representation of that kind. For example* ([10], (1.11.24)):

$$\mathfrak{M}\left\{{}_p\Psi_q\left[\begin{array}{c}(a_i, A_i)_1^p\\(b_j, B_j)_1^q\end{array}\bigg|\,-z\right]; s\right\} = \frac{\Gamma(s)\prod_{i=1}^p \Gamma(a_i - sA_i)}{\prod_{j=1}^q \Gamma(b_j - sB_j)}.$$

For variations of results, one can use in addition the relations (see, e.g., in ([28], (2.6)–(2.8))):

$$\mathfrak{M}\{f(\lambda z); s\} = \lambda^{-s} F^*(s + \gamma), \ \lambda > 0; \quad \mathfrak{M}\{z^\gamma f(z); s\} = F^*(s + \gamma); \quad \mathfrak{M}\{f(z^\mu); s\} = \frac{1}{\mu}F^*\left(\frac{s}{\mu}\right), \ \mu > 0.$$

Examples for the use of the Mellin transform in this respect are given (among many others works) in: Luchko and Kiryakova ([28], Section 4) (general scheme and examples with the M-L and Wright functions), Agarwal, Rogosin, and Trujillo [29] and Paneva-Konovska and Kiryakova [30] (images for multi-index M-L functions and their particular cases).

2.2. Operators of Generalized Fractional Calculus

In fractional calculus (FC), meant as a theory of the integration and differentiation of arbitrary (including fractional, not obligatorily integer) order, there are several almost equivalent definitions for "fractional" integrals and derivatives, applied in various functional spaces. Here we are interested in evaluating FC operator images of special functions, defined by power series, most of which are entire functions, or at least analytic ones inside/outside disks in a complex plane. Therefore, we restrict our statements to such functions, although they hold also for spaces of weighted continuous or Lebesgue integrable functions on the real half-line.

For the basic background on FC theory and related topics as SF, integral transforms, generalizations and applications, we can refer to the books (among many others, for a longer list see, e.g., Machado and Kiryakova [1]), such as those by: Samko, Kilbas, and Marichev [31], Podlubny [11], Kilbas, Srivastava, and Trujillo [10], Yakubovich and Luchko [12], including the author's one, Kiryakova [6] and a recent one, Sandev and Tomovski [32].

We state results for the Riemann-Liouville (R-L) operator for integration R^δ of order $\delta > 0$, the corresponding R-L fractional derivative D^δ, and its counterpart $*D^\delta$ in the Caputo sense, that is only the left-hand sided operators of FC (and skip similar details for the Weyl-type, right-hand sided operators).

The main operator of fractional integration we consider is the Erdélyi–Kober operator (E-K) *of integration* of order $\delta > 0$, depending on two additional parameters $\gamma \in \mathbb{R}$ and $\beta > 0$,

$$I_\beta^{\gamma,\delta} f(z) = \frac{1}{\Gamma(\delta)} \int_0^1 \sigma^\gamma (1-\sigma)^{\delta-1} f(z\sigma^{\frac{1}{\beta}}) d\sigma = \frac{z^{-\beta(\gamma+\delta)}}{\Gamma(\delta)} \int_0^z (z^\beta - \xi^\beta)^{\delta-1} \xi^{\beta\gamma} f(\xi) d(\xi^\beta), \quad (12)$$

note it is the identity for $\delta = 0$. Especially for functional spaces of weighted analytic functions of the form $f(z) = z^\mu \tilde{f}(z)$, $\mu \geq 0$ (see beginning of Section 2), to be preserved by this operator, we require $\gamma > -1 - \frac{\mu}{\beta}$, in addition to $\delta \geq 0, \beta > 0$. This operator, more general than the R-L integral, and having many more applications, was introduced in Sneddon's works, such as [33], and is considered in books ([6,10,31], Ch.2), and recently in many other works on fractional order models. The Erdélyi–Kober-type fractional integrals, or briefly Erdélyi–Kober integrals, of the form

$$If(z) = z^{\delta_0} I_\beta^{\gamma,\delta} f(z), \quad \text{with} \quad \delta_0 \geq 0. \quad (13)$$

are basic in our studies, and are called classical fractional integrals, and we consider their commutable compositions that are presented as our generalized fractional integrals, Kiryakova [6,34].

The Erdélyi–Kober operator (13) reduces to the R-L operator of integration for $\gamma=0, \beta=1, \delta_0 = \delta$,

$$R_{0+,z}^\delta f(z) := R^\delta f(z) = z^\delta I_1^{0,\delta} f(z); \quad \text{and conversely,} \quad I_1^{\gamma,\delta} f(z) = z^{-\gamma-\delta} R^\delta z^\gamma f(z). \quad (14)$$

Note that some authors often refer to the Erdélyi–Kober integral (12) as Euler integral transformation, when they are to handle various integral transforms of special functions.

The fractional order derivative of R-L type corresponding to the E-K integral (12), called E-K *fractional derivative* $D_\beta^{\gamma,\delta}$, is an extension of the R-L fractional derivative $D^\delta f(z) := \left(\frac{d}{dz}\right)^n R^{n-\delta} f(z)$. Instead of $(d/dz)^n$, a suitably chosen auxiliary differential operator D_n of integer order is used, a polynomial of the Euler differential operator $(z\,d/dz)$. It has been introduced and studied in the works of Kiryakova and Luchko et al., ([6], Ch.2) and ([12], Ch.3) and in the next ones, as [35],

$$D_\beta^{\gamma,\delta} f(z) = D_n I_\beta^{\gamma+\delta,n-\delta} f(z) = \prod_{j=1}^n \left(\frac{1}{\beta} z \frac{d}{dz} + \gamma + j\right) I_\beta^{\gamma+\delta,n-\delta} f(z), \quad n-1 < \delta \leq n, n \in \mathbb{N}. \quad (15)$$

The more formal representation ([6], Ch.1, Equation (1.6.7))

$$D_\beta^{\gamma,\delta} f(z) = \left[z^{-\gamma} D^\delta z^{\gamma+\delta} f(z^{1/\beta}) \right]_{z \mapsto z^\beta} \tag{16}$$

serves to provide a better understanding on the structure and nature of (15).

The Caputo-type analogs of the R-L and E-K fractional derivatives are defined in the same way but with exchanged order of the nonnegative order integration and the integer order differentiation, see, for example, [36], also [35], namely,

$$*D_\beta^{\gamma,\delta} f(z) = I_\beta^{\gamma+\delta, n-\delta} D_n f(z).$$

The notion of generalized fractional integration operators was introduced by S. Kalla (1969–1979), who suggested the common form of such operators (see details and references in [37]),

$$I f(z) = \int_0^1 \Phi(\sigma) \sigma^\gamma f(z\sigma) d\sigma = z^{-\gamma-1} \int_0^z \Phi(\frac{\xi}{z}) \xi^\gamma f(\xi) d\xi, \tag{17}$$

where $\Phi(\sigma)$ is an arbitrary continuous (analytical) function for which the integral makes sense, most commonly a special function as the Bessel, Gauss, G- or H-function. The operators of such generalized fractional calculus (GFC) are expected to include, in particular, these of the classical FC and should satisfy the main axioms for the FC theory.

Note that for a rather general or rather narrow choice of the special function Φ, only some formal operational rules for the generalized fractional integrals (17) can be provided. Therefore, in our generalized fractional calculus (GFC), Kiryakova [6], the suitable choice of the kernel-functions Φ as $G_{m,m}^{m,0}$- and $H_{m,m}^{m,0}$-functions was crucial. In that case, the generalized fractional integrals can be decomposed into commutative products of operators of classical FC (Erdélyi–Kober operators). Thus, the tools of the special functions and the wide usage of the classical FC are combined into a GFC with developed detailed theory and many established applications.

Definition 5 (Kiryakova [6]). *The multiple E-K integral (of multiplicity $m > 1$), is defined by means of the real parameters' sets $(\delta_1 \geq 0, ..., \delta_m \geq 0)$—multi-order of integration, $(\gamma_1, ..., \gamma_m)$— multi-weight; and $(\beta_1 > 0, ..., \beta_m > 0)$—additional multi-parameter, as:*

$$I_{(\beta_k),m}^{(\gamma_k),(\delta_k)} f(z) := \int_0^1 H_{m,m}^{m,0} \left[\sigma \left| \begin{array}{c} (\gamma_k + \delta_k + 1 - \frac{1}{\beta_k}, \frac{1}{\beta_k})_1^m \\ (\gamma_k + 1 - \frac{1}{\beta_k}, \frac{1}{\beta_k})_1^m \end{array} \right. \right] f(z\sigma) d\sigma, \tag{18}$$

if $\sum_{k=1}^m \delta_k > 0$; and as the identity operator: $I_{(\beta_k),m}^{(\gamma_k),(0,...,0)} f(z) = f(z)$, if $\delta_1 = \delta_2 = \cdots = \delta_m = 0$.

Note that the above kernel $H_{m,m}^{m,0}$-function is an analytic function in the unit disk and $H_{m,m}^{m,0}(\sigma) \equiv 0$ for $|\sigma| > 1$ (Kiryakova, [6]). Specially for functional spaces of weighted analytic functions of the form $f(z) = z^\mu \tilde{f}(z)$, $\mu \geq 0$ (see beginning of Section 2), to be preserved by this operator, we require $\gamma_k > -1 - \frac{\mu}{\beta_k}$, in addition to $\delta_k \geq 0$, $\beta_k > 0$.

If all the β's are equal: $\beta_1 = \beta_2 = ... = \beta_m = \beta > 0$, then (18) has a simpler representation where the kernel is a $G_{m,m}^{m,0}$-function of Meijer, which is also analytic in unit disk and $G_{m,m}^{m,0}(\sigma) \equiv 0$ for $|\sigma| > 1$,

$$I_{(\beta,...,\beta),m}^{(\gamma_k),(\delta_k)} f(z) := I_{\beta,m}^{(\gamma_k),(\delta_k)} f(z) = \int_0^1 G_{m,m}^{m,0} \left[\sigma \left| \begin{array}{c} (\gamma_k + \delta_k)_1^m \\ (\gamma_k)_1^m \end{array} \right. \right] f(z\sigma^{1/\beta}) d\sigma = \left[\prod_{k=1}^m I_\beta^{\gamma_k, \delta_k} \right] f(z). \tag{19}$$

The operators of the form

$$\tilde{I}f(z) = z^{\delta_0} I_{(\beta_k),m}^{(\gamma_k),(\delta_k)} f(z), \quad \tilde{I}f(z) = z^{\delta_0} I_{\beta,m}^{(\gamma_k),(\beta_k)} f(z), \quad \text{with } \delta_0 \geq 0, \quad (20)$$

are both referred to shortly as generalized fractional integrals of multi-order $(\delta_1, ..., \delta_m)$.

The following decomposition property is proved in [6], etc. (see, e.g., decomposition Th.5.2.1 in [6]). It is important because the GFC integrals (18) and (19) can be represented not only by using the kernel Fox H- and G-functions, but also by means of the repeated integral representations for the commutative product of classical E-K operators (12):

$$I_{(\beta_k),m}^{(\gamma_k),(\delta_k)} f(z) := \left[\prod_{k=1}^{m} I_{\beta_k}^{\gamma_k,\delta_k} \right] f(z) = \int_0^1 \cdots \int_0^1 \left[\prod_{k=1}^{m} \frac{(1-\sigma_k)^{\delta_k-1} \sigma_k^{\gamma_k}}{\Gamma(\delta_k)} \right] f\left(z\sigma_1^{1/\beta_1} \cdots \sigma_m^{1/\beta_m}\right) d\sigma_1 \cdots d\sigma_m. \quad (21)$$

In the book Kiryakova [6] and subsequent papers, we have provided the operational properties of the operators (18) and (19) as semigroup property, formal inversion formula, reduction to identity or to the conventional integration operators for special parameters' choice. This is to justify their names as operators of GFC.

Following the idea of how the R-L and E-K fractional derivatives are defined, we have proposed the definition of the corresponding generalized fractional derivatives. To this end, the auxiliary differential operator D_η, a polynomial of $z(\frac{d}{dz})$ of degree $\eta_1 + ... + \eta_m$, is used:

$$D_\eta = \left[\prod_{r=1}^{m} \prod_{j=1}^{\eta_r} \left(\frac{1}{\beta_r} z \frac{d}{dz} + \gamma_r + j \right) \right], \quad \text{with } \eta_k := \begin{cases} [\delta_k] + 1, & \text{for noninteger } \delta_k, \\ \delta_k, & \text{for integer } \delta_k, \end{cases} \quad k = 1, \ldots, m. \quad (22)$$

Definition 6 (Kiryakova ([6], Ch.1,Ch.5), [34,35]). *The multiple (m-tuple) Erdélyi-Kober fractional derivative of R-L type of multi-order $\delta = (\delta_1 \geq 0, \ldots, \delta_m \geq 0)$ is defined by means of the differ-integral operator:*

$$D_{(\beta_k),m}^{(\gamma_k),(\delta_k)} f(z) := D_\eta I_{(\beta_k),m}^{(\gamma_k+\delta_k),(\eta_k-\delta_k)} f(z) = D_\eta \int_0^1 H_{m,m}^{m,0} \left[\sigma \left| \begin{array}{c} (\gamma_k + \eta_k + 1 - \frac{1}{\beta_k}, \frac{1}{\beta_k})_1^m \\ (\gamma_k + 1 - \frac{1}{\beta_k}, \frac{1}{\beta_k})_1^m \end{array} \right. \right] f(z\sigma) \, d\sigma. \quad (23)$$

Analogously, the Caputo-type generalized fractional derivative has been introduced in Kiryakova and Luchko [35], as

$$*D_{(\beta_k),m}^{(\gamma_k),(\delta_k)} f(z) = I_{(\beta_k),m}^{(\gamma_k+\delta_k),(\eta_k-\delta_k)} D_\eta f(z). \quad (24)$$

For all equal β's: $\beta_1 = ... = \beta_m = \beta > 0$, the R-L and Caputo-type "derivatives" corresponding to the generalized fractional integral (19) has a simpler form with Meijer G-function in the kernel:

$$D_{\beta,m}^{(\gamma_k),(\delta_k)} f(z) = D_\eta I_{\beta,m}^{(\gamma_k+\delta_k),(\eta_k-\delta_k)} f(z) = \left[\prod_{r=1}^{m} \prod_{j=1}^{\eta_r} \left(\frac{1}{\beta} z \frac{d}{dz} + \gamma_r + j \right) \right] I_{\beta,m}^{(\gamma_k+\delta_k),(\eta_k-\delta_k)} f(z),$$

$$\text{and } *D_{\beta,m}^{(\gamma_k),(\delta_k)} f(z) = I_{\beta,m}^{(\gamma_k+\delta_k),(\eta_k-\delta_k)} D_\eta f(z). \quad (25)$$

Under generalized (multiple, multi-order) fractional derivatives of the R-L type, resp. of the Caputo type, we have in mind all the differ-integral/integro-differential operators of the form

$$\tilde{D}f(z) = D_{(\beta_k),m}^{(\gamma_k),(\delta_k)} z^{-\delta_0} f(z) = z^{-\delta_0} D_{(\beta_k),m}^{(\gamma_k - \frac{\delta_0}{\beta}),(\delta_k)} f(z), \quad *\tilde{D}f(z) = *D_{(\beta_k),m}^{(\gamma_k),(\delta_k)} z^{-\delta_0} f(z) \text{ with } \delta_0 \geq 0. \quad (26)$$

A basic formula for the image of a power function in the general case of (18) and (19) (say from Kiryakova [6]) reads as

$$I_{(\beta_k),m}^{(\gamma_k),(\delta_k)} \{z^p\} = c_p z^p, \quad \text{with } c_p = \prod_{i=1}^{m} \frac{\Gamma(\gamma_i + 1 + p/\beta_i)}{\Gamma(\gamma_i + \delta_i + 1 + p/\beta_i)}, \quad \delta_k \geq 0, \, p > -\beta_k(\gamma_k + 1), \, k = 1, ..., m, \quad (27)$$

and a similar one holds for the generalized fractional derivatives, both analogous to the same formulas for the classical Erdélyi–Kober operators. These results are in the base of the standard techniques applied by other authors for the evaluation of FC operators of special functions, in various particular cases. Using (27), in particular for $p = k = 0, 1, 2, \ldots, n, \ldots$, then interchanging the integration and summation of the power series for a particular special function, the authors of mentioned papers obtain a new power series to be recognized as another special function (in the successful cases, or the result is useless, left just as such series or as some ${}_p\Psi_q$-function). Our general result states as follows.

Theorem 1 (Kiryakova, since 1988, see, e.g., ([6,38], Ch.5)). *Let the conditions $\beta_k(\gamma_k + 1) > -\mu$, $\delta_k \geq 0$, $\beta_k > 0$, $k = 1, \ldots, m$, be satisfied for the parameters of the multiple E-K integral (18). Then, it preserves the class of weighted analytic functions $f(z)$ in a disk Δ_R, denoted by $\mathcal{H}_\mu(\Delta_R = \{|z| < R\})$:*

$$f(z) = z^\mu \sum_{k=0}^{\infty} a_k z^k = z^\mu (a_0 + a_1 z + \ldots) \in \mathcal{H}_\mu(\Delta_R) \quad \text{with} \quad R = \left\{ \limsup_{k \to \infty} \sqrt[k]{|a_k|} \right\}^{-1}. \tag{28}$$

Namely, the images of such functions have the same form:

$$I_{(\beta_k),m}^{(\gamma_k),(\delta_k)} f(z) = z^\mu \sum_{k=0}^{\infty} a_k b_k z^k \in \mathcal{H}_\mu(\Delta_R), \quad \text{with} \quad b_k = \left\{ \prod_{i=1}^{m} \frac{\Gamma(\gamma_i + \frac{k+\mu}{\beta_i} + 1)}{\Gamma(\gamma_i + \delta_i + \frac{k+\mu}{\beta_i} + 1)} \right\} > 0, \tag{29}$$

with the same radius of convergence $R > 0$ and the same signs of the coefficients in their series expansions.

2.3. Some Special Cases of GFC Operators

We emphasize here only some operators of FC that are recently exploited very often in publications on FC operators of SF. In [6] and the author's other papers as well as in works by other authors, there many other particular cases of linear integral and differential operators provided and used with applications in geometric (univalent) function theory, in differential and integral equations of integer and fractional order, operational calculus, transmutation theory, special functions theory, mathematical models of phenomena of fractional order, etc.

For $m = 1$ the kernel-functions of the generalized fractional integrals and derivatives (18) and (23) can be represented as

$$H_{1,1}^{1,0}\left[\sigma \left|\begin{array}{c}(\gamma+\delta, 1/\beta)\\(\gamma, 1/\beta)\end{array}\right.\right] = \beta \sigma^{\beta-1} G_{1,1}^{1,0}\left[\sigma^\beta \left|\begin{array}{c}\gamma+\delta\\\gamma\end{array}\right.\right] = \beta \frac{\sigma^{\beta\gamma+\beta-1}(1-\sigma^\beta)^{\delta-1}}{\Gamma(\delta)},$$

therefore we have the E-K and R-L ($\gamma = 0, \beta = 1$) operators of classical FC. Many other integration and differentiation operators introduced and used by different authors appear as special cases of $I_{\beta,1}^{\gamma,\delta} = I_\beta^{\gamma,\delta}$, $D_{\beta,1}^{\gamma,\delta} = D_\beta^{\gamma,\delta}$, R^δ and D^δ.

When $m = 2$, the kernels $H_{2,2}^{2,0}$ and $G_{2,2}^{2,0}$ reduce to a Gauss hypergeometric lfunction:

$$H_{2,2}^{2,0}\left[\sigma \left|\begin{array}{c}(\gamma_1+\delta_1+1-\frac{1}{\beta}, \frac{1}{\beta}), (\gamma_2+\delta_2+1-\frac{1}{\beta}, \frac{1}{\beta})\\(\gamma_1+1-\frac{1}{\beta}, \frac{1}{\beta}), (\gamma_2+1-\frac{1}{\beta}, \frac{1}{\beta})\end{array}\right.\right] = G_{2,2}^{2,0}\left[\sigma^\beta \left|\begin{array}{c}\gamma_1+\delta_1, \gamma_2+\delta_2\\\gamma_1, \gamma_2\end{array}\right.\right]$$

$$= \frac{\sigma^{\beta\gamma_2}(1-\sigma^\beta)^{\delta_1+\delta_2-1}}{\Gamma(\delta_1+\delta_2)} {}_2F_1(\gamma_2+\delta_2-\gamma_1, \delta_1; \delta_1+\delta_2; 1-\sigma^\beta). \tag{30}$$

In this case, the generalized fractional integrals are known as hypergeometric fractional integrals, and some of them are introduced and studied by Love, Saxena, Kalla, Saigo (see in next Section 6), Hohlov, etc.

In the case $m = 3$, a recently very popular example is with the Marichev–Saigo–Maeda (M-S-M) operators. These FC integration operators are introduced and studied by Marichev [39] and by Saigo et al. [40]. Their kernel-function, the Appel F_3 function (Horn function)

$$F_3\left(a, a', b, b', c, z, \xi\right) = \sum_{m,n=0}^{\infty} \frac{(a)_m (a')_n (b)_m (b')_n}{(c)_{m+n}} \frac{z^m \xi^n}{m! \, n!}, \quad |z| < 1, |\xi| < 1 \text{ (see, e.g., [7,15])},$$

appears as a case of the $G_{3,3}^{3,0}$-function and of $H_{3,3}^{3,0}$-function. Indeed, according to [7], p.727, (2); and as observed in Kiryakova [6], p.21:

$$\frac{(1-\sigma)^{c-1}}{\Gamma(c)} F_3\left(a, a', b, b', c, 1 - \tfrac{1}{\sigma}, 1 - \sigma\right)$$
$$= G_{3,3}^{3,0}\left[\sigma \left| \begin{array}{c} a+b, c-a', c-b' \\ a, b, c-a'-b' \end{array}\right.\right] = H_{3,3}^{3,0}\left[\sigma \left| \begin{array}{c} (a+b,1), (c-a',1), (c-b',1) \\ (a,1), (b,1), (c-a'-b',1) \end{array}\right.\right], \quad \operatorname{Re} c > 0. \tag{31}$$

Therefore, our generalized fractional integrals reduce in this case to the M-S-M integral operators.

Let $m \geq 1$ be an arbitrary integer, but all δ's are integers, say $\delta_1 = \ldots = \delta_m = 1$. Then we have the Bessel type integral and differential operators of arbitrary (higher) integer order, introduced by Dimovski [41] (see also [42]) and named as hyper-Bessel operators by Kiryakova ([6], Ch.3), as shown related to the hyper-Bessel functions of Delerue [43] as their eigenfunctions (see Example 16, in next Section 8). The studies on these operators gave rise to our GFC, since they appeared as "fractional" integrals and derivatives of integer multi-orders $(1, 1, \ldots, 1)$ and for $\lambda > 0$ their fractional powers have multi-orders $(\lambda, \lambda, \ldots, \lambda)$. In Section 8, we will discuss also the Gelfond–Leontiev operators generated by the multi-index M-L functions, as more general operators of arbitrary multiplicity $m > 1$ and arbitrary fractional multi-order.

As mentioned, here we stress on only a few particular examples of GFC operators $I_{(\beta_k),m}^{(\gamma_k),(\delta_k)}$, $D_{(\beta_k),m}^{(\gamma_k),(\delta_k)}$, that are involved in results on to the topic of this survey. This is because many other authors' works handle the evaluation of images of various elementary or special functions under the classical or some "generalized operators of FC"—such as the operators of R-L, E-K, Saigo, Marichev-Saigo-Maeda. Say, one takes first the cosine or Bessel function, later the generalized Bessel (Bessel-Maitland) function, then an M-L or generalized M-L function, etc., so as to produce new publications by same standard techniques. Very rarely observed, or mostly is ignored, the fact from relation (21) that these are 2-tuple ($m = 2$), respectively 3-tuple ($m = 3$), or m-tuple (arbitrary $m > 1$) compositions of Erdélyi–Kober operators. Therefore, the task can be done at once, if one knows how an E-K operator acts on such special functions, all being cases of Wright g.h.f. (4), and then applying the procedure a suitable number of times (2-, 3-, or m). Thus, the result can be predicted in advance, having in mind the general statements in the next sections.

3. Erdélyi-Kober and Riemann–Liouville Images of $_p\Psi_q$, $_pF_q$ and Simpler Special Functions

Some basic classical results on the topic exist from the previous century that should not be forgotten and on which our approach was built. Namely, the image of a generalized hypergeometric function $_pF_q$, with $p \leq q + 1$, under the R-L fractional integral/derivative is shown to be the same special function with indices p and q increased by 1:

$$R^\delta \left\{ z^{\nu-1} \, _pF_q(a_1, \ldots, a_p; b_1, \ldots, b_q; \lambda z) \right\} = \frac{\Gamma(\nu)}{\Gamma(\delta + \nu)} z^{\delta + \nu - 1} \, _{p+1}F_{q+1}(a_1, \ldots, a_p, \nu; b_1, \ldots, b_q, \delta + \nu; \lambda z), \tag{32}$$

with $\operatorname{Re} \delta > 0$, $\operatorname{Re} \nu > 0$, $p \leq q + 1$; $\lambda \neq 0$, $z \in \mathbb{C}$ and if $p = q + 1$: $|\lambda z| < 1$ is additionally required. For this, we can refer to Erdélyi et al. ([44], Vol.2), Ch. XIII, Equations (95)–(97); Askey ([45], p.19), and emphasize the survey by Lavoie-Osler-Tremblay [46], a table on p.261. Then, to make use of (32), in our older works on the topic since 1984-1985, we started from the R-L images of the $_pF_q$ functions with the lowest possible indices: $_1F_0$, $_1F_1$ and $_0F_1$, given by three basic elementary functions, see for example in Kiryakova ([6], Ch.4). Some particular illustrative cases in this direction were mentioned there, as

$$I_2^{-1,\nu+\frac{1}{2}}\{\cos z\} = R_{z^2}^{\nu+\frac{1}{2}}\left\{\frac{\cos z}{z}\right\} = \sqrt{\pi}\, 2^\nu z^{-\nu} J_\nu(z) = \sqrt{\pi}\, 2^{\nu-1} z^{-\nu+1}\, _0F_1\left(\nu+1;-\frac{z^2}{4}\right),\ \nu>-\tfrac{1}{2}, \quad (33)$$

$$R^\delta\left\{z^{\nu-1}\exp(\lambda z)\right\} = \frac{\Gamma(\nu)}{\Gamma(\delta+\nu)} z^{\delta+\nu-1}\, _1F_1(\nu;\delta+\nu;\lambda z),\ \mathrm{Re}\,\delta>0,\ \mathrm{Re}\,\nu>0, \quad (34)$$

$$R^\delta\left\{z^{\nu-1}(z-\lambda)^\mu\right\} = \frac{(-\lambda)^\mu z^{\delta+\nu-1}\Gamma(\nu)}{\Gamma(\delta+\nu)}\, _2F_1(-\mu,\nu;\delta+\nu,b_1,...,b_q;z/\lambda),\ \mathrm{Re}\,\delta>0,\ \mathrm{Re}\,\nu>0. \quad (35)$$

Note that (33) is an interpretation of the Poisson integral formula for the Bessel function, that has been generalized in [42] and ([6], Ch.4) to represent the hyper-Bessel functions $J^{(m)}_{\nu_1,...,\nu_m}$ with multi-indices $(\nu_1,...,\nu_m)$ (the Bessel function J_ν is the case for $m=q-p=1$ with one index ν), that is, to represent the $_0F_{q-p}$-functions by means of "generalized cosine" $\cos_m$.

R-L integrals/ derivatives of the most general G- and H-functions are also well known in the literature (for example, from [44]), and these are the same type of functions but with increased orders and additional parameters.

Along with the mentioned old classical results, recently, new articles are published on the evaluation of classical (R-L, E-K) or generalized FC operators of classical SF or of SF of FC almost every day (e.g., in 2020: [47,48]), and also of their *multivariate or matrix variants*. Just as one example on fractional operators for the matrix Wright hypergeometric functions (5), is a 2020 paper [49].

The classical results (32) and (33)–(35) have been extended in our works (as in ([6], Ch.4), [22,24,50]) in terms of the Erdélyi–Kober operators (12) and (15) and for *their counterparts of the GFC:* $I^{(\gamma_k),(\delta_k)}_{(\beta_k),m}$, $D^{(\gamma_k),(\delta_k)}_{(\beta_k),m}$, not only for $_pF_q$ but for $_p\Psi_q$ as well. To reduce the Wright g.h.f. $_p\Psi_q$ in the general case to three basic simplest functions with lowest indices p and q, we also apply modifications as the Wright–Erdélyi–Kober multiple operators with a Bessel–Maitland kernel-function and in general, GFC operators with H-functions like in (20) but with different parameters $1/\beta_k>0$ and $1/\lambda_k>0$ in the upper and low row. Details are in Kiryakova [24].

Now we provide some basic statements necessary for the topic of this survey, repeating in a few lines the ideas of the proofs, so as to clarify the approach used.

Lemma 1. *The image of a Wright g.h.f. $_p\Psi_q$ under the Erdélyi–Kober fractional integral (12) is the same type of function in which the indices p- and q are increased by one, and so, has two additional parameters:*

$$I^{\gamma,\delta}_\beta\left\{z^c\, _p\Psi_q\left[\begin{matrix}(a_1,A_1),...,(a_p,A_p)\\(b_1,B_1),...,(b_q,B_q)\end{matrix}\Big|\lambda z^\mu\right]\right\} = z^c\, _{p+1}\Psi_{q+1}\left[\begin{matrix}(a_i,A_i)_1^p,(\gamma+1+c/\beta,\mu/\beta)\\(b_j,B_j)_1^q,(\gamma+\delta+1+c/\beta,\mu/\beta)\end{matrix}\Big|\lambda z^\mu\right].$$
(36)

It is supposed that $\mathrm{Re}\,\delta>0$, $\mathrm{Re}\,\gamma>-1$, $\mu>0$, $\lambda\neq 0$, c is arbitrary (real), and if $p=q+1$, then $|\lambda z^\mu|<1$ is additionally required.

Proof. In a simpler case with $c=0$, this is Lemma 1 from Kiryakova [2]. There, a proof is based on the Formula (44) (Section 4) for the integral (the Mellin transform) of a product of two H-functions, since both the $_p\Psi_q$-function and the kernel of the E-K operator are cases of H-functions; compare (5) and Section 2.3. This approach will be discussed later for the more general case of GFC operators.

As a very standard technique, to prove (36), one can use term-by-term integration in series (4), similarly to that in Kilbas ([16], Th.2) for the particular case of R-L integral, with $c:=\nu-1$ there:

$$R^\delta\left\{z^c\, _p\Psi_q\left[\begin{matrix}(a_1,A_1),...,(a_p,A_p)\\(b_1,B_1),...,(b_q,B_q)\end{matrix}\Big|\lambda z^\mu\right]\right\} = z^{c+\delta}\, _{p+1}\Psi_{q+1}\left[\begin{matrix}(a_i,A_i)_1^p,(c+1,\mu)\\(b_j,B_j)_1^q,(c+\delta+1,\mu)\end{matrix}\Big|\lambda z^\mu\right]. \quad (37)$$

Note that the known simpler formula (32), written by means of E-K integrals, appears as a special case of (36) if all $A_1 = \ldots = A_p = B_1 = \ldots = B_q = 1$ and $\beta = 1$, see Equation (4.2.2′) in Kiryakova [6], for $\operatorname{Re} \delta > 0, \operatorname{Re} \gamma > -1$:

$$I_1^{\gamma,\delta} \{ {}_pF_q(a_1, \ldots, a_p; b_1, \ldots, b_q; \lambda z) \} = \frac{\Gamma(\gamma+1)}{\Gamma(\gamma+\delta+1)} {}_{p+1}F_{q+1}(a_1, \ldots, a_p, \gamma+1; b_1, \ldots, b_q, \gamma+\delta+1; \lambda z). \quad (38)$$

□

Lemma 2. *The image of a Wright g.h.f.* ${}_p\Psi_q$ *under the E-K fractional derivative* (15) *is the same kind of function but with indices p and q increased by 1, and with 2 the additional parameters:*

$$D_\beta^{\gamma,\delta} \left\{ z^c \, {}_p\Psi_q \left[\begin{array}{c} (a_1, A_1), \ldots, (a_p, A_p) \\ (b_1, B_1), \ldots, (b_q, B_q) \end{array} \Big| \lambda z^\mu \right] \right\} = z^c \, {}_{p+1}\Psi_{q+1} \left[\begin{array}{c} (a_i, A_i)_1^p, (\gamma+\delta+1+c/\beta, \mu/\beta) \\ (b_j, B_j)_1^q, (\gamma+1+c/\beta, \mu/\beta) \end{array} \Big| \lambda z^\mu \right], \quad (39)$$

provided $\operatorname{Re} \delta > 0$, $\operatorname{Re} \gamma > -1$, $\mu > 0$, $\lambda \neq 0$, *and if* $p = q+1$, *we require* $|\lambda z^\mu| < 1$.

Proof. For $c = 0$ this is Lemma 3 in Kiryakova [2], and for the case of $\gamma = 0, \beta = 1$ we have the formula for the R-L fractional derivative from Kilbas ([16], Th.4) (where $\nu - 1 := c$, $\operatorname{Re} c > -1$, and the same other conditions):

$$D^\delta \left\{ z^c \, {}_p\Psi_q \left[\begin{array}{c} (a_1, A_1), \ldots, (a_p, A_p) \\ (b_1, B_1), \ldots, (b_q, B_q) \end{array} \Big| \lambda z^\mu \right] \right\} = z^{c-\delta} \, {}_{p+1}\Psi_{q+1} \left[\begin{array}{c} (a_i, A_i)_1^p, (c+1, \mu) \\ (b_j, B_j)_1^q, (c+1-\delta, \mu) \end{array} \Big| \lambda z^\mu \right]. \quad (40)$$

The same standard term-by-term integration/differentiation technique can be used for the proof of (39).

Consider also the simplest case as an analog of (39), when $\beta = 1$, $c = 0$ and all $A_1 = \ldots = A_p = B_1 = \ldots = B_q = 1$. This is our Lemma 4.3.1 from [6] for the ${}_pF_q$-functions. Namely,

$$D_1^{\gamma,\delta} \{ {}_pF_q(a_1, \ldots, a_p; b_1, \ldots, b_q; \lambda z) \} = \frac{\Gamma(\gamma+\delta+1)}{\Gamma(\gamma+1)} {}_{p+1}F_{q+1}(a_1, \ldots, a_p, \gamma+\delta+1; b_1, \ldots, b_q, \gamma+1; \lambda z). \quad (41)$$

In the proof of (41) given in [2], we used the relation (16) between the E-K derivative (15) and the R-L derivative $D^\delta f(z) = (\frac{d}{dz})^n R^{n-\delta} f(z)$, $n = [\delta]+1$, combined with the result (32). Then, employed a formula from ([7], Section 7.2.3, (51)) for differentiation of integer order n of a generalized hypergeometric function ${}_pF_q$ with specific parameters as above.

Here we demonstrate a new proof of Lemma 2 *for the more general case of Wright function* ${}_p\Psi_q$. For simplicity, $\beta = 1$ and $\mu = 1$. Interpreting the E-K derivative (15) as in (16), we have subsequently:

$$\text{The L.H.S. of (41)} = \left[z^{-\gamma} D^\delta z^{\gamma+\delta} \right] \left\{ z^c \, {}_p\Psi_q \left[\begin{array}{c} (a_i, A_i)_1^p \\ (b_j, B_j)_1^q \end{array} \Big| \lambda z \right] \right\}$$

$$= \left[z^{-\gamma} \left(\frac{d}{dz} \right)^n R^{n-\delta} \right] \left\{ z^{\gamma+\delta+c} \, {}_p\Psi_q \left[\begin{array}{c} (a_i, A_i)_1^p \\ (b_j, B_j)_1^q \end{array} \Big| \lambda z \right] \right\}$$

(and due to (37)) $= z^{-\gamma} \left(\frac{d}{dz} \right)^n \left\{ z^{\gamma+n+c} \, {}_{p+1}\Psi_{q+1} \left[\begin{array}{c} (a_i, Ai)_1^p, (\gamma+\delta+c+1, 1) \\ (b_j, B_j)_1^q, (\gamma+n+c+1, 1) \end{array} \Big| \lambda z \right] \right\}.$

Now, we use the representation (5) of the Wright g.h.f. as an H-function, and may apply a formula for differentiation of integer order n of the H-function, say Equation (1.69) from Mathai-Saxena-Haubold [51], to continue as follows:

$$\ldots = z^{-\gamma} \left(\frac{d}{dz} \right)^n \left\{ z^{\gamma+n+c} H_{p+1,q+2}^{1,p+1} \left[-\lambda z \Big| \begin{array}{c} (1-a_i, A_i)_1^p, (-\gamma-\delta-c, 1) \\ (0,1), (1-b_j, B_j)_1^q, (-\gamma-n-c, 1) \end{array} \right] \right\}$$

$$= z^{-\gamma} z^{\gamma+c} H^{1,p+2}_{p+2,q+3} \left[-\lambda z \left| \begin{array}{c} (-\gamma - n - c, 1), (1 - a_i, A_i)_1^p, (-\gamma - \delta - c, 1) \\ (0, 1), (1 - b_j, B_j)_1^q, (-\gamma - n - c, 1), (-\gamma - c, 1) \end{array} \right. \right]$$

and because of the coincidence of the terms $(-\gamma - n - c, 1)$ in upper and low parameters' rows, according to the reduction order formula for the H-function: (1.56) in [51], see also (E.8) in [6], and [7,9]), we have

$$\ldots = z^c H^{1,p+1}_{p+1,q+2} \left[-\lambda z \left| \begin{array}{c} (1 - a_i, A_i)_1^p, (-\gamma - \delta - c, 1) \\ (0, 1), (1 - b_j, B_j)_1^q, (-\gamma - c, 1) \end{array} \right. \right],$$

which, by using again (5) to go back to a Wright g.h.f., gives the result (39). In case $\beta \neq 1$, substitution $z \mapsto z^{1/\beta}$ is necessary, and same for $\mu \neq 1$.

Yet another approach to check the validity of (39) is to use the identity $D^{\gamma,\delta}_\beta I^{\gamma,\delta}_\beta f(z) = f(z)$ for $f(z) := z^c {}_p\Psi_q \left[\begin{array}{c} (a_1, A_1), \ldots, (a_p, A_p) \\ (b_1, B_1), \ldots, (b_q, B_q) \end{array} \middle| \lambda z^\mu \right]$, the result from Lemma 1 and reduction of the intermediate result ${}_{p+2}\Psi_{q+2}$ to a ${}_{p+1}\Psi_{q+1}$-function, since the last two equal parameters in the upper and low rows of its series eliminate each other. □

Remark 2. *The corresponding result for the Caputo-type E-K derivative* $*D^{\gamma,\delta}_\beta$ *for images of the Wright ${}_p\Psi_q$-functions, and in particular also for the ${}_pF_q$-functions and for the simpler case of operators with $\beta_1 = \ldots = \beta_m = \beta > 0$, will be presented in another separate work.*

Example 1. *The classical result (34) can be extended to an Erdélyi–Kober integral if we use Lemma 1 for* $\exp(z) = {}_0F_0(-;-;z) = {}_1F_1(0;0;z) = {}_1\Psi_1 \left[\begin{array}{c} (0,1) \\ (0,1) \end{array} \middle| z \right]$ *(which is also a ${}_0\Psi_0$, $G^{1,0}_{0,1}$, and $H^{1,0}_{0,1}$-function):*

$$I^{\gamma,\delta}_\beta \{z^c \exp(\lambda z)\} = I^{\gamma,\delta}_\beta \left\{ z^c {}_1\Psi_1 \left[\begin{array}{c} (0,1) \\ (0,1) \end{array} \middle| \lambda z \right] \right\} = z^c {}_2\Psi_2 \left[\begin{array}{c} (0,1), (\gamma + 1 + c/\beta, 1/\beta) \\ (0,1), (\gamma + \delta + 1 + c/\beta, 1/\beta) \end{array} \middle| z \right]$$

$$= {}_1\Psi_1 \left[\begin{array}{c} (\gamma + 1 + c/\beta, 1/\beta) \\ (\gamma + \delta + 1 + c/\beta, 1/\beta) \end{array} \middle| \lambda z \right], \text{ reducible to (34) for } \gamma = 0, \beta = 1, \nu = c + 1 > 0. \quad (42)$$

4. Results for the Generalized Fractional Calculus Operators of Special Functions

Here we present our results on evaluating operators of generalized fractional calculus (in the sense of [6] and of Riemann–Liouville type) of wide classes of special functions as the Wright generalized hypergeometric functions ${}_p\Psi_q$ and even of the Fox H-functions (thus incorporating the SF of FC) and in particular, of the ${}_pF_q$- and Meijer G-functions (thus having general results also for the "classical" SF).

We start with the most general result, presented in Kiryakova ([2], Th.3) and mentioned in ([3], Th.4.3).

Theorem 2. *The generalized (m-tuple) fractional integral $I^{(\gamma_k),(\delta_k)}_{(\beta_k),m}$ of a H-function is again an H-function:*

$$I^{(\gamma_k),(\delta_k)}_{(\beta_k),m} \left\{ H^{s,t}_{u,v} \left[\lambda z \left| \begin{array}{c} (c_i, C_i)_1^u \\ (d_j, D_j)_1^v \end{array} \right. \right] \right\} = H^{s,t+m}_{u+m,v+m} \left[\lambda z \left| \begin{array}{c} (c_i, C_i)_1^t, (-\gamma_k)_1^m, (c_i, C_i)_{t+1}^u \\ (d_j, D_j)_1^s, (-\gamma_k - \delta_k)_1^m, (d_j, D_j)_{s+1}^v \end{array} \right. \right]. \quad (43)$$

Note that three of the orders of the H-function are increased by the multiplicity m, and additional $m+m$ parameters appear depending on those of the operator.

Proof. The following known *formula for integral* (can be seen as a Mellin transform) of product of two Fox H-functions is very important for evaluating integrals of products of special functions of general nature, because almost all of them can be presented as H-functions (([9], Section 5.1, (5.1.1)), ([7], Section 2.25, (1)), see also (E.21′) in ([6], Appendix)):

$$\int_0^\infty \sigma^{\alpha-1} H_{u,v}^{s,t}\left[\varkappa\sigma \left|\begin{matrix}(c_i,C_i)_1^u\\(d_l,D_l)_1^v\end{matrix}\right.\right] H_{p,q}^{m,n}\left[\omega\sigma^r \left|\begin{matrix}(a_j,A_j)_1^p\\(b_k,B_k)_1^q\end{matrix}\right.\right] d\sigma \qquad (44)$$

$$= \varkappa^{-\alpha} H_{p+v,q+u}^{m+t,n+s}\left[\frac{\omega}{\varkappa^r}\left|\begin{matrix}(a_j,A_j)_1^n,(1-d_l-\alpha D_l,rD_l)_1^v,(a_j,A_j)_{n+1}^p\\(b_k,B_k)_1^m,(1-c_i-\alpha C_i,rC_i)_1^u,(b_k,B_k)_{m+1}^q\end{matrix}\right.\right],$$

under the conditions $\Delta > -1$, $a^* = \Delta + 1 > 0$ (in terms of (2)).

To prove (43) we use the definition (18) of $I_{(\beta_k),m}^{(\gamma_k),(\delta_k)}$, the fact that the kernel $H_{m,m}^{m,0}$-function vanishes for $|\sigma| > 1$ and so the limits $(0,1)$ of the integral can be changed into $(0,\infty)$, and the above Formula (44):

$$I_{(\beta_k),m}^{(\gamma_k),(\delta_k)}\left\{H_{u,v}^{s,t}\left[\lambda z\left|\begin{matrix}(c_i,C_i)_1^u\\(d_j,D_j)_1^v\end{matrix}\right.\right]\right\} = \int_0^1 \ldots d\sigma = \int_0^\infty \ldots d\sigma$$

$$= \int_0^\infty H_{m,m}^{m,0}\left[\sigma\left|\begin{matrix}(\gamma_k+\delta_k+1-\frac{1}{\beta_k},\frac{1}{\beta_k})_1^m\\(\gamma_k+1-\frac{1}{\beta_k},\frac{1}{\beta_k})_1^m\end{matrix}\right.\right] H_{u,v}^{s,t}\left[\lambda z\sigma\left|\begin{matrix}(c_i,C_i)_1^u\\(d_j,D_j)_1^v\end{matrix}\right.\right] d\sigma$$

$$= H_{u+m,v+m}^{s+0,t+m}\left[\lambda z\left|\begin{matrix}(c_i,C_i)_1^t,(1-\gamma_k-1+\frac{1}{\beta_k}-\frac{1}{\beta_k},\frac{1}{\beta_k})_1^m,(c_i,C_i)_{t+1}^u\\(d_j,D_j)_1^s,(1-\gamma_k-\delta_k-1+\frac{1}{\beta_k}-\frac{1}{\beta_k},\frac{1}{\beta_k})_1^m,(d_j,D_j)_{s+1}^v\end{matrix}\right.\right]$$

$$= H_{u+m,v+m}^{s,t+m}\left[\lambda z\left|\begin{matrix}(c_i,C_i)_1^t,(-\gamma_k,\frac{1}{\beta_k})_1^m,(c_i,C_i)_{t+1}^u\\(d_j,D_j)_1^s,(-\gamma_k-\delta_k,\frac{1}{\beta_k})_1^m,(d_j,D_j)_{s+1}^v\end{matrix}\right.\right]. \qquad \square$$

For $\beta_1 = \beta_2 = \ldots = \beta_m = \beta > 0$, the image of an arbitrary G-function under the simpler GFC integrals with Meijer's $G_{m,m}^{m,0}$-kernels has been provided earlier.

Corollary 1 (Lemma 1.2.2, [6]). *The $I_{\beta,m}^{(\gamma_k),(\delta_k)}$-image of a G-function is also a G-function in which the three orders are increased by the multiplicity m and has additional $m+m$ parameters depending on those of the GFC operator:*

$$I_{\beta,m}^{(\gamma_k),(\delta_k)}\left\{G_{u,v}^{s,t}\left[\lambda z^\beta\left|\begin{matrix}(c_i)_1^u\\(d_j)_1^v\end{matrix}\right.\right]\right\} = G_{u+m,v+m}^{s,t+m}\left[\lambda z^\beta\left|\begin{matrix}(c_i)_1^t,(-\gamma_k)_1^m,(c_i)_{t+1}^u\\(d_j)_1^s,(-\gamma_k-\delta_k)_1^m,(d_j)_{s+1}^v\end{matrix}\right.\right]. \qquad (45)$$

Proof. In this case one can use a formula for the integral of product of two arbitrary G-functions, simpler than (44) (see for example, ([7], Section 2.24, (1)]) and with a proof in ([6], App., (A.29))), and be reminded again that the $G_{m,m}^{m,0}$-function vanishes outside the unit disc. Thus the integral (45)

$$I = \int_0^1 \ldots d\sigma = \int_0^\infty G_{m,m}^{m,0}\left[\sigma\left|\begin{matrix}(\gamma_k+\delta_k)_1^m\\(\gamma_k)_1^m\end{matrix}\right.\right] G_{u,v}^{s,t}\left[\lambda z^\beta\left|\begin{matrix}(c_i)_1^u\\(d_j)_1^v\end{matrix}\right.\right] d\sigma,$$

gives the required image G-function. Because the $G_{u,v}^{s,t}$-function is from the space of an analytic function in a disk centered at the origin, and has the following asymptotic behavior

$$G_{u,v}^{s,t}\left[\lambda z^\beta\right] = O\left(z^{d^*}\right) \text{ as } |z|\to 0, \text{ with } d^* = \beta \min_j d_j > \max_k[-\beta(\gamma_k+1)],$$

the conditions for the used formula to hold on are satisfied. $\square$

Formulas (43) and (45) can be used to evaluate practically all (classical and generalized) operators of FC of arbitrary SF (which are representable either as a G- or as a more general H-function).

Now we present *the main result on the topic of this survey paper*, which comes from Kiryakova [2], Th.1.

Theorem 3. *Assume that the conditions $\delta_k \geq 0, \gamma_k > -1, \beta_k > 0, k = 1,...,m$ and $\mu > 0, \lambda \neq 0$ hold. The image of a Wright g.h.f. $_p\Psi_q(z)$ by a generalized fractional integral* (20) *(multiple, m-tuple Erdélyi–Kober integral) is another Wright g.h.f. with indices p and q increased by the multiplicity m and with additional parameters coming from those of the GFC integral:*

$$I_{(\beta_k)_1^m,m}^{(\gamma_k)_1^m,(\delta_k)_1^m} \left\{ z^c \,_p\Psi_q \left[\begin{array}{c} (a_1,A_1),\ldots,(a_p,A_p) \\ (b_1,B_1),\ldots,(b_q,B_q) \end{array} \bigg| \lambda z^\mu \right] \right\}$$

$$= z^c \,_{p+m}\Psi_{q+m} \left[\begin{array}{c} (a_i,A_i)_1^p, (\gamma_i+1+\frac{c}{\beta_i},\frac{\mu}{\beta_i})_1^m \\ (b_j,B_j)_1^q, (\gamma_i+\delta_i+1+\frac{c}{\beta_i},\frac{\mu}{\beta_i})_1^m \end{array} \bigg| \lambda z^\mu \right]. \tag{46}$$

Proof. Here we briefly repeat the proof from Kiryakova [2], Th.1, *in order to exhibit the main ideas on which this survey paper is based.*

As one approach to prove (46), the general integral formula (44) can be used. This theorem can be seen also as a consequence of Theorem 2. It is because the kernel-function of the operator is a $H_{m,m}^{m,0}$-function and the $_p\Psi_q$-function is a $H_{p,q+1}^{1,p}$-function, see (5). Then, according to (45) the result will be a $H_{p+m,q+1+m}^{1,p+m}$-function that should be recognized as a $_{p+1}\Psi_{q+1}$-function, because it is reduced to a $H_{p+1,q+2}^{1,p+1}$-function in view of the coincidence of $(m-1)$ parameters in the upper and low row (use "reduction order" property of the H-function, [7], Section 8.3, 6.; [6], App. (E.8), etc.).

However, to clarify our main idea it is more instructive to refer to the decomposition property (21) presenting the generalized fractional integral (18) as a product of commuting (classical) Erdélyi–Kober operators. In the simplest case, we use subsequently m-times (36) from Lemma 1, to get:

$$I_{(\beta_k)_1^m,m}^{(\gamma_k)_1^m,(\delta_k)_1^m} \left\{ _p\Psi_q \left[\begin{array}{c} (a_i,A_i)_1^p \\ (b_j,B_j)_1^q \end{array} \bigg| \lambda z \right] \right\} = I_{(\beta_k)_1^{m-1},m-1}^{(\gamma_k)_1^{m-1},(\delta_k)_1^{m-1}} \left\{ I_{\beta_m}^{\gamma_m,\delta_m} \,_p\Psi_q \left[\begin{array}{c} (a_i,A_i)_1^p \\ (b_j,B_j)_1^q \end{array} \bigg| \lambda z \right] \right\}$$

$$= I_{(\beta_k)_1^{m-2},m-2}^{(\gamma_k)_1^{m-2},(\delta_k)_1^{m-2}} \left\{ I_{\beta_{m-1}}^{\gamma_{m-1},\delta_{m-1}} \left\{ _{p+1}\Psi_{q+1} \left[\begin{array}{c} (a_i,A_i)_1^p, (\gamma_m+1,1/\beta_m) \\ (b_j,B_j)_1^q, (\gamma_m+\delta_m+1,1/\beta_m) \end{array} \bigg| \lambda z \right] \right\} \right\}$$

$$= \ldots = I_{\beta_1}^{\gamma_1,\delta_1} \left\{ I_{\beta_2}^{\gamma_2,\delta_2} \left\{ _{p+m-2}\Psi_{q+m-2} \left[\begin{array}{c} (a_i,A_i)_1^p, (\gamma_r+1,1/\beta_r)_1^{m-2} \\ (b_j,B_j)_1^q, (\gamma_r+\delta_r+1,1/\beta_r)_1^{m-2} \end{array} \bigg| \lambda z \right] \right\} \right\} \tag{47}$$

$$= I_{\beta_1}^{\gamma_1,\delta_1} \left\{ _{p+m-1}\Psi_{q+m-1} \left[\begin{array}{c} (a_i,A_i)_1^p, (\gamma_r+1,1/\beta_r)_1^{m-1} \\ (b_j,B_j)_1^q, (\gamma_r+\delta_r+1,1/\beta_r)_1^{m-1} \end{array} \bigg| \lambda z \right] \right\}$$

$$= \,_{p+m}\Psi_{q+m} \left[\begin{array}{c} (a_i,A_i)_1^p, (\gamma_r+1,1/\beta_r)_1^m \\ (b_j,B_j)_1^q, (\gamma_r+\delta_r+1,1/\beta_r)_1^m \end{array} \bigg| \lambda z \right].$$

To derive the general relation (46) we apply to the above result the property for "generalized commutation" from Kiryakova ([6], Ch.5, (5.1.28)), namely:

$$I_{(\beta_k),m}^{(\gamma_k),(\delta_k)} z^c f(z^\mu) = z^c I_{(\frac{\beta_k}{\mu}),m}^{(\gamma_k+\frac{c}{\beta_k}),(\delta_k)} f(z^\mu), \quad \text{with} \quad \mu > 0. \tag{48}$$

□

Corollary 2. *(Lemma 4.2.1., Equations (4.2.2)–(4.2.2′) in Kiryakova ([6], Ch.4)) The image of a $_pF_q$ g.h.f. (6) under a generalized (m-tuple) fractional integral $I_{1,m}^{(\gamma_k,\delta_k)} := I_{(1,...,1),m}^{(\gamma_k,\delta_k)}$, (19) with (for simplicity) all $\beta_k = \beta = 1$, $k = 1, ..., m$, is another g.h.f. of the same kind with indices increased by the multiplicity m:*

$$I_{1,m}^{(\gamma_k)_1^m,(\delta_k)_1^m}\{{}_pF_q(a_1,...,a_p;b_1,...,b_q;\lambda z)\} = {}_{p+m}F_{q+m}(a_1,...,a_p,(\gamma_i+1)_1^m;(b_1,...,b_q,(\gamma_i+\delta_i+1)_1^m;\lambda z). \tag{49}$$

The above results (46) and (49) can be interpreted alternatively as the assertions stated in our earlier works ([6,50], Ch.4) (in the simpler case of Corollary 2), and later in [24] (in more general case of Theorem 3). That is, a $_{p+m}\Psi_{q+m}$-function (resp. a $_{p+m}F_{q+m}$-function) of the form below can be represented by means of a multiple (m-tuple) operator of GFC

$$\tilde{I} = I_{(1/\beta_i)_{i=1}^m,m}^{(a_{p+i}-1)_{i=1}^m,(b_{q+i}-a_{p+i})_{i=1}^m}$$

of a $_p\Psi_q$-function (resp. a $_pF_q$-function), with orders reduced by m, namely:

$$_{p+m}\Psi_{q+m}\left[\begin{array}{c}(a_i,A_i)_{i=1}^p;(a_{p+i},1/\beta_i)_{i=1}^m\\(b_j,B_j)_{k=1}^q;(b_{q+i},1/\beta_i)_{i=1}^m\end{array}\Big|\lambda z\right] = \tilde{I}\left\{{}_p\Psi_q\left[\begin{array}{c}(a_j,A_j)_{j=1}^p\\(b_k,B_k)_{k=1}^q\end{array}\Big|\lambda z\right]\right\}. \tag{50}$$

In the case of Wright function with arbitrary parameters $A_{p+i}, B_{q+i}, i = 1, ..., m$:

$$_{p+m}\Psi_{q+m}\left[\begin{array}{c}(a_i,A_i)_{i=1}^p;(a_{p+i},A_{p+i})_{i=1}^m\\(b_j,B_j)_{k=1}^q;(b_{q+i},B_{q+i})_{i=1}^m\end{array}\Big|z\right],$$

such kind of result is presented in [24] by means of more general operators $\tilde{I}$, the so-called *Wright-Erdélyi–Kober operators*. This means that using a suitable number of times of a procedure similar to that in proof of Theorem 3, from any $_p\Psi_q$-function (resp. $_pF_q$-function) we can go down to one of the *three basic generalized hypergeometric functions*, depending on if $p < q$, $p = q$ or $p = q+1$: $_0\Psi_{q-p}, {}_1\Psi_1, {}_2\Psi_1$; resp. to: $_0F_{q-p}$ (hyper-Bessel f. and $\cos_m$-f.), $_1F_1$ (confluent h.f. and exp-f.), $_2F_1$ (Gauss f. and beta-distribution of form $z^\alpha(1-z)^\beta$). This is the reason that we classified the g.h.f. to be of three basic types, as: "g.h.f. of Bessel/cosine type", "g.h.f. of confluent/exp type" and "g.h.f. of Gauss/beta-distribution type". Details on this approach and such a classification of the SF can be found in Kiryakova ([6,22,24,50], Ch.4).

Analogously to Theorem 3, we have also a relation (image) for the generalized fractional derivatives of g.h.f., presented as Theorem 2 in Kiryakova [2]. The more general formula (as below) is available in Kiryakova, ([3], Theorem 4.2.).

Theorem 4.

$$D_{(\beta_k)_1^m,m}^{(\gamma_k)_1^m,(\delta_k)_1^m}\left\{z^c\,{}_p\Psi_q\left[\begin{array}{c}(a_1,A_1),...,(a_p,A_p)\\(b_1,B_1),...,(b_q,B_q)\end{array}\Big|\lambda z^\mu\right]\right\}$$
$$= z^c\,{}_{p+m}\Psi_{q+m}\left[\begin{array}{c}(a_i,A_i)_1^p,(\gamma_k+\delta_k+1+\frac{c}{\beta_k},\frac{\mu}{\beta_k})_1^m\\(b_j,B_j)_1^q,(\gamma_k+1+\frac{c}{\beta_k},\frac{\mu}{\beta_k})_1^m\end{array}\Big|\lambda z^\mu\right]. \tag{51}$$

Proof. One possible approach to derive this, is to use a decomposition formula for the generalized (multiple) fractional derivatives (23),

$$D_{(\beta_k),m}^{(\gamma_k),(\delta_k)}f(z) = D_{\beta_m}^{\gamma_m,\delta_m}\left\{D_{\beta_{m-1}}^{\gamma_{m-1},\delta_{m-1}}\left[\cdots D_{\beta_1}^{\gamma_1,\delta_1}f(z)\right]\right\},$$

as sequential derivatives. Then, we apply m-times the result (39) of Lemma 2.

We can verify (51) also directly, in the same way as in the end of the proof of Lemma 2, using the basic relation $D_{(\beta_k),m}^{(\gamma_k),(\delta)_k}I_{(\beta)_k,m}^{(\gamma_k),(\delta_k)}f(z) = f(z)$. □

The case of images under the *Caputo type* generalized fractional derivatives $D_{(\beta_k)_1^m,m}^{(\gamma_k)_1^m,(\delta_k)_1^m}$ defined in (24) will be discussed in a separate work, see also Remark 2.

5. Examples of Erdélyi–Kober and Riemann–Liuoville Operators of Some Special Functions

In the beginning of Section 3 we already acknowledged the contributions by some classical authors, such as Erdélyi et al., Askey, Lavoie-Osler-Tremblay, to provide the images of some special and elementary functions under the Riemann–Liuoville fractional integral/derivative, see Formulas (32)–(35). We may refer also to works where detailed tables of images under Riemann–Liuoville operators are provided, for example the book Erdélyi et al. [44], some recent surveys, including *in this Journal*, such as by Rogosin [18] (as for M-L type functions), Garrappa-Kaslik-Popolizio [52] (images of elementary functions expressed by M-L functions).

As mentioned in Section 3, the proof of Lemma 1, a most general result for Riemann–Liuoville operators of special functions (in sense of g.h.f.) is formula (37) from Kilbas ([16], Th.2):

$$R^\delta \left\{ z^c {}_p\Psi_q \left[\begin{matrix} (a_1, A_1), \ldots, (a_p, A_p) \\ (b_1, B_1), \ldots, (b_q, B_q) \end{matrix} \middle| \lambda z^\mu \right] \right\} = z^{c+\delta} {}_{p+1}\Psi_{q+1} \left[\begin{matrix} (a_i, A_i)_1^p, (c+1, \mu) \\ (b_j, B_j)_1^q, (c+\delta+1, \mu) \end{matrix} \middle| \lambda z^\mu \right],$$

and for the R-L derivative, the corresponding result is as in Equation (40).

It may be instructive to repeat (as from [2]) some very special cases of the images (37) and (40) under the R-L integral $R^\delta f(z) = z^\delta I_1^{0,\delta} f(z)$, that have been derived by the cited authors by the standard term-by-term integration/differentiation. Naturally, these come also as specifications of our results from Lemmas 1 and 2 for the Erdélyi–Kober operators (case $m = 1$).

Example 2. *The R-L fractional integral of the weighted Bessel function, for* $\operatorname{Re}\delta > 0$, $\operatorname{Re}\nu > -1$, $\operatorname{Re}(\gamma + \nu) > 0$, *is given by Kilbas-Sebastian ([53], Cor.1, (28)), in the form*

$$R^\delta \left\{ z^{\gamma-1} J_\nu(z) \right\} = \frac{z^{\gamma+\nu+\delta-1}}{2^\nu} {}_1\Psi_2 \left[\begin{matrix} (\gamma+\nu, 2) \\ (\gamma+\nu+\delta, 2), (\nu+1, 1) \end{matrix} \middle| -\frac{1}{4}z^2 \right]. \tag{52}$$

We can use $J_\nu(z) = {}_0\Psi_1 \left[\begin{matrix} -- \\ (\nu+1, 1) \end{matrix} \middle| -\frac{1}{4}z^2 \right]$, so to see (52) as an immediate corollary of (37), and also of our E-K result (36). Because a 1-tuple fractional calculus operator (R-L, or E-K) is applied, the preliminary expectation is confirmed to have as a result a ${}_{0+1}\Psi_{1+1}$-function.

Example 3. *The R-L fractional integral of the generalized Bessel function* J_ν^κ *(usually called Bessel–Maitland function, a name that should correctly be called the Bessel-Wright function) is derived in Kilbas ([16], Th.8, (26)), and extends the above formula (52): Note the representation* $J_\nu^\kappa(z) = {}_0\Psi_1 \left[\begin{matrix} -- \\ (\nu+1, \kappa) \end{matrix} \middle| -z \right]$, *then from our result in Lemma 1, and in particular, from (37), it is expected to have the result as a ${}_1\Psi_2$-function:*

$$R^\delta \left\{ z^{\gamma-1} J_\nu^\kappa(\lambda z^\mu) \right\} = z^{\gamma+\delta-1} {}_1\Psi_2 \left[\begin{matrix} (\gamma, \mu) \\ (\gamma+\delta, \mu), (\nu+1, \kappa) \end{matrix} \middle| -\lambda z^\mu \right], \tag{53}$$

for $\operatorname{Re}\delta > 0$, $\operatorname{Re}(\gamma - 1) > -1$, $\kappa > -1$, $\mu > 0$. *The sign "minus" in the argument of RHS was missing in [16] due to a possible typo.*

The same result, in terms of the (classical) Wright function is presented in the same paper, Kilbas ([16], Th.6, (18)), with the true argument sign (we slightly change the denotations to be similar as in the first row of our (11)),

$$R^\delta \left\{ z^{\gamma-1} \phi(\kappa, \nu+1; \lambda z^\mu) \right\} = z^{\gamma+\delta-1} {}_1\Psi_2 \left[\begin{matrix} (\gamma, \mu) \\ (\gamma+\delta, \mu), (\nu+1, \kappa) \end{matrix} \middle| \lambda z^\mu \right]. \tag{54}$$

A more useful result, in the sense that the R-L integral transforms a generalized Bessel function/resp. Wright function, into same kind of function but with increased index comes if we put $\gamma - 1 = \nu, \kappa = \mu$ in (53), see, for example, ([16], Cor.8.1, (28)), wth $\text{Re}\,\delta > 0$, $\text{Re}\,\nu > -1$, $\mu > 0$:

$$R^\delta \left\{ z^\nu J_\nu^\mu(\lambda z^\mu) \right\} = z^{\nu+\delta} J_{\nu+\delta+1}^\mu(\lambda z^\mu), \quad R^\delta \left\{ z^\nu \phi(\mu, \nu+1; \lambda z^\mu) \right\} = z^{\nu+\delta} \phi(\mu, \nu+\delta+1; \lambda z^\mu), \quad \lambda \neq 0. \tag{55}$$

Next, we mention an example with the so-called generalized M-series. Namely, (M.) Sharma and Jain [54] introduced the special function ${}_p M_q^{\alpha,\beta}(z)$, as an extension of both g.h.f. ${}_pF_q(z)$ and the (2-parameters) M-L function $E_{\alpha,\beta}(z)$:

$${}_p M_q^{\alpha,\beta}(a_1,\ldots,a_p;b_1,\ldots,s,b_q;z) = \sum_{k=0}^\infty \frac{(a_1)_k \cdots (a_p)_k}{(b_1)_k \cdots (b_q)_k} \frac{z^k}{\Gamma(\alpha k + \beta)} = \kappa\, {}_{p+1}\Psi_{q+1}\left[\begin{array}{c} (a_1,1),\ldots,(a_p,1),(1,1) \\ (b_1,1),\ldots,(b_q,1),(\beta,\alpha) \end{array} \Big| z \right]. \tag{56}$$

Here $z, \alpha, \beta \in \mathbb{C}$, $\text{Re}\,\alpha > 0$, $p \leq q$ are the integer orders, and if $p = q+1$ we require additionally that $|z| < R = \alpha^\alpha$, and $\kappa := \prod_{j=1}^q \Gamma(b_j)/\prod_{i=1}^p \Gamma(a_i)$. Usually the following particular cases are always mentioned: (1) $\beta = 1$: this is the (simpler) M-series, introduced by M. Sharma (2008, in same journal as [54]); (2) $p = q = 0$ (that is, no upper and no lower parameters): this is the M-L function $E_{\alpha,\beta}(z)$; (3) $p = 0, q = 1, b_1 = 1$: one has the Wright function $\phi(\alpha,\beta,z)$, or the generalized Bessel–Maitland function; (4) $p = q = 1, a_1 = \gamma, b_1 = 1$: this is the Prabhakar M-L type function (8), (5) $\alpha = \beta = 1$: we have the g.h.f. ${}_pF_q(a_1,\ldots,a_p;b_1,\ldots,b_q;z)$, etc.

Since (56) is a ${}_{p+1}\Psi_{q+1}$-function, all FC operators of the form (20) (and their particular cases as R-L, E-K, Saigo, M-S-M) of the M-series can be evaluated using our formulas in Lemmas 1 and 2 and Theorems 3 and 4.

Example 4. *In [54], the images of the generalized M-series are derived in the case of R-L fractional integral and derivative of order $\delta > 0$:*

$$R^\delta \left\{ {}_p M_q^{\alpha,\beta}(a_1,\ldots,a_p;b_1,\ldots,b_q;z) \right\} = \frac{z^\delta}{\Gamma(1+\delta)}\, {}_{p+1}M_{q+1}^{\alpha,\beta}(a_1,\ldots,a_p,1;b_1,\ldots,b_q,1+\delta;z), \tag{57}$$

$$D^\delta \left\{ {}_p M_q^{\alpha,\beta}(a_1,\ldots,a_p;b_1,\ldots,b_q;z) \right\} = \frac{z^{-\delta}}{\Gamma(1-\delta)}\, {}_{p+1}M_{q+1}^{\alpha,\beta}(a_1,\ldots,a_p,1;b_1,\ldots,b_q,1-\delta;z), \tag{58}$$

using term-by-term integration/differentiation of the series (56). However, having in mind the representations in both sides as Wright g.h.f., one can get these results directly from the corollaries of Lemmas 1 and 2, the R-L integral (37) and derivative (40). See also in Lavault [55].

The formulas in Theorems 3 and 4 can easily be reduced to corresponding results for generalized fractional integrals and derivatives (20) and (23) of the M-series, to appear in terms of ${}_{p+m}M_{q+m}^{\alpha,\beta}(z)$, with additional parameters depending on $(\gamma_k)_1^m, (\delta_k)_1^m$. Again in view of (56), evaluation of other particular FC operators, such as E-K, Saigo, M-S-M, of the M-series can be done. For example, the M-S-M images were evaluated by Kumar and Saxena [56].

6. Saigo Hypergeometric Operators of Various Special Functions

In a series of papers since 1978, such as [57] (for more references see in [6,58]), Saigo introduced a linear integral operator with Gauss function in the kernel, and applied it first for studying BVP for PDE as the Euler-Darboux equation. Later on, this operator was used by him and collaborators in geometric function theory (classes of univalent functions). It happens that, as a case of the hypergeometric integral operators, the Saigo operator has also a role as an FC operator and this has recently become a reason for great interest for researchers in FC, and mainly to authors whose job is to evaluate images of Saigo operator(s) of various special functions. A search in *Google* for the phrase "Saigo operator"

+ "function" returns now more than 1060 results (of course some of them may also concern the more general Marichev–Saigo–Maeda, discussed in next Section 7).

First, let us remind the definition and two basic properties of the *Saigo operators*.
For complex α, β, η and $\operatorname{Re} \alpha > 0$, the *Saigo fractional integration operator* (the LHS version) is

$$I^{\alpha,\beta,\eta} f(z) = \frac{z^{-\alpha-\beta}}{\Gamma(\alpha)} \int_0^z (z-\xi)^{\alpha-1} {}_2F_1\left(\alpha+\beta, -\eta; \alpha; 1-\tfrac{\xi}{z}\right) f(\xi) d\xi \tag{59}$$

$$= \frac{z^{-\beta}}{\Gamma(\alpha)} \int_0^1 s(1-\sigma)^{\alpha-1} {}_2F_1(\alpha+\beta, -\eta; \alpha; 1-\sigma) f(z\sigma) d\sigma,$$

and we skip the discussion on the RHS versions of the Saigo integrals, as similar. The *Saigo fractional derivative* is used as: $D^{\alpha,\beta,\eta} f(z) = (d/dz)^n I^{\alpha+n,\beta-n,\eta-n} f(z)$ with $n = [-\operatorname{Re}\alpha] + 1$. For its explicit differ-integral expression, see for example in ([6], Ch.1). More details can be found in Kiryakova ([6,58], Ch.1, Ch.5) and other our papers dealing with these operators in classes of univalent functions (some of which are joined with Professor Megumi Saigo). A basic formula (*known from the original Saigo works*) that all authors use (and sometimes derive again) is for the image of a power function:

$$I^{\alpha,\beta,\eta} \{z^p\} = [\Gamma(p+1)\Gamma(p+\eta-\beta+1)/\Gamma(p-\beta+1)\Gamma(p+\alpha+\eta+1)]\, z^{p-\beta},$$

for $\operatorname{Re}\alpha > 0, \operatorname{Re}(p+1) > \max[0, \operatorname{Re}(\beta-\eta)]$.

As mentioned in Section 2.3, the Saigo operators are cases of the hypergeometric operators of FC, and of the GFC operators for $m = 2$, simply because according to (30) the Gauss kernel function is representable as the kernel of (19) and (18) with $m = 2$, $\beta = 1, \gamma_1 = \eta - \beta, \gamma_2 = 0, \delta_1 = -\eta, \delta_2 = \alpha + \eta$:

$$\frac{(1-\sigma)^{\alpha-1}}{\Gamma(\alpha)} {}_2F_1(\alpha+\beta, -\eta; \alpha; 1-\sigma) = G_{2,2}^{2,0}\left[\sigma \left|\begin{array}{c} -\beta, \alpha+\eta \\ \eta-\beta, 0 \end{array}\right.\right].$$

Thus, the Saigo operator is a generalized (2-tuple fractional integral) of the form (20) and therefore in view of (21), it is also a *commutable composition of two classical E-K fractional integrals*, see for example ([6], Ch.1):

$$I^{\alpha,\beta,\eta} f(z) = z^{-\beta} I_{(1,1),2}^{(\eta-\beta,0),(-\eta,\alpha+\eta)} f(z) = z^{-\beta} I_1^{\eta-\beta,-\eta} I_1^{0,\alpha+\eta} f(z) \tag{60}$$

$$= I_1^{\eta,-\eta} I_1^{\beta,\alpha+\eta} z^{-\beta} f(z) = I_{(1,1),2}^{(\eta,\beta),(-\eta,\alpha+\eta)} z^{-\beta} f(z) = R^{-\eta} z^{-\alpha-\beta} R^{\alpha+\eta} f(z).$$

The relation between the first and second lines follows by application of the "generalized commutation" between (multiple) Erdélyi–Kober operators and power functions (([6], Ch.1, (1.3.3)), ([34], Th.4), etc.). For particular parameters α, β, η, the Saigo operator can reduce to *one* E-K operator or an R-L operator, say for $\beta = -\alpha, \eta = 0$ it is an R-L integral; and for $\eta = -\alpha$, an E-K integral: $I^{\alpha,\beta,-\alpha} = z^{-\beta} I_1^{-\alpha-\beta,\alpha}$.

Therefore, the Saigo image of some special function, which can be represented as a Wright function ${}_p\Psi_q$, can be written as a particular case of the general formulas (46), resp. (51), or also, as a subsequent two-times application of classical E-K operators. Therefore, *the Saigo image of a ${}_p\Psi_q$-function can always be predicted to result into a ${}_{p+2}\Psi_{q+2}$-function* (unless some parameters in upper and lower rows eliminate each other, and so the indices can be reduced). Our result, as a corollary of Theorem 3 and Corollary 2 states as follows.

Lemma 3. *The images of the Wright g.h.f. ${}_p\Psi_q$, and in particular of the g.h.f. ${}_pF_q$, under the Saigo operator (59) are the same kind of functions with orders increased by 2:*

$$I^{\alpha,\beta,\eta}\left\{z^c {}_p\Psi_q\left[\begin{array}{c}(a_i, A_i)_1^p \\ (b_j, B_j)_1^q\end{array}\bigg| \lambda z^\mu\right]\right\} = z^{c-\beta} {}_{p+2}\Psi_{q+2}\left[\begin{array}{c}(a_i, A_i)_1^p, (\eta-\beta+1+c,\mu), (1+c,\mu) \\ (b_j, B_j)_1^q, (-\beta+1+c,\mu), (\alpha+\eta+1+c,\mu)\end{array}\bigg| \lambda z^\mu\right], \tag{61}$$

(for $c = 0$, $\mu = 1$, this is Cor. 3 in [2]) and

$$I^{\alpha,\beta,\eta}\left\{{}_pF_q\left(a_1,...,a_p;b_1,...,b_q;\lambda z\right)\right\}=z^{-\beta}{}_{p+2}F_{q+2}\left(a_1,...,a_p,\eta-\beta+1,1;b_1,...,b_q,-\beta+1,\alpha+\eta+1;\lambda z\right), \quad (62)$$

under the mentioned conditions in the definition of (59).

The following examples for Saigo operators of particular functions from our previous papers [2,3] are repeated here as an illustration for the general result in Lemma 3.

Example 5. *The Saigo fractional integral (59) of a weighted Bessel function was evaluated in Kilbas-Sebastian ([53], Th.1), for* $\operatorname{Re}\alpha > 0$, $\operatorname{Re}\nu > -1$, $\operatorname{Re}(\gamma+\nu) > \max[0, \operatorname{Re}(\beta-\eta)]$:

$$I^{\alpha,\beta,\eta}\left\{z^{\gamma-1}J_\nu(z)\right\} = \frac{z^{\gamma+\nu-\beta-1}}{2^\nu}\,{}_2\Psi_3\left[\begin{array}{c}(\gamma+\nu,2),(\gamma+\eta+\nu-\beta,2)\\(\gamma+\nu-\beta,2),(\gamma+\nu+\alpha+\eta,2),(\nu+1,1)\end{array}\bigg|-\tfrac{1}{4}z^2\right]. \quad (63)$$

To apply (61) from Lemma 3, let us remind the reader again that $z^{\gamma-1}J_\nu(z) = z^{\gamma-1}{}_0\Psi_1\left[\begin{array}{c}--\\(\nu,1)\end{array}\bigg|-\tfrac{1}{4}z^2\right]$ and so, the result should be expected to appear as a ${}_{0+2}\Psi_{1+2}$. Alternatively, to exhibit the use of decomposition of the Saigo operator in two R-L operators (the last relation in (60)) combined with (52) from Example 2, we may proceed as follows:

$$R^{-\eta}z^{-\alpha-\beta}R^{\alpha+\eta}\left\{z^{\gamma-1}{}_0\Psi_1\left[\begin{array}{c}--\\(\nu,1)\end{array}\bigg|-\tfrac{1}{4}z^2\right]\right\}$$

$$= R^{-\eta}z^{-\alpha-\beta}\left\{2^{-\nu}z^{\gamma+\nu+\alpha+\eta-1}{}_1\Psi_2\left[\begin{array}{c}(\gamma+\nu,2)\\(\gamma+\nu+\alpha+\eta,2),(\nu+1,1)\end{array}\bigg|-\tfrac{1}{4}z^2\right]\right\}$$

$$= 2^{-\nu}R^{-\eta}\left\{z^{\gamma-\beta+\eta+\nu-1}{}_1\Psi_2\left[\begin{array}{c}(\gamma+\nu,2)\\(\gamma+\nu+\alpha+\eta,2),(\nu+1,1)\end{array}\bigg|-\tfrac{1}{4}z^2\right]\right\} = ...\,{}_2\Psi_3\,..., \text{ as in (63).}$$

Example 6. *The more special case for Saigo fractional integral of a (weighted) cosine function is the formula from the same paper of Kilbas-Sebastian ([53], Th.5, (47)), for* $\operatorname{Re}\alpha > 0$, $\operatorname{Re}\gamma > \max[0, \operatorname{Re}(\beta-\eta)]$:

$$I^{\alpha,\beta,\eta}\left\{z^{\gamma-1}\cos z\right\} = \sqrt{\pi}\,z^{\gamma-\beta-1}\,{}_2\Psi_3\left[\begin{array}{c}(\gamma,2),(\gamma+\eta-\beta,2)\\(\gamma-\beta,2),(\gamma+\eta+\alpha,2),(\tfrac{1}{2},1)\end{array}\bigg|-\tfrac{1}{4}z^2\right]. \quad (64)$$

Note that $\cos z = \sqrt{\pi z/2}\,J_{-1/2}(z)$, and use the result (63) of Example 4 with $\nu = -1/2$. To use our general approach, we can present the cos-function as a ${}_0\Psi_1$-function, and predict the result to be a ${}_{0+2}\Psi_{1+2}$-function in (64).

Next, we consider a case with a more general special function, called generalized K-series. In [59] (K.) Sharma introduced an extension of both a g.h.f. ${}_pF_q(z)$ and Prabhakar (three-parameter Mittag-Leffler) function $E^\gamma_{\alpha,\beta}(z)$ (see (8)):

$$_pK^{\alpha,\beta;\gamma}_q\left(a_1,\ldots,a_p;b_1,\ldots,b_q;z\right) := {}_pK^{\alpha,\beta;\gamma}_q(z) = \sum_{k=0}^\infty \frac{(a_1)_k\ldots(a_p)_k}{(b_1)_k\ldots(b_q)_k}\frac{(\gamma)_k\,z^k}{\Gamma(\alpha k+\beta)}, \quad (65)$$

with $z,\alpha,\beta \in \mathbb{C}$, $\operatorname{Re}\alpha > 0$, integers $p \leq q$ (and additional requirement $|z| < R = \alpha^\alpha$ if $p = q+1$). When $\gamma = 1$ it reduces to the *(generalized) M-series* (56) by Sharma-Jain [54], Example 4.

Example 7. *Recently, Lavault [55] represented the above K-series in terms of a Wright g.h.f.:*

$$
{}_p K_q^{\alpha,\beta;\gamma}(a_1,\ldots,a_p;b_1,\ldots,b_q;z) = \frac{\prod_{j=1}^{q}\Gamma(b_j)}{\Gamma(\gamma)\prod_{i=1}^{p}\Gamma(a_i)} {}_{p+2}\Psi_{q+2}\left[\begin{array}{c}(a_1,1),\ldots(a_p,1),(\gamma,1),(1,1)\\(b_1,1),\ldots,(b_q,1),(1,1),(\beta,\alpha)\end{array}\bigg| z\right], \quad (66)
$$

and calculated some of its FC operators, as the R-L, Saigo and M-S-M operators. As should be expected, the image of a ${}_p K_q^{\alpha,\beta;\gamma}$-function under the R-L integral is a ${}_{p+1}K_{q+1}^{\alpha,\beta;\gamma}$-function (Th. 4.1 there), similarly to Example 4 for the M-series.

The Saigo operator is also derived by Lavault in [55]: for the M-series—in Th. 4.2, and for the K-series—in Cor. 4.3. Namely, Equation (4.10), [55] reads as:

$$
I^{\alpha,\beta,\gamma}\left\{t^{\sigma-1}{}_pK_q^{\xi,\eta;\nu}(cz^\mu)\right\} = \frac{\prod_1^q\Gamma(b_j)}{\prod_1^p\Gamma(a_i)}\frac{z^{\sigma-\beta-1}}{\Gamma(\nu)}{}_{p+3}\Psi_{q+3}\left[\begin{array}{c}(a_i,1)_1^p,(\sigma,\mu),(-\beta+\gamma+\sigma,\mu),(\nu,1)\\(b_j)_1^q,(\beta+\sigma,\mu),(\alpha+\gamma+\sigma,\mu),(\eta,\xi)\end{array}\bigg| cz^\mu\right]. \quad (67)
$$

Let us note that the K-series is a ${}_{p+2}\Psi_{q+2}$-function (66), and from our Lemma 3 the expected result should be a ${}_{p+4}\Psi_{q+4}$-function, with indices increased by two. However, pairs of upper and lower rows' parameters appear the same and eliminate each other, therefore the result reduces to a ${}_{p+3}\Psi_{q+3}$, as above.

7. Marichev–Saigo–Maeda (M-S-M) Operators of Various Special Functions

As mentioned in Section 2.3, there is an interesting particular case of the GFC operators (20) and (23) for $m = 3$, often abbreviated as M-S-M (MSM) operators. These operators have also become very popular in works dedicated to evaluate FC images of special functions. A search in *Google* for "Marichev–Saigo–Maeda" returns at least 2430 results, and for "MSM operator"—some 2670 results.

This operator appeared in a paper by Marichev of 1974, [39], see also in the book ([31], Section 8.4.51); and further was introduced and studied by Saigo, Saigo and Maeda in 1996, see [40], also by Saigo and Saxena (1996, 1998, 2001), details on references are in ([6,37,58,60], Ch.1), etc.

For complex parameters a, a', b, b', c, Re $c > 0$, the *Marichev–Saigo–Maeda (M-S-M) integral operator*, of which the kernel is the *Appel function, or Horn's function F_3* (see ([15], Vol.1), also [7])

$$
F_3(a,a',b,b',c,z,\xi) = \sum_{m,n=0}^{\infty}\frac{(a)_m(a')_n(b)_m(b')_n}{(c)_{m+n}}\frac{z^m\xi^n}{m!n!}, \quad |z|<1, |\xi|<1,
$$

is defined as the linear integral operator

$$
\begin{aligned}
I^{a,a',b,b',c}f(z) &= \frac{z^{-a}}{\Gamma(c)}\int_0^z(z-\xi)^{c-1}\xi^{-a'}F_3\left(a,a',b,b';c;1-\frac{\xi}{z},1-\frac{z}{\xi}\right)f(\xi)d\xi \\
&= z^{c-a-a'}\int_0^1\frac{(1-\sigma)^{c-1}}{\Gamma(c)}\sigma^{-a'}F_3\left(a,a',b,b';c;1-\sigma,1-\frac{1}{\sigma}\right)f(z\sigma)d\sigma.
\end{aligned} \quad (68)
$$

Observing the representation (31) of the kernel F_3-function as a kernel of the generalized fractional integrals (20) (see Section 2.3), it is evident that the M-S-M operator is nothing but their special case for $m = 3$. Then, in view of (21), it is also a *composition of three commutable classical E-K integrals* (see Kiryakova [6,37,58]). This fact seems to be *unknown* to the other authors (or is continuously *ignored*):

$$I^{a,a',b,b',c} f(z) = z^{c-a-a'} \int_0^1 \sigma^{-a'} G_{3,3}^{3,0} \left[\sigma \left| \begin{array}{c} a+b, c-a', c-b' \\ a, b, c-a'-b' \end{array} \right. \right] f(z\sigma) d\sigma$$

$$= z^{c-a-a'} \int_0^1 G_{3,3}^{3,0} \left[\sigma \left| \begin{array}{c} a-a'+b, c-2a', c-a'-b' \\ a-a', b-a, c-2a'-b' \end{array} \right. \right] f(z\sigma) d\sigma \quad (69)$$

$$= z^{c-a-a'} \int_0^1 H_{3,3}^{3,0} \left[\sigma \left| \begin{array}{c} (a-a'+b,1), (c-2a',1), (c-a'-b',1) \\ (a-a',1), (b-a,1), (c-2a'-b',1) \end{array} \right. \right] f(z\sigma) d\sigma$$

$$= z^{c-a-a'} I_{(1,1,1),3}^{(a-a',b-a',c-2a'-b'),(b,c-a'-b,a')} f(z)$$

$$= z^{c-a-a'} I_1^{a-a',b} I_1^{b-a',c-a'-b} I_1^{c-2a'-b',a'} f(z).$$

The relations (31) and (69) have been recently denied and argued in the *Response of authors* [61] to our critical *Commentary* [5] to their paper [62]. Then, I needed to support (by my footnote remark to [61]) the truth of (69) as appearing also in the basic FC book by Samko-Kilbas-Marichev [31], see there Equation (10.38) (for decomposition of Saigo operator) and Equation (10.46), p. 193 (for decomposition of the M-S-M operator).

For the above reasons, to evaluate M-S-M images of special functions, which are representable as Wright g.h.f., one can use the general result of Theorem 3. Thus we have:

Lemma 4. *The image of a Wright g.h.f. under the M-S-M fractional integral is given by the formula*

$$I^{a,a',b,b',c} \left\{ z^\nu {}_p \Psi_q \left[\begin{array}{c} (a_i, A_i)_1^p \\ (b_j, B_j)_1^q \end{array} \bigg| \lambda z^\mu \right] \right\}$$

$$= z^{c-a-a'} {}_{p+3} \Psi_{q+3} \left[\begin{array}{c} (a_i, A_i)_1^p, (a-a'+1+\nu, 1), (b-a'+1+\nu, 1), (c-2a'-b'+1+\nu, 1) \\ (b_j, B_j)_1^q, (a-a'+b+1+\nu, 1), (c-2a'+1+\nu, 1), (c-a'-b'+1+\nu, 1) \end{array} \bigg| \lambda z^\mu \right]. \quad (70)$$

The corresponding simpler result for $\nu = 0$ is given by Corollary 4 in Kiryakova [2].

The M-S-M fractional derivatives $D^{a,a',b,b',c}$, denoted also by $I^{a,a',b,b',c}$ with $\operatorname{Re} c \leq 0$, are considered by Saigo and Maeda and by the next authors as originally defined by analogy with the Saigo derivatives $D^{\alpha,\beta,\eta}$. In view of (69), they can be considered also as special cases of the generalized fractional derivatives (23) with $m = 3$, namely as $D^{a,a',b,b',c} = D_{(1,1,1),3}^{(a-a',b-a',c-2a'-b'),(b,c-a'-b,a')} z^{-c}$.

The authors after Saigo-Maeda use to derive first a formula for the M-S-M image of a power function z^p, ignoring the fact that it exists in the original paper (1996) (and follows also as a particular case of our (27) in Section 2). Then, to find the M-S-M fractional integral or derivative of a particular special function, they use the standard techniques of term-by-term integration/or differentiation of the corresponding powers series. However, our general approach says that we know in advance the image of a ${}_p\Psi_q$-function expected as a ${}_{p+3}\Psi_{q+3}$, see (70).

We provide a few illustrative examples for other authors' results, mentioned also in Kiryakova [2].

Example 8. *The formula for the M-S-M generalized fractional integral of a weighted Bessel function:*

$$I^{a,a',b,b',c} \{z^{\gamma-1} J_\nu(z)\} = \frac{z^{\gamma+\nu-a-a'+c-1}}{2^\nu}$$

$$\times {}_3\Psi_4 \left[\begin{array}{c} (\gamma+\nu, 2), (\gamma+\nu+c-a-a'-b, 2), (\gamma+\nu+b'-a', 2) \\ (\gamma+\nu+b', 2), (\gamma+\nu+c-a-a', 2), (\gamma+\nu+c-a'-b, 2), (\nu+1, 1) \end{array} \bigg| -\frac{z^2}{4} \right], \quad (71)$$

can be found in Purohit-Suthar-Kalla ([60], Th.2.1, (10)). It is supposed that $\mathrm{Re}\, c > 0$, $\mathrm{Re}\, \nu > -1$, $\mathrm{Re}\,(\gamma + \nu) > \max[0, \mathrm{Re}\,(a + a' + b - c), \mathrm{Re}\,(a' - b')]$. The same result, however, can be obtained in the way as discussed in Example 5, using the M-S-M operator's representation (69) and the result from Lemma 4. Then, by analogy with Example 6, one can derive the particular result from ([60], Cor.3.1, (24)) for the M-S-M image of $z^{\gamma-1} \cos z$, again in terms of $_3\Psi_4(-\frac{z^2}{4})$.

Example 9. Mondal-Nisar ([63], Th.3, (11)) evaluated the M-S-M integral (68) of the so-called generalized Bessel function

$$W_{p,\beta,\gamma}(z) = \sum_{k=0}^{\infty} \frac{(-1)^k \gamma^k}{\Gamma(p + \frac{\beta}{2} + \frac{1}{2} + k)\, k!} \left(\frac{z}{2}\right)^{2k+p}.$$

Evidently, it is a variant (up to variable substitution) of the Bessel–Maitland-Wright function $J_\nu^\kappa(z)$ and of the Wright function $\phi(z)$, representable as $_0\Psi_1$-function of $(z^2/4)$. Then as well expected, the result comes as a $_3\Psi_4$-function, since the indices are increased by 3.

Example 10. Next, in Nisar-Mondal-Agrawal ([64], Th.1), the authors derive the M-S-M operator of the Bessel-Struve function, which is representable as a 2×2-indices (multi-index, $m = 2$) Mittag-Leffler function (9), see Examples (11) mentioned in Section 2.1, as well as a Wright g.h.f. $_1\Psi_1$,

$$S_\nu(z) = \frac{\Gamma(\nu)}{\sqrt{\pi}} \sum_{k=0}^{\infty} \frac{\Gamma(\frac{1}{2} + \frac{k}{2})}{\Gamma(\nu + 1 + \frac{k}{2})} \frac{z^k}{k!} = \frac{\Gamma(\nu)}{\sqrt{\pi}} {}_1\Psi_1 \left[\begin{matrix} (\frac{1}{2}, \frac{1}{2}) \\ (\nu + 1, \frac{1}{2}) \end{matrix} \middle| z \right].$$

Then, the result for $I^{a,a',b,b',c}\{t^{\gamma-1} S_\nu(\lambda z)\}$ is expected, written in terms of a $_{1+3}\Psi_{1+3}(\lambda z) = {}_4\Psi_4(\lambda z)$, with parameters following from the general scheme.

Example 11. The M-S-M operator (68) of a generalized multi-index Mittag-Leffler function

$$E^{\gamma,\kappa}_{(\alpha_j,\beta_j)_1^m}(z) = \sum_{k=0}^{\infty} \frac{(\gamma)_{\kappa k}\, z^k}{\prod_{j=1}^{m} \Gamma(\alpha_j k + \beta_j)} \frac{1}{k!} = \frac{1}{\Gamma(\gamma)} {}_1\Psi_m \left[\begin{matrix} (\gamma, \kappa) \\ (\beta_j, \alpha_j)_1^m \end{matrix} \middle| z \right],$$

is handled in Agarwal-Rogosin-Trujillo [29]. When $m = 1$ it was studied also by Srivastava-Tomovski [65]. Note that for $\gamma = \kappa = 1$ the above function reduces to the $(2m)$ multi-index Mittag-Leffler function (9). This appeared also in Saxena-Nishimoto [66] and was studied in Saxena-Pogany-Ram-Daiya [67]. The result from ([29], Th.3.1, (3.2)) is the following:

$$I^{a,a',b,b',c}\left\{z^{\rho-1} E^{\gamma,\kappa}_{(\alpha_j,\beta_j)_1^m}(\lambda z^\mu)\right\} = \frac{z^{\rho+c-a-a'-1}}{\Gamma(\gamma)} \tag{72}$$

$$\times {}_4\Psi_{m+3} \left[\begin{matrix} (\gamma, \kappa), (\rho, \mu), (\rho + c - a - a' - b, \mu), (\rho + b' - a', \mu) \\ (\alpha_j, \beta_j)_1^m, (\rho + b', \mu), (\rho + c - a - a', \mu), (\rho + c - b - a', \mu) \end{matrix} \middle| \lambda z^\mu \right].$$

Using the representations of the M-S-M operator as three-tuple generalized fractional integral (69) and of this special function as a $_1\Psi_m$-function, the same formula can be evaluated by the general result in Theorem 3, that is the image is again a Wright g.h.f. but its indices are increased by three.

Example 12. We were stuck on a paper by Kumar-Gupta-Rawat [68] (very fast accepted and published with a lot of typographical problems). The authors there aim to "establish certain generalized fractional differentiation involving M-L type function with four parameters, recently introduced by Garg et al. (2016)". Namely, they have evaluated its image under the Marichev–Saigo–Maeda derivative $D^{a,a',b,b',c}$, corresponding to the integral operator (68). Their result, Theorem 1 (p.205), reads as follows:

$$D^{a,a',b,b',c}\left\{t^{\rho-1}{}_{\xi,\gamma}E_{\mu,\nu}(\lambda z^\sigma)\right\} = \frac{z^{\rho+a+a'-c-1}}{\Gamma(\xi)}$$

$$\times {}_5\Psi_4 \left[\begin{array}{c} (\xi,\gamma),(\rho,\sigma),(\rho+a+a'+b'-c,\sigma),(\rho+a-b,\sigma),(1,1) \\ (\rho-b,\sigma),(\rho+a+a'-c,\sigma),(\nu,\mu),(\rho+a+b'-c,\sigma) \end{array} \bigg| \lambda z^\sigma \right].$$

The authors did not observe the fact that the considered M-L type function is a case of the generalized Wright function (4), the definition of which is also given in the mentioned paper, namely:

$$\xi,\gamma E_{\mu,\nu}(z) = \sum_{k=0}^{\infty} \frac{(\xi)_{\gamma k}}{\Gamma(\mu k + \nu)} z^k = \frac{1}{\Gamma(\xi)} \sum_{k=0}^{\infty} \frac{\Gamma(\xi+\gamma k)\Gamma(k+1)}{\Gamma(\mu k+\nu)} \frac{z^k}{k!} = \frac{1}{\Gamma(\xi)} {}_2\Psi_1 \left[\begin{array}{c} (\xi,\gamma),(1,1) \\ (\nu,\mu) \end{array} \bigg| z \right].$$

Then, its image under the M-S-M differentiation, as a three-tuple generalized fractional derivative, is well expected to be the Wright function ${}_{2+3}\Psi_{1+3}$, in view of our general results as (51) and (70).

Example 13. *In [62], Agarwal-Jain-Baleanu considered the M-S-M images of the generalized Lommel-Wright function*

$$J^{\varphi,m}_{\omega,\theta}(z) = \left(\frac{z}{2}\right)^{\omega+2\theta} \sum_{k=0}^{\infty} \frac{(-1)^k (\frac{z}{2})^{2k}}{(\Gamma(\theta+k+1))^m \Gamma(\omega+k\varphi+\theta+1)} \tag{73}$$

$$= \left(\frac{z}{2}\right)^{\omega+2\theta} {}_1\Psi_{m+1}\left[(1,1);(\theta+1,1),...,(\theta+1,1),(\omega+\theta+1,\varphi);-z^2/4\right],$$

which is a Wright g.h.f. (see Equation (1.1) there). We can note that it is also example of the multi-index M-L function (9), namely $J^{\varphi,m}_{\omega,\theta}(z) = \left(\frac{z}{2}\right)^{\omega+2\theta} \left(\frac{z}{2}\right)^{\omega+2\theta} E^{(m+1)}_{(1,...,1,\varphi),(\theta+1,...,\theta+1,\omega+\theta+1)}\left(-\left(\frac{z}{2}\right)^2\right)$. Then the result, as calculated by the authors, follows directly from Theorem 3 and especially from Lemma 4 (below, $A := \chi + \omega + 2\theta$, $\varphi > 0$):

$$I^{\xi,\xi',\rho,\rho',\varkappa}_{0+}\left[t^{\chi-1}J^{\varphi,m}_{\omega,\theta}(tz)\right](x) = x^{A-\xi-\xi'+\varkappa-1}\left(\frac{z}{2}\right)^{\omega+2\theta} \tag{74}$$

$$\times {}_4\Psi_{4+m}\left[\begin{array}{c}(A,2),(A+\varkappa-\xi-\xi'-\rho,2),(A+\rho'-\xi',2),(1,1) \\ (A+\rho',2),(A+\varkappa-\xi-\xi',2),(A+\varkappa-\xi'-\rho,2),(\omega+\theta+1,\varphi),(\theta+1,1)\end{array}\bigg| -\frac{(tz)^2}{4}\right],$$

to be again a Wright g.h.f. but with indices increased by three, that is, a ${}_4\Psi_{m+4}$-function. In [62] also many special cases are derived, such as Beta-transform (that is E-K integral), Saigo operator, path integral, of the function (73) and of its particular cases. As in the Commentary [5] we discussed the possibilities to use our unified approach, the authors tried to argue with the facts in their Response [61]. The curious readers are recommended to read Kiryakova's footnote comments at the bottom to this Response [61].

8. Multiple Gel'fond-Leontiev Operators of Multi-Index Mittag-Leffler Functions; Hyper-Bessel Operators and Functions

We consider now GFC images of the multi-index M-L functions (9).

Lemma 5. *Taking in general $m \neq n$ (m-tuple operators of GFC and 2n-indexed M-L functions), we have*

$$I^{(\gamma_k),(\delta_k)}_{(\beta_k),m}\left\{z^c E^{(n)}_{(\alpha_i),(\nu_i)}(\lambda z^\mu)\right\} = I^{(\gamma_k),(\delta_k)}_{(\beta_k),m}\left\{z^c {}_1\Psi_n\left[\begin{array}{c}(1,1) \\ (\nu_i,\alpha_i)_1^n\end{array}\bigg| \lambda z^\mu\right]\right\}$$

$$= z^c {}_{1+m}\Psi_{n+m}\left[\begin{array}{c}(1,1),(\gamma_k+1+c/\beta_k,\mu/\beta_k)_1^m \\ (\nu_i,\alpha_i)_1^n,(\gamma_k+\delta_k+1+c/\beta_k,\mu/\beta_k)_1^m\end{array}\bigg| \lambda z^m\right]. \tag{75}$$

This is an easy corollary of Theorem 3. In particular, for $c = 0$, $\mu = 1$, $m = n$, and for GFC parameters taken to be $\gamma_k = \nu_k - 1$, $\beta_k = 1/\alpha_k$, $k = 1, 2, ..., m$, it happens that the parameters $(\gamma_k + 1, 1/\beta_k)_1^m$ in the upper row and $(\nu_i, \alpha_i)_1^m$ in bottom row appear equal and cancel each other, and then the W. g.h.f. ${}_{1+m}\Psi_{m+m}$ reduces to ${}_1\Psi_m$, again a multi-index M-L function!

Then, as proved (in other direct way) in our previous papers, we have:

Example 14 (Kiryakova, [21,22]). *For each fixed $j = 1, ..., m$, a (classical) E-K integral of (9) reads as:*

$$I_{1/\alpha_j}^{\beta_j-1,\delta_j} E_{(\alpha_i),(\beta_i)}(\lambda z) = E_{(\alpha_i),(\beta_1,...,\beta_{j-1},\beta_j+\delta_j,\beta_{j+1},...,\beta_m)}(\lambda z). \quad (76)$$

This is an extension of the result for an E-K integral (12) of an M-L function: $I_{1/\alpha}^{\beta-1,\delta} E_{\alpha,\beta}(z) = E_{\alpha,\beta+\delta}(z)$, where its second index is increased by the order of fractional integral. After m-times application of the above relation with respect to each $j = 1, ..., m$, we obtain that a GFC integral (18) with suitably chosen parameters transforms a multi-index M-L function into the same kind of multi-index M-L function of which the indices of the second set are increased by the multi-order of fractional integration:

$$I_{(1/\alpha_k),m}^{(\beta_k-1),(\delta_k)} E_{(\alpha_i)_1^m,(\beta_i)_1^m}(\lambda z) = E_{(\alpha_i)_1^m,(\beta_i+\delta_i)_1^m}(\lambda z), \text{ with } \operatorname{Re}\delta_k > 0, \gamma_k > -1, \alpha_k > 0, k = 1, ..., m, \lambda \neq 0. \quad (77)$$

If we take $\delta_k = \alpha_k$, $k = 1, ..., m$, formula (77) has the form (Kiryakova, [21,22])

$$I_{(1/\alpha_k),m}^{(\beta_k-1),(\alpha_k)} E_{(\alpha_i),(\beta_i)}(\lambda z) = E_{(\alpha_i),(\beta_i+\alpha_i)}(\lambda z) = (\lambda z)^{-1}\left[E_{(\alpha_i),(\beta_i)}(\lambda z) - \frac{1}{\prod_{i=1}^m \Gamma(\beta_i)}\right].$$

According to the operational rules of the GFC (([6], Ch.5)) and rewriting the above relation for the generalized fractional derivative $D_{(1/\alpha_k),m}^{(\beta_k-1-\alpha_k),(\alpha_k)}$ defined as in (23), we have

$$D_{(1/\alpha_k),m}^{(\beta_k-1-\alpha_k),(\alpha_k)} E_{(\alpha_i),(\beta_i)}(\lambda z) = (\lambda z) E_{(\alpha_i),(\beta_i)}(\lambda z) + \left[\prod_{i=1}^m \Gamma(\beta_i - \alpha_i)\right]^{-1}.$$

Here one can see an analogy with the results (10.6), (10.9) from Haubold-Mathai-Saxena [17] for the R-L operators (in the case $m = 1$).

Next, let us consider the *special cases of GFC operators* for which the multi-index M-L functions (9) appear as eigenfunctions, that is, these special functions are transformed into the same kind of functions with the same multi-indices.

Example 15. *The so-called Gelfond–Leontiev (G-L) operators are operators of generalized integration and differentiation, defined for functions $f(z) = \sum_{j=0}^\infty a_j z_j$ analytic in a disk $|z| < R$, and are generated by the coefficients of a given entire function $\varphi(\sigma)$, used as a multipliers' sequence. They were introduced in a paper by Gelfond and Leontiev of 1951 (for details and references see our works, such as [6,20,21]). In the case when $\varphi(\sigma) = E_{(\alpha_i),(\beta_i)}(\sigma)$ is the multi-index M-L function (9), these operators were considered by Kiryakova [20], see also [21], etc., and called (multiple) Dzrbashjan–Gelfond–Leontiev (D-G-L) operators. We defined them as follows:*

$$\mathfrak{D}f(z) = \sum_{j=1}^\infty a_j \frac{\Gamma(\beta_1 + j\alpha_1)...\Gamma(\beta_m + j\alpha_m)}{\Gamma(\beta_1 + (j-1)\alpha_1)...\Gamma(\beta_m + (j-1)\alpha_m)} z^{j-1},$$
$$\mathfrak{L}f(z) = \sum_{j=0}^\infty a_j \frac{\Gamma(\beta_1 + j\alpha_1)...\Gamma(\beta_m + j\alpha_m)}{\Gamma(\beta_1 + (j+1)\alpha_1)...\Gamma(\beta_m + (j+1)\alpha_m)} z^{j+1}, \quad (78)$$

and noted that the image functions $\mathfrak{D}f(z)$, $\mathfrak{L}f(z)$ are also analytic functions in the same disk $|z| < R$.

We have shown (e.g., in [21,22]) that the operators (78) can be analytically extended (outside a disk, to holomorphic functions in starlike domain) to operators of GFC, namely to generalized integrals and derivatives of fractional multi-order $(\alpha_1, ..., \alpha_m)$ as follows:

$$\mathfrak{L}f(z) = z^1 I_{(1/\alpha_k),m}^{(\beta_k-1),(\alpha_k)} f(z), \quad \mathfrak{D}f(z) = z^{-1} D_{(1/\alpha_k),m}^{(\beta_k-1-\alpha_k),(\alpha_k)} f(z) - z^{-1} f(0) \left[\prod_{k=1}^m \frac{\Gamma(\beta_k)}{\Gamma(\beta_k - \alpha_k)}\right].$$

Then, as proved in [21,22], etc., and seen also in the end of Example 14, the multi-index M-L function (9) is a solution of the differential equation of fractional multi-order

$$\mathfrak{D} E_{(\alpha_i),(\beta_i)}(\lambda z) = E_{(\alpha_i),(\beta_i)}(\lambda z), \text{ that is, } E_{(\alpha_i),(\beta_i)}(z) \text{ is an eigenfunction of the operator } \mathfrak{D}. \tag{79}$$

In view of this relation, the multi-index M-L function serves as an eigenfunction of the D-G-L differentiation generated by its own coefficients!

In the case of *the 3m-parametric M-L type functions* (multi-index Prabhakar functions, [26])

$$E_{(\alpha_i),(\beta_i)}^{(\gamma_i),m}(z) = \sum_{k=0}^{\infty} \frac{(\gamma_1)_k \cdots (\gamma_m)_k}{\Gamma(\alpha_1 k + \beta_1) \cdots \Gamma(\alpha_m k + \beta_m)} \frac{z^k}{(k!)^m}, \text{ with Pochhamer symbols } (\gamma_i)_k := \frac{\Gamma(\gamma_i + k)}{\Gamma(\gamma_i)},$$

the R-L, classical E-K operators and some multiple E-K operators are evaluated in the works of Paneva-Konovska, for example, the book [19].

Example 16. *In 1966, and his later works, Dimovski [41] introduced a very general class of differential operators of arbitrary (integer) order generalizing the Bessel operators of second order. His aims were to develop operational calculus for these operators, both via a Laplace-type integral transform and by the Mikusinski algebraical approach. These operators have the alternative representations*

$$Bf(t) = t^{\alpha_0} \frac{d}{dt} t^{\alpha_1} \frac{d}{dt} \cdots t^{\alpha_{m-1}} \frac{d}{dt} t^{\alpha_m} f(t) = t^{-\beta} P_m\left(t\frac{d}{dt}\right) f(t) = t^{-\beta} \prod_{k=1}^{m}\left(t\frac{d}{dt} + \beta \gamma_k\right) f(t), \ t > 0, \tag{80}$$

with arbitrary parameters α_0; α_k, γ_k, $k = 1, ..., m$; $\beta > 0$; P_m a polynomial of degree m, and their different cases were studied by many authors as appearing in various equations of mathematical physics, problems in analysis, etc., disciplines. The name "hyper-Bessel differential operator" for (80) *was introduced by Kiryakova in the further studies on the topic, for example ([6], Ch.3), and next ones as [69]. For the linear right inverse operator denoted by L and called hyper-Bessel integral operator (such that $BLf(t) = f(t)$), we have found a representation by an integral operator with Meijer's $G_{m,m}^{m,0}$-function in the kernel, and later, the same kind of integral representation also for its fractional powers L^{λ}, $\lambda > 0$. These results were the hint of how to introduce the operators of GFC: the generalized integration and differentiation* (20) *and* (26) *of arbitrary fractional multi-order $(\delta_1, \delta_2, ..., \delta_m)$ instead of the multi-order $(\lambda, \lambda, ..., \lambda)$ for L^{λ}. The story is explained in [69]. Due to the representation of the hyper-Bessel operators in the form: $B = t^{-1} D_{(\beta,\beta,...,\beta),m}^{(\gamma_k-1),(1,1,...,1)}$, $L = t I_{(\beta,\beta,...,\beta),m}^{(\gamma_k),(1,1,...,1)}$, these operators are important examples of the generalized "fractional" derivatives and also of the Gelfond–Leontiev operators* (78). *Indeed, for simplicity we take $\beta = 1$ and $\gamma_m = 0$, then we have that $B = t^{-1} D_{(1)_1^m, m}^{(\gamma_k-1)_1^m, (1)_1^m}$ is a particular case of the operator denoted by $\mathfrak{D}$ in the previous Example 15, with modified denotations.*

Consider the m-th order (that is, of multi-order $(1, 1, ..., 1)$) hyper-Bessel differential equation

$$By(t) = \lambda y(t), \quad \lambda \neq 0. \tag{81}$$

In ([6], Ch.3, Th.3.4.3) and the next corollaries (see also [37]), we proved that the functions, $j = 1, ..., m$:

$$y_j(t) = G_{0,m}^{1,0}\left[-\lambda t \left| \begin{array}{c} -- \\ -\gamma_j, -\gamma_1, ..., -\gamma_{j-1}, -\gamma_{j+1}, ..., -\gamma_{m-1}, 0 \end{array}\right.\right] \tag{82}$$

$$= \frac{(\lambda t)^{-\gamma_j}}{\prod_{k=1}^{m} \Gamma(\gamma_k+1)} {}_0F_{m-1}\left((1+\gamma_i-\gamma_j)_{i\neq j}; \lambda t\right) := J_{(1+\gamma_i-\gamma_j)_{i\neq j}}^{(m-1)}(\lambda t), \text{ the hyper-Bessel functions,}$$

form a fundamental system of solutions of equation (81) in a neighborhood of origin $t = +0$. Under the assumptions of Th.3.4.3 in [6], $\gamma_1 < ... < \gamma_m < \gamma_1 + 1$ and $\gamma_m = 0$, we have that $-\gamma_j \in (0,1)$ for all $j = 1, ..., m$, and one of these solutions, for $j = m$, can be written as

$$y_m(t) = \left[\prod_{k=1}^{m}(\gamma_k+1)\right]^{-1} {}_0F_{m-1}\left((1+\gamma_i)_1^{m-1}; \lambda t\right) = \left[\prod_{k=1}^{m}(\gamma_k+1)\right]^{-1} {}_1F_m\left(1; (1+\gamma_i)_1^{m-1}, 1; \lambda t\right)$$

$$= {}_1\Psi_m\left[\begin{array}{c}(1,1)\\(1+\gamma_i)_1^{m-1}, 1\end{array}\middle| \lambda z\right] = E_{(1,1,...,1),(1+\gamma_i)_1^m}(\lambda z), \quad \text{a case of the multi-index M-L functions.}$$

Therefore, Example 16 appears a special case of Example 15, and shows that the multi-index Mittag-Leffler functions (9) can be seen also as "fractional indices" analogs, extensions of the hyper-Bessel functions (82), which themselves are multi-index variants of the classical Bessel function.

9. Some "New" Special Functions and Their FC Images

Recently, some authors claimed to introduce and consider "new" SF. Among these are examples of the so-called k-analogs of the Bessel and Mittag-Leffler functions, some generalized multi-index Bessel and Mittag-Leffer functions, and some S-functions. The mentioned k-analogs are based on the use of the k-Γ-function, which, however, can be rewritten in terms of the "classical" Γ-function:

$$\Gamma_k(s) = \int_0^\infty \exp(-\frac{t^k}{k}) t^{s-1} dt = k^{\frac{s}{k}-1}\Gamma(\frac{s}{k}), \quad s \in \mathbb{C}, \text{Re}(s) > 0, k > 0; \quad \Gamma(.) \text{ the Gamma-function.} \quad (83)$$

Usually, the denotations include also the k-analogs of the Pochhamer symbol:

$$(\eta)_{\nu,\kappa} := \Gamma_k(\eta + \nu\kappa)/\Gamma_k(\lambda), \quad \eta \in \mathbb{C} \setminus \{0\}, \nu \in \mathbb{C}, \quad (84)$$

and in view of (83) are representable again by means of classical Gamma-functions.

Then, one can easily observe that such "new SF" are just cases of the Wright generalized hypergeometric function ${}_p\Psi_q$. Therefore, all the results provided by the mentioned authors to evaluate FC operators of these special functions follow from our general ones, say from Theorems 3 and 4, or the special cases as Lemmas 1 and 2 (for E-K operators, incl. R-L ones), Lemma 3 (for Saigo operators), Lemma 4 (for M-S-M operators), and so on. As an illustration, we repeat some examples from Kiryakova [4].

Example 17. *A generalization of the Bessel function, called generalized k-Bessel function was introduced by Gehlot [70] and studied by Mondal [71], Shaktawat et al. [72], defined as*

$$W_{\nu,c}^k(z) = \sum_{n=0}^{\infty} \frac{(-c)^n}{\Gamma_k(nk+\nu+k)} \cdot \frac{(z/2)^{2n+\frac{\nu}{k}}}{n!}, \quad z \in \mathbb{C}, \ k > 0, \ \text{Re}(\nu) > -1, \ c \in \mathbb{C}. \quad (85)$$

Lets us note that this function is practically a Wright g.h.f. ${}_0\Psi_1$, and even a simpler g.h.f. ${}_0F_1$ of the same type as the classical Bessel function:

$$W_{\nu,c}^k(z) = (z/2)^{\nu/k} \sum_{n=0}^{\infty} \frac{[-c(z/2)^2]^n}{k^{n+1+(\nu/k)}\Gamma(n+1+(\nu/k))\Gamma(n+1)} = \frac{(z/2)^{\nu/k}}{k^{1+(\nu/k)}} \sum_{n=0}^{\infty} \frac{[-c(z/2)^2]^n}{k^n \Gamma(1+(\nu/k)+n.1)\Gamma(1+n.1)}$$

$$= \frac{(z/2)^{\nu/k}}{k^{1+(\nu/k)}} \sum_{n=0}^{\infty} \frac{[-(c/k)(z/2)^2]^n}{\Gamma(1+(\nu/k)+n.1)\Gamma(1+n.1)} = \frac{(z/2)^{\nu/k}}{k^{1+(\nu/k)}} {}_1\Psi_2\left[\begin{array}{c}(1,1)\\(1+\frac{\nu}{k},1),(1,1)\end{array}\middle| -\frac{c}{k}\left(\frac{z}{2}\right)^2\right] \quad (86)$$

$$= \frac{(z/2)^{\nu/k}}{k^{1+(\nu/k)}} {}_0\Psi_1\left[\begin{array}{c}--\\(1+\frac{\nu}{k},1)\end{array}\middle| -\frac{c}{k}\left(\frac{z}{2}\right)^2\right] = \frac{(z/2)^{\nu/k}}{k^{1+(\nu/k)}\Gamma(1+\nu)} {}_0F_1\left(-;1+\frac{\nu}{k};-\frac{cz^2}{k\,4}\right).$$

Naturally, for $k = 1$, $c = 1$, (85) becomes the classical Bessel function:

$$W_{\nu,1}^1(z) = \frac{(z/2)^\nu}{\Gamma(1+\nu)} \,_0F_1\left(-;1+\nu;-\frac{z^2}{4}\right).$$

In the case $c = 1$, Gehlot [70] considered (85) as a solution of a k-Bessel differential equation. Mondal [71] studied its properties for complex $c \in \mathbb{C}$. Shaktawat et al. [72] evaluated the M-S-M operators of FC of this function. In view of Lemma 4, the result there (Th.1, (18)) is well expected to appear in terms of the $_3\Psi_4$-function (because the 3-tuple FC integral increases by three the indices of the initial $_0\Psi_1$-function).

Example 18. *The simplest k-analogs of the M-L function are considered by Dorrego-Cerruti [73] and Gupta and Parihar [74], and very recently (2020/2021) studied also by Ali et al. [47]:*

$$E_{k,\alpha,\beta}(z) = \sum_{n=0}^{\infty} \frac{z^n}{\Gamma_k(\alpha n + \beta)}, \quad resp. \quad E_{k;\nu,\rho}^{\delta} = \sum_{n=0}^{\infty} \frac{(\delta)_{n,k} z^n}{\Gamma_k(\nu n + \rho)\, n!}.$$

Various further extensions appeared, as the generalized k-Mittag-Leffler function, studied by Gupta and Parihar [74] and Nisar-Eata-Dhaifalla-Choi [75] in the form (note that the index p was missing in these authors' original denotation):

$$E_{\kappa,\alpha,\beta}^{\eta,\delta,p,q}(z) := \sum_{n=0}^{\infty} \frac{(\eta)_{qn,\kappa}}{\Gamma_k(\alpha n + \beta)\,(\delta)_{pn,\kappa}} z^n, \quad \kappa, p, q \in \mathbb{R}_+; \ \alpha,\beta,\eta,\delta \in \mathbb{C}, \quad (87)$$

with $\min\{Re(\alpha), Re(\beta), Re(\eta), Re(\delta)\} > 0$; $q \le Re(\alpha) + p$; *the k-Pochhammer symbol as in (84).*

Again, the function (87) can be rewritten as a Wright g.h.f., now as $_2\Psi_2$. Using the representations for (83) and (84) we have, respectively:

$$(\eta)_{qn,\kappa} = \ldots = \kappa^{qn\kappa/k}\Gamma\left(\frac{\eta + qn\kappa}{k}\right)/\Gamma\left(\frac{\eta}{k}\right); \quad (\delta)_{pn,\kappa} = \ldots = \kappa^{pn\kappa/k}\Gamma\left(\frac{\delta + pn\kappa}{k}\right)/\Gamma\left(\frac{\delta}{k}\right);$$

$$and \quad \Gamma_k(\alpha n + \beta) = k^{\alpha n + \beta/k} \cdot k^{-1}\Gamma\left(\frac{\alpha n + \beta}{k}\right).$$

Then,

$$E_{\kappa,\alpha,\beta}^{\eta,\delta,p,q}(z) = k^{1-\frac{\beta}{k}} \frac{\Gamma(\delta/k)}{\Gamma(\eta/k)} \sum_{n=0}^{\infty} \frac{\Gamma(\frac{\eta}{k} + n \cdot \frac{q\kappa}{k})\Gamma(1+n.1)}{\Gamma(\frac{\delta}{k} + n \cdot \frac{p\kappa}{k})\Gamma(\frac{\beta}{k} + n \cdot \frac{\alpha}{k})} \cdot \frac{\left[k^{\frac{q\kappa - p\kappa - \alpha}{k}} z\right]^n}{n!}$$

$$= k^{1-\frac{\beta}{k}} \frac{\Gamma(\delta/k)}{\Gamma(\eta/k)} \,_2\Psi_2\left[\begin{array}{c}(\frac{\eta}{k},\frac{q\kappa}{k}),(1,1)\\(\frac{\delta}{k},\frac{p\kappa}{k}),(\frac{\beta}{k},\frac{\alpha}{k})\end{array}\,\bigg|\, k^{\frac{(q-p)\kappa - \alpha}{k}} z\right].$$

By the standard techniques, Nisar-Eata-Dhaifalla-Choi [75] evaluated FC operators of the functions (87). In view of our general results, as expected, the results are $_5\Psi_5$-functions (for the MSM operators, Ths. 1–2, 3–4) there), and in particular, $_4\Psi_4$-functions (for the Saigo operators, Cor. 3.1–3.2, there). Also, the pathway integrals (that are related to E-K integrals) are calculated.

Example 19. *The so-called multi-index Bessel function:*

$$J_{(\beta_j)_m,\kappa,b}^{(\alpha_j)_m,\gamma,c}(z) = \sum_{k=0}^{\infty} \frac{c^k\,(\gamma)_{\kappa k}}{\prod_{j=1}^{m} \Gamma(\alpha_j k + \beta_j + \frac{b+1}{2})} \cdot \frac{z^k}{k!}, \quad m = 1, 2, 3, \ldots, \quad (88)$$

with the Pochhammer symbol $(\gamma)_{\kappa k}$, were introduced and studied in a series of papers by Nisar at al., see, for example, Nisar-Purohit-Parmar [76]. The authors proposed a result for the R-L fractional integral of (88), unfortunately written wrongly in their Theorem 1, Equation (2.4) of [76] as a $_2\Psi_2$-function, although it is evidently a $_2\Psi_{m+1}$-function having $(m+1)$ parameters in the low row. The true result should be

$$R^\lambda \left\{ t^{\delta-1} J_{(\beta_j)_m,\varkappa,b}^{(\alpha_j)_m,\gamma,c}(z) \right\} = \frac{1}{\Gamma(\gamma)} z^{\delta+\lambda-1} {}_2\Psi_{m+1} \left[\begin{array}{c} (\gamma,\varkappa),(\delta,1) \\ (\beta_j + \frac{b+1}{2}, \alpha_j)_1^m, (\lambda+\delta,1) \end{array} ; cz \right]. \quad (89)$$

However, it is easily seen that (88) is: $J_{(\beta_j)_m,\varkappa,b}^{(\alpha_j)_m,\gamma,c}(z) = \frac{1}{\Gamma(\gamma)} {}_1\Psi_m \left[\begin{array}{c} (\gamma,\varkappa) \\ (\beta_j + \frac{b+1}{2}, \alpha_j)_{j=1}^m \end{array} ; cz \right]$, so this result follows directly from our Lemma 1. Note that the function (88) is also a special case of the generalized multi-index M-L function in Example 11 with $\beta_j \mapsto \beta_j + (b+1)/2$, and then the results for its images under FC operators follow from these in Agarwal-Rogosin-Trujillo [29].

Very recently (published 24 September 2020), in Mubeen etal. [48], the authors considered integral transforms, including FC operators, of yet more general SF called "extended generalized multi-index Bessel function" introduced by Kamarujjama-Khan-Khan (2019) with an additional member $(\delta)_k$ in the denominator of the series, as:

$$J_{(\beta_j)_m,\varkappa,b,\delta}^{(\alpha_j)_m,\gamma,c}(z) = \sum_{k=0}^\infty \frac{(\gamma)_{\varkappa k}(-cz)^k}{(\delta)_k \prod_{j=1}^m \Gamma(\alpha_j k + \beta_j + \frac{b+1}{2})}. \quad (90)$$

Following similar manipulations as we did in [4] (Section 5.3, Equation (48)) for the case of (88), one can show that

$$J_{(\beta_j)_m,\varkappa,b,\delta}^{(\alpha_j)_m,\gamma,c}(z) = \frac{\Gamma(\delta)}{\Gamma(\gamma)} {}_2\Psi_{m+1} \left[\begin{array}{c} (\gamma,\varkappa),(1,1) \\ (\delta,1), (\beta_j + \frac{b+1}{2}, \alpha_j)_1^m \end{array} \bigg| -cz \right],$$

and is evidently reducible to (88) for $\delta = 1$. Therefore, the M-S-M fractional integral will be a $_5\Psi_{m+4}$-function—presented as an explicit SF, instead of the authors' hardly visible result in form of some unknown complicated series, compared with Th.5.3, [48] for the extension (1.17) of the function (90).

Example 20. The S-function was introduced in Saxena-Daiya [77] as a "new" special function extending the M-L function ($p = q = 0, k = 1$), the Prabhakar function (8), the M-series (56) of Sharma and Jain [54] ($\gamma = 1$, $k = 1$), etc., by

$$S[z] := S_{(p,q)}^{\alpha,\beta,\gamma,\tau,k}(a_1,...,a_p;b_1,...,b_q;z) = \sum_{n=0}^\infty \frac{(a_1)_n...(a_p)_n \cdot (\gamma)_{n\tau,k}}{(b_1)_n...(b_q)_n \cdot \Gamma_k(n\alpha+\beta)} \cdot \frac{z^n}{n!}, \quad (91)$$

with $k \in \mathbb{R}, ; \alpha, \beta, \gamma, \tau \in \mathbb{C}; Re(\alpha) > 0; Re(\alpha) > k, Re(\tau), p < q+1$. For $p = q = 0$ it reduces to the generalized k-Mittag-Leffler function $E_{k,\alpha,\beta}^{\gamma,\tau}(z)$, see in Example 18, the simplest case by Gupta and Parihar [74].

However, as shown in Kiryakova ([4], Section 5.4), this S-function (91) appears to be a Wright g.h.f. (4) of the form $_{p+1}\Psi_{q+1}\left(zk^{\tau-\frac{\alpha}{k}}\right)$, namely:

$$S[z] = k^{1-\frac{\beta}{k}} \frac{\Gamma(b_1)...\Gamma(b_q)}{\Gamma(a_1)...\Gamma(a_p) \cdot \Gamma(\frac{\gamma}{k})} {}_{p+1}\Psi_{q+1} \left[\begin{array}{c} (a_1,1),...,(a_p,1),(\frac{\gamma}{k},\tau) \\ (b_1,1),...,(b_q,1),(\frac{\beta}{k},\frac{\alpha}{k}) \end{array} ; zk^{\tau-\frac{\alpha}{k}} \right].$$

Unfortunately, this fact has not been observed neither by the authors of [77] introducing it, nor by their numerous followers. Then, all results for images of FC operators, such as R-L, E-K, Saigo, M-S-M, the Euler-transform (which is in fact E-K operator), Laplace transform, follow as images of the Wright function according to our general results, say Theorem 4.1. Then, as evaluated in [77], Th. 2.10, (32), the Euler transform is a function $_{p+2}\Psi_{q+2}(zk^{\tau-\frac{\alpha}{k}})$, because the indices are increased by one for the E-K operator; the Saigo operators will increase the indices by two; the M-S-M integral will be in terms of Wright function with indices increased by three, namely: $_{(p+1)+3}\Psi_{(q+1)+3}$, etc.

Special cases of (91) are the generalized K-series (65) and M-series (56), see Examples 4 and 7.

Example 21. *The generalized k-Wright function (multi-parametric k-M-L function) is introduced by Purohit and Badguzer [78] as a k-extension of the Wright g.h.f. (4):*

$$
{}_p\Psi_q^k(z) = {}_p\Psi_q^k \left[\begin{matrix} (a_1, A_1), \ldots, (a_p, A_p) \\ (b_1, B_1), \ldots, (b_q, B_q) \end{matrix} \bigg| z \right] = \sum_{n=0}^{\infty} \frac{\Gamma_k(a_1 + nA_1) \ldots \Gamma_k(a_p + nA_p)}{\Gamma_k(b_1 + nB_1) \ldots \Gamma_k(b_q + nB_q)} \frac{z^n}{n!}, \quad k > 0. \quad (92)
$$

However, from the representation (83), it is seen that this "new" function is again a Wright generalized hypergeometric function, namely:

$$
\text{const } {}_{p+1}\Psi_{q+1} \left[\begin{matrix} (a_i/k, A_i/k)_{i=1}^p \\ (b_j/k, B_j/k)_{j=1}^q \end{matrix} \bigg| k^{(A_1+\ldots+A_p-B_1-\ldots-B_q)/k} \cdot z \right].
$$

Then the M-S-M operators evaluated for (92) by these authors can appear directly from our general results (say, Lemma 4) in terms of ${}_{p+4}\Psi_{q+4}$-functions.

10. Conclusions

10.1. The researchers on the topic can be advised *to follow a procedure like this*:

(1) Check if the considered SF can be presented as a Wright g.h.f. ${}_p\Psi_q$ or as simpler ${}_pF_q$-function; in more complicated cases, or if it is a Fox H-function or a Meijer G-function;
(2) Check if the operator of FC to be evaluated is some special case of the GFC operators, that is, if it can be presented as a composition of classical R-L or E-K operators (also in the form (20) and (26));
(3) Then, apply a general result like Theorem 3, Theorem 4 (or more generally, Theorem 2) and their special cases (Lemmas 1–4) and the examples, provided in this survey.

10.2. In Section 3 we first give the images of the SF (the generalized hypergeometric functions ${}_p\Psi_q$, ${}_pF_q$ and their simplest cases) for the classical FC operators: E-K and R-L, and show that these are the same kinds of functions of which the indices p, q are increased by 1. Then, the images under the GFC operators are obtained by m-times application of these results, in Section 4. Our result states that *the image of a ${}_p\Psi_q$-function (resp. ${}_pF_q$-function) under any (m-tuple) GFC operator can be predicted by Theorems 3 and 4 to be a ${}_{p+m}\Psi_{q+m}$-function (resp. ${}_{p+m}F_{q+m}$-function)* with additional parameters depending on those of the FC operators. Using this general approach, one can avoid application of the standard term-by-term integration/differentiation of the power series for each particular special function the authors choose to treat.

10.3. For the *proofs* of Theorems 3 and 4 and their corollaries, see Kiryakova [2–4], and for their *alternative interpretations*—in other our works as ([6,22,24,50], Ch.4), [27]. The basic idea is that by means of a multiple (m-tuple) operator of GFC each ${}_{p+m}\Psi_{q+m}$-function (resp. a ${}_{p+m}F_{q+m}$-function) can be reduced to a ${}_p\Psi_q$-function (resp. a ${}_pF_q$-function), see comments and formula (50) before Theorem 4. Thus, by a suitable number of steps, from any ${}_p\Psi_q$-function (resp. ${}_pF_q$-function) we can reach to one of the *three basic generalized hypergeometric functions*, depending on either $p < q$, $p = q$ or $p = q+1$: ${}_0\Psi_{q-p}$, ${}_1\Psi_1$, ${}_2\Psi_1$ (resp. ${}_0F_{q-p}$, ${}_1F_1$, ${}_2F_1$). Additionally, by an Erdélyi–Kober operator these are reducible to one of the three elementary functions: $\cos z$, $z^\alpha \exp(z)$ or $(1-z)^\alpha z^\beta$. Details are given in Kiryakova [27], submitted to this Journal. As a conclusion, we have classified the g.h.f., that is the SF, in three basic classes: "g.h.f. of Bessel/cosine type", "g.h.f. of confluent/exp type" and "g.h.f. of Gauss/beta-distribution type", each of these classes with own specific properties. Thus, the title of Kiryakova [50] appeared as: *"All the special functions are fractional differintegrals of elementary functions"*.

10.4. In some papers, the authors evaluate an operator of FC of a particular special function in terms of *another* special function. Or even, the final result is written *only as a series not recognized as some SF*. However, for both theoretical reasons and possible applications, the results can be useful when a GFC operator transforms a special function from some class *into a special function of the same class*, although with changed (increased/decreased) indices and additional parameters. Such are our Theorems 3 and 4, showing that a $_p\Psi_q$-function (resp. a $_pF_q$-function) is transformed into a $_{p+m}\Psi_{q+m}$-function (resp. a $_{p+m}F_{q+m}$-function). We discussed similar results also for some particular cases of SF, such as for the (classical) Wright function $\phi(\alpha,\beta,z)$, the M-series $_p M_q^{\alpha,\beta}(z)$. Among the *illustrative examples for SF having FC images of the same class*, are our formulas: (32), (36), (37), (39), (40), (41), (46), (49), (51), and their corollaries like (55)–(58),(61), (62), (70), (77).

10.5. Next goal: the most useful results on the topic are when we can *specify an operator of GFC corresponding to the considered special function, so that this function is to be its eigenfunction*. In this survey we give such examples, say for the multi-index Mittag-Leffler functions $E_{(\alpha_i),(\beta_i)}(z)$ and the hyper-Bessel functions of Delerue $_0F_{m-1}\left((1+\gamma_i)_1^{m-1};z\right)$. These are the formulas (79), (81) and (82), etc. Another example, for the eigenfunction of the simplest fractional order differential equation, is the Rabotnov function $z^{\alpha-1}E_{\alpha,\alpha}(z^\alpha)$ (called also fractional exponent), namely:

$$D^\alpha y_\alpha(z) = \lambda y_\alpha(z), \text{ where } y_\alpha(z) = z^{\alpha-1}E_{\alpha,\alpha}(\lambda z^\alpha), \quad \alpha > 0, \lambda \neq 0. \tag{93}$$

10.6. Many authors are publishing results on the *images of particular special functions under some integral transforms* like the *Laplace transform, Mellin transform, Euler (Beta) transform, Whittaker transform*. Observe that the *Euler transform* (called so after the Euler integral formula for the Gauss function) is just a case of the Erdélyi–Kober fractional integral (12), as an extension of the Riemann–Liuoville fractional integral (14). Therefore, there is no need to separately evaluate these two transforms (Euler transform and Riemann–Liouville operator), and what is more, to repeat such calculations for each particular special function. One can just apply the general result, as in Lemma 1. Note that the so-called *pathway-transform* is also closely related to the E-K integral. To evaluate a *Laplace transform*, say for any special function which is an *H*-function, one can use the general integral formula (44) and the representation of the kernel exponential function as a Wright g.h.f. (see (42), Example 1), then also as a H-function: $\exp(-z) = H_{0,1}^{1,0}\left[z \left|\begin{array}{c}--\\(0,1)\end{array}\right.\right]$. As already mentioned in Remark 1, a basic approach to evaluate integral transforms (also FC operators) of special functions relies on their images under the *Mellin transform* in terms of Gamma-functions, to which the fundamental book by Marichev [8] is devoted.

Author Contributions: The ideas and results in this paper survey and reflect the author's (V. K.) sole contributions, resulting from more than 30 years of research on the topic. The author haves read and agreed to the published version of the manuscript.

Funding: This research received no financial funding.

Acknowledgments: This paper is done under the working programs on bilateral collaboration contracts of Bulgarian Academy of Sciences with Serbian and Macedonian Academies of Sciences and Arts, and under the COST program, COST Action CA15225 'Fractional'.

Conflicts of Interest: The author declares no conflict of interest.

References

1. Machado, J.A.T.; Kiryakova, V. Recent history of the fractional calculus: Data and statistics. In *Handbook of Fractional Calculus with Applications. Volume 1: Basic Theory*; Kochubei, A., Luchko, Y., Eds.; De Gryuter: Berlin, Germany, 2019; pp. 1–21. [CrossRef]
2. Kiryakova, V. Fractional calculus operators of special functions?—The result is well predictable! *Chaos Solitons Fractals* **2017**, *102*, 2–15. [CrossRef]
3. Kiryakova, V. Use of fractional calculus to evaluate some improper integrals of special functions. *AIP Conf. Proc.* **2017**, *1910*, 050012. [CrossRef]
4. Kiryakova, V. Fractional calculus of some "new" but not new special functions: k-, multi-index-, and S-analogues. *AIP Conf. Proc.* **2019**, *2172*, 0500088. [CrossRef]

5. Kiryakova, V. Commentary: "A remark on the fractional integral operators and the image formulas of generalized Lommel-Wright function". *Front. Phys.* **2019**, *7*, 145. [CrossRef]
6. Kiryakova, V. *Generalized Fractional Calculus and Applications*; Longman—J. Wiley: Harlow, UK; New York, NY, USA, 1994.
7. Prudnikov, A.P.; Brychkov, Y.; Marichev, O.I. *Integrals and Series, Vol. 3: More Special Functions*; Gordon and Breach Sci. Publ.: New York, NY, USA; London, UK; Paris, France; Tokyo, Japan, 1992.
8. Marichev, O.I. *Handbook of Integral Transforms of Higher Transcendental Functions, Theory and Algorithmic Tables*; Ellis Horwood: Chichester, UK, 1983; Translated from Russian; Method of Evaluation of Integrals of Special Functions (In Russian); Nauka i Teknika, Minsk, Belarus, 1978.
9. Srivastava, H.M.; Gupta, K.S.; Goyal, S.P. *The H-Functions of One and Two Variables with Applications*; South Asian Publs: New Delhi, India, 1982.
10. Kilbas, A.A.; Srivastava, H.M.; Trujillo, J.J. *Theory and Applications of Fractional Differential Equations*; Elsevier: Amsterdam, The Netherlands, 2006.
11. Podlubny, I. *Fractional Differential Equations*; Acad. Press: Boston, MA, USA, 1999.
12. Yakubovich, S.; Luchko, Y. *The Hypergeometric Approach to Integral Transforms and Convolutions*; Ser. Mathematics and Its Applications 287; Kluwer Acad. Publ.: Dordrecht, The Netherlands; Boston, MA, USA; London, UK, 1994.
13. Mathai, A.M.; Haubold, H.J. *Special Functions for Applied Scientists*; Springer: Berlin/Heidelberg, Germany, 2008.
14. Gorenflo, R.; Kilbas, A.; Mainardi, F.; Rogosin, S. *Mittag-Leffler Functions, Related Topics and Applications*, 2nd ed.; Springer: Berlin/Heidelberg, Germany, 2014. [CrossRef]
15. Erdélyi, A. (Ed.) *Higher Transcendental Functions*; McGraw Hill: New York, NY, USA, 1953; Volume 1–3.
16. Kilbas, A.A. Fractional calculus of the generalized Wright function. *Fract. Calc. Appl. Anal.* **2005**, *8*, 113–126.
17. Haubold, H.J.; Mathai, A.M.; Saxena, R.K. Mittag-Leffler functions and their applications. *J. Appl. Math.* **2011**, *51*, 298628. [CrossRef]
18. Rogosin, S. The role of the Mittag-Leffler function in fractional modeling. *Mathematics* **2015**, *3*, 368–381. [CrossRef]
19. Paneva-Konovska, J. *From Bessel to Multi-Index Mittag-Leffler Functions: Enumerable Families, Series in Them and Convergence*; World Scientific Publishing: London, UK, 2016.
20. Kiryakova, V. Multiple Dzrbashjan-Gelfond-Leontiev fractional differintegrals. *Recent Adv. Appl. Math.* **1996**, *96*, 281–294.
21. Kiryakova, V. Multiple (multiindex) Mittag-Leffler functions and relations to generalized fractional calculus. *J. Comput. Appl. Math.* **2000**, *118*, 241–259. [CrossRef]
22. Kiryakova, V. The multi-index Mittag-Leffler functions as important class of special functions of fractional calculus. *Comput. Math. Appl.* **2010**, *59*, 1885–1895. [CrossRef]
23. Kilbas, A.A.; Koroleva, A.A.; Rogosin, S.V. Multi-parametric Mittag-Leffler functions and their extension. *Fract. Calc. Appl. Anal.* **2013**, *16*, 378–404. [CrossRef]
24. Kiryakova, V. The special functions of fractional calculus as generalized fractional calculus operators of some basic functions. *Comput. Math. Appl.* **2010**, *59*, 1128–1141. [CrossRef]
25. Dzrbashjan, M.M. On the integral transformations generated by the generalized Mittag-Leffler function. *Izv. Arm. SSR* **1960**, *13*, 21–63. (In Russian)
26. Paneva-Konovska, J. Multi-index (3m-parametric) Mittag-Leffler functions and fractional calculus. *Compt. Rend. Acad. Bulg. Sci.* **2011**, *64*, 1089–1098.
27. Kiryakova, V. A guide to special functions in fractional calculus. *Math. Spec. Issue Spec. Funct. Math. Phys. Part II* **2020**, Submitted.
28. Luchko, Y.; Kiryakova, V. The Mellin integral transform in fractional calculus. *Fract. Calc. Appl. Anal.* **2013**, *16*, 405–430. [CrossRef]
29. Agarwal, P.; Rogosin, S.V.; Trujillo, J.J. Certain fractional integral operators and the generalized multi-index Mittag-Leffler functions. *Proc. Indian Acad. Sci. (Math. Sci.)* **2015**, *125*, 291–306. [CrossRef]
30. Paneva-Konovska, J.; Kiryakova, V. On the multi-index Mittag-Leffler functions and their Mellin transforms. *Int. J. Appl. Math.* **2020**, *33*, 549–571. [CrossRef]
31. Samko, S.; Kilbas, A.; Marichev, O. *Fractional Integrals and Derivatives: Theory and Applications*; Gordon and Breach: Yverdon, Switzerland, 1993.

32. Sandev, T.; Tomovski, Ž. *Fractional Equations and Models (Theory and Applications)*; Springer: Berlin/Heidelberg, Germany, 2019. [CrossRef]
33. Sneddon, I.N. The use in mathematical analysis of Erdélyi-Kober operators and some of their applications. In *Fractional Calculus and Its Applications (Proc. Internat. Conf. Held in New Haven)*; Ross, B., Ed.; Lecture Notes in Math. 457; Springer: New York, NY, USA, 1975; pp. 37–79.
34. Kiryakova, V. Generalized fractional calculus operators with special functions. In *Handbook of Fractional Calculus with Applications. Volume 1: Basic Theory*; Kochubei, A., Luchko, Y., Eds.; De Gryuter: Berlin, Germany, 2019; pp. 87–110. [CrossRef]
35. Kiryakova, V.; Luchko, Y. Riemann-Liouville and Caputo type multiple Erdélyi-Kober operators. *Cent. Eur. J. Phys.* **2013**, *11*, 1314–1336. [CrossRef]
36. Luchko, Y.; Trujillo, J.J. Caputo type modification of the Erdélyi-Kober fractional derivative. *Fract. Calc. Appl. Anal.* **2007**, *10*, 249–267.
37. Kiryakova, V. A brief story about the operators of the generalized fractional calculus. *Fract. Calc. Appl. Anal.* **2008**, *11*, 203–220.
38. Kiryakova, V. Gel'fond-Leont'ev integration operators of fractional (multi-)order generated by some special functions. *AIP Conf. Proc.* **2018**, *2048*, 050016. [CrossRef]
39. Marichev, O.I. Volterra equation of Mellin convolutional type with a Horn function in the kernel. *Izv. AN BSSR, Ser. Fiz.-Mat. Nauk* **1974**, *1*, 128–129. (In Russian)
40. Saigo, M.; Maeda, N. More generalization of fractional calculus. In *Transform Methods & Special Functions, Varna'96 (Proc. Second In- ternat. Workshop)*; Rusev, P., Dimovski, I., Kiryakova, V., Eds.; Science Culture Technology Publishing: Singapore, 1998; pp. 386–400.
41. Dimovski, I. Operational calculus for a class of differential operators. *C.R. Acad. Bulg. Sci.* **1966**, *19*, 1111–1114.
42. Dimovski, I.; Kiryakova, V. Generalized Poisson transmutations and corresponding representations of hyper-Bessel functions. *C. R. Acad. Bulg. Sci.* **1986**, *39*, 29–32.
43. Delerue, P. Sur le calcul symbolique à n variables et fonctions hyperbesseliennes (II). *Annales Soc. Sci. Bruxelles, Ser. 1* **1953**, *3*, 229–274.
44. Erdélyi, A. (Ed.) *Tables of Integral Transforms*; McGraw Hill: New York, NY, USA, 1954; Volume 1–2.
45. Askey, R. *Orthogonal Polynomials and Special Functions*; SIAM: Philadelphia, PA, USA, 1975.
46. Lavoie, J.L.; Osler, T.J.; Tremblay, R. Fractional derivatives and special functions. *SIAM Rev.* **1976**, *18*, 240–268. [CrossRef]
47. Ali, R.S.; Mubeen, S.; Ahmad, M.M. A class of fractional integral operators with multi-index Mittag-Leffler k-function and Bessel k-function of first kind. *J. Math. Comput. Sci.* **2020**, *22*, 266–281. [CrossRef]
48. Mubeen, S.; Ali, R.S.; Nayab, I.; Rahman, G.; Abdeljavad, T.; Nisar, K.S. Integral transforms of an extended generalized multi-index Bessel function. *AIMS Math.* **2020**, *5*, 7531–7546. [CrossRef]
49. Abdalla, M. Fractional operators for the Wright hypergeometric matrix functions. *Adv. Differ. Equ.* **2020**, *2020*, 246. [CrossRef]
50. Kiryakova, V. All the special functions are fractional differintegrals of elementary functions. *J. Phys. A Math. Gen.* **1997**, *30*, 5085–5103. [CrossRef]
51. Mathai, A.M.; Saxena, R.K.; Haubold, H.J. *The H-Function*; Springer: Berlin/Heidelberg, Germany, 2010.
52. Garrappa, R.; Kaslik, E.; Popolizio, M. Evaluation of fractional integrals and derivatives of elementary functions: Overview and tutorial. *Mathematics* **2019**, *7*, 407. [CrossRef]
53. Kilbas, A.A.; Sebastian, N. Generalized fractional integration of Bessel function of first kind. *Integr. Transf. Spec. Funct.* **2008**, *19*, 869–883. [CrossRef]
54. Sharma, M.; Jain, R. A note on a generalized M-series as a special function of fractional calculus. *Fract. Calc. Appl. Anal.* **2009**, *12*, 449–452.
55. Lavault, C. Fractional calculus and generalized Mittag-Leffler type functions. *arXiv* **2017**, arXiv:1703.01912.
56. Kumar D.; Saxena, R.K. Generalized fractional calculus of the M-Series involving F_3 hypergeometric function. *Sohag. J. Math.* **2015**, *2*, 17–22. [CrossRef]
57. Saigo, M. A remark on integral operators involving the Gauss hypergeometric functions. *Math. Rep. Coll. Gen. Ed. Kyushu Univ.* **1978**, *11*, 135–143.
58. Kiryakova, V. On two Saigo's fractional integral operators in the class of univalent functions. *Fract. Calc. Appl. Anal.* **2006**, *9*, 159–176.

59. Sharma, K. An introduction to the generalized fractional integration. *Bol. Soc. Paran. Math.* **2012**, *30*, 85–90. [CrossRef]
60. Purohit, S.D.; Suthar, D.L.; Kalla, S.L. Marichev-Saigo-Maeda fractional integration operators of the Bessel functions. *Le Mat.* **2012**, *LXVII*, 21–32. [CrossRef]
61. Agarwal, R.; Jain, S.; Agarwal, R.P.; Baleanu, D. Response: Commentary: A remark on the fractional integral operators and the image formulas of generalized Lommel-Wright function. *Front. Phys.* **2020**, *8*, 72. [CrossRef]
62. Agarwal, R.; Jain, S.; Agarwal, R.P.; Baleanu, D. A remark on the fractional integral operators and the image formulas of generalized Lommel-Wright function. *Front. Phys.* **2018**, *6*, 79. [CrossRef]
63. Mondal, S.R.; Nisar, K.S. Marichev-Saigo-Maeda fractional integration operators involving generalized Bessel functions. *Math. Probl. Eng.* **2014**, *11*, 274093. [CrossRef]
64. Nisar, K.S.; Mondal, S.R.; Agarwal, P. Composition formulas of Bessel-Struve kernel function. *arXiv* **2016**, arXiv:1602.00279v1.
65. Srivastava, H.M.; Tomovski, Ž. Fractional calculus with an integral operator containing generalized Mittag-Leffler function in the kernel. *Appl. Math. Comput.* **2009**, *211*, 198–210. [CrossRef]
66. Saxena, R.K.; Nishimoto, K. N-fractional calculus of generalized Mittag-Leffler functions. *J. Fract. Calc.* **2010**, *37*, 43–52.
67. Saxena, R.K.; Pogany, T.K.; Ram, J.; Daiya, J. Dirichlet averages of generalized multi-index Mittag-Leffler functions. *Armen. J. Math.* **2010**, *3*, 174–187.
68. Kumar, D.; Kumar Gupta, R.; Singh Rawat, D. Marichev-Saigo-Maeda fractional differential operator involving mittag-Leffler type function with four parameters. *J. Chem. Biol. Phys. Sci. Sect. C: Phys. Sci.* **2017**, *7*, 201–210.
69. Kiryakova, V. From the hyper-Bessel operators of Dimovski to the generalized fractional calculus. *Fract. Calc. Appl. Anal.* **2014**, *17*, 977–1000. [CrossRef]
70. Gehlot, K.S. Differential equation of k-Bessel's function and its properties. *Nonl. Anal. Differ. Equ.* **2014**, *2*, 61–67. [CrossRef]
71. Mondal, S.R. Representation formulae and monotonicity of the generalized k-Bessel functions. *arXiv* **2016**, arXiv:1611.07499.
72. Shaktawat, B.S.; Rawat, D.S.; Gupta, R.K. On generalized fractional calculus of the generalized k-Bessel function. *J. Rajasthan Acad. Phys. Sci.* **2017**, *16*, 9–19.
73. Dorrego, G.A.; Cerruti, R.A. The k-Mittag-Leffler function. *Int. J. Contemp. Math. Sci.* **2012**, *7*, 705–716.
74. Gupta, A.; Parihar, C.L. k-New generalized Mittag-Leffler function. *J. Fract. Calc. Appl.* **2014**, *5*, 165–176.
75. Nisar, K.S.; Eata, A.F.; Al-Dhaifallah, M.; Choi, J. Fractional calculus of generalized k-Mittag-Leffler function and its applications to statistical distribution. *Adv. Differ. Equ.* **2016**, *304*. [CrossRef]
76. Nisar, K.S.; Purohit, S.D.; Parmar, R.K. Fractional calculus and certain integrals of generalized multiindex Bessel function. *arXiv* **2017**, arXiv:1706.08039.
77. Saxena, R.K.; Daiya, J. Integral transforms of S-functions. *Le Mat.* **2015**, *LXX*, 147–159.
78. Purohit, M.; Badguzer, A. MSM fractional integration and differentiation operators of multi-parametric K-Mittag Leffler function and generalized multi-index Bessel function. *Intern. J. Stat. Appl. Math.* **2018**, *3*, 156–161.

Publisher's Note: MDPI stays neutral with regard to jurisdictional claims in published maps and institutional affiliations.

© 2020 by the author. Licensee MDPI, Basel, Switzerland. This article is an open access article distributed under the terms and conditions of the Creative Commons Attribution (CC BY) license (http://creativecommons.org/licenses/by/4.0/).

Article

Numerical Approaches to Fractional Integrals and Derivatives: A Review

Min Cai and Changpin Li *

Department of Mathematics, Shanghai University, Shanghai 200444, China; caimin_69@163.com
* Correspondence: lcp@shu.edu.cn

Received: 30 November 2019; Accepted: 19 December 2019; Published: 1 January 2020

Abstract: Fractional calculus, albeit a synonym of fractional integrals and derivatives which have two main characteristics—singularity and nonlocality—has attracted increasing interest due to its potential applications in the real world. This mathematical concept reveals underlying principles that govern the behavior of nature. The present paper focuses on numerical approximations to fractional integrals and derivatives. Almost all the results in this respect are included. Existing results, along with some remarks are summarized for the applied scientists and engineering community of fractional calculus.

Keywords: fractional integral; fractional derivative; numerical approximation

1. Introduction

1.1. Historical Review

The primary attempt, which was recorded in history to discuss the idea of generalizing the integer-order differentiation $\frac{d^n f(t)}{dt^n}$ to $\frac{d^\alpha f(t)}{dt^\alpha}$ with non-integer α, was contained in the correspondence of Leibniz [1]. Some remarks were made on the possibility of considering differentials and derivatives of order one-half. Then, the formulation for derivative of non-integer orders was considered by Euler [2] and Fourier [3]. At the end of the 19-th century, the theory of more-or-less complete form appeared for fractional calculus, primarily due to Liouville [4–11], Riemann [12], Grünwald [13], Letnikov [14–17], and Marchaud [18,19].

Theoretical analysis of fractional calculus has been booming since the 20-th century. Results in this respect are fruitful, for example, in mapping properties of fractional integration and integro–differentiation [20], Leibniz rule for the generalized differentiation [21], formulae for fractional integration by parts [22], and the Bernstein-type inequality for fractional integration and differentiation operators [23,24], et al.

It is believed that the proper history of fractional calculus began in the realm of physics, with the papers by Abel [25,26]. In those two papers the integral equation

$$\int_a^x \frac{\varphi(t) dt}{(x-t)^\mu} = f(x), \ x > a, \ 0 < \mu < 1, \tag{1}$$

was solved in connection with the tautochrone problem. Although Abel did not intend to generalize differentiation, the left-hand side of the integral equation leads to the fractional integral operator of order $1 - \mu$. Fractional integro–differentiation in such a form was sharpened somewhat later. It was not until the recent few decades that scholars came to realize the importance of fractional calculus for applied sciences, such as rheology, continuum mechanics, porous media, thermodynamics, electrodynamics, quantum mechanics, plasma dynamics, and cosmic rays [27]. Achievements in this regard were also presented in Refs. [28–31].

1.2. Current Situations

For the time being, the most frequently utilized fractional integrals and derivatives in applications are the left- and right-sided Riemann–Liouville integrals [32] (fractional integrals for short),

$$_{RL}D_{a,x}^{-\alpha}f(x) = \frac{1}{\Gamma(\alpha)} \int_a^x \frac{f(t)dt}{(x-t)^{1-\alpha}}, \quad \alpha > 0, \tag{2}$$

$$_{RL}D_{x,b}^{-\alpha}f(x) = \frac{1}{\Gamma(\alpha)} \int_x^b \frac{f(t)dt}{(t-x)^{1-\alpha}}, \quad \alpha > 0, \tag{3}$$

the left- and right-sided Riemann-Liouville derivatives [32],

$$_{RL}D_{a,x}^{\alpha}f(x) = \frac{1}{\Gamma(m-\alpha)} \frac{d^m}{dx^m} \int_a^x \frac{f(t)dt}{(x-t)^{\alpha+1-m}}, \quad m-1 \le \alpha < m \in \mathbb{Z}^+, \tag{4}$$

$$_{RL}D_{x,b}^{\alpha}f(x) = \frac{(-1)^m}{\Gamma(m-\alpha)} \frac{d^m}{dx^m} \int_x^b \frac{f(t)dt}{(t-x)^{\alpha+1-m}}, \quad m-1 \le \alpha < m \in \mathbb{Z}^+, \tag{5}$$

the left- and right-sided Caputo derivatives [32],

$$_{C}D_{a,x}^{\alpha}f(x) = \frac{1}{\Gamma(m-\alpha)} \int_a^x \frac{f^{(m)}(t)dt}{(x-t)^{\alpha+1-m}}, \quad m-1 \le \alpha < m \in \mathbb{Z}^+, \tag{6}$$

$$_{C}D_{x,b}^{\alpha}f(x) = \frac{(-1)^m}{\Gamma(m-\alpha)} \int_x^b \frac{f^{(m)}(t)dt}{(t-x)^{\alpha+1-m}}, \quad m-1 \le \alpha < m \in \mathbb{Z}^+, \tag{7}$$

and Riesz derivative [33]

$$_{RZ}D_x^{\alpha}f(x) = \Psi_\alpha \left[_{RL}D_{a,x}^{\alpha}f(x) + _{RL}D_{x,b}^{\alpha}f(x) \right], \quad 0 < \alpha \ne 1, 3, 5, \ldots, \tag{8}$$

where $\Psi_\alpha = -\frac{1}{2\cos(\frac{\alpha\pi}{2})}$. They are the subjects of this paper. Fractional integrals and derivatives of other kinds such as ones in [34] and the very newly defined ones in [35,36] and their approximations are omitted here.

Fractional calculus which has two main characteristics—singularity and nonlocality from its origin, is a generalization of the classical one to some extent. However, these two concepts are different. First of all, fractional integral and Riemann–Liouville derivatives coincide with the integer-order ones while Caputo derivative and Riesz derivative fail to be consistent with integer-order derivatives in general cases. Besides, semigroup property is valid for fractional integral while is invalid for the case with fractional derivatives. Fractional derivatives of periodic functions are not in the same form of those in the integer-order case, either. For example, the α-th order Riemann–Liouville derivative of $\sin x$ and $\cos x$ with $\alpha > 0$ are not $\sin(x + \frac{\alpha\pi}{2})$ and $\cos(x + \frac{\alpha\pi}{2})$ unless one adopts the new axiom system proposed in Ref. [37], which differs from the commonly used one. Last but not least, definite conditions for fractional differential equations are in general different from the integer-order case. Especially, in the case with fractional derivatives, boundary and/or initial conditions usually contain fractional derivatives/integrals at the terminals or integer-order derivatives/integrals at points close to the terminals [38,39]. The behavior of the solutions to fractional differential equations may also differ from that of the solutions to the general class of difference equations presented in [40].

In light of potential applications of fractional integration and differentiation operators, there is a substantial demand for efficient algorithms for their numerical handling. Discretizing fractional integrals and derivatives gives a series of quadrature formulae. Different choices of nodes and coefficients give distinct accuracies. Numerical approximations to fractional integrals and derivatives are mainly derived from three distinct paths. Based on the polynomial interpolation, numerical schemes can be obtained with accuracy generally depending on the order of integration and differentiation, for example, the L1, L2, and L2C methods. Convolution quadratures, which preserve properties of fractional integrals and derivatives can be viewed as numerical evaluations of fractional integrals and derivatives with integer-order accuracy. For instance, the fractional multistep methods, among which the fractional backward difference formulae are mostly used, are of integer-order accuracy independent of the order of integration and differentiation. These methods can be verified through Fourier analysis. So do the Grünwald-Letnikov type approximations and fractional centered difference methods. Reformulating fractional integrals and derivatives as infinite integrals of solutions to integer-order ordinary differential equations, the diffusive approximation for fractional calculus can be obtained. Those numerical approximations are discussed in the coming sections. Without specific clarification, the introduced methods are in the setting of uniformed mesh with $h = \frac{b-a}{N}$, $N \in \mathbb{Z}^+$, and $x_j = a + jh$, $j = 0, 1, 2, \ldots, N$.

2. Numerical Approximations to Fractional Integral

The weak singularities in Equations (2) and (3) often make it difficult to calculate fractional integrals directly. In the following, several kinds of numerical methods are introduced.

2.1. Numerical Methods Based on Polynomial Interpolation

Assume that $f(x)$ is suitably smooth on $[a, b]$. Then the α-th order fractional integral of $f(x)$ at $x = x_j$ with $1 \leq j \leq N$ can be expressed as

$$\left[{}_{RL}D_{a,x}^{-\alpha}f(x)\right]_{x=x_j} = \frac{1}{\Gamma(\alpha)} \sum_{k=0}^{j-1} \int_{x_k}^{x_{k+1}} (x_j - t)^{\alpha-1} f(t) dt. \tag{9}$$

It is reasonable to utilize an interpolate function $\widetilde{f}(x)$ to approximate $f(x)$ on each subinterval, such that the integral $\int_{x_k}^{x_{k+1}} (x_j - t)^{\alpha-1} \widetilde{f}(t) dt$ can be calculated exactly. This idea yields a series of numerical formulae in the form

$$\left[{}_{RL}D_{a,x}^{-\alpha}f(x)\right]_{x=x_j} \approx \sum_{k=0}^{j} \omega_k f(x_k), \ 1 \leq j \leq N, \tag{10}$$

where ω_k ($k = 0, 1, \ldots, j$) are the corresponding coefficients. To better understand this method, we retrospect some specific formulae with their brief derivations.

If $f(x) \in C[a, b]$ on the right-hand side of Equation (9) is approximated by a piecewise constant function

$$\widetilde{f}(x) = f(x_k), \ x \in [x_k, x_{k+1}), \ 0 \leq k \leq j-1, \tag{11}$$

then there holds

$$\left[{}_{RL}D_{a,x}^{-\alpha}f(x)\right]_{x=x_j} \approx \frac{1}{\Gamma(\alpha)} \sum_{k=0}^{j-1} \int_{x_k}^{x_{k+1}} (x_j - t)^{\alpha-1} f(x_k) dt. \tag{12}$$

This yields the left fractional rectangular formula [38]

$$\left[{}_{RL}D_{a,x}^{-\alpha}f(x)\right]_{x=x_j} \approx \sum_{k=0}^{j-1} b_{j-k-1} f(x_k), \tag{13}$$

where the convolution coefficients b_k ($0 \leq k \leq j-1$) are given by

$$b_k = \frac{1}{\Gamma(\alpha)} \int_{x_k}^{x_{k+1}} (x_j - t)^{\alpha-1} dt = \frac{h^\alpha}{\Gamma(\alpha+1)} \left[(k+1)^\alpha - k^\alpha\right]. \tag{14}$$

Similarly, if the function $f(x)$ on the right-hand side of Equation (9) is replaced by

$$\tilde{f}(x) = f(x_{k+1}), \ x \in (x_k, x_{k+1}], \tag{15}$$

then we have the right fractional rectangular formula [38]

$$\left[{}_{RL}D_{a,x}^{-\alpha}f(x)\right]_{x=x_j} \approx \sum_{k=0}^{j-1} b_{j-k-1} f(x_{k+1}), \tag{16}$$

with b_k ($0 \leq k \leq j-1$) given by Equation (14). Based on the left and right rectangular formulae, the weighted fractional rectangular formula [38] yields

$$\left[{}_{RL}D_{a,x}^{-\alpha}f(x)\right]_{x=x_j} \approx \sum_{k=0}^{j-1} b_{j-k-1} \left[\theta f(x_k) + (1-\theta) f(x_{k+1})\right], \ 0 \leq \theta \leq 1, \tag{17}$$

or the similar form

$$\left[{}_{RL}D_{a,x}^{-\alpha}f(x)\right]_{x=x_j} \approx \sum_{k=0}^{j-1} b_{j-k-1} f\left(x_k + (1-\theta)h\right), \ 0 \leq \theta \leq 1. \tag{18}$$

Remark 1. (I) *The left fractional rectangular formula (13) and the right fractional rectangular formula (16) will be recovered if $\theta = 1$ and $\theta = 0$, respectively. In addition, the weighted rectangular formula (17) (or (18)) is reduced to the composite trapezoidal formula (or midpoint formula) [41] for the classical integral provided that $\alpha = 1$ and $\theta = \frac{1}{2}$.*

(II) Leading terms of remainders for left- and right-rectangular formulae generally can not be canceled out by introducing weights as the remainders depend on $f'(\xi_k)(t-x_k), 0 \leq k \leq j-1$ and $f'(\eta_{k+1})(t-x_{k+1}), 0 \leq k \leq j-1$, respectively. Therefore, the accuracy of fractional rectangular formulae are of first order accuracy for all $0 \leq \theta \leq 1$. And all the above fractional rectangular formulae are of the first order accuracy.

Assume that $f(x) \in C[a,b]$. Replacing $f(x)$ in Equation (9) with the piecewise linear polynomial

$$\tilde{f}(x) = \frac{x_{k+1} - x}{x_{k+1} - x_k} f(x_k) + \frac{x - x_k}{x_{k+1} - x_k} f(x_{k+1}), \ x \in [x_k, x_{k+1}], \tag{19}$$

we obtain the fractional trapezoidal formula [38]

$$\left[{}_{RL}D_{a,x}^{-\alpha}f(x)\right]_{x=x_j} \approx \frac{h^\alpha}{\Gamma(\alpha+2)} \sum_{k=0}^{j} a_{k,j} f(x_k). \tag{20}$$

Here the coefficients $a_{k,j}$ ($0 \leq k \leq j$) are given by

$$\begin{cases} a_{0,j} = (j-1)^{\alpha+1} - (j-1-\alpha) j^\alpha, \\ a_{k,j} = (j-k+1)^{\alpha+1} - 2(j-k)^{\alpha+1} + (j-k-1)^{\alpha+1}, \ 1 \leq k \leq j-1, \\ a_{j,j} = 1. \end{cases} \tag{21}$$

Suppose that $f(x) \in C[a,b]$. For $0 \leq k \leq j-1$, let $\{l_{k,i}(x)\}$ be Lagrangian functions defined on the grid points $\{x_{k+s}, s \in S\}$ with $S = \{0, \frac{1}{2}, 1\}$, which are given by

$$l_{k,i}(x) = \prod_{s \in S, s \neq i} \frac{x - x_{k+s}}{x_{k+i} - x_{k+s}}, \ i \in S. \tag{22}$$

Denote $x_{k+\frac{1}{2}} = \frac{x_k + x_{k+1}}{2}$ and utilize the piecewise quadratic polynomial

$$\tilde{f}(x) = \sum_{i \in S} l_{k,i}(x) f(x_{k+i}), \quad x \in [x_k, x_{k+1}]. \tag{23}$$

Then we obtain the following fractional Simpson's formula [38]

$$[_{RL}D_{a,x}^{-\alpha} f(x)]_{x=x_j} \approx \frac{h^\alpha}{\Gamma(\alpha+3)} \left[\sum_{k=0}^{j} c_{k,j} f(x_k) + \sum_{k=0}^{j-1} \hat{c}_{k,j} f(x_{k+\frac{1}{2}}) \right], \tag{24}$$

in which

$$\hat{c}_{k,j} = 4(\alpha+2) \left[(j-k)^{1+\alpha} + (j-k-1)^{1+\alpha} \right] - 8 \left[(j-k)^{2+\alpha} - (j-k-1)^{2+\alpha} \right], \quad 0 \le k \le j-1, \tag{25}$$

and

$$\begin{cases} c_{0,j} = 4 \left[j^{2+\alpha} - (j-1)^{2+\alpha} \right] - (\alpha+2) \left[3j^{1+\alpha} + (j-1)^{1+\alpha} \right] + (\alpha+2)(\alpha+1)j^\alpha, \\ c_{k,j} = -(\alpha+2) \left[(j-k+1)^{1+\alpha} + (j-k-1)^{1+\alpha} + 6(j-k)^{1+\alpha} \right] \\ \qquad + 4 \left[(j-k+1)^{2+\alpha} - (j-k-1)^{2+\alpha} \right], \quad 1 \le k \le j-1, \\ c_{j,j} = 2 - \alpha. \end{cases} \tag{26}$$

Assume that $f(x) \in C[a,b]$. Let $f(x)$ be approximated by the following r-th degree polynomial on the grid points $\{x_k = x_0^{(k)}, x_1^{(k)}, \ldots, x_{r-1}^{(k)}, x_r^{(k)} = x_{k+1}\}$,

$$p_{k,r}(x) = \sum_{i=0}^{r} l_{k,i}(x) f(x_i^{(k)}), \quad x \in [x_k, x_{k+1}], \tag{27}$$

with

$$l_{k,i}(x) = \prod_{\substack{n=0 \\ n \ne i}}^{r} \frac{x - x_n^{(k)}}{x_i^{(k)} - x_n^{(k)}}. \tag{28}$$

Then we obtain the fractional Newton–Cotes formula [38]

$$[_{RL}D_{a,x}^{-\alpha} f(x)]_{x=x_j} \approx [_{RL}D_{a,x}^{-\alpha} p_{k,r}(x)]_{x=x_j} = \sum_{k=0}^{j-1} \sum_{i=0}^{r} A_{i,j}^{(k)} f(x_i^{(k)}), \tag{29}$$

with the coefficients being calculated by

$$A_{i,j}^{(k)} = \frac{1}{\Gamma(\alpha)} \int_{x_k}^{x_{k+1}} (x_j - t)^{\alpha-1} l_{k,i}(t) dt. \tag{30}$$

Remark 2. *It has been demonstrated in Ref. [38] that the error estimate of Equation (29) is $\mathcal{O}(h^{r+1})$ provided that $f \in C^{r+1}([a,b])$. The error estimate does not equal that for the classical one, which is $\mathcal{O}(h^{r+2})$. This inconsistency may be due to the asymmetry of the weakly singular kernel $(x_j - x)^{\alpha-1}$, which leads to the non-symmetry of the remainder term $(x_j - x)^{\alpha-1} \prod_{i=0}^{r}(x - x_i^{(k)})$ in the integrand. Note that formulae (13), (16), (20), and (24) are special cases of Equation (29). Therefore, they are of the first, second, and third-order accuracy, respectively.*

Consider the function $f(x)$ in the Sobolev space $H^r[a,b]$ with $r \geq 1$ being an integer. Generalizing the above approaches, we can derive spectral approximations [42]. For $f(x)$ defined on $[-1,1]$, we consider the interpolation function

$$p_N(x) = \sum_{j=0}^{N} \widetilde{p}_j^{u,v} P_j^{u,v}(x), \tag{31}$$

based on the Jacobi polynomials $\{P_j^{u,v}(x)\}_{j=0}^{N}$ $(u, v > -1)$. Here

$$\widetilde{p}_j^{u,v} = \frac{1}{\delta_j^{u,v}} \sum_{k=0}^{N} f(x_k) P_j^{u,v}(x_k) \omega_k, \ j = 0, 1, \ldots, N, \tag{32}$$

with $\{x_k\}_{k=0}^{N}$ and $\{\omega_k\}_{k=0}^{N}$ being the collocation points and the corresponding quadrature weights [43]. The constants $\delta_j^{u,v}$ are given by

$$\delta_j^{u,v} = \begin{cases} \gamma_j^{u,v}, & j = 0, 1, \ldots, N-1, \\ (2 + \frac{u+v+1}{N}) \gamma_N^{u,v}, & j = N, \end{cases} \tag{33}$$

with $\gamma_j^{u,v}$ being defined by

$$\gamma_j^{u,v} = \frac{2^{u+v+1} \Gamma(j+u+1) \Gamma(j+v+1)}{(2j+u+v+1) j! \Gamma(j+u+v+1)}. \tag{34}$$

In this case, we have the following spectral approximation based on Jacobi polynomials

$$_{RL}D_{-1,x}^{-\alpha} f(x) \approx {}_{RL}D_{-1,x}^{-\alpha} p_N(x) = \sum_{j=0}^{N} \widetilde{p}_j^{u,v} \widehat{P}_j^{u,v,\alpha}(x), \ x \in [-1,1]. \tag{35}$$

Here $\widehat{P}_j^{u,v,\alpha}(x) = \frac{1}{\Gamma(\alpha)} \int_{-1}^{x} (x-t)^{\alpha-1} P_j^{u,v}(t) dt$ can be explicitly calculated by the recurrence formula

$$\begin{cases} \widehat{P}_0^{u,v,\alpha}(x) = \frac{(x+1)^\alpha}{\Gamma(\alpha+1)}, \\ \widehat{P}_1^{u,v,\alpha}(x) = \frac{u+v+2}{2} \left(\frac{x(x+1)^\alpha}{\Gamma(\alpha+1)} - \frac{\alpha(x+1)^{\alpha+1}}{\Gamma(\alpha+2)} \right) + \frac{u-v}{2} \widehat{P}_0^{u,v,\alpha}(x), \\ \widehat{P}_{j+1}^{u,v,\alpha}(x) = \frac{A_j^{u,v} x - B_j^{u,v} - \alpha A_j^{u,v} \widehat{B}_j^{u,v}}{1 + \alpha A_j^{u,v} \widehat{C}_j^{u,v}} \widehat{P}_j^{u,v,\alpha}(x) \\ \qquad + \frac{\alpha \left(\widehat{A}_j^{u,v} P_{j-1}^{u,v}(-1) + \widehat{B}_j^{u,v} P_j^{u,v}(-1) + \widehat{C}_j^{u,v} P_{j+1}^{u,v}(-1) \right)}{\Gamma(\alpha+1)(1 + \alpha A_j^{u,v} \widehat{C}_j^{u,v})} A_j^{u,v} (x+1)^\alpha \\ \qquad - \frac{C_j^{u,v} + \alpha A_j^{u,v} \widehat{A}_j^{u,v}}{1 + \alpha A_j^{u,v} \widehat{C}_j^{u,v}} \widehat{P}_{j-1}^{u,v,\alpha}(x), \ j \geq 1, \end{cases} \tag{36}$$

which follows from the three-term recurrence relation of the Jacobi polynomials. Here the coefficients are given by

$$\begin{cases} A_j^{u,v} = \dfrac{(2j+u+v+1)(2j+u+v+2)}{2(j+1)(j+u+v+1)}, \\ B_j^{u,v} = \dfrac{(v^2-u^2)(2j+u+v+1)}{2(j+1)(j+u+v+1)(2j+u+v)}, \\ C_j^{u,v} = \dfrac{(j+u)(j+v)(2j+u+v+2)}{(j+1)(j+u+v+1)(2j+u+v)}, \end{cases} \quad (37)$$

and

$$\begin{cases} \widehat{A}_j^{u,v} = \dfrac{-2(j+u)(j+v)}{(j+u+v)(2j+u+v)(2j+u+v+1)}, \\ \widehat{B}_j^{u,v} = \dfrac{2(u-v)}{(2j+u+v)(2j+u+v+2)}, \\ \widehat{C}_j^{u,v} = \dfrac{2(j+u+v+1)}{(2j+u+v+1)(2j+u+v+2)}, \end{cases} \quad (38)$$

and $\widehat{A}_j^{u,v} = 0$ if $j = 1$. For $f(x)$ defined on an arbitrary interval $[a,b]$, it follows from the affine transformation $\widehat{x} = \frac{2x-a-b}{b-a} \in [-1,1]$ that

$$_{RL}D_{a,x}^{-\alpha}f(x) \approx \left(\frac{b-a}{2}\right)^\alpha {}_{RL}D_{a,\widehat{x}}^{-\alpha} p_N(\widehat{x}) = \left(\frac{b-a}{2}\right)^\alpha \sum_{j=0}^N \widetilde{p}_j^{u,v} \widehat{P}_j^{u,v,\alpha}(\widehat{x}). \quad (39)$$

Let $u = v = 0$. Then the Jacobi polynomials $\{P_j^{u,v}\}_{j=0}^N$ reduce to the Legendre polynomials $\{L_j(x)\}_{j=0}^N$. Consequently, numerical scheme (35) becomes the spectral approximation based on Legendre polynomials

$$_{RL}D_{-1,x}^{-\alpha}f(x) \approx \sum_{j=0}^N \widetilde{l}_j \widehat{L}_j^\alpha(x), \quad x \in [-1,1]. \quad (40)$$

Here the coefficients are given by

$$\widetilde{l}_j = \frac{1}{\widetilde{\gamma}_j} \sum_{k=0}^N f(x_k) L_j(x_k) \omega_k, \quad (41)$$

with $\widetilde{\gamma}_j = \frac{2}{2j+1}$ for $0 \leq j \leq N-1$, $\widetilde{\gamma}_N = \frac{2}{N}$, and $\{\omega_k\}_{k=0}^N$ being the corresponding quadrature weights. Recurrence formula for $\widehat{L}_j^\alpha(x)$ is in the form

$$\begin{cases} \widehat{L}_0^\alpha(x) = \dfrac{(x+1)^\alpha}{\Gamma(\alpha+1)}, \\ \widehat{L}_1^\alpha(x) = \dfrac{x(x+1)^\alpha}{\Gamma(\alpha+1)} - \dfrac{\alpha(x+1)^{\alpha+1}}{\Gamma(\alpha+2)}, \\ \widehat{L}_{j+1}^\alpha(x) = \dfrac{(2j+1)x\widehat{L}_j^\alpha(x) - (j-\alpha)\widehat{L}_{j-1}^\alpha(x)}{j+1+\alpha}, \quad j \geq 1. \end{cases} \quad (42)$$

Correspondingly, for the case with arbitrary interval $[a,b]$, we also have

$$_{RL}D_{a,x}^{-\alpha}f(x) \approx \left(\frac{b-a}{2}\right)^\alpha \sum_{j=0}^n \widetilde{l}_j \widehat{L}_j^\alpha(\widehat{x}), \quad \widehat{x} = \frac{2x-a-b}{b-a}. \quad (43)$$

Let $u = v = -\frac{1}{2}$ in Equation (39). We obtain the spectral approximation based on Chebyshev polynomials

$$_{RL}D_{a,x}^{-\alpha}f(x) \approx \left(\frac{b-a}{2}\right)^\alpha \sum_{j=0}^{N} \tilde{t}_j \widehat{T}_j^\alpha(\hat{x}), \quad \hat{x} = \frac{2x-a-b}{b-a} \in [-1,1], \tag{44}$$

which follows from the relation $P_j^{-\frac{1}{2},-\frac{1}{2}}(x) = \frac{\Gamma(j+\frac{1}{2})}{j!\sqrt{\pi}} T_j(x)$ with $\{T_j(x)\}_{j=0}^{N}$ being the Chebyshev polynomials. Here $\widehat{T}_j^\alpha(x) = \frac{1}{\Gamma(\alpha)}\int_{-1}^{x}(x-s)^{\alpha-1}T_j(s)ds$ can be computed by the recurrence formula

$$\begin{cases}
\widehat{T}_0^\alpha(x) = \dfrac{(x+1)^\alpha}{\Gamma(\alpha+1)}, \\
\widehat{T}_1^\alpha(x) = \dfrac{x(x+1)^\alpha}{\Gamma(\alpha+1)} - \dfrac{\alpha(x+1)^{\alpha+1}}{\Gamma(\alpha+2)}, \\
\widehat{T}_2^\alpha(x) = \dfrac{4x}{2+\alpha}\widehat{T}_1^\alpha(x) - \dfrac{2}{2+\alpha}\widehat{T}_0^\alpha(x) + \dfrac{\alpha(x+1)^\alpha}{(2+\alpha)\Gamma(\alpha+1)}, \\
\widehat{T}_{j+1}^\alpha(x) = \dfrac{2(j+1)x}{j+1+\alpha}\widehat{T}_j^\alpha(x) - \dfrac{(j+1)(j-1-\alpha)}{(j+1+\alpha)(j-1)}\widehat{T}_{j-1}^\alpha(x) \\
\qquad + \dfrac{2(-1)^j \alpha(x+1)^\alpha}{\Gamma(\alpha+1)(j+1+\alpha)(j-1)}, \quad j \geq 2.
\end{cases} \tag{45}$$

The coefficients $\tilde{t}_j$ $(0 \leq j \leq N)$ are determined by

$$\tilde{t}_j = \frac{1}{\sigma_j} \sum_{k=0}^{N} f(x_k) T_j(x_k) \omega_k, \tag{46}$$

with

$$\sigma_j = \begin{cases} \gamma_j^{-\frac{1}{2},-\frac{1}{2}}, & j = 0, 1, \ldots, N-1, \\ 2\gamma_N^{-\frac{1}{2},-\frac{1}{2}}, & j = N, \end{cases} \tag{47}$$

and $\{\omega_k\}_{k=0}^{N}$ being the corresponding quadrature weights.

The above spectral approximations can be rewritten in matrix forms. For differential matrices for fractional integrals and derivatives, see Refs. [42,44] for more details. Here we present numerical examples given by Ref. [42] to verify the spectral accuracy of spectral approximations.

Example 1. Let $f(x) = x^\mu$, $x \in [0, 1]$. Apply scheme (39) to evaluating its fractional integral. Table 1 shows the absolute maximum errors at the Jacobi–Gauss–Lobatto points. The spectral accuracy is visible.

Table 1. The absolute errors for Example 1.

			$u = v = 0$, $\mu = 3.5$			
n	$\alpha = 0.2$	$\alpha = 0.5$	$\alpha = 0.8$	$\alpha = 1.2$	$\alpha = 1.5$	$\alpha = 1.8$
10	4.57×10^{-8}	3.57×10^{-8}	1.78×10^{-8}	5.18×10^{-9}	1.67×10^{-9}	6.04×10^{-10}
20	2.89×10^{-10}	1.52×10^{-10}	5.37×10^{-11}	9.88×10^{-12}	2.54×10^{-12}	1.31×10^{-12}
40	1.82×10^{-12}	6.36×10^{-13}	1.52×10^{-13}	1.74×10^{-14}	3.68×10^{-15}	2.77×10^{-15}
80	1.12×10^{-14}	2.59×10^{-15}	4.11×10^{-16}	1.67×10^{-16}	1.67×10^{-16}	1.18×10^{-16}
			$u = v = -\frac{1}{2}$, $\mu = 3.5$			
n	$\alpha = 0.2$	$\alpha = 0.5$	$\alpha = 0.8$	$\alpha = 1.2$	$\alpha = 1.5$	$\alpha = 1.8$
10	5.49×10^{-8}	4.59×10^{-8}	2.54×10^{-8}	8.33×10^{-9}	3.09×10^{-9}	1.67×10^{-9}
20	3.08×10^{-10}	1.96×10^{-10}	7.62×10^{-11}	1.70×10^{-11}	4.97×10^{-12}	2.93×10^{-12}
40	1.81×10^{-12}	7.79×10^{-13}	2.14×10^{-13}	3.23×10^{-14}	8.09×10^{-15}	5.61×10^{-15}
80	1.06×10^{-14}	3.05×10^{-15}	5.66×10^{-16}	3.33×10^{-16}	1.80×10^{-16}	1.73×10^{-16}

Example 2. Let $f(x) = \sin x$, $x \in [0,1]$. Utilize scheme (39) to evaluate its fractional integral. Table 2 displays the absolute maximum errors of the spectral approximations to the fractional integral with $u = v = 0$ and $u = v = -\frac{1}{2}$. We can observe that satisfactory results are obtained as well.

Table 2. The absolute errors for Example 2.

n	$\alpha = 0.2$	$\alpha = 0.5$	$\alpha = 0.8$	$\alpha = 1.2$	$\alpha = 1.5$	$\alpha = 1.8$
			$u = v = 0$			
4	4.46×10^{-6}	5.40×10^{-6}	3.88×10^{-6}	1.71×10^{-6}	6.94×10^{-7}	2.90×10^{-7}
8	4.79×10^{-12}	4.72×10^{-12}	2.73×10^{-12}	8.94×10^{-13}	2.88×10^{-13}	6.96×10^{-14}
16	6.66×10^{-16}	2.22×10^{-16}	2.22×10^{-16}	1.67×10^{-16}	1.67×10^{-16}	8.33×10^{-17}
			$u = v = -\frac{1}{2}$			
n	$\alpha = 0.2$	$\alpha = 0.5$	$\alpha = 0.8$	$\alpha = 1.2$	$\alpha = 1.5$	$\alpha = 1.8$
4	6.08×10^{-6}	7.91×10^{-6}	6.26×10^{-6}	3.41×10^{-6}	1.87×10^{-6}	1.27×10^{-6}
8	6.58×10^{-12}	6.63×10^{-12}	3.96×10^{-12}	1.38×10^{-12}	4.97×10^{-13}	1.36×10^{-13}
16	3.33×10^{-16}	3.33×10^{-16}	2.22×10^{-16}	5.55×10^{-17}	2.22×10^{-16}	5.55×10^{-17}

2.2. Fractional Linear Multistep Method

In Ref. [45], the convolution quadrature

$$I_h^\alpha f(x) = h^\alpha \sum_{j=0}^{n} \omega_{\ell, n-j} f(jh) + h^\alpha \sum_{j=0}^{s} w_{n,j} f(jh), \quad x = nh, \tag{48}$$

is utilized to evaluate fractional integrals (with $\alpha > 0$) and derivatives (with $\alpha < 0$). Here the convolution quadrature weights $\omega_{\ell, j}$ ($j \geq 0$) and the starting quadrature weights $w_{n,j}$ ($n \geq 0, j = 0, \ldots, s$; s fixed) do not depend on h.

On the basis of Dahlquist's theorem on linear multistep methods [46], the proposed convolution quadrature was proved to be convergent of order ℓ if and only if it is stable and consistent of order ℓ. An easy way of obtaining such a convolution quadrature is by using an ℓ-th order linear multistep method to the power α, which gives fractional linear multistep methods. The widely used one is fractional backward difference formula (the fractional BDF), whose implementations are as follows.

Theorem 1 ([45,47,48]). *The convolution quadrature* (48) *approximates the fractional integral* $_{RL}D_{0,x}^{-\alpha} f(x)$ *with accuracy order* $\mathcal{O}(h^\ell)$, *i.e.*,

$$_{RL}D_{0,x}^{-\alpha} f(x) = h^\alpha \sum_{j=0}^{n} \omega_{\ell, n-j} f(jh) + h^\alpha \sum_{j=0}^{s} w_{n,j} f(jh) + \mathcal{O}(h^\ell), \quad x = a + nh, \tag{49}$$

where s is a fixed integer with $s \leq n$. Here the convolution coefficients $\omega_{\ell, j}$ are respectively those of the Taylor series expansions of the corresponding generating functions

$$W_\ell^{(\alpha)}(z) = \left(\sum_{k=1}^{\ell} \frac{1}{k} (1-z)^k \right)^\alpha = \sum_{j=0}^{\infty} \omega_{\ell, j} z^j, \quad |z| < 1, \tag{50}$$

with ℓ being the order of consistency. Technically all the coefficients $\omega_{\ell, j}$ can be computed by using any implementation of the fast Fourier transform. For the starting weights $w_{n,j}$, we can consider the fixed $s = 0$. In this case, we obtain the Lubich formulae for fractional integrals

$$_{RL}D_{0,x}^{-\alpha} f(x) = h^\alpha \sum_{j=0}^{n} \omega_{\ell, n-j} f(jh) + \mathcal{O}(h^\ell), \quad x = a + nh, \tag{51}$$

when $f(0) = 0$. For $s \neq 0$, the coefficients $w_{n,j}$ can be constructed such that Equation (49) exactly holds for power functions. Therefore, we recover

$$\sum_{j=0}^{s} w_{n,j} j^q = \frac{\Gamma(q+1)}{\Gamma(q+\alpha+1)} n^{q+\alpha} - \sum_{j=0}^{n} \omega_{\ell,n-j} j^q, \quad q = 0, \ldots, s, \quad (52)$$

and it makes sense to choose $s = \ell - 1$.

In the case with the lower terminal $a \neq 0$, we can readily adopt affine transform to modify the fractional linear multistep methods.

Remark 3. *Apart from the choice given by Equation (50), which corresponds to the fractional BDF, there are alternatives for the generating functions of the convolution coefficients. When we choose*

$$W_2^\alpha(z) = \left(\frac{1}{2}\frac{1+z}{1-z}\right)^\alpha \quad (53)$$

as the generating function, the fractional trapezoidal rule with second order accuracy for $\alpha \geq 0$ is obtained.
Let γ_i $(i = 0, 1, 2, \ldots)$ denote the coefficients of

$$\sum_{i=0}^{\infty} \gamma_i (1-z)^i = \left(\frac{\ln z}{z-1}\right)^{-\alpha}, \quad (54)$$

and set

$$\widetilde{W}_\ell^\alpha = (1-z)^{-\alpha} \left[\gamma_0 + \gamma_1(1-z) + \cdots + \gamma_{\ell-1}(1-z)^{\ell-1}\right], \quad \ell = 1, 2, \ldots \quad (55)$$

Then we obtain the generating function for the coefficients of the generalized Newton-Gregory formulae, which is convergent of order ℓ. Direct calculation gives $\gamma_0 = 1$ and $\gamma_1 = -\frac{\alpha}{2}$. Then the corresponding generating function for the second order generalized Newton–Gregory formula is given by

$$\widetilde{W}_2^\alpha = (1-z)^{-\alpha} \left[1 - \frac{\alpha}{2}(1-z)\right]. \quad (56)$$

More details for generating functions can be found in Refs. [45,49–51].

2.3. Diffusive Approximation

The above numerical methods may result in expensive computational costs. To eliminate this deficiency, the diffusive approximation reformulates the model containing the fractional integral as a system of differential equations.

Recalling the relations

$$\Gamma(\alpha) = \int_0^\infty e^{-z} z^{\alpha-1} dz, \quad (57)$$

and

$$\Gamma(1-\alpha)\Gamma(\alpha) = \frac{\pi}{\sin(\pi\alpha)}, \quad (58)$$

the fractional integral with $0 < \alpha < 1$ can be rewritten as [52]

$$_{RL}D_{0,x}^{-\alpha} f(x) = \frac{\sin(\pi\alpha)}{\pi} \int_0^x \left[\int_0^\infty e^{-z} \left(\frac{z}{x-t}\right)^{1-\alpha} \frac{dz}{z}\right] f(t) dt. \quad (59)$$

Define the variable transformation $z = (x-t)\omega^2$, $\omega \geq 0$. The Fubini's Theorem yields

$$_{RL}D_{0,x}^{-\alpha} f(x) = \frac{2\sin(\pi\alpha)}{\pi} \int_0^\infty \omega^{1-2\alpha} \left(\int_0^x e^{-(x-t)\omega^2} f(t) dt\right) d\omega. \quad (60)$$

Introducing the auxiliary function

$$\phi(\omega, x) = \frac{2\sin(\pi\alpha)}{\pi} \omega^{1-2\alpha} \int_0^x e^{-(x-t)\omega^2} f(t) dt, \qquad (61)$$

we have

$$_{RL}D_{0,x}^{-\alpha} f(x) = \int_0^\infty \phi(\omega, x) d\omega, \quad 0 < \alpha < 1. \qquad (62)$$

It follows from the definition of $\phi(\omega, x)$ that the auxiliary function satisfies

$$\begin{cases} \dfrac{\partial}{\partial x}\phi(\omega, x) = \dfrac{2\sin(\pi\alpha)}{\pi}\omega^{1-2\alpha} f(x) - \omega^2 \phi(\omega, x), \\ \phi(\omega, 0) = 0. \end{cases} \qquad (63)$$

In this case, evaluating Riemann–Liouville integral $_{RL}D_{0,x}^{-\alpha} f(x)$ consists of two steps: solving the first order differential equation (63), and computing the infinite integral in Equation (62) via suitable quadratures.

Instead of utilizing the properties of the Gamma function, Chatterjee adopted a popular integral representation [53,54]

$$x^{\alpha-1} = \frac{1}{\Gamma(1-\alpha)} \int_0^\infty e^{-zx} z^{-\alpha} dz. \qquad (64)$$

Consequently, the Fubini's Theorem gives

$$\begin{aligned} _{RL}D_{0,x}^{-\alpha} f(x) &= \frac{1}{\Gamma(\alpha)}\frac{1}{\Gamma(1-\alpha)} \int_0^\infty \left(\int_0^x e^{-z(x-t)} f(t) dt\right) \frac{dz}{z^\alpha} \\ &= \frac{\sin(\pi\alpha)}{\pi} \int_0^\infty g(z,x) z^{-\alpha} dz, \end{aligned} \qquad (65)$$

where $g(z, x)$ is defined as

$$g(z, x) = \int_0^x e^{-z(x-t)} f(t) dt. \qquad (66)$$

In order to generate nonreflecting boundary conditions [55] and accelerate convolutions with the heat kernel [56], literatures such as Ref. [57] usually recognize

$$g(z, x) = e^{-z\Delta x} g(z, x - \Delta x) + \Psi(z, x, \Delta x), \qquad (67)$$

where

$$\Psi(z, x, \Delta x) = \int_{x-\Delta x}^x e^{-z(x-t)} f(t) dt. \qquad (68)$$

Alternatively, other literatures have regarded $g(z, x)$ as the solution of a first order ordinary differential (ODE) equation [52,53,58,59],

$$\frac{dg(z, x)}{dx} = -zg(z, x) + f(x), \quad g(z, 0) = 0. \qquad (69)$$

Any approximate method for ODEs can be used to obtain $g(z, x), x = \Delta x, 2\Delta x, \ldots$, in an amount of work that is linear.

The principle difficulty of implementing both approaches lies in the discretization of the integrals on the right-hand side of Equations (62) and (65). The choices of quadrature nodes and corresponding weights have been investigated in several literatures, see Refs. [57,60,61] for more details.

3. Numerical Approximations to Fractional Derivatives

We introduce the existing numerical evaluations to Caputo, Riemann–Liouville, and Riesz derivatives in this section. The basic ideas of these methods are presented as well.

3.1. Numerical Caputo Differentiation

Caputo derivatives in Equations (6) and (7) can be viewed as Riemann–Liouville fractional integrals of integer-order derivatives. As a result, most of the numerical evaluations of Caputo derivatives follow from those of fractional integrals. We derive numerical evaluations of the Caputo derivative as follows.

3.1.1. L1, L2, and L2C Methods

The well-known L1 method was originally introduced in Ref. [62] to evaluate Riemann–Liouville derivative with $0 < \alpha < 1$, which equivalently reads as

$$_{RL}D^\alpha_{a,x} f(x) = \frac{(x-a)^{-\alpha}}{\Gamma(1-\alpha)} f(a) + \frac{1}{\Gamma(1-\alpha)} \int_a^x (x-t)^{-\alpha} f'(t) dt. \tag{70}$$

Note that the second term on the right-hand side happens to be Caputo derivative with $0 < \alpha < 1$. That is the reason why we introduce the L1 method when considering numerical approximations to the Caputo derivative.

Let $f(x) \in C^2[a,b]$. On the setting of uniform grids $\{x_k\}_{k=0}^N$, utilizing the constant $\frac{f(x_{k+1})-f(x_k)}{h}$ to approximate $f'(x)$ on each interval $[x_k, x_{k+1}]$ yields the following L1 method on uniform grids for Caputo derivative [62]

$$\begin{aligned}[_{C}D^\alpha_{a,x} f(x)]_{x=x_j} &= \frac{1}{\Gamma(1-\alpha)} \sum_{k=0}^{j-1} \int_{x_k}^{x_{k+1}} (x_j - t)^{-\alpha} \frac{f(x_{k+1}) - f(x_k)}{h} dt + \mathcal{O}(h^{2-\alpha}) \\ &= \sum_{k=0}^{j-1} b_{j-k-1} [f(x_{k+1}) - f(x_k)] + \mathcal{O}(h^{2-\alpha}),\ 0 < \alpha < 1,\ 1 \le j \le N.\end{aligned} \tag{71}$$

Here the coefficients are given by

$$b_k = \frac{h^{-\alpha}}{\Gamma(2-\alpha)} \left[(k+1)^{1-\alpha} - k^{1-\alpha}\right],\ 0 \le k \le j-1. \tag{72}$$

Normally, the L1 method can lead to unconditionally stable algorithms [63–69]. Therefore, it is frequently used in the discretization of time fractional differential equations. Since the proof for this scheme available is not very direct or a little cryptic, it is necessary to present clear proof of its truncated error for reference as it is mostly used.

Theorem 2. *Let $0 < \alpha < 1$ and $f(x) \in C^2[a,b]$. Denote by*

$$[\delta^\alpha_x f(x)]_{x=x_j} = \sum_{k=0}^{j-1} b_{j-k-1} [f(x_{k+1}) - f(x_k)],\ 1 \le j \le N. \tag{73}$$

Then it holds that

$$\left| [\delta^\alpha_x f(x)]_{x=x_j} - [_{C}D^\alpha_{a,x} f(x)]_{x=x_j} \right| \le C h^{2-\alpha}, \tag{74}$$

where C is a positive constant given by

$$C = \frac{1}{\Gamma(2-\alpha)} \left[\frac{1-\alpha}{12} + \frac{2^{2-\alpha}}{2-\alpha} - (2^{-\alpha} + 1) \right] \max_{x_0 \le x \le x_j} |f''(x)|. \tag{75}$$

Proof. Denote

$$A = \sum_{k=0}^{j-1} \frac{f(x_{k+1}) - f(x_k)}{h} \int_{x_k}^{x_{k+1}} \frac{dt}{(x_j - t)^\alpha} - \int_{x_0}^{x_j} \frac{f'(t)}{(x_j - t)^\alpha} dt. \qquad (76)$$

Then it immediately follows that

$$\frac{|A|}{\Gamma(1-\alpha)} = \left| \sum_{k=0}^{j-1} b_{j-k-1} \left[f(x_{k+1}) - f(x_k) \right] - \left[{}_C D^\alpha_{a,x} f(x) \right]_{x=x_j} \right|. \qquad (77)$$

Using the Taylor expansion with integral remainder, we have for $t \in [x_k, x_{k+1}]$,

$$f'(t) - \frac{f(x_{k+1}) - f(x_k)}{h} = \frac{1}{h} \left[\int_{x_k}^{t} f''(s)(s - x_k) ds - \int_{t}^{x_{k+1}} f''(s)(x_{k+1} - s) ds \right], \qquad (78)$$

which yields

$$A = \sum_{k=0}^{j-1} \int_{x_k}^{x_{k+1}} \left[f'(t) - \frac{f(x_{k+1}) - f(x_k)}{h} \right] (x_j - t)^{-\alpha} dt$$

$$= \frac{1}{h} \sum_{k=0}^{j-1} \int_{x_k}^{x_{k+1}} \left[\int_{x_k}^{t} f''(s)(s - x_k) ds - \int_{t}^{x_{k+1}} f''(s)(x_{k+1} - s) ds \right] \frac{dt}{(x_j - t)^\alpha}. \qquad (79)$$

Exchanging the order of integration gives

$$A = \frac{1}{h} \sum_{k=0}^{j-1} \int_{x_k}^{x_{k+1}} \left[\int_{x_k}^{t} f''(s)(s - x_k) ds - \int_{t}^{x_{k+1}} f''(s)(x_{k+1} - s) ds \right] \frac{dt}{(x_j - t)^\alpha}$$

$$= \frac{1}{h} \sum_{k=0}^{j-1} \left[\int_{x_k}^{x_{k+1}} f''(s)(s - x_k) \int_{s}^{x_{k+1}} (x_j - t)^{-\alpha} dt ds - \int_{x_k}^{x_{k+1}} f''(s)(x_{k+1} - s) \int_{x_k}^{s} (x_j - t)^{-\alpha} dt ds \right]$$

$$= \frac{1}{1-\alpha} \sum_{k=0}^{j-1} \left[\int_{x_k}^{x_{k+1}} f''(s) \frac{s - x_k}{h} \left[(x_j - s)^{1-\alpha} - (x_j - x_{k+1})^{1-\alpha} \right] ds \right. \qquad (80)$$

$$\left. - \int_{x_k}^{x_{k+1}} f''(s) \frac{x_{k+1} - s}{h} \left[(x_j - x_k)^{1-\alpha} - (x_j - s)^{1-\alpha} \right] ds \right]$$

$$= \frac{1}{1-\alpha} \sum_{k=0}^{j-1} \int_{x_k}^{x_{k+1}} f''(s) \left\{ (x_j - s)^{1-\alpha} - \left[\frac{s - x_k}{h(x_j - x_{k+1})^{\alpha-1}} + \frac{x_{k+1} - s}{h(x_j - x_k)^{\alpha-1}} \right] \right\} ds.$$

In the following, we show that when $0 < \alpha < 1$,

$$\int_{x_k}^{x_{k+1}} \left\{ (x_j - s)^{1-\alpha} - \left[\frac{s - x_k}{h(x_j - x_{k+1})^{\alpha-1}} + \frac{x_{k+1} - s}{h(x_j - x_k)^{\alpha-1}} \right] \right\} ds \geq 0 \qquad (81)$$

for $0 \leq k \leq j - 1$, and

$$\sum_{k=0}^{j-1} \int_{x_k}^{x_{k+1}} \left\{ (x_j - s)^{1-\alpha} - \left[\frac{s - x_k}{h(x_j - x_{k+1})^{\alpha-1}} + \frac{x_{k+1} - s}{h(x_j - x_k)^{\alpha-1}} \right] \right\} ds < +\infty. \qquad (82)$$

Denote $g(s) = (x_j - s)^{1-\alpha}$. Then it holds for any $s \in (x_k, x_{k+1})$ that

$$g(s) - \left[\frac{s - x_k}{h} g(x_{k+1}) + \frac{x_{k+1} - s}{h} g(x_k) \right] = \frac{(1-\alpha)(-\alpha)(s - x_k)(s - x_{k+1})}{2(x_j - \xi_k)^{\alpha+1}} \geq 0, \qquad (83)$$

with certain $\xi_k \in (x_k, x_{k+1})$. As a result, inequality (81) holds. For the inequality (82), one has

$$\sum_{k=0}^{j-3} \int_{x_k}^{x_{k+1}} \left\{ g(s) - \left[\frac{s - x_k}{h} g(x_{k+1}) + \frac{x_{k+1} - s}{h} g(x_k) \right] \right\} ds$$

$$= \sum_{k=0}^{j-3} \int_{x_k}^{x_{k+1}} \frac{\alpha(1-\alpha)(s-x_k)(x_{k+1}-s)}{2(x_j - \xi_k)^{\alpha+1}} ds$$

$$\leq \frac{\alpha(1-\alpha)}{2} \sum_{k=0}^{j-3} (x_j - x_{k+1})^{-\alpha-1} \int_{x_k}^{x_{k+1}} (s - x_k)(x_{k+1} - s) ds \quad (84)$$

$$\leq \frac{h^2}{12} \alpha(1-\alpha) \sum_{k=0}^{j-3} \int_{x_{k+1}}^{x_{k+2}} (x_j - s)^{-\alpha-1} ds$$

$$\leq \frac{1-\alpha}{12} h^{2-\alpha},$$

and

$$\sum_{k=j-2}^{j-1} \int_{x_k}^{x_{k+1}} \left\{ g(s) - \left[\frac{s - x_k}{h} g(x_{k+1}) + \frac{x_{k+1} - s}{h} g(x_k) \right] \right\} ds$$

$$= \int_{x_{j-2}}^{x_j} g(s) ds - \left[\frac{g(x_{j-2})}{2} + g(x_{j-1}) + \frac{g(x_j)}{2} \right] h$$

$$= \int_{x_{j-2}}^{x_j} g(s) ds - \left[\frac{g(x_{j-2})}{2} + g(x_{j-1}) \right] h \quad (85)$$

$$= \int_{x_{j-2}}^{x_j} (x_j - s)^{1-\alpha} ds - \left[\frac{(x_j - x_{j-2})^{1-\alpha}}{2} + (x_j - x_{j-1})^{1-\alpha} \right] h$$

$$= \left[\frac{2^{2-\alpha}}{2-\alpha} - (2^{-\alpha} + 1) \right] h^{2-\alpha}.$$

The above two equalities yield that Equation (82) holds.
Combining the above analysis, one has

$$0 \leq \sum_{k=0}^{j-1} \int_{x_k}^{x_{k+1}} \left\{ (x_j - s)^{1-\alpha} - \left[\frac{s - x_k}{h(x_j - x_{k+1})^{\alpha-1}} + \frac{x_{k+1} - s}{h(x_j - x_k)^{\alpha-1}} \right] \right\} ds$$

$$\leq \left[\frac{1-\alpha}{12} + \frac{2^{2-\alpha}}{2-\alpha} - (2^{-\alpha} + 1) \right] h^{2-\alpha}. \quad (86)$$

Inserting the above estimate into Equation (80) gives

$$|A| \leq \frac{1}{1-\alpha} \left[\frac{1-\alpha}{12} + \frac{2^{2-\alpha}}{2-\alpha} - (2^{-\alpha} + 1) \right] \max_{x_0 \leq x \leq x_j} |f''(x)| h^{2-\alpha}. \quad (87)$$

All this ends the proof. □

Remark 4. *The idea of proving Theorem 2 is borrowed from Ref. [70] where the case with $\alpha \in (1,2)$ was considered. Such an estimate was also considered in [71].*

Let $\{\tilde{x}_i\}$ be any division of $[a,b]$ with $a = \tilde{x}_0 < \tilde{x}_1 < \ldots < \tilde{x}_{N-1} < \tilde{x}_N = b$. Then the classical L1 method is generalized into the L1 method on nonuniform grids for Caputo derivative [72]

$$[{}_CD^\alpha_{a,x} f(x)]_{x=\tilde{x}_j} = \sum_{k=0}^{j-1} b^j_{k+1} [f(\tilde{x}_{k+1}) - f(\tilde{x}_k)] + \mathcal{O}(\tilde{h}^{2-\alpha}_{max}), \tag{88}$$

provided that $\dfrac{\max\limits_{0\le k\le j-1} \tilde{h}_k}{\min\limits_{0\le k\le j-1} \tilde{h}_k} \le C$ with C being a positive constant. Here $\tilde{h}_k = \tilde{x}_{k+1} - \tilde{x}_k$, $\tilde{h}_{max} = \max\limits_{0\le k\le j-1} \tilde{h}_k$, and the coefficients are given by

$$b^j_{k+1} = \frac{1}{\Gamma(2-\alpha)\tilde{h}_k} \left[(\tilde{x}_j - \tilde{x}_k)^{1-\alpha} - (\tilde{x}_j - \tilde{x}_{k+1})^{1-\alpha}\right]. \tag{89}$$

In the special case of nonuniform grids with $\tilde{x}_0 = x_0$, $\tilde{x}_j = x_{j-\frac{1}{2}} = \frac{x_j + x_{j-1}}{2}$, $j = 1, 2, \ldots$, scheme (88) is reduced to

$$[{}_CD^\alpha_{a,x} f(x)]_{x=\tilde{x}_{j+1}} = b_0 f(\tilde{x}_{j+1}) - \sum_{k=1}^{j} (b_{j-k} - b_{j-k+1}) f(\tilde{x}_k) - B_j f(\tilde{x}_0) + \mathcal{O}(h^{2-\alpha}). \tag{90}$$

Here the coefficients are given by

$$\begin{cases} b_k = \dfrac{(k+1)^{1-\alpha} - k^{1-\alpha}}{\Gamma(2-\alpha) h^\alpha}, & 0 \le k \le j, \\[2mm] B_j = \dfrac{2\left(j + \frac{1}{2}\right)^{1-\alpha} - 2j^{1-\alpha}}{\Gamma(2-\alpha) h^\alpha}, & 0 \le j \le N. \end{cases} \tag{91}$$

Replacing $f(\tilde{x}_k) = f(\frac{x_{k-1}+x_k}{2})$ with $\frac{f(x_k)+f(x_{k-1})}{2}$ yields the following modified L1 method for Caputo derivative [38]

$$[{}_CD^\alpha_{a,x} f(x)]_{x=x_{j+\frac{1}{2}}} = -\frac{1}{2}\sum_{k=1}^{j} (b_{j-k} - b_{j-k+1})[f(x_{k-1}) + f(x_k)] \\ + \frac{b_0}{2}[f(x_{j+1}) + f(x_j)] - B_j f(x_0) + \mathcal{O}(h^{2-\alpha}). \tag{92}$$

Remark 5. (I) *The modified L1 method* (92) *is useful to obtain the Crank–Nicolson scheme for the time-fractional subdiffusion equation* [73,74], *which can be regarded as a natural extension of the classical Crank–Nicolson scheme* [75].
(II) *The (weak) singularity makes it difficult to evaluate fractional derivatives. In this case, approximations such as Equations* (88) *and* (92) *on nonuniform meshes or graded meshes can be utilized. One can refer to* [72,76,77] *for more details in this respect.*

For the case with $1 < \alpha < 2$ and the lower terminal $a = 0$, there holds

$$[{}_CD^\alpha_{0,x} f(x)]_{x=x_j} = \frac{1}{\Gamma(2-\alpha)} \sum_{k=0}^{j-1} \int_{x_k}^{x_{k+1}} t^{1-\alpha} f''(x_j - t) dt. \tag{93}$$

Suppose that $f(x) \in C^3[a,b]$. Utilizing the central difference scheme

$$\frac{f(x_j - x_{k+1}) - 2f(x_j - x_k) + f(x_j - x_{k-1})}{h^2}, \tag{94}$$

to approximate $f''(x_j - t)$ on each interval $[x_k, x_{k+1}]$, we have the following L2 method for Caputo derivative [62]

$$[{}_cD_{0,x}^\alpha f(x)]_{x=x_j} = \frac{1}{\Gamma(3-\alpha)h^\alpha} \sum_{k=-1}^{j} W_k f(x_{j-k}) + \mathcal{O}(h^{3-\alpha}), \tag{95}$$

in which

$$\begin{cases} W_{-1} = 1, \quad W_0 = 2^{2-\alpha} - 3, \\ W_k = \left[(k+2)^{2-\alpha} - 3(k+1)^{2-\alpha} + 3k^{2-\alpha} - (k-1)^{2-\alpha}\right], \ 1 \leq k \leq j-2, \\ W_{j-1} = -2j^{2-\alpha} + 3(j-1)^{2-\alpha} - (j-2)^{2-\alpha}, \\ W_j = j^{2-\alpha} - (j-1)^{2-\alpha}. \end{cases} \tag{96}$$

In Ref. [78], the integral $\int_{x_k}^{x_{k+1}} t^{1-\alpha} f''(x_j - t) dt$ was evaluated in a more symmetric form. For $t \in [x_{k-1}, x_k]$, if we replace $f''(x_j - t)$ with the difference

$$\frac{f(x_j - x_{k+2}) - f(x_j - x_{k+1}) + f(x_j - x_{k-1}) - f(x_j - x_k)}{2h^2}, \tag{97}$$

then the L2C method for Caputo derivative

$$[{}_cD_{0,x}^\alpha f(x)]_{x=x_j} = \frac{1}{2\Gamma(3-\alpha)h^\alpha} \sum_{k=-1}^{j+1} \widehat{W}_k f(x_{j-k}) + \mathcal{O}(h^{3-\alpha}) \tag{98}$$

is obtained. Here the coefficients are given by

$$\begin{cases} \widehat{W}_{-1} = 1, \quad \widehat{W}_0 = 2^{2-\alpha} - 2, \quad \widehat{W}_1 = 3^{2-\alpha} - 2^{3-\alpha}, \\ \widehat{W}_k = \left[(k+2)^{2-\alpha} - 2(k+1)^{2-\alpha} + 2(k-1)^{2-\alpha} - (k-2)^{2-\alpha}\right], \ 2 \leq k \leq j-2, \\ \widehat{W}_{j-1} = 2(j-2)^{2-\alpha} - j^{2-\alpha} - (j-3)^{2-\alpha}, \\ \widehat{W}_j = 2(j-1)^{2-\alpha} - j^{2-\alpha} - (j-2)^{2-\alpha}, \\ \widehat{W}_{j+1} = j^{2-\alpha} - (j-1)^{2-\alpha}. \end{cases} \tag{99}$$

Note that in the above two schemes the value of $f(x_{-1})$ is needed. We can set $f(x_{-1}) = f(x_1)$ when the condition $f'(0) = 0$ is met. For the case with lower terminal $a \neq 0$, we can utilize affine transformation before applying the L2 and L2C methods.

Remark 6. *The L2 and L2C methods reduce to the backward difference method and the central difference method for the first order derivative, respectively, when $\alpha = 1$. If $\alpha = 2$, the L2 method reduces to the central difference method for the second order derivative and the L2C method reduces to*

$$\frac{d^2 f(x_k)}{dx^2} \approx \frac{f(x_{k+2}) - f(x_k) + f(x_{k-1}) - f(x_{k+1})}{2h^2} \tag{100}$$

with the first order accuracy. As a matter of fact, the error bound for the L2 method is $\mathcal{O}(h^{3-\alpha})$. Numerical experiments indicate that the L2 method is more accurate than the L2C method for $1.5 < \alpha < 2$, while the opposite result appears when $1 < \alpha < 1.5$. And these two methods behave in almost the same way near $\alpha = 1.5$ [78].

3.1.2. Numerical Methods Based on Polynomial Interpolation

It is evident that the higher-order accuracy can be achieved by utilizing the higher-order interpolation, provided that $f(x)$ is suitably smooth. In the following, we introduce numerical approximations in this respect.

(I) $(3-\alpha)$-th order approximations

Let $f(x) \in C^3[a,b]$. For $0 \leq k \leq j$ and $0 < x - x_k < h$, it follows from Taylor expansions of $f(x_{k+1})$, $f(x_k)$, and $f(x_{k-1})$ at x that

$$f'(x) = \frac{f(x_{k+1}) - f(x_{k-1})}{2h} + \frac{f(x_{k+1}) - 2f(x_k) + f(x_{k-1})}{h^2}(x - x_k)$$

$$- \frac{f^{(3)}(x_k)}{3!}h^2 + \frac{f^{(3)}(x_k)}{2!}(x - x_k)^2 + \mathcal{O}\left((x-x_k)^3\right). \quad (101)$$

In this case, we have the following $(3-\alpha)$-th order approximation [79],

$$[{}_cD^\alpha_{a,x}f(x)]_{x=x_j} = \frac{1}{\Gamma(1-\alpha)} \int_{x_0}^{x_j} (x_j - t)^{-\alpha} f'(t)dt \quad (102)$$

$$= \frac{h^{-\alpha}}{\Gamma(3-\alpha)} \sum_{k=0}^{j-1} \left\{ \omega_{1,j-k} [f(x_{k+1}) - f(x_{k-1})] + \omega_{2,j-k} [f(x_{k+1}) - 2f(x_k) + f(x_{k-1})] \right\} + R^j,$$

where $0 < \alpha < 1$, R^j denotes the truncated error, and the coefficients are given by

$$\begin{cases} \omega_{1,j-k} = \dfrac{2-\alpha}{2}\left[(j-k)^{1-\alpha} - (j-k-1)^{1-\alpha}\right], \\ \omega_{2,j-k} = (j-k)^{2-\alpha} - (j-k-1)^{2-\alpha} - (2-\alpha)(j-k-1)^{1-\alpha}, \end{cases} \quad (103)$$

with $0 \leq k \leq j-1$ and $1 \leq j \leq N$.

Since the above $(3-\alpha)$-th order method is also widely used, we estimate its truncated error in detail.

Theorem 3 ([80]). *Let $0 < \alpha < 1$ and $f(x) \in C^3[a,b]$. For the truncated error R^j of approximation (102), it holds that*

$$|R^j| \leq ch^{3-\alpha}, \ 1 \leq j \leq N, \quad (104)$$

with c being a positive constant and $f(x_{-1})$ in Equation (102) being used.

Proof. It is clear that the truncated error is given by

$$R^j = -\frac{1}{\Gamma(1-\alpha)} \sum_{k=0}^{j-1} \int_{x_k}^{x_{k+1}} (x_j - t)^{-\alpha} \left\{ \frac{1}{2! \cdot 2h}\left[\int_t^{x_{k+1}} (x_{k+1} - s)^2 f^{(3)}(s)ds \right. \right.$$

$$\left. - \int_t^{x_{k-1}} (x_{k-1} - s)^2 f^{(3)}(s)ds \right]$$

$$+ \frac{(t-x_k)}{2h^2}\left[\int_t^{x_{k+1}} (x_{k+1} - s)^2 f^{(3)}(s)ds - 2\int_t^{x_k}(x_k - s)^2 f^{(3)}(s)ds \right. \quad (105)$$

$$\left. \left. + \int_t^{x_{k-1}} (x_{k-1} - s)^2 f^{(3)}(s)ds \right] \right\} dt.$$

Interchanging the order of integrations yields

$$R^j = \frac{1}{2h\Gamma(2-\alpha)} \sum_{k=0}^{j-1} \left\{ \int_{x_k}^{x_{k+1}} (x_{k+1}-s)^2 f^{(3)}(s) \left[\frac{(x_j-s)^{1-\alpha} - (x_j-x_k)^{1-\alpha}}{2} \right. \right.$$

$$\left. + \frac{(x_j-s)^{1-\alpha}(s-x_k)}{h} + \frac{(x_j-s)^{2-\alpha} - (x_j-x_k)^{2-\alpha}}{h(2-\alpha)} \right] ds$$

$$+ \frac{2}{h} \int_{x_k}^{x_{k+1}} (x_k-s)^2 f^{(3)}(s) \left[(x_j - x_{k+1})^{1-\alpha} h - (x_j-s)^{1-\alpha}(s-x_k) \right.$$

$$\left. + \frac{(x_j - x_{k+1})^{2-\alpha} - (x_j-s)^{2-\alpha}}{2-\alpha} \right] ds$$

$$+ \int_{x_k}^{x_{k+1}} (x_{k-1}-s)^2 f^{(3)}(s) \left[\frac{(x_j - x_{k+1})^{1-\alpha} - (x_j-s)^{1-\alpha}}{2} - (x_j - x_{k+1})^{1-\alpha} \right. \tag{106}$$

$$\left. + \frac{(x_j-s)^{1-\alpha}(s-x_k)}{h} - \frac{(x_j - x_{k+1})^{2-\alpha} - (x_j-s)^{2-\alpha}}{h(2-\alpha)} \right] ds$$

$$+ \int_{x_{k-1}}^{x_k} (x_{k-1}-s)^2 f^{(3)}(s) \left[\frac{(x_j-x_{k+1})^{1-\alpha} - (x_j-x_k)^{1-\alpha}}{2} \right.$$

$$\left. - \frac{(x_j-x_{k+1})^{2-\alpha} - (x_j-x_k)^{2-\alpha}}{h(2-\alpha)} - (x_j - x_{k+1})^{1-\alpha} \right] ds \biggr\}$$

$$= \frac{1}{2h\Gamma(2-\alpha)} \sum_{k=0}^{j-1} S_k.$$

For $k = 0, 1, \ldots, j-1$, denote

$$B_k = \int_{x_{k-1}}^{x_k} (x_{k-1}-s)^2 f^{(3)}(s) \left[\frac{(x_j-x_{k+1})^{1-\alpha} - (x_j-x_k)^{1-\alpha}}{2} \right.$$

$$\left. - \frac{(x_j-x_{k+1})^{2-\alpha} - (x_j-x_k)^{2-\alpha}}{h(2-\alpha)} - (x_j-x_{k+1})^{1-\alpha} \right] ds, \tag{107}$$

and
$$A_k = S_k - B_k, \tag{108}$$

where the expression of A_k can be derived from Equations (106) and (107) so is left out due to lengthiness.

Let $l = j - k$, $k = 0, 1, \ldots, j-1$. The affine transformation $s = x_{k-1} + \xi h$ with $\xi \in [0,1]$ yields

$$B_{j-l} = h^{4-\alpha} \int_0^1 \xi^2 f^{(3)}(x_{j-l-1} + \xi h) \left[\frac{(l-1)^{1-\alpha} - l^{1-\alpha}}{2} - (l-1)^{1-\alpha} - \frac{(l-1)^{2-\alpha} - l^{2-\alpha}}{2-\alpha} \right] d\xi$$

$$= h^{4-\alpha} b_l \int_0^1 \xi^2 f^{(3)}(x_{j-l-1} + \xi h) d\xi, \quad 1 \le l \le j. \tag{109}$$

It is evident that $b_1 = \frac{1}{2-\alpha} - \frac{1}{2}$, and for $l \ge 2$,

$$b_l = l^{1-\alpha} \sum_{n=2}^{\infty} \frac{1}{l^n} \left(\frac{1}{2n!} - \frac{1}{(n+1)!} \right) (-\alpha+1)\alpha(\alpha+1) \cdots (\alpha+n-2) \ge 0. \tag{110}$$

Thus, it holds for $l \geq 2$ that

$$|B_{j-l}| = h^{4-\alpha} \left| b_l \int_0^1 \xi^2 f^{(3)}(x_{j-l-1} + \xi h) d\xi \right|$$

$$\leq \frac{h^{4-\alpha}}{3} \max_{x \in [x_{j-l-1}, x_{j-l}]} \left| f^{(3)}(x) \right| l^{1-\alpha} \sum_{n=2}^{\infty} \frac{1}{l^n} \left(\frac{1}{2n!} - \frac{1}{(n+1)!} \right) (1-\alpha)\alpha(\alpha+1)\cdots(\alpha+n-2)$$

$$\leq \frac{h^{4-\alpha}}{3} \max_{x \in [x_{j-l-1}, x_{j-l}]} \left| f^{(3)}(x) \right| l^{1-\alpha} \sum_{n=2}^{\infty} \frac{1}{l^n} \left(\frac{1}{2} - \frac{1}{n+1} \right) \frac{1-\alpha}{n}$$

$$\leq \frac{h^{4-\alpha}}{3} \max_{x \in [x_{j-l-1}, x_{j-l}]} \left| f^{(3)}(x) \right| l^{-1-\alpha} \sum_{n=2}^{\infty} \frac{1}{l^{n-2}} \cdot \frac{1}{2} \cdot \frac{1-\alpha}{2} \qquad (111)$$

$$\leq \frac{h^{4-\alpha}(1-\alpha)}{12} l^{-1-\alpha} \frac{l}{l-1} \max_{x \in [x_{j-l-1}, x_{j-l}]} \left| f^{(3)}(x) \right|$$

$$\leq \frac{h^{4-\alpha}}{6} \max_{x \in [x_{j-l-1}, x_{j-l}]} \left| f^{(3)}(x) \right| \frac{1-\alpha}{l^{1+\alpha}}.$$

As a result,

$$\left| \sum_{k=0}^{j-1} B_k \right| = \left| \sum_{l=1}^{j} B_{j-l} \right| \leq h^{4-\alpha} \left\{ \left| b_1 \int_0^1 \xi^2 f^{(3)}(x_{j-2} + \xi h) d\xi \right| + \sum_{l=2}^{j} |B_l| \right\}$$

$$\leq h^{4-\alpha} \max_{x \in [x_{-1}, x_{j-1}]} \left| f^{(3)}(x) \right| \left[\frac{1}{3} \left(\frac{1}{2-\alpha} - \frac{1}{2} \right) + \frac{1-\alpha}{12} \sum_{l=2}^{j} l^{-1-\alpha} \right] \qquad (112)$$

$$\leq C_2 \max_{x \in [x_{-1}, x_{j-1}]} \left| f^{(3)}(x) \right| h^{4-\alpha}$$

with $C_2 > 0$ being a constant.

Note that A_k contains all the terms in Equation (106) with the form of integrals over $[x_k, x_{k+1}]$. Then the affine transformation $s = x_k + \xi h$, $\xi \in [0,1]$ and $l = j-k$, $k = 0, 1, \ldots, j-1$ yield

$$A_{j-l} = h^{4-\alpha} \int_0^1 f^{(3)}(x_{j-l} + \xi h) \left\{ -\frac{2}{2-\alpha} \left[(l-1)^{2-\alpha} - (l-\xi)^{2-\alpha} \right] \right.$$

$$+ \frac{(1-\xi)^2}{2-\alpha} \left[(l-1)^{2-\alpha} - l^{2-\alpha} \right] + 2 \left[(l-\xi)^{1-\alpha} \xi - (l-1)^{1-\alpha} \right]$$

$$+ (1-\xi)^2 (l-1)^{1-\alpha} + \frac{(1-\xi)^2}{2} \left[(l-\xi)^{1-\alpha} - l^{1-\alpha} \right] \qquad (113)$$

$$\left. + \frac{(\xi+1)^2}{2} \left[(l-1)^{1-\alpha} - (l-\xi)^{1-\alpha} \right] \right\} d\xi$$

$$= h^{4-\alpha} \int_0^1 f^{(3)}(x_{j-l} + \xi h) a_l(\xi) d\xi.$$

Rewrite $a_l(\xi)$ in the form

$$a_l(\xi) = l^{1-\alpha} \sum_{n=2}^{\infty} \frac{1}{l^n} \tilde{a}_n(\xi)(-\alpha+1)\alpha(\alpha+1)\cdots(\alpha+n-2), \qquad (114)$$

with

$$\tilde{a}_n(\xi) = \frac{2\xi^{n+1} - 2 + (1-\xi)^2}{(n+1)!} + \frac{2 - (1-\xi)^2 - \frac{1}{2}(1+\xi)^2}{n!}, \quad n \geq 2. \qquad (115)$$

For $n \geq 2$, we have $\tilde{a}_n(\xi) \geq 0$ for arbitrary $\xi \in [0,1]$. To see this, recall that

$$\begin{cases} \tilde{a}'_n(\xi) = \dfrac{2\xi^n}{n!} - \dfrac{2(1-\xi)}{(n+1)!} + \dfrac{1-3\xi}{n!}, \\ \tilde{a}''_n(\xi) = \dfrac{2\xi^{n-1}}{(n-1)!} + \dfrac{2}{(n+1)!} - \dfrac{3}{n!}. \end{cases} \quad (116)$$

When $\xi_0 = \left(\dfrac{1}{2n} + \dfrac{1}{n+1}\right)^{\frac{1}{n-1}} \in (0,1)$, there hold

$$\tilde{a}''_n(\xi_0) = 0, \quad (117)$$

and

$$\tilde{a}''_n(\xi) \begin{cases} < 0, & \xi \in [0, \xi_0), \\ \geq 0, & \xi \in [\xi_0, 1], \end{cases} \quad (118)$$

Note that

$$\begin{cases} \tilde{a}'_n(1) = 0, \\ \tilde{a}'_n(0) = \dfrac{1}{n!} - \dfrac{2}{(n+1)!} > 0. \end{cases} \quad (119)$$

One has $\tilde{a}'_n(\xi_0) < \tilde{a}'_n(1) = 0$, and there exits $\xi_1 \in (0, \xi_0)$ such that $\tilde{a}'_n(\xi_1) = 0$ since $\tilde{a}'_n(0) > 0$. Therefore,

$$\tilde{a}'_n(\xi) \begin{cases} > 0, & \xi \in [0, \xi_1), \\ \leq 0, & \xi \in [\xi_1, 1]. \end{cases} \quad (120)$$

Since

$$\begin{cases} \tilde{a}_n(1) = 0, \\ \tilde{a}_n(0) = \dfrac{1}{2n!} - \dfrac{1}{(n+1)!} > 0, \end{cases} \quad (121)$$

it holds that $\tilde{a}_n(\xi) \geq 0$ for arbitrary $\xi \in [0,1]$ when $n \geq 2$. As a result,

$$a_l(\xi) = l^{1-\alpha} \sum_{n=2}^{\infty} \dfrac{1}{l^n} \tilde{a}_n(\xi)(1-\alpha)\alpha(\alpha+1)\cdots(\alpha+n-1) \geq 0,\ 2 \leq l \leq j. \quad (122)$$

Furthermore,

$$a_l(\xi) = l^{1-\alpha} \sum_{n=2}^{\infty} \dfrac{1}{l^n} \left[\dfrac{2\xi^{n+1} - 2 + (1-\xi)^2}{(n+1)!} + \dfrac{2 - (1-\xi)^2 - \frac{1}{2}(1+\xi)^2}{n!} \right] (1-\alpha)\alpha(\alpha+1)\cdots(\alpha+n-1)$$

$$\leq l^{1-\alpha} \sum_{n=2}^{\infty} \dfrac{1}{l^n} \left\{ \dfrac{2\xi^3 - 2 + (1-\xi)^2}{n+1} + \left[2 - (1-\xi)^2 - \dfrac{1}{2}(1+\xi)^2 \right] \right\} \dfrac{1-\alpha}{n}$$

$$\leq l^{1-\alpha} \sum_{n=2}^{\infty} \dfrac{1}{l^n} \left(\dfrac{1}{n+1} \cdot \dfrac{35 + 13\sqrt{13}}{54} + \dfrac{2}{3} \right) \dfrac{1-\alpha}{n} \quad (123)$$

$$\leq l^{-1-\alpha}(1-\alpha) \dfrac{143 + 13\sqrt{13}}{324} \cdot \dfrac{1}{1 - \frac{1}{l}}.$$

Especially, for $l \geq 2$,

$$a_l(\xi) \leq l^{-1-\alpha}(1-\alpha) \dfrac{143 + 13\sqrt{13}}{162}. \quad (124)$$

As a result, it holds that

$$\left|\sum_{l=1}^{j} A_l\right| \leq h^{4-\alpha}\left\{\left|\int_0^1 a_1(\xi) f^{(3)}(x_{j-1} + \xi h) d\xi\right| + \sum_{l=2}^{j}|A_l|\right\}$$

$$\leq h^{4-\alpha} \max_{x\in[x_0,x_j]}\left|f^{(3)}(x)\right|\left\{\int_0^1 a_1(\xi)d\xi + \sum_{l=2}^{j}\int_0^1 a_l(\xi)d\xi\right\} \qquad (125)$$

$$\leq h^{4-\alpha} \max_{x\in[x_0,x_j]}\left|f^{(3)}(x)\right|\left[\frac{2}{(2-\alpha)(3-\alpha)} - \frac{1}{3(2-\alpha)} - \frac{1}{6} + \sum_{l=2}^{j} l^{-1-\alpha}(1-\alpha)\frac{31\sqrt{13}+125}{162}\right]$$

$$\leq C_1 \max_{x\in[x_0,x_j]}\left|f^{(3)}(x)\right| h^{4-\alpha},$$

with $C_1 > 0$ being a constant. Consequently, the truncated error has the bound

$$\left|R^j\right| = \frac{1}{2h\Gamma(2-\alpha)}\left|\sum_{k=0}^{j-1}(A_k + B_k)\right|$$

$$\leq \frac{1}{2h\Gamma(2-\alpha)}\left\{\left|\sum_{l=1}^{j} A_{j-l}\right| + \left|\sum_{l=1}^{j} B_{j-l}\right|\right\} \qquad (126)$$

$$\leq \frac{h^{3-\alpha}}{2\Gamma(2-\alpha)}\left\{C_1 \max_{x\in[x_0,x_j]}\left|f^{(3)}(x)\right| + C_2 \max_{x\in[x_{-1},x_{j-2}]}\left|f^{(3)}(x)\right|\right\}.$$

Note that the derivative $f^{(3)}(x)$ with $x \in [x_{-1}, x_0]$ is needed in the above inequality. In this case, $f^{(3)}(x_k)$ with $k \geq 0$ can be utilized to approximate $f^{(3)}(x)$ when $x \in [x_{-1}, x_0]$ and then $f^{(3)}(x)$ is bounded on $[x_{-1}, x_0]$. Consequently, the desired estimate is obtained. □

Remark 7. *In formula (102), $f(x_{-1})$ is defined outside of $[a,b]$. In numerical calculation, we can approximate $f(x_{-1})$ based on the relation $f(x_{-1}) = f(a) - hf'(a) + \frac{h^2}{2}f''(a) + \mathcal{O}(h^3)$. When $f'(a) = f''(a) = 0$, then $f(x_{-1}) = f(a) + \mathcal{O}(h^3)$, and we have $R^j = \mathcal{O}(h^{3-\alpha})$. When $f'(a) = 0$ and $f''(a) \neq 0$, then $f(x_{-1}) = f(a) + \frac{h^2}{2}f''(a) + \mathcal{O}(h^3)$, and $R^j = \mathcal{O}(h^2)$. If $f'(a) \neq 0$, then $R^j = \mathcal{O}(h)$.*

Example 3 ([79]). *Consider the function $f(x) = x^4$, $x \in [0,1]$. Evaluate its Caputo derivative at $x = 1$ by formula (102). Absolute error (AE) and convergence order (CO) are shown in Table 3. It is obvious that the convergence order is $(3-\alpha)$, which is in line with the theoretical analysis.*

Table 3. Numerical results for Example 3.

α	h	AE	CO	α	h	AE	CO
	$\frac{1}{10}$	0.0015	-		$\frac{1}{10}$	0.0139	-
0.2	$\frac{1}{40}$	3.7575×10^{-5}	2.6809	0.6	$\frac{1}{40}$	5.4517×10^{-4}	2.3606
	$\frac{1}{160}$	8.6640×10^{-7}	2.7289		$\frac{1}{160}$	2.0146×10^{-5}	2.3846
	$\frac{1}{10}$	0.0052	-		$\frac{1}{10}$	0.0331	-
0.4	$\frac{1}{40}$	1.6158×10^{-4}	2.5282	0.8	$\frac{1}{40}$	0.0017	2.1414
	$\frac{1}{160}$	4.6455×10^{-6}	2.5676		$\frac{1}{160}$	8.1011×10^{-5}	2.1910

In Ref. [81], another $(3-\alpha)$-th order approximation was proposed. Denote

$$\begin{cases} \delta_x f_{j-\frac{1}{2}} = \dfrac{f(x_j) - f(x_{j-1})}{h}, & j \geq 1 \\ \delta_x^2 f_j = \dfrac{1}{h}\left(\delta_x f_{j+\frac{1}{2}} - \delta_x f_{j-\frac{1}{2}}\right), & j \geq 1. \end{cases} \quad (127)$$

Let $f(x) \in C^3[a,b]$ and $0 < \alpha < 1$. We utilize the linear interpolation

$$P_{1,k}(x) = f(x_{k-1})\frac{x_k - x}{h} + f(x_k)\frac{x - x_{k-1}}{h}, \quad (128)$$

on the first interval $[x_0, x_1]$, and the quadratic interpolation

$$P_{2,k}(x) = \sum_{l=0}^{2} f(x_{k-l}) \prod_{\substack{i=0 \\ i \neq l}}^{2} \frac{x - x_{k-i}}{x_{k-l} - x_{k-i}} = P_{1,k}(x) + \frac{1}{2}\left(\delta_x^2 f_{k-1}\right)(x - x_{k-1})(x - x_k) \quad (129)$$

on the remaining intervals $[x_{k-1}, x_k]$ ($k \geq 2$) to approximate $f(x)$. Denote $x_{j+\frac{1}{2}} = \frac{x_{j+1}+x_j}{2}$, $j \geq 0$. We obtain the following L1-2 formula [81]

$$\begin{aligned}
\left[{}_cD_{a,x}^\alpha f(x)\right]_{x=x_j} &= \frac{1}{\Gamma(1-\alpha)}\left[\int_{x_0}^{x_1}\frac{(P_{1,1}(t))'}{(x_j-t)^\alpha}dt + \sum_{k=2}^{j}\int_{x_{k-1}}^{x_k}\frac{(P_{2,k}(t))'}{(x_j-t)^\alpha}dt\right] + R^j \\
&= \frac{1}{\Gamma(1-\alpha)}\left\{\sum_{k=2}^{j}\int_{x_{k-1}}^{x_k}\frac{\delta_x f_{k-\frac{1}{2}} + (\delta_x^2 f_{k-1})(t-x_{k-\frac{1}{2}})}{(x_j-t)^\alpha}dt + \delta_x f_{\frac{1}{2}}\int_{x_0}^{x_1}\frac{dt}{(x_j-t)^\alpha}\right\} + R^j \quad (130) \\
&= \frac{h^{-\alpha}}{\Gamma(2-\alpha)}\left[c_0^{(\alpha)}f(x_j) - \sum_{k=1}^{j-1}\left(c_{j-k-1}^{(\alpha)} - c_{j-k}^{(\alpha)}\right)f(x_k) - c_{j-1}^{(\alpha)}f(x_0)\right] + R^j,
\end{aligned}$$

with the truncated errors $R^1 = \mathcal{O}(h^{2-\alpha})$ and $R^j = \mathcal{O}(h^{3-\alpha})$, $j \geq 2$. The coefficient $c_0^{(\alpha)} = 1$ when $j = 1$. For $j \geq 2$, the coefficients are give by

$$c_k^{(\alpha)} = \begin{cases} a_0^{(\alpha)} + b_0^{(\alpha)}, & k = 0, \\ a_k^{(\alpha)} + b_k^{(\alpha)} - b_{k-1}^{(\alpha)}, & 1 \leq k \leq j-2, \\ a_k^{(\alpha)} - b_{k-1}^{(\alpha)}, & k = j-1 \end{cases} \quad (131)$$

with

$$a_k^{(\alpha)} = (k+1)^{1-\alpha} - k^{1-\alpha}, \ 0 \leq k \leq j-1, \quad (132)$$

and

$$b_k^{(\alpha)} = \frac{(k+1)^{2-\alpha} - k^{2-\alpha}}{2-\alpha} - \frac{(k+1)^{1-\alpha} + k^{1-\alpha}}{2}, \ 0 \leq k \leq j-2. \quad (133)$$

Numerical results in Ref. [81] imply that the computational errors given by the L1-2 formula are obviously much smaller than those of the L1 formula.

Modifying the above L1-2 formula, Alikhanov proposed an overall $(3-\alpha)$-th order approximation. Let $\sigma = 1 - \frac{\alpha}{2}$ with $0 < \alpha < 1$, then the Caputo derivative of $f(x) \in C^3[a,b]$ at $x_{j+\sigma} = a + (j+\sigma)h$ with $0 \leq j \leq N-1$ can be expressed by

$$\left[{}_cD_{0,x}^\alpha f(x)\right]_{x=x_{j+\sigma}} = \frac{1}{\Gamma(1-\alpha)}\left[\sum_{k=1}^{j}\int_{x_{k-1}}^{x_k}\frac{f'(t)dt}{(x_{j+\sigma}-t)^\alpha} + \int_{x_j}^{x_{j+\sigma}}\frac{f'(t)dt}{(x_{j+\sigma}-t)^\alpha}\right]. \quad (134)$$

Applying the quadratic interpolation

$$\Pi_{2,k}f(x) = f(x_{k-1})\frac{(x-x_k)(x-x_{k+1})}{2h^2} - f(x_k)\frac{(x-x_{k-1})(x-x_{k+1})}{h^2} \qquad (135)$$
$$+ f(x_{k+1})\frac{(x-x_{k-1})(x-x_k)}{2h^2}, \quad x \in [x_{k-1}, x_k], \ 1 \le k \le j,$$

which is different from the one defined in Equation (129) to approximating $f(x)$, and utilizing the expression $f'(t) \approx \frac{f(x_{j+1}) - f(x_j)}{h}$ on the interval $[x_j, x_{j+\sigma}]$, we obtain the L2-1_σ formula [82]

$$[{}_cD_{a,x}^\alpha f(x)]_{x=x_{j+\sigma}} = \frac{h^{-\alpha}}{\Gamma(2-\alpha)} \sum_{k=0}^{j} c_{j-k}^{(\alpha,\sigma)} [f(x_{k+1}) - f(x_k)] + \mathcal{O}(h^{3-\alpha}) \qquad (136)$$

with $0 \le j \le N-1$. Here $c_0^{(\alpha,\sigma)} = a_0^{(\alpha,\sigma)}$ when $j=0$, and for $j \ge 1$,

$$c_k^{(\alpha,\sigma)} = \begin{cases} a_0^{(\alpha,\sigma)} + b_1^{(\alpha,\sigma)}, & k=0, \\ a_k^{(\alpha,\sigma)} + b_{k+1}^{(\alpha,\sigma)} - b_k^{(\alpha,\sigma)}, & 1 \le k \le j-1, \\ a_j^{(\alpha,\sigma)} - b_j^{(\alpha,\sigma)}, & k=j, \end{cases} \qquad (137)$$

with $a_k^{(\alpha,\sigma)}$ and $b_k^{(\alpha,\sigma)}$ given by

$$\begin{cases} a_0^{(\alpha,\sigma)} = \sigma^{1-\alpha}, \\ a_k^{(\alpha,\sigma)} = (k+\sigma)^{1-\alpha} - (k+\sigma-1)^{1-\alpha}, \ k \ge 1, \end{cases} \qquad (138)$$

and

$$b_k^{(\alpha,\sigma)} = \frac{(k+\sigma)^{2-\alpha} - (k+\sigma-1)^{2-\alpha}}{2-\alpha} - \frac{(k+\sigma)^{1-\alpha} + (k+\sigma-1)^{1-\alpha}}{2}. \qquad (139)$$

The comparison between the L2-1_σ and L1-2 methods in Ref. [82] shows that the L2-1_σ formula refines the accuracy indeed.

Remark 8 ([80]). *The L2-1_σ formula for the right-sided Caputo derivative can be derived in a similar manner. In this case, the parameter should be chosen as $\sigma = \frac{\alpha}{2}$, $\alpha \in (0,1)$. The corresponding approximation is given by*

$$[{}_cD_{x,b}^\alpha f(x)]_{x=x_{j+\sigma}} = \frac{h^{-\alpha}}{\Gamma(2-\alpha)} \sum_{k=j}^{N-1} \tilde{c}_{k-j}^{(\alpha,\sigma)} [f(x_k) - f(x_{k+1})] + \mathcal{O}(h^{3-\alpha}) \qquad (140)$$

with $0 \le j \le N-1$. Here the coefficients are given by

$$\tilde{c}_0^{(\alpha,\sigma)} = -\tilde{a}_0^{(\alpha,\sigma)}, \qquad (141)$$

if $j = N-1$, and for $0 \le j < N-1$,

$$\tilde{c}_k^{(\alpha,\sigma)} = \begin{cases} \tilde{b}_1^{(\alpha,\sigma)}, & k=0, \\ \tilde{b}_{k+1}^{(\alpha,\sigma)} - \tilde{b}_k^{(\alpha,\sigma)}, & 1 \le k \le N-j-2, \\ -\tilde{a}_{N-j-1}^{(\alpha,\sigma)} - \tilde{b}_{N-j-1}^{(\alpha,\sigma)}, & k = N-j-1, \end{cases} \qquad (142)$$

where

$$\tilde{a}_k^{(\alpha,\sigma)} = (k+1-\sigma)^{1-\alpha}, \qquad (143)$$

and
$$\tilde{b}_k^{(\alpha,\sigma)} = \frac{(k+1-\sigma)^{1-\alpha} - (k-\sigma)^{1-\alpha}}{2} - \frac{(k+1-\sigma)^{2-\alpha} - (k-\sigma)^{2-\alpha}}{2-\alpha}. \quad (144)$$

For other $(3-\alpha)$-th order approximations to Caputo derivative based on interpolation, one may refer to Refs. [83,84].

(II) $(4-\alpha)$-th order approximation

Let $0 < \alpha < 1$ and $f(x) \in C^4[x_0, x_j]$. A linear interpolation of $f(x)$ on the first subinterval $[x_0, x_1]$ yields

$$\int_{x_0}^{x_1} (x_j - t)^{-\alpha} f'(t) dt \approx \frac{f(x_1) - f(x_0)}{h} \int_{x_0}^{x_1} (x_j - t)^{-\alpha} dt = \frac{a_{j-1}}{h^\alpha (1-\alpha)} [f(x_1) - f(x_0)] \quad (145)$$

with $a_{j-1} = j^{1-\alpha} - (j-1)^{1-\alpha}$. On the second subinterval $[x_1, x_2]$, we similarly obtain

$$\int_{x_1}^{x_2} (x_j - t)^{-\alpha} f'(t) dt \approx \frac{h^{-\alpha}}{1-\alpha} \left[(a_{j-2} + b_{j-2}) f(x_2) - (a_{j-2} + 2b_{j-2}) f(x_1) + b_{j-2} f(x_0) \right] \quad (146)$$

through the quadratic interpolation, where

$$\begin{cases} a_{j-2} = (j-1)^{1-\alpha} - (j-2)^{1-\alpha}, \\ b_{j-2} = \dfrac{(j-1)^{2-\alpha} - (j-2)^{2-\alpha}}{2-\alpha} - \dfrac{(j-1)^{1-\alpha} + (j-2)^{1-\alpha}}{2}. \end{cases} \quad (147)$$

For the remaining subintervals, we use the cubic interpolation function

$$p_3(x) = \sum_{l=0}^{3} f(x_{k-l}) \prod_{i=0, i \neq l}^{3} \frac{x - x_{k-i}}{x_{k-l} - x_{k-i}}, \quad x \in [x_{k-1}, x_k], \; k \geq 3, \quad (148)$$

to approximate $f(x)$. Consequently, it holds that

$$\frac{1}{\Gamma(1-\alpha)} \sum_{k=3}^{j} \int_{x_{k-1}}^{x_k} \frac{f'(t)}{(x_j - t)^\alpha} dt \approx \frac{1}{\Gamma(1-\alpha)} \sum_{k=3}^{j} \int_{x_{k-1}}^{x_k} (x_j - t)^{-\alpha} p_3'(t) dt$$
$$= \frac{h^{-\alpha}}{\Gamma(2-\alpha)} \sum_{k=3}^{j} \left[\omega_{1,j-k} f(x_k) + \omega_{2,j-k} f(x_{k-1}) + \omega_{3,j-k} f(x_{k-2}) + \omega_{4,j-k} f(x_{k-3}) \right], \; j \geq 3, \quad (149)$$

where the coefficients are given by

$$\omega_{1,j-k} = \frac{2(j-k+1)^{1-\alpha} - 11(j-k)^{1-\alpha}}{6} - \frac{2(j-k)^{2-\alpha} - (j-k+1)^{2-\alpha}}{2-\alpha} - \frac{(j-k)^{3-\alpha} - (j-k+1)^{3-\alpha}}{(2-\alpha)(3-\alpha)},$$

$$\omega_{2,j-k} = \frac{6(j-k)^{1-\alpha} + (j-k+1)^{1-\alpha}}{2} + \frac{5(j-k)^{2-\alpha} - 2(j-k+1)^{2-\alpha}}{2-\alpha} + \frac{3(j-k)^{3-\alpha} - 3(j-k+1)^{3-\alpha}}{(2-\alpha)(3-\alpha)},$$

$$\omega_{3,j-k} = -\frac{3(j-k)^{1-\alpha} + 2(j-k+1)^{1-\alpha}}{2} - \frac{4(j-k)^{2-\alpha} - (j-k+1)^{2-\alpha}}{2-\alpha} - \frac{(j-k)^{3-\alpha} - (j-k+1)^{3-\alpha}}{(2-\alpha)(3-\alpha)},$$

$$\omega_{4,j-k} = \frac{2(j-k)^{1-\alpha} + (j-k+1)^{1-\alpha}}{6} + \frac{(j-k)^{2-\alpha}}{2-\alpha} + \frac{(j-k)^{3-\alpha} - (j-k+1)^{3-\alpha}}{(2-\alpha)(3-\alpha)}, \; 3 \leq j \leq N.$$

In view of the above analysis, we obtain the numerical approximation [85]

$$[{}_C D_{a,x}^\alpha f(x)]_{x=x_j} = \frac{1}{\Gamma(1-\alpha)} \sum_{k=1}^{j} \int_{x_{k-1}}^{x_k} \frac{f'(t)}{(x_j-t)^\alpha} dt = \frac{h^{-\alpha}}{\Gamma(2-\alpha)} \sum_{k=0}^{j} g_k f(x_{j-k}) + R^j, \; 0 < \alpha < 1. \quad (150)$$

The coefficients g_k have different values for different j. When $j = 1$,

$$g_0 = a_0, \quad g_1 = -a_0. \tag{151}$$

When $j = 2$,

$$\begin{cases} g_0 = a_0 + b_0, \\ g_1 = a_1 - a_0 - 2b_0, \\ g_2 = b_0 - a_1. \end{cases} \tag{152}$$

When $j = 3$,

$$\begin{cases} g_0 = w_{1,0}, \quad g_1 = w_{2,0} + a_1 + b_1, \\ g_2 = w_{3,0} + a_2 - a_1 - 2b_1, \\ g_3 = w_{4,0} - a_2 + b_1. \end{cases} \tag{153}$$

When $j = 4$,

$$\begin{cases} g_0 = w_{1,0}, \quad g_1 = w_{1,1} + w_{2,0}, \\ g_2 = w_{2,1} + w_{3,0} + a_2 + b_2, \\ g_3 = w_{3,1} + w_{4,0} + a_3 - a_2 - 2b_2, \\ g_4 = w_{4,1} - a_3 + b_2. \end{cases} \tag{154}$$

When $j = 5$,

$$\begin{cases} g_0 = w_{1,0}, \quad g_1 = w_{1,1} + w_{2,0}, \\ g_2 = w_{1,2} + w_{2,1} + w_{3,0}, \\ g_3 = w_{2,2} + w_{3,1} + w_{4,0} + a_3 + b_3, \\ g_4 = w_{3,2} + w_{4,1} + a_4 - a_3 - 2b_3, \\ g_5 = w_{4,2} - a_4 + b_3. \end{cases} \tag{155}$$

When $j \geq 6$,

$$\begin{cases} g_0 = w_{1,0}, \quad g_1 = w_{1,1} + w_{2,0}, \\ g_2 = w_{1,2} + w_{2,1} + w_{3,0}, \\ g_k = w_{1,k} + w_{2,k-1} + w_{3,k-2} + w_{4,k-3}, \ 3 \leq k \leq j-3, \\ g_{j-2} = a_{j-2} + b_{j-2} + w_{2,j-3} + w_{3,j-4} + w_{4,j-5}, \\ g_{j-1} = w_{3,j-3} + w_{4,j-4} + a_{j-1} - a_{j-2} - 2b_{j-2}, \\ g_j = w_{4,j-3} - a_{j-1} + b_{j-2}. \end{cases} \tag{156}$$

If $f(x) \in C^4[x_0, x_j]$ and $\alpha \in (0,1)$, the truncated error R^j in Equation (150) satisfies

$$\begin{cases} \left|R^1\right| \leq c_1 \max_{x_0 \leq x \leq x_1} \left|f''(x)\right| h^{2-\alpha}, \ c_1 > 0, \\ \left|R^2\right| \leq c_2 \max_{x_0 \leq x \leq x_2} \left|f'''(x)\right| h^{3-\alpha}, \ c_2 > 0, \\ \left|R^j\right| \leq \dfrac{1}{\Gamma(1-\alpha)} \left\{ \dfrac{2\alpha}{3} \max_{x_0 \leq x \leq x_2} \left|f'''(x)\right| (x_j - x_2)^{-\alpha-1} h^4 \right. \\ \left. + \left[\dfrac{1}{12} + \dfrac{3\alpha^2}{2(1-\alpha)(2-\alpha)}\right] \max_{x_0 \leq x \leq x_j} \left|f^{(4)}(x)\right| h^{4-\alpha} \right\}, \ j \geq 3. \end{cases} \tag{157}$$

Numerical examples in Ref. [85] verify the above theoretical results.

Example 4. *Suppose that $0 < \alpha < 1$ and $f(x) = x^4$. Evaluate the α-th order Caputo derivative of $f(x)$ at $x = 1$ by Equation (150). Maximum errors (ME) and convergence order (CO) are presented in Table 4.*

Table 4. Numerical results for Example 4.

α	h	ME	CO	α	h	ME	CO
0.2	$\frac{1}{10}$	1.2176×10^{-4}	-	0.6	$\frac{1}{10}$	1.0943×10^{-3}	-
	$\frac{1}{40}$	6.9376×10^{-7}	3.7336		$\frac{1}{40}$	9.9598×10^{-6}	3.3918
	$\frac{1}{160}$	3.8404×10^{-9}	3.7528		$\frac{1}{160}$	8.9946×10^{-8}	3.3963
0.4	$\frac{1}{10}$	4.1401×10^{-4}	-	0.8	$\frac{1}{10}$	2.6315×10^{-3}	-
	$\frac{1}{40}$	2.9349×10^{-6}	3.5741		$\frac{1}{40}$	3.1265×10^{-5}	3.1982
	$\frac{1}{160}$	2.0437×10^{-8}	3.5855		$\frac{1}{160}$	3.7065×10^{-7}	3.1994

Example 5. *Let $f(x) = e^{2x}$. We evaluate Caputo derivative of $f(x)$ at $x = 1$ by utilizing Equation (150). The maximum errors (ME) and convergence order (CO) are shown in Table 5.*

Table 5. Numerical results for Example 5.

α	h	ME	CO	α	h	ME	CO
0.2	$\frac{1}{10}$	4.9025×10^{-4}	-	0.6	$\frac{1}{10}$	4.2309×10^{-3}	-
	$\frac{1}{40}$	3.8638×10^{-6}	3.5125		$\frac{1}{40}$	4.6839×10^{-5}	3.2902
	$\frac{1}{160}$	3.1440×10^{-8}	3.4447		$\frac{1}{160}$	4.5469×10^{-7}	3.3536
0.4	$\frac{1}{10}$	1.6156×10^{-3}	-	0.8	$\frac{1}{10}$	1.0190×10^{-2}	-
	$\frac{1}{40}$	1.4478×10^{-5}	3.4371		$\frac{1}{40}$	1.4521×10^{-4}	3.1086
	$\frac{1}{160}$	1.1851×10^{-7}	3.4669		$\frac{1}{160}$	1.8089×10^{-6}	3.1747

(III) $(r + 1 - \alpha)$-th order approximation

Generalizing the above $(4 - \alpha)$-th order approximation, an $(r + 1 - \alpha)$-th order approximation was proposed in Ref. [86] by virtue of the Lagrange polynomials of degree r. Let $f(x) \in C^r[a,b]$ ($r \geq 4$) and $0 < \alpha < 1$. On the subintervals $[x_{k-1}, x_k]$, $j \geq k \geq r$, $N \geq j \geq r$, we utilize the Lagrange polynomial

$$p_r(x) = \sum_{l=0}^{r} f(x_{k-l}) \prod_{i=0, i \neq l}^{r} \frac{x - x_{k-i}}{x_{k-l} - x_{k-i}}, \quad x \in [x_{k-1}, x_k], \tag{158}$$

to approximate $f(x)$. Denote

$$I_k[p_r(x)] = \frac{1}{\Gamma(1-\alpha)} \int_{x_{k-1}}^{x_k} (x_j - t)^{-\alpha} p_r'(t) dt. \tag{159}$$

Then it holds that

$$\frac{1}{\Gamma(1-\alpha)} \int_{x_{k-1}}^{x_k} \frac{f'(t)}{(x_j - t)^\alpha} dt \approx I_k[p_r(x)]$$

$$= \frac{1}{\Gamma(1-\alpha)} \sum_{l=0}^{r} \frac{(-1)^l f(x_{k-l})}{l!(r-l)!h^r} \int_{x_{k-1}}^{x_k} (x_j - t)^{-\alpha} \left[\prod_{i=0, i \neq l}^{r} (t - x_{k-i}) \right]' dt \tag{160}$$

$$= \frac{h^{-\alpha}}{\Gamma(1-\alpha)} \sum_{l=0}^{r} \omega_{l,j-k}^r f(x_{k-l}).$$

To compute the coefficients $w_{i,j-k}^r$, we denote by

$$a_k^s = s^{k-\alpha}, \quad b_k^s = (s+1)^{k-\alpha}, \quad p_k = \prod_{l=1}^{k}(l-\alpha), \quad (161)$$

and

$$\alpha_{j,i}^k = \begin{cases} \phi_{j,i}^k, & k \neq 0, \\ 1, & k = 0, \end{cases} \quad \beta_{j,i}^k = \begin{cases} \psi_{j,i}^k, & k \neq 0, \\ 1, & k = 0. \end{cases} \quad (162)$$

Here $\phi_{j,i}^k$ and $\psi_{j,i}^k$ are the sums of products of all different combinations of k elements in the sets $A_{j,i} = \{\bar{a}|\bar{a} \in [0, j-1], \bar{a} \neq i, \bar{a} \in \mathbb{Z}\}$, and $B_{j,i} = \{\bar{b}|\bar{b} \in [-1, j-2], \bar{b} \neq i-1, \bar{b} \in \mathbb{Z}\}$, respectively. Then

$$w_{i,j-k}^r = \frac{(-1)^{i+1}}{i!(r-i)!} \sum_{l=1}^{r} \left[\frac{l!}{p_l} \left(\alpha_{r+1,i}^{r-l} a_l^{j-k} - \beta_{r+1,i}^{r-l} b_l^{j-k} \right) \right], \quad 0 \leq i \leq r-1. \quad (163)$$

On the subinterval $[x_{k-1}, x_k]$, $1 < k < r$, $1 \leq j \leq N$, there are no enough nodes to obtain an r-th degree Lagrange polynomial. In this case, we use $I_k[p_k(x)]$ to approximate the integral $\frac{1}{\Gamma(1-\alpha)} \int_{x_{k-1}}^{x_k} (x_j - t)^{-\alpha} f'(t) dt$. In summary, we obtain the following approximation [86]

$$[{}_c D_{a,x}^\alpha f(x)]_{x=x_j} = \begin{cases} \sum_{k=1}^{j} I_k[p_k(x)] + R_r^j, & j < r, \\ \sum_{k=1}^{r-1} I_k[p_k(x)] + \sum_{k=r}^{j} I_k[p_r(x)] + R_r^j, & r \leq j \leq N, \end{cases} \quad (164)$$

with R_r^j being the truncated error. It has been proved that when $f(x) \in C^r[a,b]$ ($r \geq 4$), the truncation error satisfies

(1) $|R_r^1| \leq c_1 \max_{x_0 \leq x \leq x_1} |f^{(2)}(x)| h^{2-\alpha}$, $c_1 > 0$;

(2) $|R_r^2| \leq c_2 \max_{x_0 \leq x \leq x_2} |f^{(3)}(x)| h^{3-\alpha}$, $c_2 > 0$;

(3) If $f^{(1)}(a) = f^{(2)}(a) = 0$, then $|R_r^3| \leq c_3 \max_{x_0 \leq x \leq x_3} |f^{(4)}(x)| h^{4-\alpha}$, $c_3 > 0$;

(4) Provided that $f^{(k)}(a) = 0$ for $0 < k \leq j$, then

$$|R_r^j| \leq c_j \max_{x_0 \leq x \leq x_j} |f^{(j+1)}(x)| h^{j+1-\alpha}, \quad 2 < j < r, \; c_j > 0;$$

(5) Provided that $f^{(k)}(a) = 0$ for $0 \leq k \leq r-1$, then

$$|R_r^j| = \frac{\alpha}{\Gamma(1-\alpha)} \left[\frac{1}{r+1} \left(\frac{1}{\alpha} + \frac{1}{(1-\alpha)(2-\alpha)} \right) \max_{x_0 \leq x \leq x_j} |f^{(r+1)}(x)| h^{r+1-\alpha} \right.$$
$$\left. + \sum_{k=1}^{r-1} \frac{(x_j - x_{r-1})^{-\alpha-1}(r-1)^{r-1}}{(r-k-1)!(k+1)} \max_{x_0 \leq x \leq x_j} |f^{(r)}(x)| h^{r+1} \right], \; j \geq r.$$

Numerical examples in Ref. [86] verify the above theoretical results.

Example 6. Suppose $0 < \alpha < 1$, and let $f(x) = x^6$, $x \in [0, 1]$. Use scheme (164) to compute Caputo derivative of $f(x)$ at $x = 1$ with different stepsizes. Table 6 lists the computational errors and convergence orders at $x = 1$ with different values for α, and $r = 4, 5$. It can be observed that the numerical convergence order of the utilized scheme is $(r + 1 - \alpha)$.

Table 6. Numerical results for Example 6.

α	h	Errors ($r=4$)	Order	Errors ($r=5$)	Order
0.2	$\frac{1}{40}$	3.4260×10^{-7}	-	7.5843×10^{-9}	-
	$\frac{1}{50}$	1.2040×10^{-7}	4.6900	2.0999×10^{-9}	5.7583
	$\frac{1}{60}$	5.1128×10^{-8}	4.7001	7.3082×10^{-10}	5.8067
0.4	$\frac{1}{40}$	1.5377×10^{-6}	-	3.3066×10^{-8}	-
	$\frac{1}{50}$	5.5973×10^{-7}	4.5319	9.5097×10^{-9}	5.5862
	$\frac{1}{60}$	2.4468×10^{-7}	4.5408	3.4281×10^{-9}	5.6014
0.8	$\frac{1}{40}$	1.7551×10^{-5}	-	3.7712×10^{-7}	-
	$\frac{1}{50}$	6.9402×10^{-6}	4.1604	1.1820×10^{-7}	5.1992
	$\frac{1}{60}$	3.2473×10^{-6}	4.1673	4.5817×10^{-8}	5.1977

Example 7. Suppose $0<\alpha<1$, and consider the function $f(x)=e^{2x}-2x-2x^2-\frac{4}{3}x^3-\frac{2}{3}x^4$, $x\in[0,1]$. Table 7 lists the numerical results with different values for α, and $r=4,5$. It is evident that scheme (164) can reach $(r+1-\alpha)$-th order accuracy.

Table 7. Numerical results for Example 7.

α	h	Errors ($r=4$)	Order	Errors ($r=5$)	Order
0.2	$\frac{1}{26}$	7.7151×10^{-7}	-	4.7095×10^{-8}	-
	$\frac{1}{28}$	5.4884×10^{-7}	4.5977	3.1083×10^{-8}	5.6001
	$\frac{1}{30}$	3.9948×10^{-7}	4.6064	2.1014×10^{-8}	5.7337
0.4	$\frac{1}{26}$	3.2710×10^{-6}	-	1.9686×10^{-7}	-
	$\frac{1}{28}$	2.3542×10^{-6}	4.4402	1.3170×10^{-7}	5.4275
	$\frac{1}{30}$	1.7322×10^{-6}	4.4490	9.0617×10^{-8}	5.4056
0.8	$\frac{1}{26}$	3.2210×10^{-5}	-	1.9644×10^{-6}	-
	$\frac{1}{28}$	2.3819×10^{-5}	4.0750	1.3521×10^{-6}	5.0431
	$\frac{1}{30}$	1.7973×10^{-5}	4.0832	9.5398×10^{-7}	5.0602

Remark 9. The $(3-\alpha)$-th, $(4-\alpha)$-th, and $(r+1-\alpha)$-th order numerical schemes established in Refs. [79,85,86] are of unconditional stability in the practical sense when solving fractional differential equations. In other words, numerical schemes for fractional differential equations based on these approximations are stable only if α lies in their respective subsets of the interval $(0,1)$. On the other hand, there are some other interesting methods in this respect. See [87,88] for more details.

(IV) Spectral approximations

Let $m-1<\alpha<m\in\mathbb{Z}^+$ and $f(x)\in H^r[a,b]$ with $r\geq 2m$. Now we introduce spectral approximations to Caputo derivative [42,44]. Here we take the Jacobi approximation as a representative example since the others such as Chebyshev approximation are special cases of the Jacobi one. Let the polynomial

$$p_N(x)=\sum_{j=0}^{N}\widetilde{p}_j^{u,v}P_j^{u,v}(x),\ x\in[-1,1] \tag{165}$$

be an approximation of $f(x)$ based on the Jacobi polynomials. Recall that

$$\begin{cases}\dfrac{d^m}{dx^m}P_j^{u,v}(x)=d_{j,m}^{u,v}P_{j-m}^{u+m,v+m}(x),\ j\geq m\in\mathbb{Z}^+,\\ d_{j,m}^{u,v}=\dfrac{\Gamma(j+m+u+v+1)}{2^m\Gamma(j+u+v+2)}.\end{cases} \tag{166}$$

It holds that

$$_cD^\alpha_{-1,x} p_N(x) = \frac{1}{\Gamma(m-\alpha)} \int_{-1}^{x} (x-t)^{m-\alpha-1} \sum_{j=m}^{N} \widetilde{p}_j^{u,v} d_{j,m}^{u,v} \widehat{P}_{j-m}^{u+m,v+m}(t) dt \qquad (167)$$

$$= \sum_{j=m}^{N} \widetilde{p}_j^{u,v} d_{j,m}^{u,v} \widehat{P}_{j-m}^{u+m,v+m,m-\alpha}(x),$$

where $\widetilde{p}_j^{u,v}$ and $\widehat{P}_j^{u+m,v+m,m-\alpha}(x)$ are defined by Equations (32) and (36). Denote

$$D_j^{u,v,\alpha,m}(x) = d_{j,m}^{u,v} \widehat{P}_{j-m}^{u+m,v+m,m-\alpha}(x) \qquad (168)$$

with $D_j^{u,v,\alpha,m}(x) = 0$ for $0 \le j \le m-1$. Then it holds that

$$_cD^\alpha_{-1,x} f(x) \approx {_cD^\alpha_{-1,x}} p_N(x) = \sum_{j=m}^{n} \widetilde{p}_j^{u,v} D_j^{u,v,\alpha,m}(x), \; x \in [-1,1]. \qquad (169)$$

The affine transformation $\widehat{x} = \frac{2x-a-b}{b-a}$ with $x \in [a,b]$ yields

$$_cD^\alpha_{a,x} f(x) \approx \left(\frac{b-a}{2}\right)^{-\alpha} \sum_{j=m}^{N} \widetilde{p}_j^{u,v} D_j^{u,v,\alpha,m}(\widehat{x}). \qquad (170)$$

For the corresponding differential matrix, see Ref. [44] for more details.
The following numerical examples verify the efficiency of the spectral approximation.

Example 8 ([42]). *Let $f(x) = x^\mu$, $x \in [0,1]$. We use formula (170) to compute $_cD^\alpha_{0,x} f(x)$. Table 8 shows the absolute maximum errors at the Jacobi-Gauss-Lobatto points. The spectral accuracy is obtained.*

Table 8. The absolute errors for Example 8.

			$u = v = 0, \mu = 3.5$			
n	$\alpha = 0.2$	$\alpha = 0.5$	$\alpha = 0.8$	$\alpha = 1.2$	$\alpha = 1.5$	$\alpha = 1.8$
20	2.49×10^{-9}	2.90×10^{-8}	1.99×10^{-7}	6.63×10^{-6}	1.92×10^{-5}	2.67×10^{-5}
40	2.70×10^{-11}	4.73×10^{-10}	4.88×10^{-9}	2.81×10^{-7}	1.22×10^{-6}	2.55×10^{-6}
80	2.88×10^{-13}	7.62×10^{-12}	1.19×10^{-10}	1.18×10^{-8}	7.77×10^{-8}	2.44×10^{-7}
			$u = v = -\frac{1}{2}, \mu = 3.5$			
n	$\alpha = 0.2$	$\alpha = 0.5$	$\alpha = 0.8$	$\alpha = 1.2$	$\alpha = 1.5$	$\alpha = 1.8$
20	2.12×10^{-9}	2.11×10^{-8}	1.62×10^{-7}	5.37×10^{-6}	1.86×10^{-5}	3.95×10^{-5}
40	2.15×10^{-11}	3.22×10^{-10}	3.77×10^{-9}	2.17×10^{-7}	1.14×10^{-6}	3.66×10^{-6}
80	2.20×10^{-13}	5.00×10^{-12}	8.89×10^{-11}	8.92×10^{-9}	7.10×10^{-8}	3.45×10^{-7}

Example 9 ([42]). *Let $f(x) = \sin x$, $x \in [0,1]$. We utilized Equation (170) to evaluate $_cD^\alpha_{0,x} f(x)$. Table 9 presents the absolute maximum errors for the cases of $u = v = 0$ and $u = v = -\frac{1}{2}$. The expected results can be observed.*

Table 9. The absolute errors for Example 9.

			$u = v = 0$			
n	$\alpha = 0.2$	$\alpha = 0.5$	$\alpha = 0.8$	$\alpha = 1.2$	$\alpha = 1.5$	$\alpha = 1.8$
4	1.05×10^{-5}	6.17×10^{-5}	2.31×10^{-4}	1.02×10^{-3}	2.18×10^{-3}	5.66×10^{-3}
8	1.48×10^{-11}	1.08×10^{-10}	5.96×10^{-10}	4.14×10^{-9}	1.47×10^{-8}	5.33×10^{-8}
16	3.22×10^{-15}	1.91×10^{-14}	9.29×10^{-14}	6.39×10^{-13}	2.45×10^{-12}	9.03×10^{-12}
			$u = v = -\frac{1}{2}$			
n	$\alpha = 0.2$	$\alpha = 0.5$	$\alpha = 0.8$	$\alpha = 1.2$	$\alpha = 1.5$	$\alpha = 1.8$
4	1.34×10^{-5}	5.66×10^{-5}	1.95×10^{-4}	9.93×10^{-4}	2.05×10^{-3}	5.40×10^{-3}
8	1.98×10^{-11}	1.04×10^{-10}	3.92×10^{-10}	3.34×10^{-9}	1.15×10^{-8}	4.34×10^{-8}
16	4.44×10^{-16}	1.22×10^{-15}	7.55×10^{-15}	4.40×10^{-14}	2.32×10^{-13}	9.82×10^{-13}

(V) Radial basis function discretization

Being a natural generalization of univariate polynomial splines to a multivariate setting, radial basis functions work for arbitrary geometry with high dimensions and it does not require a mesh at all [89]. Numerically solving fractional differential equations based on radial basis functions has attracted sustained attention in engineering and science community. See [90–93] and references cited therein. In [94], radial basis functions are utilized to evaluate fractional differential operators. In the following, we introduce the basic idea of this method.

Take the one-dimensional case as an example. Let x_j $(j = 1, 2, \ldots, N)$ be the collocation points in the interval $[a, b]$. An radial basis function interpolant of a given function $f(x)$ is defined in the form

$$f(x) \approx S(x) = \sum_{j=1}^{N} \lambda_j \phi(|x - x_j|). \tag{171}$$

In order to take the values $f(x_i)$, $i = 1, 2, \ldots, N$, the expansion coefficients λ_j are required to satisfy the matrix form

$$\mathbf{A}\vec{\lambda} = \vec{f} \tag{172}$$

with $\vec{\lambda} = (\lambda_1, \lambda_2, \ldots, \lambda_N)^\top$, $\vec{f} = (f(x_1), f(x_2), \ldots, f(x_N))^\top$, and $\mathbf{A}_{ij} = \phi(|x_i - x_j|)$. Here $\phi(\cdot)$ is the radial basis function. Some popular choices of radial basis function are cubic ($\phi(r) = r^3$), multiquadrics ($\phi(r) = \sqrt{r^2 + c^2}$), and Gaussian ($\phi(r) = e^{-cr^2}$), where the free parameter c is called the shape parameter for the radial basis function. The smooth radial functions (such as multiquadrics and Gaussian) give rise to spectrally accurate function representation while the piecewise smooth radial functions (such as cubic) only produce algebraically accurate representations [94]. Applying the Caputo differentiation operator to (171) yields

$$\sum_{j=1}^{N} \lambda_j \, {}_cD_{a,x}^\alpha \phi(|x - x_j|)\big|_{x=x_i} \approx g(x_i), \ 1 \leq i \leq N, \tag{173}$$

which can be written in the matrix form

$$\mathbf{B}\vec{\lambda} \approx \vec{g} \tag{174}$$

with $\mathbf{B}_{i,j} = {}_cD_{a,x}^\alpha \phi(|x - x_j|)\big|_{x=x_i}$ and $\vec{g} = (g(x_1), g(x_2), \ldots, g(x_N))^\top$. Here $g(x_i)$ is the value of ${}_cD_{a,x}^\alpha f(x)$ at the point $x = x_i$. Note that the collocation matrix $\mathbf{A}$ is unconditionally nonsingular [94]. Combing equations (172) and (174) gives

$$\vec{g} \approx \mathbf{B}\mathbf{A}^{-1}\vec{f}. \tag{175}$$

Therefore, the differential matrix $\mathbf{D} = \mathbf{B}\mathbf{A}^{-1}$ yields an radial basis function discretiazation of the operator ${}_cD_{a,x}^\alpha$.

Remark 10. (I) *The above procedure of deriving differential matrix based radial basis functions is applicable for other fractional differentiation operators as well.*

(II) *Finding a closed form analytical expression for the fractional derivative of a given function may be challenging. In practice, one has to represent the radial basis function in the form of Taylor series before applying fractional differentiation operator term by term. Then the infinite sum can be truncated once the terms are smaller in magnitude than machine precision.*

(III) *The standard radial basis function methods may result in ill-conditioning which often impairs the convergence. To offset this deficiency, the so-called RBF-QR method can be utilized instead of the standard one. See Ref. [94] for more details.*

3.1.3. Fractional Backward Difference Formulae

It has been mentioned in Ref. [45] that the fractional linear multistep method is applicable for numerical Riemann-Liouville differentiation, provided that the generating functions are properly chosen. We can therefore derive fractional backward multistep methods for Caputo derivative, on the basis of the relation

$$\begin{cases} {}_cD^\alpha_{a,x}f(x) = {}_{RL}D^\alpha_{a,x}f(x) - \sum_{k=0}^{m-1} \dfrac{f^{(k)}(a)(x-a)^{k-\alpha}}{\Gamma(1+k-\alpha)}, \\ {}_cD^\alpha_{x,b}f(x) = {}_{RL}D^\alpha_{x,b}f(x) - \sum_{k=0}^{m-1} \dfrac{(-1)^k f^{(k)}(b)(b-x)^{k-\alpha}}{\Gamma(1+k-\alpha)}, \end{cases} \quad (176)$$

where $m-1 < \alpha < m \in \mathbb{Z}^+$. In Ref. [95], shifted fractional backward difference formulae for Caputo derivative were derived through three steps. Shifted Lubich formulae for Riemann–Liouville derivative on bounded domain were first introduced for the case with homogeneous conditions. At that stage, generating functions of the coefficients were constructed to maintain high-order accuracy. By virtue of adopting suitable auxiliary functions, the shifted formulae were modified for the case with inhomogeneous conditions. Finally, the shifted formulae for Caputo derivative are obtained based on relation (176). Theoretical results which can be proved through Fourier analysis are as follows.

Theorem 4. *Suppose that $f(x) \in C^{[\alpha]+3}[a,b]$, and that the derivatives of $f(x)$ up to order $[\alpha]+4$ belong to $L^1[a,b]$. Then there hold*

$$\left[{}_cD^\alpha_{a,x}f(x)\right]_{x=x_j} = \frac{1}{h^\alpha}\sum_{k=0}^{j+p}\zeta^{(\alpha)}_{2,p,k}f(x_{j-k+p}) + \mathcal{O}(h^2), \quad 0 < \alpha < 1, \quad (177)$$

and

$$\left[{}_cD^\alpha_{a,x}f(x)\right]_{x=x_j} = \frac{1}{h^\alpha}\sum_{k=0}^{j+p}\xi^{(\alpha)}_{2,p,k}f(x_{j-k+p}) + \mathcal{O}(h^2), \quad 1 < \alpha < 2, \quad (178)$$

where the weights $\zeta^{(\alpha)}_{2,p,k}$ and $\xi^{(\alpha)}_{2,p,k}$ ($k = 0, 1, \ldots$) are given by

$$\zeta^{(\alpha)}_{2,p,k} = \begin{cases} \omega^{(\alpha)}_{2,p,k'} & k = 0, 1, \ldots, j+p-1, \\ -\sum_{l=0}^{j+p-1}\omega^{(\alpha)}_{2,p,l'} & k = j+p, \end{cases} \quad (179)$$

and,

$$\varsigma_{2,p,k}^{(\alpha)} = \begin{cases} \omega_{2,p,k}^{(\alpha)}, & k = 0,1,\ldots,j+p-3, \\ \omega_{2,p,j+p-2}^{(\alpha)} + \frac{1}{2}\sum_{l=0}^{j+p}\omega_{2,p,l}^{(\alpha)}(j-l+p), & k = j+p-2, \\ \omega_{2,p,j+p-1}^{(\alpha)} - 2\sum_{l=0}^{j+p}\omega_{2,p,l}^{(\alpha)}(j-l+p), & k = j+p-1, \\ -\sum_{l=0}^{j+p-1}\omega_{2,p,l}^{(\alpha)} + \frac{3}{2}\sum_{l=0}^{j+p}\omega_{2,p,l}^{(\alpha)}(j-l+p), & k = j+p. \end{cases} \qquad (180)$$

Here the shift $p \leq \frac{3\alpha}{2}$, and the coefficients $\omega_{2,p,k}^{(\alpha)}$ ($0 \leq k \leq j+p$) are given by

$$\omega_{2,p}^{(\alpha)}(z) = \left(\frac{3\alpha - 2p}{2\alpha} - \frac{2(\alpha-p)}{\alpha}z + \frac{\alpha - 2p}{2\alpha}z^2\right)^\alpha = \sum_{k=0}^\infty \omega_{2,p,k}^{(\alpha)} z^k, \ |z| < 1. \qquad (181)$$

Theorem 5. Suppose that $f(x) \in C^{[\alpha]+4}[a,b]$, and that the derivatives of $f(x)$ up to order $[\alpha]+5$ belong to $L^1[a,b]$. Then the third order schemes are given by

$$\left[{}_C D_{a,x}^\alpha f(x)\right]_{x=x_j} = \frac{1}{h^\alpha}\sum_{k=0}^{j+p}\varsigma_{3,p,k}^{(\alpha)} f(x_{j-k+p}) + \mathcal{O}(h^3), \ 0 < \alpha < 1, \qquad (182)$$

and

$$\left[{}_C D_{a,x}^\alpha f(x)\right]_{x=x_j} = \frac{1}{h^\alpha}\sum_{k=0}^{j+p}\varsigma_{3,p,k}^{(\alpha)} f(x_{j-k+p}) + \mathcal{O}(h^3), \ 1 < \alpha < 2, \qquad (183)$$

where the weights $\varsigma_{3,p,k}^{(\alpha)}$ and $\xi_{3,p,k}^{(\alpha)}$ ($k = 0,1,\ldots$) are

$$\varsigma_{3,p,k}^{(\alpha)} = \begin{cases} \omega_{3,p,k}^{(\alpha)}, & k = 0,1,\ldots,j+p-1, \\ -\sum_{l=0}^{j+p-1}\omega_{3,p,l}^{(\alpha)}, & k = j+p, \end{cases} \qquad (184)$$

and,

$$\xi_{3,p,k}^{(\alpha)} = \begin{cases} \omega_{3,p,k}^{(\alpha)}, & k = 0,1,\ldots,j+p-4, \\ \omega_{3,p,j+p-3}^{(\alpha)} - \frac{1}{3}\sum_{l=0}^{j+p}\omega_{3,p,l}^{(\alpha)}(j-l+p), & k = j+p-3, \\ \omega_{3,p,j+p-2}^{(\alpha)} + \frac{3}{2}\sum_{l=0}^{j+p}\omega_{3,p,l}^{(\alpha)}(j-l+p), & k = j+p-2, \\ \omega_{3,p,j+p-1}^{(\alpha)} - 3\sum_{l=0}^{j+p}\omega_{3,p,l}^{(\alpha)}(j-l+p), & k = j+p-1, \\ -\sum_{l=0}^{j+p-1}\omega_{3,p,l}^{(\alpha)} + \frac{11}{6}\sum_{l=0}^{j+p}\omega_{3,p,l}^{(\alpha)}(j-l+p), & k = j+p. \end{cases} \qquad (185)$$

Here the shift p satisfies $3p^2 - 12\alpha p + 11\alpha^2 \geq 0$, and $\omega_{3,p,k}^{(\alpha)}$ ($0 \leq k \leq j+p$) are given by

$$\omega_{3,p}^{(\alpha)}(z) = \left(a_0 + a_1 z + a_2 z^2 + a_3 z^3\right)^\alpha = \sum_{k=0}^\infty \omega_{3,p,k}^{(\alpha)} z^k, \ |z| < 1, \qquad (186)$$

with

$$\begin{cases} a_0 = \dfrac{11\alpha^2 - 12\alpha p + 3p^2}{6\alpha^2}, & a_1 = \dfrac{-18\alpha^2 + 30\alpha p - 9p^2}{6\alpha^2}, \\ a_2 = \dfrac{9\alpha^2 - 24\alpha p + 9p^2}{6\alpha^2}, & a_3 = \dfrac{-2\alpha^2 + 6\alpha p - 3p^2}{6\alpha^2}. \end{cases} \qquad (187)$$

Theorem 6. Suppose that $f(x) \in C^{[\alpha]+5}[a,b]$, and that the derivatives of $f(x)$ up to order $[\alpha]+6$ belong to $L^1[a,b]$. Then there hold

$$[{}_C D^\alpha_{a,x} f(x)]_{x=x_j} = \frac{1}{h^\alpha} \sum_{k=0}^{j+p} \zeta^{(\alpha)}_{4,p,k} f(x_{j-k+p}) + \mathcal{O}(h^4), \quad 0 < \alpha < 1, \tag{188}$$

and

$$[{}_C D^\alpha_{a,x} f(x)]_{x=x_j} = \frac{1}{h^\alpha} \sum_{k=0}^{j+p} \zeta^{(\alpha)}_{4,p,k} f(x_{j-k+p}) + \mathcal{O}(h^4), \quad 1 < \alpha < 2, \tag{189}$$

where the weights $\zeta^{(\alpha)}_{4,p,k}$ and $\tilde{\zeta}^{(\alpha)}_{4,p,k}$ ($k = 0, 1, \ldots$) are defined by

$$\zeta^{(\alpha)}_{4,p,k} = \begin{cases} \omega^{(\alpha)}_{4,p,k}, & k = 0, 1, \ldots, j+p-1, \\ -\sum_{l=0}^{j+p-1} \omega^{(\alpha)}_{4,p,l}, & k = j+p, \end{cases} \tag{190}$$

and,

$$\tilde{\zeta}^{(\alpha)}_{4,p,k} = \begin{cases} \omega^{(\alpha)}_{4,p,k}, & k = 0, 1, \ldots, j+p-5, \\ \omega^{(\alpha)}_{4,p,j+p-4} + \frac{1}{4}\sum_{l=0}^{j+p} \omega^{(\alpha)}_{4,p,l}(j-l+p), & k = j+p-4, \\ \omega^{(\alpha)}_{4,p,j+p-3} - \frac{4}{3}\sum_{l=0}^{j+p} \omega^{(\alpha)}_{4,p,l}(j-l+p), & k = j+p-3, \\ \omega^{(\alpha)}_{4,p,j+p-2} + 3\sum_{l=0}^{j+p} \omega^{(\alpha)}_{4,p,l}(j-l+p), & k = j+p-2, \\ \omega^{(\alpha)}_{4,p,j+p-1} - 4\sum_{l=0}^{j+p} \omega^{(\alpha)}_{4,p,l}(j-l+p), & k = j+p-1, \\ -\sum_{l=0}^{j+p-1} \omega^{(\alpha)}_{4,p,l} + \frac{25}{12}\sum_{l=0}^{j+p} \omega^{(\alpha)}_{4,p,l}(j-l+p), & k = j+p. \end{cases} \tag{191}$$

Here the shift p satisfies $25\alpha^3 - 35\alpha^2 p + 15\alpha p^2 - 2p^3 \geq 0$, and $\omega^{(\alpha)}_{4,p,k}$ ($0 \leq k \leq j+p$) are given by

$$\omega^{(\alpha)}_{4,p}(z) = \left(b_0 + b_1 z + b_2 z^2 + b_3 z^3 + b_4 z^4\right)^\alpha = \sum_{k=0}^\infty \omega^{(\alpha)}_{2,p,k} z^k, \quad |z| < 1, \tag{192}$$

with

$$\begin{cases} b_0 = \dfrac{25\alpha^3 - 35\alpha^2 p + 15\alpha p^2 - 2p^3}{12\alpha^3}, \\ b_1 = \dfrac{-48\alpha^3 + 104\alpha^2 p - 54\alpha p^2 + 8p^3}{12\alpha^3}, \\ b_2 = \dfrac{36\alpha^3 - 114\alpha^2 p + 72\alpha p^2 - 12p^3}{12\alpha^3}, \\ b_3 = \dfrac{-16\alpha^3 + 56\alpha^2 p - 42\alpha p^2 + 8p^3}{12\alpha^3}, \\ b_4 = \dfrac{3\alpha^3 - 11\alpha^2 p + 9\alpha p^2 - 2p^3}{12\alpha^3}. \end{cases} \tag{193}$$

Different from numerical algorithms based on the polynomial interpolation, in which the corresponding accuracy depends on the derivative order α and homogenous conditions are needed, the formulae presented in Theorems 4–6 for Caputo derivatives have no restriction on the initial conditions, and are of integer-order accuracy. Here we display two numerical examples to verify these arguments.

Example 10 ([95]). *Consider the function $f(x) = x^{6+\alpha}$, $x \in [0,1]$. We utilize schemes (178), (183), and (189) to evaluate the α-th order Caputo derivative at $x = 1$. The absolute errors and convergence orders for $p = 0$ and $p = 1$ are shown in Table 10. The experiment convergence orders are consistent with theoretical analysis. Furthermore, the shifted numerical methods are more efficient than the unshifted one when $1 < \alpha < 2$.*

Table 10. The absolute errors and convergence orders of Example 10.

		$p=0$					
α	h	$n=2$		$n=3$		$n=4$	
		$E(h)$	rate	$E(h)$	rate	$E(h)$	rate
1.70	$\frac{1}{20}$	1.0791	-	1.5714×10^{-1}	-	1.9056×10^{-2}	-
	$\frac{1}{80}$	7.5767×10^{-2}	1.94	2.8201×10^{-3}	2.93	8.4873×10^{-5}	3.94
	$\frac{1}{320}$	4.8717×10^{-3}	1.99	4.5587×10^{-5}	2.99	3.4203×10^{-7}	3.96
1.80	$\frac{1}{20}$	1.4104	-	2.0548×10^{-1}	-	2.4921×10^{-2}	-
	$\frac{1}{80}$	9.9074×10^{-2}	1.94	3.6878×10^{-3}	2.93	1.1098×10^{-4}	3.94
	$\frac{1}{320}$	6.3705×10^{-3}	1.99	5.9612×10^{-5}	2.98	4.4862×10^{-7}	3.98
1.85	$\frac{1}{20}$	1.6112	-	2.3480×10^{-1}	-	2.8478×10^{-2}	-
	$\frac{1}{80}$	1.1321×10^{-1}	1.94	4.2140×10^{-3}	2.93	1.2682×10^{-4}	3.94
	$\frac{1}{320}$	7.2796×10^{-3}	1.99	6.8119×10^{-5}	2.98	5.1220×10^{-7}	3.98
		$p=1$					
α	h	$n=2$		$n=3$		$n=4$	
		$E(h)$	rate	$E(h)$	rate	$E(h)$	rate
1.70	$\frac{1}{20}$	3.0173×10^{-1}	-	4.3788×10^{-2}	-	5.0963×10^{-3}	-
	$\frac{1}{80}$	1.9203×10^{-2}	1.99	7.3362×10^{-4}	2.97	2.1650×10^{-5}	3.96
	$\frac{1}{320}$	1.2061×10^{-3}	2.00	1.1665×10^{-5}	2.99	8.6359×10^{-8}	4.00
1.80	$\frac{1}{20}$	3.2844×10^{-1}	-	5.0679×10^{-2}	-	6.0965×10^{-3}	-
	$\frac{1}{80}$	2.0842×10^{-2}	1.99	8.4958×10^{-4}	2.97	2.5938×10^{-5}	3.96
	$\frac{1}{320}$	1.3081×10^{-3}	2.00	1.3511×10^{-5}	2.99	1.0348×10^{-7}	3.99
1.85	$\frac{1}{20}$	3.3890×10^{-1}	-	5.4135×10^{-2}	-	6.6280×10^{-3}	-
	$\frac{1}{80}$	2.1454×10^{-2}	1.99	9.0750×10^{-4}	2.97	2.8214×10^{-5}	3.96
	$\frac{1}{320}$	1.3456×10^{-3}	2.00	1.4432×10^{-5}	2.99	1.1252×10^{-7}	3.99

Example 11 ([95]). *Consider $f(x) = x^{6+\alpha} + (x+1)^2$, $x \in [0,1]$. In this case, $f(0) \neq 0$. We utilize schemes (178), (183), and (189) to evaluate its α-th order Caputo derivative at $x = 1$. Numerical results are presented in Table 11. These results imply that the numerical approximations can be used to compute Caputo derivatives of suitably smooth functions with inhomogeneous conditions at the initial time.*

Table 11. The absolute errors and convergence orders of Example 11 with $p = 1$.

α	h	$n=2$		$n=3$		$n=4$	
		$E(h)$	rate	$E(h)$	rate	$E(h)$	rate
1.70	$\frac{1}{20}$	3.0157×10^{-1}	-	4.3799×10^{-2}	-	5.0897×10^{-3}	-
	$\frac{1}{80}$	1.9193×10^{-2}	1.99	7.3380×10^{-4}	2.97	2.1622×10^{-5}	3.96
	$\frac{1}{320}$	1.2055×10^{-3}	2.00	1.1668×10^{-5}	2.99	8.6891×10^{-8}	3.98
1.80	$\frac{1}{20}$	3.2833×10^{-1}	-	5.0688×10^{-2}	-	6.0949×10^{-3}	-
	$\frac{1}{80}$	2.0836×10^{-2}	1.99	8.4970×10^{-4}	2.97	2.5935×10^{-5}	3.96
	$\frac{1}{320}$	1.3077×10^{-3}	2.00	1.3513×10^{-5}	2.99	1.0350×10^{-7}	3.99
1.85	$\frac{1}{20}$	3.3882×10^{-1}	-	5.4142×10^{-2}	-	6.6269×10^{-3}	-
	$\frac{1}{80}$	2.1449×10^{-2}	1.99	9.0760×10^{-4}	2.97	2.8211×10^{-5}	3.96
	$\frac{1}{320}$	1.3453×10^{-3}	2.00	1.4433×10^{-5}	2.99	1.1225×10^{-7}	3.99

3.1.4. Diffusive Approximation

Recall that Caputo derivative is defined as the Riemann–Liouville integral of an integer-order derivative, i.e.,

$$_C D_{a,x}^\alpha f(x) = {}_{RL}D_{0,x}^{-(m-\alpha)} f^{(m)}(x), \quad m-1 < \alpha < m \in \mathbb{Z}^+. \tag{194}$$

In this case, Equation (62) implies

$$_C D_{0,x}^\alpha f(x) = \int_0^\infty \phi(\omega, x) d\omega, \tag{195}$$

with $m - 1 < \alpha < m \in \mathbb{Z}^+$ and $\phi : (0,\infty) \times [0,x] \to \mathbb{R}$ being the auxiliary bivariate function defined by

$$\phi(\omega, x) = (-1)^{m-1} \frac{2\sin(\pi\alpha)}{\pi} \omega^{2\alpha - 2m + 1} \int_0^x f^{(m)}(t) e^{-(x-t)\omega^2} dt. \tag{196}$$

For fixed $\omega > 0$, the function $\phi(\omega, \cdot)$ satisfies the differential equation

$$\frac{\partial}{\partial x}\phi(\omega, x) = (-1)^{m-1} \frac{2\sin(\pi\alpha)}{\pi} \omega^{2\alpha - 2m + 1} f^{(m)}(x) - \omega^2 \phi(\omega, x) \tag{197}$$

subject to the initial condition $\phi(\omega, 0) = 0$. Consequently, any implementation solving ODEs and suitable quadratures approximating the infinite integral (196) yield numerical approximations to Caputo derivative.

For more details of discussions, modifications, and applications of the diffusive approximation, see Refs. [60,61].

3.2. Numerical Riemann-Liouville Differentiation

Now we consider numerical approximations to Riemann-Liouville derivatives. The relation (176) indicates that numerical evaluations of Riemann–Liouville derivative can be readily obtained based on those of Caputo derivative. Numerical approximations derived from evaluations of Caputo derivative are therefore omitted in this section. Here we present alternative approaches.

3.2.1. Numerical Methods Based on Linear Spline Interpolation

Let $f(x) \in C^4[a,b]$ and $1 < \alpha < 2$. For $j = 1, 2, \ldots, N-1$, there holds

$$\begin{aligned}
\left[{}_{RL}D_{a,x}^\alpha f(x)\right]_{x=x_j} &= \frac{1}{\Gamma(2-\alpha)} \left[\frac{d^2}{dx^2} \int_a^x (x-t)^{1-\alpha} f(t) dt\right]_{x=x_j} \\
&\approx \frac{h^{-2}}{\Gamma(2-\alpha)} \left[\mathcal{I}_\alpha^l(x_{j-1}) - 2\mathcal{I}_\alpha^l(x_j) + \mathcal{I}_\alpha^l(x_{j+1})\right],
\end{aligned} \tag{198}$$

where $\mathcal{I}_\alpha^l(x_j) = \int_a^{x_j} (x_j - t)^{1-\alpha} f(t) dt$. Approximate $f(x)$ with the linear spline

$$s_j^l(x) = \sum_{k=0}^{j} f(x_k) s_{j,k}^l(x), \tag{199}$$

where

$$s_{j,k}^l(x) = \begin{cases} \dfrac{x - x_{k-1}}{x_k - x_{k-1}}, & x_{k-1} \le x \le x_k, \\ \dfrac{x_{k+1} - x}{x_{k+1} - x_k}, & x_k \le x \le x_{k+1}, \\ 0, & \text{otherwise,} \end{cases} \tag{200}$$

for $1 \leq k \leq j-1$,

$$s_{j,0}^l(x) = \begin{cases} \dfrac{x_1 - x}{x_1 - x_0}, & x_0 \leq x \leq x_1, \\ 0, & \text{otherwise,} \end{cases} \tag{201}$$

and

$$s_{j,j}^l(x) = \begin{cases} \dfrac{x - x_{j-1}}{x_j - x_{j-1}}, & x_{j-1} \leq x \leq x_j, \\ 0, & \text{otherwise.} \end{cases} \tag{202}$$

Then we obtain an approximation to $\mathcal{I}_a^l(x_j)$ given by

$$\mathcal{I}_a^l(x_j) = \int_a^{x_j}(x_j - t)^{1-\alpha} s_j^l(t) dt = \frac{h^{2-\alpha}}{(2-\alpha)(3-\alpha)} \sum_{k=0}^{j} a_{j,k}^l f(x_k) \tag{203}$$

with

$$a_{j,k}^l = \begin{cases} (3-\alpha)j^{2-\alpha} + (j-1)^{3-\alpha} - j^{3-\alpha}, & k = 0, \\ (j-k+1)^{3-\alpha} - 2(j-k)^{3-\alpha} + (j-k-1)^{3-\alpha}, & 1 \leq k \leq j-1, \\ 1, & k = j. \end{cases} \tag{204}$$

As a result, there holds [96]

$$\left[{}_{RL}D_{a,x}^\alpha f(x)\right]_{x=x_j}$$

$$\approx \frac{h^{-\alpha}}{\Gamma(4-\alpha)} \left[\sum_{k=0}^{j-1} a_{j-1,k}^l f(x_k) - 2\sum_{k=0}^{j} a_{j,k}^l f(x_k) + \sum_{k=0}^{j+1} a_{j+1,k}^l f(x_k) \right] \tag{205}$$

$$= \frac{h^{-\alpha}}{\Gamma(4-\alpha)} \left[\sum_{k=0}^{j-1} \left(a_{j-1,k}^l - 2a_{j,k}^l + a_{j+1,k}^l\right) f(x_k) + \left(a_{j+1,j}^l - 2a_{j,j}^l\right) f(x_j) + a_{j+1,j+1}^l f(x_{j+1}) \right].$$

Similarly, the right-sided Riemann–Liouville derivative can be approximated by [96]

$$\left[{}_{RL}D_{x,b}^\alpha f(x)\right]_{x=x_j} \approx \frac{h^{-\alpha}}{\Gamma(4-\alpha)} \left[\sum_{k=j+1}^{N} \left(a_{j-1,k}^r - 2a_{j,k}^r + a_{j+1,k}^r\right) f(x_k) \right. \tag{206}$$

$$\left. + a_{j-1,j-1}^r f(x_{j-1}) + \left(a_{j-1,j}^r - 2a_{j,j}^r\right) f(x_j) \right],$$

where

$$\begin{cases} a_{j,N}^r = (3-\alpha)(N-j)^{2-\alpha} + (N-j-1)^{3-\alpha} - (N-j)^{3-\alpha}, \\ a_{j,k}^r = (k-j+1)^{3-\alpha} - 2(k-j)^{3-\alpha} + (k-j-1)^{3-\alpha}, \ j+1 \leq k \leq N-1, \\ a_{j,j}^r = 1. \end{cases} \tag{207}$$

The truncated error of this approach has been proved in Ref. [96] to be $\mathcal{O}(h^2)$ provided that $f^{(4)}(x)$ has compact support on $[a,b]$.

In the particular cases with $a = -\infty$ and $b = +\infty$, Equations (205) and (206) can be written as [96–98]

$$\left[{}_{RL}D_{-\infty,x}^\alpha f(x)\right]_{x=x_j} \approx \frac{h^{-\alpha}}{\Gamma(4-\alpha)} \sum_{m=-1}^{\infty} q_m f(x_{j-m}), \tag{208}$$

and

$$\left[{}_{RL}D_{x,\infty}^\alpha f(x)\right]_{x=x_j} \approx \frac{h^{-\alpha}}{\Gamma(4-\alpha)} \sum_{m=-1}^{\infty} q_m f(x_{j+m}), \tag{209}$$

with

$$q_m = \begin{cases} a_{m-1} - 2a_m + a_{m+1}, & m \geq 1, \\ -2a_0 + a_1, & m = 0, \\ a_0, & m = -1. \end{cases} \quad (210)$$

Here

$$a_m = \begin{cases} 1, & m = 0, \\ (m+1)^{3-\alpha} - 2m^{3-\alpha} + (m-1)^{3-\alpha}, & m \geq 1. \end{cases} \quad (211)$$

Both series on the right-hand side of Equations (208) and (209) converge absolutely for $1 < \alpha < 2$ if $f(x)$ is bounded [96]. When $\alpha = 1$, Equations (208) and (209) reduce to the second order finite difference formula for the first order derivative. When $\alpha = 2$, Equations (208) and (209) are consistent with the central difference formula for the second order derivative.

3.2.2. Grünwald-Letnikov Type Approximations

Let $m - 1 < \alpha < m \in \mathbb{Z}^+$ and $f(x)$ be m times continuously differentiable. It is known that when $a = -\infty$ or $f(a) = 0$, the Grünwald–Letnikov derivative

$$_{GL}D^\alpha_{a,x}f(x) = \lim_{h \to 0} \frac{1}{h^\alpha} \sum_{l=0}^{\left[\frac{x-a}{h}\right]} (-1)^l \binom{\alpha}{l} f(x - lh), \quad (212)$$

can approximate the α-th order Riemann–Liouville derivative with first order accuracy [48], i.e.,

$$_{RL}D^\alpha_{a,x}f(x) = \frac{1}{h^\alpha} \sum_{l=0}^{j} \omega_l^{(\alpha)} f(x - lh) + \mathcal{O}(h), \quad jh = x - a, \quad (213)$$

where $\omega_l^{(\alpha)} = (-1)^l \binom{\alpha}{l}$. Equation (213) can be verified through the Fourier transform. The above numerical approximation, which is called the classical Grünwald–Letnikov formula, is warmly applied to solving fractional differential equations. However, this approximation is not suitable for the discretization of fractional differential equations when $\alpha \in (1,2)$ since it leads to unstable numerical schemes [99].

One way to construct stable schemes for fractional differential equations is to make the corresponding coefficient matrix diagonally dominated via replacing $f(x - lh)$ in Equation (213) by $f(x - (l - p)h)$ with $p \in \mathbb{Z}$ being the shift. In this case, we obtain the shifted Grünwald–Letnikov formula

$$_{RL}D^\alpha_{a,x}f(x) = \frac{1}{h^\alpha} \sum_{l=0}^{[j+p]} \omega_l^{(\alpha)} f(x - (l - p)h) + \mathcal{O}(h), \quad jh = x - a. \quad (214)$$

It turns out that the best performances of the shifted Grünwald–Letnikov method come from minimizing $|p - \frac{\alpha}{2}|$. And it coincides with the central difference of the classical second order differentiation. If the shift is chosen to be non-integers, numerical method (214) may have superconvergent behaviors [100–102].

Introducing integer shifts to the classical Grünwald-Letnikov approximation may eliminate the instability indeed, while the truncated error is still $\mathcal{O}(h)$. A modification called the weighted and shifted Grünwald-Letnikov formulae was proposed in Ref. [103], in the spirit of eliminating the first order terms in the truncated errors of the shifted Grünwald–Letnikov formulae. In the following, we introduce the basic idea in details.

Let $f(x) \in L^1(\mathbb{R})$, $_{RL}D^{\alpha+1}_{-\infty,x}f$, and its Fourier transform belong to $L^1(\mathbb{R})$. It can be verified through the Fourier analysis that the shifted Grünwald–Letnikov difference operator

$$A^\alpha_{h,p}f(x) = h^{-\alpha} \sum_{k=0}^{\infty} \omega_k^{(\alpha)} f(x-(k-p)h), \tag{215}$$

with $p \in \mathbb{Z}$ approximates the left-sided Riemann-Liouville derivative $_{RL}D^\alpha_{-\infty,x}f(x)$ with first order accuracy. Assume that the following weighted and shifted Grünwald–Letnikov difference (WSGD) operator

$$B^\alpha_{h,p,q}f(x) = a_1 A^\alpha_{h,p}f(x) + b_1 A^\alpha_{h,q}f(x), \quad a_1, b_1 \in \mathbb{R}, \tag{216}$$

approximates the Riemann–Liouville derivative with second order accuracy. Then the Fourier transform gives

$$\mathcal{F}\left\{B^\alpha_{h,p,q}f(x); \xi\right\} = \frac{1}{h^\alpha} \sum_{k=0}^{\infty} \omega_k^{(\alpha)} \left(a_1 e^{-i(k-p)h\xi} + b_1 e^{-i(k-q)h\xi}\right) \mathcal{F}\{f(x); \xi\}$$

$$= \frac{1}{h^\alpha} \left(a_1 (1-e^{-ih\xi})^\alpha e^{iph\xi} + b_1 (1-e^{-ih\xi})^\alpha e^{iqh\xi}\right) \mathcal{F}\{f(x); \xi\} \tag{217}$$

$$= \left(\frac{1-e^{ih\xi}}{ih\xi}\right)^\alpha \left[a_1 e^{iph\xi} + b_1 e^{iqh\xi}\right] (i\xi)^\alpha \mathcal{F}\{f(x); \xi\}$$

with $i^2 = -1$. Note that

$$\mathcal{F}\{_{RL}D^\alpha_{-\infty,x}f(x); \xi\} = (i\xi)^\alpha \mathcal{F}\{f(x); \xi\}, \tag{218}$$

and

$$\left(\frac{1-e^z}{z}\right)^\alpha e^{zr} = 1 + (r - \frac{\alpha}{2})z + \mathcal{O}(|z|^2). \tag{219}$$

Therefore, a_1 and b_1 need to satisfy

$$\begin{cases} a_1 + b_1 = 1, \\ a_1(p - \frac{\alpha}{2}) + b_1(q - \frac{\alpha}{2}) = 0, \end{cases} \tag{220}$$

to assure that $B^\alpha_{h,p,q}f(x)$ is of second order accuracy. In other words,

$$B^\alpha_{h,p,q}f(x) = {}_{RL}D^\alpha_{-\infty,x}f(x) + \mathcal{O}(h^2) \tag{221}$$

holds uniformly when $a_1 = \frac{\alpha-2q}{2p-2q}$ and $b_1 = \frac{2p-\alpha}{2p-2q}$ [103].

Remark 11. *The relevant academic literature has revealed that numerical approximation (221) results in unstable schemes for fractional partial differential equations when the shift paring $(p,q) = (0,-1)$ [99]. The corresponding schemes are stable when $(p,q) = (1,0)$ and $(p,q) = (1,-1)$. Furthermore, the WSGD operator reduces to the centred difference approximation for the classical second order differentiation when $(p,q) = (1,0)$ and $(p,q) = (1,-1)$ in the case with $\alpha = 2$, while for the classical first order one when $(p,q) = (1,0)$ if $\alpha = 1$.*

A third order WSGD operator was also proposed in Ref. [103]. Nevertheless, it fails to obtain stable numerical schemes when $\alpha \in (1,2)$. To offset this situation, the compact-WSGD operator [104] was introduced through combining WSGD operators with Taylor expansions of the shifted Grünwald formula for sufficiently smooth function $f(x)$ that satisfies homogeneous initial conditions.

The construction of the WSGD operators implies the possibility of deriving higher-order numerical approximations to Riemann-Liouville derivative by imposing various weights and shifts on higher-order Lubich formulae. In Ref. [105], numerical algorithms with second, third, and fourth order accuracy are proposed based on the second order Lubich formula.

3.2.3. Fractional Backward Difference Formulae and Their Modifications

It is evident that the classical Grünwald–Letnikov approximation coincides with the first order Lubich formula (51) with $\alpha > 0$ replaced by $-\alpha$. In fact, Lubich formulae (51) are applicable for evaluating Riemann–Liouville derivative indeed.

Let $f^{(k)}(a+) = 0$ $(k = 0, 1, \ldots, \ell - 1)$. We have the classical Lubich formulae [106]

$$_{RL}D_{a,x}^\alpha f(x) = \frac{1}{h^\alpha} \sum_{l=0}^{[\frac{x-a}{h}]} \omega_{\ell,l}^{(\alpha)} f(x - lh) + \mathcal{O}(h^\ell), \quad \alpha > 0, \tag{222}$$

in which h is the stepsize. The convolution coefficients $\omega_{\ell,l}^{(\alpha)}$ are generated by $W_\ell^{(\alpha)}(z)$ defined in Equation (50). This can be readily verified by the Fourier transform.

Remark 12. *In Ref. [106], the coefficients of high-order approximations (till 10-th order) for Riemann–Liouville derivative were computed. Furthermore, a conjecture on coefficients of the third, fourth, and fifth order schemes was proposed by Li and Ding and was rephrased on Page 80 of Ref. [38], stated as follows,*

(1) *If $0 < \alpha < 1$, then $\omega_{3,l}^{(\alpha)} \leq \omega_{3,l+1}^{(\alpha)}$ for $l \geq 4$, $\omega_{4,l}^{(\alpha)} \leq \omega_{4,l+1}^{(\alpha)}$ for $l \geq 7$, and $\omega_{5,l}^{(\alpha)} \leq \omega_{5,l+1}^{(\alpha)}$ for $l \geq 12$;*
(2) *If $1 < \alpha < 2$, then $\omega_{3,l}^{(\alpha)} \geq \omega_{3,l+1}^{(\alpha)}$ for $l \geq 7$, $\omega_{4,l}^{(\alpha)} \geq \omega_{4,l+1}^{(\alpha)}$ for $l \geq 12$, and $\omega_{5,l}^{(\alpha)} \geq \omega_{5,l+1}^{(\alpha)}$ for $l \geq 16$.*

Recently, the above conjecture for $\omega_{3,l}^{(\alpha)}$ with $0 < \alpha < 1$ has been proved in Ref. [107].

Similarly to the case of the classical Grünwald-Letnikov approximation, the classical Lubich formulae may produce unstable numerical schemes for fractional differential equations due to the eigenvalue issue [105]. In this case, we often introduce shifts. To maintain the high-order accuracy, the corresponding generating functions need modifying. The shifted fractional backward difference formulae [108], which can be proved via the Fourier transform method, are presented as follows.

Theorem 7. *Suppose that $f(x) \in C^{[\alpha]+3}(\mathbb{R})$, and all the derivatives of $f(x)$ up to order $[\alpha] + 4$ belong to $L^1(\mathbb{R})$. Then we have*

$$_{RL}D_{-\infty,x}^\alpha f(x) = \frac{1}{h^\alpha} \sum_{l=0}^\infty k_{2,l}^{(\alpha)} f(x - (l-1)h) + \mathcal{O}(h^2), \tag{223}$$

and

$$_{RL}D_{x,+\infty}^\alpha f(x) = \frac{1}{h^\alpha} \sum_{l=0}^\infty k_{2,l}^{(\alpha)} f(x + (l-1)h) + \mathcal{O}(h^2), \tag{224}$$

as $h \to 0$. Here $k_{2,l}^{(\alpha)}$ $(l = 0, 1, \ldots)$ are generated by

$$\widetilde{W}_2(z) = \left(\frac{3\alpha - 2}{2\alpha} - \frac{2(\alpha-1)}{\alpha} z + \frac{\alpha - 2}{2\alpha} z^2\right)^\alpha = \sum_{l=0}^\infty k_{2,l}^{(\alpha)} z^l, \quad |z| < 1. \tag{225}$$

Theorem 8. *Let $p \geq 3$, $f(x) \in C^{[\alpha]+p+1}(\mathbb{R})$, and all the derivatives of $f(x)$ up to order $[\alpha] + p + 2$ belong to $L^1(\mathbb{R})$. Then there hold*

$$_{RL}D_{-\infty,x}^\alpha f(x) = \frac{1}{h^\alpha} \sum_{l=0}^\infty k_{p,l}^{(\alpha)} f(x - (l-1)h) + \mathcal{O}(h^p), \quad p \geq 3, \tag{226}$$

and

$$_{RL}D_{x,+\infty}^\alpha f(x) = \frac{1}{h^\alpha} \sum_{l=0}^\infty k_{p,l}^{(\alpha)} f(x + (l-1)h) + \mathcal{O}(h^p), \quad p \geq 3. \tag{227}$$

Here the generating functions of coefficients $k_{p,l}^{(\alpha)}$ ($l = 0, 1, \ldots$) with $p \geq 3$ are

$$\widetilde{W}_p(z) = \left((1-z) + \frac{\alpha-2}{2\alpha}(1-z)^2 + \sum_{k=3}^{p} \frac{\lambda_{k-1,k-1}^{(\alpha)}}{\alpha}(1-z)^k \right)^\alpha, \tag{228}$$

in which the parameters $\lambda_{k-1,k-1}^{(\alpha)}$ ($k = 3, 4, \ldots$) can be determined by the relation

$$\widetilde{W}_k(e^{-z})\frac{e^z}{z^\alpha} = 1 - \sum_{l=k}^{\infty} \lambda_{k,l}^{(\alpha)} z^l, \quad k = 2, 3, \ldots \tag{229}$$

Introducing suitable weights and shifts to the classical Lubich operators, we can obtain weighted and shifted Lubich formulae [105], which are not only of high-order accuracy but also stable when $\alpha \in (1,2)$.

Define the operator

$$\mathcal{A}_p^\alpha = \frac{1}{h^\alpha} \sum_{k=0}^{\infty} \omega_{2,k}^{(\alpha)} f(x - (k-p)h), \tag{230}$$

where the shift p is an integer. The coefficients $\omega_{2,k}^{(\alpha)}$ can be calculated by Equation (50) with $\ell = 2$. The weighted and shifted Lubich formulae, whose convergence and accuracy can be verified through the Fourier transform, are given as follows.

Theorem 9. Let $f(x)$, $_{RL}D_{-\infty,x}^{\alpha+1}f(x)$ (or $_{RL}D_{-\infty,x}^{\alpha+2}f(x)$) with $1 < \alpha < 2$ and their Fourier transforms belong to $L^1(\mathbb{R})$. Then we have

$$\begin{cases} _{RL}D_{-\infty,x}^\alpha f(x) = \mathcal{A}_p^\alpha f(x) + \mathcal{O}(h), \ p \neq 0, \\ _{RL}D_{-\infty,x}^\alpha f(x) = \mathcal{A}_p^\alpha f(x) + \mathcal{O}(h^2), \ p = 0, \end{cases} \tag{231}$$

where $\mathcal{A}_p^\alpha$ is given by Equation (230)

Theorem 10. When $f(x)$, $_{RL}D_{-\infty,x}^{\alpha+2}f(x)$ with $1 < \alpha < 2$, and their Fourier transforms belong to $L^1(\mathbb{R})$, there holds

$$_{RL}D_{-\infty,x}^\alpha f(x) = \mathcal{A}_{p,q}^\alpha f(x) + \mathcal{O}(h^2). \tag{232}$$

Here

$$\mathcal{A}_{p,q}^\alpha f(x) = W_p \, \mathcal{A}_p^\alpha f(x) + W_q \, \mathcal{A}_q^\alpha f(x) \tag{233}$$

with $\mathcal{A}_p^\alpha$, $\mathcal{A}_q^\alpha$ being defined in Equation (230), $W_p = \frac{q}{q-p}$, $W_q = \frac{p}{p-q}$, $p \neq q$, and p, q being integers.

It was proved in Ref. [105] that the approximation (232) with $1 < \alpha < 2$ works well for space fractional differential equations when the pair $(p, q) = (1, q)$ with $|q| \geq 2$.

Theorem 11. Assume that $f(x)$, $_{RL}D_{-\infty,x}^{\alpha+3}f(x)$ with $1 < \alpha < 2$, and their Fourier transforms belong to $L^1(\mathbb{R})$. Then there holds

$$_{RL}D_{-\infty,x}^\alpha f(x) = \mathcal{A}_{p,q,r,s}^\alpha f(x) + \mathcal{O}(h^3), \tag{234}$$

$$\mathcal{A}_{p,q,r,s}^\alpha f(x) = W_{p,q} \, \mathcal{A}_{p,q}^\alpha f(x) + W_{r,s} \, \mathcal{A}_{r,s}^\alpha f(x), \tag{235}$$

where $\mathcal{A}_{p,q}^\alpha$ and $\mathcal{A}_{r,s}^\alpha$ are defined in Equation (233), $W_{p,q} = \frac{3rs+2\alpha}{3(rs-pq)}$, $W_{r,s} = \frac{3pq+2\alpha}{3(pq-rs)}$, $rs \neq pq$, and p, q, r, s are integers.

When $(p, q, r, s) = (1, q, 1, s)$, $|q| \geq 2, |s| \geq 2$, and $qs < 0$, the approximation (234) with $1 < \alpha < 2$ works well for space fractional differential equations.

For higher-order weighted and shifted Lubich formulae such as the fourth order one, see Ref. [105] for more details.

Remark 13. *All the above weighted and shifted Lubich formulae are applicable to the bounded domain (a,b) through performing zero extension, whenever the zero extended function satisfies the corresponding assumptions of the approximations.*

An alternative approach modifying the Lubich formulae is to introduce compact operators, which gives the following fractional-compact formulae [109]. The corresponding accuracy can be proved by the Fourier transform.

Theorem 12. *Define the following two difference operators,*

$$^L\mathcal{B}_2^\alpha f(x+sh) = \frac{1}{h^\alpha} \sum_{l=0}^\infty k_{2,l}^{(\alpha)} f(x-(l-s-1)h), \tag{236}$$

and

$$^R\mathcal{B}_2^\alpha f(x+sh) = \frac{1}{h^\alpha} \sum_{l=0}^\infty k_{2,l}^{(\alpha)} f(x+(l-s-1)h), \tag{237}$$

where the coefficients $k_{2,l}^{(\alpha)}$ are given by the function

$$\widehat{W}_2(z) = \left(\frac{3\alpha-1}{2\alpha} - \frac{2(\alpha-1)}{\alpha}z + \frac{\alpha-2}{2\alpha}z^2\right)^\alpha = \sum_{l=0}^\infty k_{2,l}^{(\alpha)} z^l, \quad |z| < 1. \tag{238}$$

If we introduce the fractional-compact difference operator

$$\mathcal{L} f(x) = \left(1 - \frac{2\alpha^2 - 6\alpha + 3}{6\alpha}\delta_x^2\right) f(x), \tag{239}$$

with δ_x^2 being a second order central difference operator defined by $\delta_x^2 f(x) = f(x+h) - 2f(x) + f(x+h)$, then equalities

$$^L\mathcal{B}_2^\alpha f(x) = \mathcal{L}\,_{RL}D_{-\infty,x}^\alpha f(x) + \mathcal{O}(h^3) \tag{240}$$

and

$$^R\mathcal{B}_2^\alpha f(x) = \mathcal{L}\,_{RL}D_{x,\infty}^\alpha f(x) + \mathcal{O}(h^3) \tag{241}$$

hold uniformly for $x \in \mathbb{R}$, provided that $f(x) \in C^{[\alpha]+4}(\mathbb{R})$ and all derivatives of $f(x)$ up to order $[\alpha]+5$ belong to $L^1(\mathbb{R})$.

Theorem 13. *Choose the generating function as*

$$\widetilde{W}_2(z) = \left(\frac{3\alpha+2}{2\alpha} - \frac{2(\alpha+1)}{\alpha}z + \frac{\alpha+2}{2\alpha}z^2\right)^\alpha = \sum_{l=0}^\infty \widetilde{k}_{2,l}^{(\alpha)} z^l, \quad |z| < 1, \tag{242}$$

and define the following difference operators

$$^L\widetilde{\mathcal{B}}_2^\alpha f(x+sh) = \frac{1}{h^\alpha} \sum_{l=0}^\infty \widetilde{k}_{2,l}^{(\alpha)} f(x-(l-s+1)h), \tag{243}$$

$$^R\widetilde{\mathcal{B}}_2^\alpha f(x+sh) = \frac{1}{h^\alpha} \sum_{l=0}^\infty \widetilde{k}_{2,l}^{(\alpha)} f(x+(l-s+1)h). \tag{244}$$

Then the equalities
$$^{L}\widetilde{\mathcal{B}}_2^\alpha f(x) = \widetilde{\mathcal{L}}\,_{RL}D^\alpha_{-\infty,x}f(x) + \mathcal{O}(h^3), \tag{245}$$

and
$$^{R}\widetilde{\mathcal{B}}_2^\alpha f(x) = \widetilde{\mathcal{L}}\,_{RL}D^\alpha_{x,\infty}f(x) + \mathcal{O}(h^3) \tag{246}$$

hold uniformly for $x \in \mathbb{R}$, provided that $f(x) \in C^{[\alpha]+4}(\mathbb{R})$ and all derivatives of $f(x)$ up to order $[\alpha]+5$ belong to $L^1(\mathbb{R})$. Here the fractional-compact difference operator $\widetilde{\mathcal{L}}$ is given by

$$\widetilde{\mathcal{L}} f(x) = \left(1 - \frac{2\alpha^2 + 6\alpha + 3}{6\alpha}\delta_x^2\right) f(x). \tag{247}$$

Theorem 14. *Let $f(x) \in C^{[\alpha]+5}(\mathbb{R})$ and all derivatives of $f(x)$ up to order $[\alpha]+6$ belong to $L^1(\mathbb{R})$. Define the fractional-compact operator as*

$$\mathcal{H}f(x) = \left[\left(\widetilde{\sigma}_{3,0}^{(\alpha)} - \sigma_{3,0}^{(\alpha)}\right) + \left(\sigma_{2,0}^{(\alpha)}\widetilde{\sigma}_{3,0}^{(\alpha)} - \widetilde{\sigma}_{2,0}^{(\alpha)}\sigma_{3,0}^{(\alpha)}\right)\delta_x^2\right] f(x), \tag{248}$$

where
$$\begin{cases} \sigma_{2,0}^{(\alpha)} = -\dfrac{2\alpha^2 - 6\alpha + 3}{6\alpha}, & \widetilde{\sigma}_{2,0}^{(\alpha)} = -\dfrac{2\alpha^2 + 6\alpha + 3}{6\alpha}, \\[6pt] \sigma_{3,0}^{(\alpha)} = \dfrac{3\alpha^3 - 11\alpha^2 + 12\alpha - 4}{12\alpha^2}, & \widetilde{\sigma}_{3,0}^{(\alpha)} = \dfrac{3\alpha^3 + 11\alpha^2 + 12\alpha + 4}{12\alpha^2}. \end{cases} \tag{249}$$

Then
$$\widetilde{\sigma}_{3,0}^{(\alpha)}\,^{L}\mathcal{B}_2^\alpha f(x) - \sigma_{3,0}^{(\alpha)}\,^{L}\widetilde{\mathcal{B}}_2^\alpha f(x) = \mathcal{H}\,_{RL}D^\alpha_{-\infty,x}f(x) + \mathcal{O}(h^4), \tag{250}$$

and
$$\widetilde{\sigma}_{3,0}^{(\alpha)}\,^{R}\mathcal{B}_2^\alpha f(x) - \sigma_{3,0}^{(\alpha)}\,^{R}\widetilde{\mathcal{B}}_2^\alpha f(x) = \mathcal{H}\,_{RL}D^\alpha_{x,\infty}f(x) + \mathcal{O}(h^4) \tag{251}$$

hold uniformly on $\mathbb{R}$.

The idea of the above approximations can be applied to evaluating tempered fractional derivatives, see Ref. [110] for more details.

3.2.4. Fractional Average Central Difference Method

In Ref. [111], a shifted operator of the form

$$_C\Delta^\alpha_{-h}f(x) = \sum_{k=0}^{\infty}(-1)^k \binom{\alpha}{k} f\left(x - \left(k - \frac{\alpha}{2}\right)h\right) \tag{252}$$

with $h > 0$ was proposed to approximate Riemann–Liouville derivative. This difference operator reduces to the standard central difference operator when α is a positive integer. It can be verified through the Fourier transform that the equality

$$\frac{1}{h^\alpha}\,_C\Delta^\alpha_{-h}f(x) = \,_{RL}D^\alpha_{-\infty,x}f(x) + \mathcal{O}(h^2), \tag{253}$$

holds uniformly for $x \in \mathbb{R}$ as $h \to 0$, provided that $f \in C^{[\alpha]+3}(\mathbb{R})$ and all derivative of f up to the order $[\alpha]+4$ exist and belong to $L^1(\mathbb{R})$.

Recently, the above shifted operator has been modified to obtain higher-order approximations, which are based on the fractional left and right average central difference operators [112]

$$_{AC}\Delta^\alpha_{-h}f(x) = \frac{1}{2}\sum_{k=0}^{\infty}(-1)^k\binom{\alpha}{k}\left[f(x - kh) + f(x - (k-\alpha)h)\right], \tag{254}$$

and
$$_{AC}\Delta^\alpha_{+h}f(x) = \frac{1}{2}\sum_{k=0}^\infty (-1)^k \binom{\alpha}{k}[f(x+kh)+f(x+(k-\alpha)h)]. \tag{255}$$

The main results, which can be verified through Fourier transform, are presented as follows.

Theorem 15. *Assume that $f(x)$, the Fourier transform of $_{RL}D^{\alpha+2}_{-\infty,x}f(x)$ and $_{RL}D^{\alpha+2}_{x,+\infty}f(x)$ are in $L^1(\mathbb{R})$. Then the equalities*

$$_{RL}D^\alpha_{-\infty,x}f(x) = \frac{_{AC}\Delta^\alpha_{-h}f(x)}{h^\alpha} + \mathcal{O}(h^2), \tag{256}$$

$$_{RL}D^\alpha_{x,+\infty}f(x) = \frac{_{AC}\Delta^\alpha_{+h}f(x)}{h^\alpha} + \mathcal{O}(h^2) \tag{257}$$

hold uniformly on $\mathbb{R}$.

Theorem 16. *When $f(x)$ and the Fourier transforms of $_{RL}D^{\alpha+4}_{-\infty,x}f(x)$ and $_{RL}D^{\alpha+4}_{x,+\infty}f(x)$ are in $L^1(\mathbb{R})$, then the relations*

$$\left[1 + \frac{\alpha(3\alpha+1)}{24}\delta_x^2\right]{_{RL}D^\alpha_{-\infty,x}f(x)} = \frac{1}{h^\alpha}{_{AC}\Delta^\alpha_{-h}f(x)} + \mathcal{O}(h^4), \tag{258}$$

and

$$\left[1 + \frac{\alpha(3\alpha+1)}{24}\delta_x^2\right]{_{RL}D^\alpha_{x,+\infty}f(x)} = \frac{1}{h^\alpha}{_{AC}\Delta^\alpha_{+h}f(x)} + \mathcal{O}(h^4) \tag{259}$$

hold uniformly on $\mathbb{R}$. Here δ_x^2 denotes the second order central difference operator defined by $\delta_x^2 f(x_j) = f(x_{j+1}) - 2f(x_j) + f(x_{j-1})$.

For functions defined on $[a,b]$, the fractional average central difference formulae can be modified through suitable extensions.

3.3. Numerical Riesz Differentation

Since Riesz derivative can be viewed as a linear combination of the left- and right-sided Riemann–Liouville derivatives. Several numerical approximations to Riesz derivative can be readily obtained based on the aforementioned methods of Riemann–Liouville derivative. Here we only present the one based on L2-1_σ formulae when introducing indirect evaluations of Riesz derivative. For more details of these indirect approaches, one can refer to Refs. [108,109,112,113]. Then we focus on introducing some schemes evaluating Riesz derivative in direct ways.

3.3.1. Approximation Based on L2-1_σ Formulae

In Ref. [107], the L2-1_σ formulae are reformulated in the following forms,

$$[_{c}D^\alpha_{a,x}f(x)]_{x=x_{j+\sigma}} = \frac{h^{-\alpha}}{\Gamma(2-\alpha)}\sum_{k=0}^j d_{j-k}^{(\alpha,\sigma)}[f(x_{k+1})-f(x_k)] + \mathcal{O}(h^{3-\alpha}), \tag{260}$$

and

$$[_{c}D^\alpha_{x,b}f(x)]_{x=x_{j+\sigma'}} = -\frac{h^{-\alpha}}{\Gamma(2-\alpha)}\sum_{k=j}^{N-1} \tilde{d}_{k-j}^{(\alpha,\sigma')}[f(x_{k+1})-f(x_k)] + \mathcal{O}(h^{3-\alpha}), \tag{261}$$

with $0 \leq j \leq N-1$, $\sigma = 1 - \frac{\alpha}{2}$, and $\sigma' = \frac{\alpha}{2}$. When $j = 0$, $d_0^{(\alpha,\sigma)} = c_0^{(\alpha,\sigma)}$. When $j = N-1$, $\tilde{d}_0^{(\alpha,\sigma)} = c_0^{(\alpha,\sigma')}$. For $j \geq 1$, the coefficients are given by

$$d_k^{(\alpha,\sigma)} = \begin{cases} c_0^{(\alpha,\sigma)} + \tilde{c}_1^{(\alpha,\sigma)}, & k = 0, \\ c_k^{(\alpha,\sigma)} + \tilde{c}_{k+1}^{(\alpha,\sigma)} - \tilde{c}_k^{(\alpha,\sigma)}, & 1 \leq k \leq j-1, \\ c_j^{(\alpha,\sigma)} - \tilde{c}_j^{(\alpha,\sigma)}, & k = j, \end{cases} \quad (262)$$

in which the second case of the right-hand side of Equation (262) should be removed if $j = 1$, and

$$\tilde{d}_k^{(\alpha,\sigma')} = \begin{cases} c_0^{(\alpha,\sigma')} + \tilde{c}_2^{(\alpha,\sigma')}, & k = 0, \\ c_{k+1}^{(\alpha,\sigma')} - \tilde{c}_{k+1}^{(\alpha,\sigma')} + \tilde{c}_{k+2}^{(\alpha,\sigma')}, & 1 \leq k \leq N-2-j, \\ c_{k+1}^{(\alpha,\sigma')} - \tilde{c}_{k+1}^{(\alpha,\sigma')}, & k = N-1-j, \end{cases} \quad (263)$$

in which the second case of the right-hand side of Equation (263) should be removed if $j = N - 2$. Here

$$\begin{cases} c_0^{(\alpha,\sigma)} = \sigma^{1-\alpha}, \\ c_k^{(\alpha,\sigma)} = (k+\sigma)^{1-\alpha} - (k-1+\sigma)^{1-\alpha}, \ k \geq 1, \\ \tilde{c}_k^{(\alpha,\sigma)} = \frac{1}{2-\alpha}\left[(k+\sigma)^{2-\alpha} - (k-1+\sigma)^{2-\alpha}\right] \\ \qquad - \frac{1}{2}\left[(k+\sigma)^{1-\alpha} + (k-1+\sigma)^{1-\alpha}\right], \ k \geq 1, \end{cases} \quad (264)$$

and

$$\begin{cases} c_0^{(\alpha,\sigma')} = (1-\sigma')^{1-\alpha}, \\ c_k^{(\alpha,\sigma')} = (k-\sigma')^{1-\alpha} - (k-1-\sigma')^{1-\alpha}, \ k \geq 1, \\ \tilde{c}_k^{(\alpha,\sigma')} = \frac{1}{2-\alpha}\left[(k-\sigma')^{2-\alpha} - (k-1-\sigma')^{2-\alpha}\right] \\ \qquad - \frac{1}{2}\left[(k-\sigma')^{1-\alpha} + (k-1-\sigma')^{1-\alpha}\right], \ k \geq 1. \end{cases} \quad (265)$$

In order to combine Equations (260) and (261), $\sigma + \sigma' = 1$ should be satisfied such that $2\sigma - 2 + \alpha + 2\sigma' - \alpha = 0$. Furthermore, assume that $\sigma = \sigma'$, i.e., $\sigma = \sigma' = \frac{1}{2}$, then $x_{j+\sigma} = x_{j+\frac{1}{2}} = x_{j+\sigma'}$. One can get the following $(3-\alpha)$-th order scheme for Riesz derivatives at $x = x_{j+\frac{1}{2}}$ with $0 \leq j \leq N-1$ [107],

$$[_{RZ}D_x^\alpha f(x)]_{x=x_{j+\frac{1}{2}}} = -\frac{1}{2\cos(\frac{\pi\alpha}{2})}\left[\frac{(x_{j+\frac{1}{2}} - a)^{-\alpha}f(a)}{\Gamma(1-\alpha)} + \frac{(b - x_{j+\frac{1}{2}})^{-\alpha}f(b)}{\Gamma(1-\alpha)}\right.$$

$$\left. + \frac{h^{-\alpha}}{\Gamma(2-\alpha)}\left(\sum_{k=0}^j d_{j-k}^{(\alpha,\frac{1}{2})}[f(x_{k+1}) - f(x_k)]\right.\right. \quad (266)$$

$$\left.\left. - \sum_{k=j}^{N-1} \tilde{d}_{k-j}^{(\alpha,\frac{1}{2})}[f(x_{k+1}) - f(x_k)]\right)\right] + \mathcal{O}(h^{3-\alpha}),$$

in which $d_{j-k}^{(\alpha,\frac{1}{2})}$ and $\tilde{d}_{k-j}^{(\alpha,\frac{1}{2})}$ are defined by Equations (262) and (263) with $\sigma = \sigma' = \frac{1}{2}$, respectively.

3.3.2. Asymmetric Centred Difference Operators

Slightly different from the fractional average central difference operators in the previous section, the symmetric fractional centred difference operator is defined as

$$\Delta_h^\alpha f(x) = \sum_{k=-\infty}^{+\infty} g_k^{(\alpha)} f(x - kh), \qquad (267)$$

with the coefficients being given by

$$g_k^{(\alpha)} = \frac{(-1)^k \Gamma(\alpha+1)}{\Gamma(\frac{\alpha}{2} - k + 1)\Gamma(\frac{\alpha}{2} + k + 1)}, \quad k = 0, \pm 1, \pm 2, \ldots \qquad (268)$$

It can be verified through the Fourier analysis that the relation [114]

$$_{RZ}D_x^\alpha f(x) = -\frac{1}{h^\alpha} \Delta_h^\alpha f(x) + \mathcal{O}(h^2), \quad 1 < \alpha \le 2, \qquad (269)$$

holds uniformly for $x \in \mathbb{R}$ as $h \to 0^+$, provided that $f \in C^5(\mathbb{R})$ and all of its derivatives up to order five belong to $L^1(\mathbb{R})$. It has been pointed out by Ref. [115] that Equation (269) also holds for $0 < \alpha \le 1$.

3.3.3. Weighted and Shifted Centred Difference Operators

Introducing shifts to the symmetric fractional centred difference operator in Equation (267), the shifted centred difference operators [112]

$$\mathcal{L}_\theta f(x) = \sum_{k=-\infty}^{\infty} g_k^{(\alpha)} f(x - (k+\theta)h), \quad |\theta| = 0, 1, 2, \ldots \qquad (270)$$

can be obtained. To achieve high-order accuracy, the following high-order approximations to Riesz derivative can be derived through combining these shifted operators with suitable weights.

Theorem 17 ([116]). *If $f(x)$ lies in $C^7(\mathbb{R})$ with all the derivatives up to order 7 in $L^1(\mathbb{R})$, then the relation*

$$_{RZ}D_x^\alpha f(x) = \frac{1}{h^\alpha} \left[\frac{\alpha}{24} \mathcal{L}_{-1} f(x) - \left(1 + \frac{\alpha}{12}\right) \mathcal{L}_0 f(x) + \frac{\alpha}{24} \mathcal{L}_1 f(x) \right] + \mathcal{O}(h^4) \qquad (271)$$

holds uniformly for $x \in \mathbb{R}$.

Theorem 18 ([112]). *Assume that $f(x) \in C^9(\mathbb{R})$ with all the derivatives up to order 9 in $L^1(\mathbb{R})$. Then*

$$_{RZ}D_x^\alpha f(x) = \frac{1}{h^\alpha} \left[A_1 \mathcal{L}_{-2} f(x) + A_2 \mathcal{L}_{-1} f(x) + A_3 \mathcal{L}_0 f(x) + A_2 \mathcal{L}_1 f(x) + A_1 \mathcal{L}_2 f(x) \right] + \mathcal{O}(h^6) \qquad (272)$$

holds uniformly for $x \in \mathbb{R}$, in which

$$\begin{cases} A_1 = -\left(\dfrac{\alpha}{1152} + \dfrac{11}{2880}\right)\alpha, \\[4pt] A_2 = \left(\dfrac{\alpha}{288} + \dfrac{41}{720}\right)\alpha, \\[4pt] A_3 = -\left(\dfrac{\alpha^2}{192} + \dfrac{17\alpha}{160} + 1\right). \end{cases} \qquad (273)$$

Theorem 19 ([112]). *Assume that $f(x)$ lies in $C^{11}(\mathbb{R})$ with all the derivatives up to order 11 in $L^1(\mathbb{R})$. Then*

$$_{RZ}D_x^\alpha f(x) = \frac{1}{h^\alpha}[\mathcal{B}_1 \mathcal{L}_{-3}f(x) + \mathcal{B}_2 \mathcal{L}_{-2}f(x) + \mathcal{B}_3 \mathcal{L}_{-1}f(x) + \mathcal{B}_4 \mathcal{L}_0 f(x) \qquad (274)$$
$$+ \mathcal{B}_3 \mathcal{L}_1 f(x) + \mathcal{B}_2 \mathcal{L}_2 f(x) + \mathcal{B}_1 \mathcal{L}_3 f(x)] + \mathcal{O}(h^8)$$

holds uniformly for $x \in \mathbb{R}$. Here

$$\begin{cases} \mathcal{B}_1 = \left(\dfrac{\alpha^2}{82944} + \dfrac{11\alpha}{69120} + \dfrac{191}{362880}\right)\alpha, \; \mathcal{B}_2 = -\left(\dfrac{\alpha^2}{13824} + \dfrac{7\alpha}{3840} + \dfrac{211}{30240}\right)\alpha, \\ \mathcal{B}_3 = \left(\dfrac{5\alpha^2}{27648} + \dfrac{3\alpha}{512} + \dfrac{7843}{120960}\right)\alpha, \; \mathcal{B}_4 = -\left(\dfrac{5\alpha^3}{20736} + \dfrac{29\alpha^2}{3456} + \dfrac{5297\alpha}{45360} + 1\right). \end{cases} \qquad (275)$$

For much higher-order difference operators in this respect, see Ref. [112].

3.3.4. Compact Centred Difference Operators

As another variant of the centred difference operator, compact centred difference operators are based on the idea of introducing compact operators to maintain even-order accuracy.

Theorem 20 ([117]). *Suppose that $f(x) \in C^{2n+3}(\mathbb{R})$, and all the derivatives of $f(x)$ up to order $2n+4$ exist and belong to $L^1(\mathbb{R})$. Then*

$$\left(\delta_x^0 - b_{n-1}\delta_x^{2n-2}\right){}_{RZ}D_x^\alpha f(x) = \left(\sum_{l=0}^{n-2} b_l \delta_x^{2l}\right)\left(-\frac{\Delta_h^\alpha f(x)}{h^\alpha}\right) + \mathcal{O}(h^{2n}), \; n \in \mathbb{Z}^+, \qquad (276)$$

where

$$\delta_x^{2l} f(x_j) = \sum_{s=0}^{2l} (-1)^s \binom{2l}{s} f(x_{l+j-s}), \; l \geq 0. \qquad (277)$$

Specifically, δ_x^0 is the identity operator, i.e., $\delta_x^0 f(x_j) = f(x_j)$. The coefficients b_l ($l = 0, 1, \ldots, n-2$) satisfy the following equation

$$\sum_{l=0}^{n-2} b_l \left(2 \sum_{s=0}^{l-1} \sum_{q=0}^{n-1} \sum_{p=0}^{n-1-q} \frac{(-1)^{s+q}(l-s)^{2q}\binom{2l}{s} a_p}{(2q)!} |\omega h|^{2(p+q)} + (-1)^l \binom{2l}{l} \sum_{p=0}^{n-1} a_p |\omega h|^{2p}\right) \qquad (278)$$

$$= 1 - b_{n-1}\left(\sum_{s=0}^{n-2}(-1)^s\binom{2n-2}{s}\frac{2(n-1-s)^{2n-2}}{(2n-2)!}\right)(-1)^{n-1}|\omega h|^{2n-2},$$

and a_p ($p = 0, 1, \ldots$) satisfy the equation

$$\sum_{p=0}^\infty a_p |\omega h|^{2p} = \left|\frac{2\sin\left(\frac{\omega h}{2}\right)}{\omega h}\right|^\alpha = \left[1 - \frac{\alpha}{24}|\omega h|^2 + \left(\frac{1}{1920} + \frac{\alpha-1}{1152}\right)\alpha|\omega h|^4 \right. \qquad (279)$$
$$\left. -\left(\frac{1}{322560} + \frac{\alpha-1}{46080} + \frac{(\alpha-1)(\alpha-2)}{82944}\right)\alpha|\omega h|^6 + \cdots\right].$$

Remark 14. *In view of proofs in Refs. [111,118,119], conditions in Theorem 20 can be weakened as $f(x) \in \mathcal{L}^{2n+\alpha}(\mathbb{R})$, where*

$$\mathcal{L}^{2n+\alpha}(\mathbb{R}) = \left\{f \,\Big|\, f \in L^1(\mathbb{R}), \text{ and } \int_\mathbb{R}(1+|\omega|)^{2n+\alpha}|\widehat{f}(\omega)|d\omega < +\infty\right\}. \qquad (280)$$

Remark 15. *One should bear in mind that some suitable smooth conditions for $f(x)$ are necessary and cannot be dropped. Once these conditions are violated, the expected accuracy cannot be achieved. Example 13 will verify this assertion latter.*

For function $f(x)$ defined on the bounded interval $[a,b]$ with $f(a) = f(b) = 0$, we can extend $f(x)$ by zero outside of the domain. When the conditions in Theorem 20 or Remark 14 are satisfied, the compact centred difference formula can be written in the following form,

$$(\delta_x^0 - b_{n-1}\delta_x^{2n-2})\frac{\partial^\alpha f(x)}{\partial |x|^\alpha} = \left(-\frac{1}{h^\alpha}\sum_{l=0}^{n-2}b_l\delta_x^{2l}\sum_{k=[-\frac{b-x}{h}]+l}^{[\frac{x-a}{h}]-l}g_k^{(\alpha)}f(x-kh)\right) + \mathcal{O}(h^{2n}). \tag{281}$$

The following numerical examples demonstrate the accuracy of the fractional-compact centred formula (281) and the assertion in Remark 15.

Example 12 ([117]). *Consider the function $f_n(x) = x^{2n}(1-x)^{2n}$, $x \in [0,1]$, $n = 2,3,4,5$. Utilize numerical scheme (281) with $n = 2,3,4,5$ to compute the Riesz derivative of $f(x)$ at $x = 0.5$. The absolute error (AE) and experimental convergence order (CO) displayed in Table 12 are in line with the theoretical analysis.*

Table 12. The absolute error and the experimental convergence order of function $f_2(x)$ in Example 12 by numerical scheme (281) with $n = 2, 3, 4, 5$.

α	h	AE	CO	α	h	AE	CO
				$n = 2$			
1.1	$\frac{1}{20}$	1.985528×10^{-6}	-	1.7	$\frac{1}{20}$	9.316621×10^{-6}	-
	$\frac{1}{80}$	7.806460×10^{-9}	3.9981		$\frac{1}{80}$	3.661963×10^{-8}	3.9982
	$\frac{1}{320}$	3.050456×10^{-11}	4.0000		$\frac{1}{320}$	1.431370×10^{-10}	3.9995
1.3	$\frac{1}{20}$	3.418165×10^{-6}	-	1.9	$\frac{1}{20}$	1.486627×10^{-5}	-
	$\frac{1}{80}$	1.343913×10^{-8}	3.9981		$\frac{1}{80}$	5.841643×10^{-8}	3.9983
	$\frac{1}{320}$	5.251979×10^{-11}	3.9998		$\frac{1}{320}$	2.284402×10^{-10}	3.9988
				$n = 3$			
1.1	$\frac{1}{20}$	3.120201×10^{-8}	-	1.7	$\frac{1}{20}$	2.053203×10^{-7}	-
	$\frac{1}{28}$	4.223802×10^{-9}	5.9537		$\frac{1}{28}$	2.780411×10^{-8}	5.9527
	$\frac{1}{36}$	9.422903×10^{-10}	5.9735		$\frac{1}{36}$	6.204232×10^{-9}	5.9727
1.3	$\frac{1}{20}$	6.008620×10^{-8}	-	1.9	$\frac{1}{20}$	3.675466×10^{-7}	-
	$\frac{1}{28}$	8.135715×10^{-9}	5.9531		$\frac{1}{28}$	4.976658×10^{-8}	5.9530
	$\frac{1}{33}$	1.815230×10^{-9}	5.9730		$\frac{1}{36}$	1.110461×10^{-8}	5.9727
				$n = 4$			
1.1	$\frac{1}{30}$	3.442344×10^{-11}	-	1.7	$\frac{1}{30}$	2.889640×10^{-10}	-
	$\frac{1}{38}$	5.303531×10^{-12}	7.9206		$\frac{1}{38}$	4.449502×10^{-11}	7.9249
	$\frac{1}{46}$	1.164521×10^{-12}	7.9393		$\frac{1}{46}$	9.752953×10^{-12}	7.9502
1.3	$\frac{1}{30}$	7.195825×10^{-11}	-	1.9	$\frac{1}{30}$	5.601931×10^{-10}	-
	$\frac{1}{38}$	1.108608×10^{-11}	7.9213		$\frac{1}{38}$	8.624075×10^{-11}	7.9260
	$\frac{1}{46}$	2.433728×10^{-12}	7.9402		$\frac{1}{46}$	1.890060×10^{-11}	7.9506
				$n = 5$			
1.1	$\frac{1}{30}$	2.669378×10^{-12}	-	1.7	$\frac{1}{30}$	2.756366×10^{-11}	-
	$\frac{1}{38}$	2.057061×10^{-13}	9.5521		$\frac{1}{38}$	2.007171×10^{-12}	9.8190
	$\frac{1}{46}$	2.983513×10^{-14}	8.8684		$\frac{1}{46}$	2.368039×10^{-13}	10.2781
1.3	$\frac{1}{30}$	6.076893×10^{-12}	-	1.9	$\frac{1}{30}$	5.640743×10^{-11}	-
	$\frac{1}{38}$	4.563785×10^{-13}	9.6687		$\frac{1}{38}$	4.097488×10^{-12}	9.8250
	$\frac{1}{46}$	6.127250×10^{-14}	9.3784		$\frac{1}{46}$	4.868649×10^{-13}	10.2198

Example 13 ([117]). *Consider the function $f(x) = x(1-x)$, $x \in [0,1]$. This function fails to meet prerequisites for Equation (281) and Remark 14. We numerically compute its Riesz derivative at $x = 0.5$ by using scheme (281) with $n = 2$. The absolute error (AE) and experimental convergence order (CO) are displayed in Table 13. One can see that the expected fourth order accuracy is not achieved, which verifies the assertion in Remark 15.*

Table 13. Numerical results of Example 13 by using scheme (281) with $n = 2$.

α	h	AE	CO	α	h	AE	CO
1.1	$\frac{1}{10}$	5.900848×10^{-4}	-	1.7	$\frac{1}{10}$	6.101600×10^{-4}	-
	$\frac{1}{40}$	3.674047×10^{-5}	2.0011		$\frac{1}{40}$	3.779590×10^{-5}	2.0026
	$\frac{1}{160}$	2.295736×10^{-6}	2.0001		$\frac{1}{160}$	2.360928×10^{-6}	2.0002
1.3	$\frac{1}{10}$	6.828118×10^{-4}	-	1.9	$\frac{1}{10}$	2.863584×10^{-4}	-
	$\frac{1}{40}$	4.245071×10^{-5}	2.0015		$\frac{1}{40}$	1.770032×10^{-5}	2.0032
	$\frac{1}{160}$	2.652296×10^{-6}	2.0001		$\frac{1}{160}$	1.105502×10^{-6}	2.0002

4. Conclusions

In this paper we focus on numerical approximations to fractional integrals and derivatives, which are essential for solving fractional differential equations. This work is targeted at systematically clarifying basic ideas of the existing numerical evaluations, which provides the readers with comprehensive understanding of numerical methods for fractional calculus.

As the experimental advances further reveal nonlocality, memory, and hereditary properties of numerous materials and processes, the importance of the fractional calculus is becoming obvious. We hope that this work, which is designated to compressively review numerical approximations to fractional calculus, will become the first step in elucidating underlying principles and results of a wider variety of fractional dynamics.

Author Contributions: Investigation, M.C. and C.L.; writing—original draft preparation, M.C. and C.L.; writing—review and editing, M.C. and C.L. These two authors contributed equally to this paper. All authors have read and agreed to the published version of the manuscript.

Funding: The present job was partially supported by the National Natural Science Foundation under grant no. 11872234.

Acknowledgments: The authors wish to thank Yuri Luchko for his cordial invitation.

Conflicts of Interest: The authors declare no conflict of interest.

References

1. Leibniz, G.W. *Mathematische Schiften*; Georg Olms Verlagsbuchhandlung: Hildesheim, Germany, 1962.
2. Euler, L. De progressionibvs transcendentibvs, sev qvarvm termini generales algebraice dari negvevnt. *Comment Acad. Sci. Imperialis Petropolitanae* **1738**, *5*, 36–57.
3. Fourier, J. *Analytical Theory of Heat*; Dover Publishers: New York, NY, USA, 1955.
4. Liouville, J. Mémories sur quelques questions de géométrie et de mécanique, et sur un nouveau genre de calcul pour résoudre ces questions. *J. l'Ecole Roy. Polytéchn.* **1832**, *13*, 1–69.
5. Liouville, J. Mémories sur le calcul des différentielles à indices quelconques. *J. l'Ecole Roy. Polytéchn.* **1832**, *13*, 71–162.
6. Liouville, J. Mémories sur l'integration de l'équation: $(mx^2 + nx + p)d^2y/dx^2 + (qx + pr)dy/dx + sy = 0$ á l'aide des différentielles indices quelconques. *J. l'Ecole Roy. Polytéchn.* **1832**, *13*, 163–186.
7. Liouville, J. Mémorie sur le théoréme des fonctions complémentaries. *J. für Reine und Ungew. Math.* **1834**, *11*, 1–19.
8. Liouville, J. Mémorie sur une formule d'analys. *J. für Reine und Ungew. Math.* **1834**, *12*, 273–287.
9. Liouville, J. Mémorie sur l'usage que l'on peut faire de la formule de Fourier, dans le calcul des differentielles à indices quelconques. *J. für Reine und Ungew. Math.* **1835**, *13*, 219–232.

10. Liouville, J. Mémorie sur le changement de la variable indépandante dans le calcul de differentielles indices quelconques. *J. l'Ecole Roy. Polytéchn.* **1835**, *15*, 17–54.
11. Liouville, J. Mémorie sur l'intégration des équetions différentilles à indices fractionnaries. *J. l'Ecole Roy. Polytéchn.* **1837**, *15*, 58–84.
12. Riemann, B. Versuch einer allgemeinen Auffassung der Integration und Differentiation. In *Gesammelte Mathematische Werke und Wissenschaftlicher*; Teubner: Leipzig, Germany, 1876.
13. Grünwald, A.K. Uber "begrenzte" Derivationen und deren Anwendung. *Z. Angew. Math. und Phys.* **1867**, *12*, 441–480.
14. Letnikov, A.V. Theory of differentiation with an arbitrary index (Russian). *Mat. Sb.* **1868**, *3*, 1–66.
15. Letnikov, A.V. On historical development of differentiation theory with an arbitrary index (Russian). *Mat. Sb.* **1868**, *3*, 85–112.
16. Letnikov, A.V. On explanation of the main propositions of differentiation theory with an arbitrary index (Russian). *Mat. Sb.* **1872**, *6*, 413–445.
17. Letnikov, A.V. Investigations on the theory of integrals the form $\int_a^x (x-u)^{p-1} f(u) du$ (Russian). *Mat. Sb.* **1874**, *7*, 5–205.
18. Ferrari, F. Weyl and Marchaud derivatives: A forgotten history. *Mathematics* **2018**, *6*, 6. [CrossRef]
19. Rogosin, S.; Dubatovskaya, M. Letnikov vs. Marchaud: A survey on two prominent constructions of fractional derivatives. *Mathematics* **2018**, *6*, 3. [CrossRef]
20. Hardy, G.H.; Littlewood, J.E. Some properties of fractional integrals, I. *Math. Z.* **1928**, *39*, 565–606. [CrossRef]
21. Watanabe, J. On some properties of fractional powers of linear operators. *Proc. Jpn. Acad.* **1961**, *37*, 273–275. [CrossRef]
22. Love, E.R.; Yong, L.C. In fractional integration by parts. *Proc. Lond. Math. Soc.* **1938**, *44*, 1–35. [CrossRef]
23. Bang, T. Une inégalité de Kolmogoroff et les fonctions presque-périodiques. *Det. Kgl. Danske Vid. Selskab. Math.-Fys. Medd. Kobenhavn* **1941**, *19*, 3–28.
24. Civin, P. Inequalities for trigonometric integrals. *Duke Math. J.* **1941**, *8*, 656–665. [CrossRef]
25. Abel, N.H. Solution de quelques problèmes à l'aide d'integreales définies. In *Gesammelte Mathematische Werke*; Teubner: Leipzig, Germany, 1881.
26. Abel, N.H. Auflösung einer mechanischen Aufgabe. *J. für Die Reine und Angew. Math.* **1826**, *1*, 153–157.
27. Uchaikin, V.V. *Fractional Derivatives for Physicists and Engineers: Application*; Higher Education Press: Beijing, China, 2013.
28. Metzier, R.; Klafter, J. The random walk's guide to anomalous diffusion: a fractional dynamics approach. *Phys. Rep.* **2000**, *339*, 1–77. [CrossRef]
29. Zaslavsky, G.M. Chaos, fractional kinetics, and anomalous transport. *Phys. Rep.* **2002**, *37*, 461–580. [CrossRef]
30. Kärger, J.; Stallmach, F. *Diffusion in Condensed Matter: Methods, Materials, Models*; Heitjans, P., Kärger, J., Eds.; Springer: New York, NY, USA, 2005.
31. Machado, J.A.T.; Mainardi, F.; Kiryakova, V.; Atanacković, T. Fractional calculus: D'où venons-nous? Que sommes-nous? Où allons-nous? *Fract. Calc. Appl. Anal.* **2016**, *19*, 1074–1104. [CrossRef]
32. Podlubny, I. *Fractional Differential Equations*; Academic Press: San Diego, CA, USA, 1999.
33. Cai, M.; Li, C.P. On Riesz derivative. *Fract. Calc. Appl. Anal.* **2019**, *22*, 287–301. [CrossRef]
34. Yin, C.T.; Li, C.P.; Bi, Q.S. Approximation to Hadamard derivative via the finite part integral. *Entropy* **2018**, *20*, 983. [CrossRef]
35. Khalil, R.; Horani, A.A.; Yousef, A.; Sababheh, M. A new definition of fractional derivative. *J. Comput. Appl. Math.* **2014**, *264*, 65–70. [CrossRef]
36. Atangana, A.; Baleanu, D. New fractional derivatives with non-local and non-singular kernel theory and application to heat transfer model. *Ther. Sci.* **2016**, *20*, 763–769. [CrossRef]
37. Trenčevski, K.; Tomovski, Ž. On fractional derivatives of some functions of exponential type. *Univ. Beograd. Publ. Elektrotein. Fak.* **2002**, *13*, 77–84. [CrossRef]
38. Li, C.P.; Zeng, F.H. *Numerical Methods for Fractional Calculus*; CRC Press: Boca Raton, FL, USA, 2015.
39. Li, C.P.; Ma, L. Well-posedness of fractional differential equations. In Proceedings of the AMSE International Design Engineering Technical Conference and Computers and Information in Engineering Conference, Cleveland, OH, USA, 6–9 August 2017; p 9.

40. Moaaz, O.; Chalishajar, D.; Bazighifan, O. Some qualitative behavior of solutions of general class of difference equations. *Mathematics* **2019**, *7*, 585. [CrossRef]
41. Quarteroni, A.; Sacco, R.; Saleri, F. *Numerical Mathematics*; Springer: New York, NY, USA, 2000.
42. Li, C.P.; Zeng, F.H.; Liu, F.W. Spectral approximations to the fractional integral and derivative. *Fract. Calc. Appl. Anal.* **2012**, *15*, 383–406. [CrossRef]
43. Shen, J.; Tang, T.; Wang, L.L. *Spectral Methods: Algorithms, Analysis and Applications*; Springer: Berlin, Germany, 2011.
44. Zeng, F.H.; Li, C.P. Fractional differentiation matrices with applications. *arXiv* **2014**, arXiv:1404.4429v2.
45. Lubich, C. Discretized fractional calculus. *SIAM J. Math. Anal.* **1986**, *17*, 704–719. [CrossRef]
46. Dahlquist, G. Convergence and stability in the numerical integration of ordinary differential equations. *Math. Scand.* **1956**, *19*, 33–53. [CrossRef]
47. Diethelm, K.; Ford, J.M.; Ford, N.J.; Weilbeer, M. Pitfalls in fast numerical solvers for fractional differential equations. *J. Comput. Appl. Math.* **2006**, *186*, 482–503. [CrossRef]
48. Sousa, E. How to approximate the fractional derivatives of order $1 < \alpha \leq 2$. *Int. J. Bifurcat. Chaos* **2012**, *22*, 1250075.
49. Zeng, F.H.; Li, C.P.; Liu, F.W.; Turner, I. The use of finite difference/element approaches for solving the time-fractional subdiffusion equation. *SIAM J. Sci. Comput.* **2013**, *35*, A2976–A3000. [CrossRef]
50. Zeng, F.H.; Li, C.P.; Liu, F.W.; Turner, I. Numerical algorithms for time-fractional subdiffusion equation with second-order accuracy. *SIAM J. Sci. Comput.* **2015**, *37*, A55–A78. [CrossRef]
51. Zeng, F.H.; Liu, F.W.; Li, C.P.; Burrage, K.; Turner, I. A Crank-Nicolson ADI spectral method for a two-dimensional Riesz space fractional nonlinear reaction-diffusion equation. *SIAM J. Numer. Anal.* **2014**, *52*, 2599–2622. [CrossRef]
52. Yuan, L.X.; Agrawal, O.P. A numerical scheme for dynamic systems containing fractional derivatives. *J. Vib. Acoust.* **2002**, *124*, 321–324. [CrossRef]
53. Chatterjee, A. Statistical origins of fractional derivatives in viscoelasticity. *J. Sound. Vib.* **2005**, *284*, 1239–1245. [CrossRef]
54. Singh, S.J.; Chatterjee, A. Galerkin projections and finite elements for fractional order derivatives. *Nonlinear Dyn.* **2006**, *45*, 183–206. [CrossRef]
55. Alpert, B.; Greengard, L.; Hagstrom, T. Rapid evaluation of nonreflecting boundary kernels for time-domain wave propagation. *SIAM J. Numer. Anal.* **2000**, *37*, 1138–1164. [CrossRef]
56. Greengard, L.; Strain, J. A fast algorithm for the evaluation of heat potentials. *Commun. Pur. Appl. Math.* **1990**, *43*, 949–963. [CrossRef]
57. Li, J.R. A fast time stepping method for evaluating fractional integrals. *SIAM J. Sci. Comput.* **2010**, *31*, 4696–4714. [CrossRef]
58. Hélie, T.; Matignon, D. Diffusive representations for the analysis and simulation of flared acoustic pipes with visco-thermal losses. *Math. Mod. Meth. Appl. Sci.* **2006**, *16*, 503–536. [CrossRef]
59. Montseny, G.; Audounet, J.; Matignon, D. Diffusive representation for pseudodifferentially damped nonlinear systems, Nonlinear Control in the Year 2000, Vol. 2. In *Lecture Notes in Control and Information Sciences*; Springer: London, UK, 2001.
60. Diethelm, K. An investigation of some nonclassical methods for the numerical approximation of Caputo-type fractional derivatives. *Numer. Algor.* **2008**, *47*, 361–390. [CrossRef]
61. Lombard, B.; Matignon, D. Diffusive approximation of a time-fractional Burger's equation in nonlinear acoustics. *SIAM J. Appl. Math.* **2016**, *76*, 1765–1791. [CrossRef]
62. Oldham, K.B.; Spanier, J. *The Fractional Calculus*; Acdemic Press: New York, NY, USA, 1974; Renewed 2002.
63. Gao, G.H.; Sun, Z.Z. A compact finite difference scheme for the fractional subdiffusion equations. *J. Comput. Phys.* **2011**, *230*, 586–595. [CrossRef]
64. Jiang, Y.J.; Ma, J.T. High-order finite element methods for time-fractional partial differential equations. *J. Comput. Appl. Math.* **2011**, *253*, 3285–3290. [CrossRef]
65. Jin, B.; Lazarov, R.; Zhou, Z. Error estimates for a semidiscrete finite element method for fractional order parabolic equations. *SIAM J. Numer. Anal.* **2013**, *51*, 445–466. [CrossRef]
66. Langlands, T.A.M.; Henry, B.I. The accuracy and stability of an implicit solution method for the fractional diffusion equation. *J. Comput. Phys.* **2005**, *205*, 719–736. [CrossRef]

67. Ren, J.C.; Sun, Z.Z.; Zhao, X. Compact difference scheme for the fractional sub-diffusion equation with Neumann boundary conditions. *J. Comput. Phys.* **2013**, *232*, 456–467. [CrossRef]
68. Sun, Z.Z. *The Method of Order Reduction and Its Application to the Numerical Solutions of Partial Differential Equations*; Science Press: Beijing, China, 2009.
69. Zhao, X.; Sun, Z.Z. A box-type scheme for fractional sub-diffusion equation with Nuemann boundary conditions. *J. Comput. Phys.* **2011**, *230*, 6061–6074. [CrossRef]
70. Sun, Z.Z.; Wu, X.N. A fully discrete difference scheme for a diffusion-wave sysytem. *Appl. Numer. Math.* **2006**, *56*, 193–209. [CrossRef]
71. Lin, Y.M.; Xu, C.J. Finite difference/spectral approximations for the time-fractional diffusion equation. *J. Comput. Phys.* **2007**, *225*, 1533–1552. [CrossRef]
72. Zhang, Y.N.; Sun, Z.Z.; Liao, H.L. Finite difference methos for the time fractional diffusion equation on non-uniform meshes. *J. Comput. Phys.* **2014**, *265*, 195–210. [CrossRef]
73. Zhang, Y.N.; Sun, Z.Z.; Wu, H.W. Error estimates of Crank-Nicolson-type difference schemes for the sub-diffusion equation. *SIAM J. Numer. Anal.* **2011**, *49*, 2302–2322. [CrossRef]
74. Zhang, Y.N.; Sun, Z.Z. Error analysis of a compact ADI scheme for the 2D fractional subdiffusion equation. *J. Sci. Comput.* **2014**, *59*, 104–128. [CrossRef]
75. Chen, A.; Li, C.P. A novel compact ADI scheme for the time-fractional subdiffusion equation in two space dimensions. *Int. J. Comput. Math.* **2016**, *93*, 889–914. [CrossRef]
76. Stynes, M.; O'Riordan, E.; Gracia, J.L. Error analysis of a finite difference method on graded meshes for a time-fractional diffusion equation. *SIAM J. Numer. Anal.* **2017**, *55*, 1057–1079. [CrossRef]
77. Liao, H.L.; Li, D.F.; Zhang, J.W. Sharp error estimate of the nonuniform L1 formula for linear reaction-subdiffusion equations. *SIAM J. Numer. Anal.* **2018**, *56*, 1112–1133. [CrossRef]
78. Lynch, V.E.; Carreras, B.A.; del Castillo-Negrete, D.; Ferreira-Mejias, K.M.; Hicks, H.R. Numerical methods for the solution of partial differential equations of fractional order. *J. Comput. Phys.* **2003**, *192*, 406–421. [CrossRef]
79. Li, C.P.; Wu, R.F.; Ding, H.F. High-order approximation to Caputo derivatives and Caputo-type advection-diffusion equations. *Commun. Appl. Ind. Math.* **2015**, *6*, e-536.
80. Li, C.P.; Cai, M. *Theory and Numerical Approximations of Fractional Integrals and Derivatives*; SIAM: Philadelphia, PA, USA, 2019.
81. Gao, G.H.; Sun, Z.Z.; Zhang, H.W. A new fractional numerical differentiation formula to approximate the Caputo derivative and its applications. *J. Comput. Phys.* **2014**, *259*, 33–50. [CrossRef]
82. Alikhanov, A.A. A new difference scheme for the time fractional diffusion equation. *J. Comput. Phys.* **2015**, *280*, 424–438. [CrossRef]
83. Lv, C.W.; Xu, C.J. Error analysis of a high order method for time-fractional diffusion equations. *SIAM J. Sci. Comput.* **2016**, *38*, A2699–A2724. [CrossRef]
84. Luo, W.H.; Li, C.P.; Huang, T.Z.; Gu, X.M.; Wu, G.C. A high-order accurate numerical scheme for the Caputo derivative with an application to fractional diffusion problems. *Numer. Funct. Anal. Optim.* **2018**, *39*, 600–622. [CrossRef]
85. Cao, J.X.; Li, C.P.; Chen, Y.Q. High-order approximation to Caputo derivatives and Caputo-type advection-diffusion equations (II). *Fract. Calc. Appl. Anal.* **2015**, *18*, 735–761. [CrossRef]
86. Li, H.F.; Cao, J.X.; Li, C.P. High-order approximation to Caputo derivatives and Caputo-type advection-diffusion equations (III). *J. Comput. Appl. Math.* **2016**, *299*, 159–175. [CrossRef]
87. Liu, Y.Z.; Roberts, J.; Yan, Y. A note on finite difference methods for nonlinear fractional differential equations with non-uniform meshes. *Int. J. Comput. Math.* **2018**, *95*, 1151–1169. [CrossRef]
88. Du, R.L.; Yan, Y.; Liang, Z.Q. A high-order scheme to approximate the Caputo fractional derivative and its application to solve the fractional diffusion wave equation. *J. Comput. Phys.* **2019**, *376*, 1312–1330. [CrossRef]
89. Mohebbi, A.; Abbaszadeh, M.; Dehghan, M. The use of a meshless technique based on collocation and radial basis functions for solving the time fractional nonlinear Schrödinger equation arising inquantum mechanics. *Eng. Anal. Bound. Elem.* **2013**, *37*, 475–485. [CrossRef]
90. Chen, W.; Ye, L.J.; Sun, H.G. Fractional diffusion equations by the Kansa method. *Comput. Math. Appl.* **2010**, *59*, 1614–1620. [CrossRef]
91. Hosseini, V.R.; Chen, W.; Avazzadeh, Z. Numerical solution of fractional telegraph equation by using radial basis functions. *Eng. Anal. Bound. Elem.* **2014**, *38*, 31–39. [CrossRef]

92. Mohebbi, A.; Abbaszadeh, M.; Dehghan, M. Solution of two-dimensional modified anomalous fractional sub-diffusion equation via radial basis functions (RBF) meshless method. *Eng. Anal. Bound. Elem.* **2014**, *38*, 72–82. [CrossRef]
93. Aslefallah, M.; Shivanian, E. Nonlinear fractional integro-differential reaction-diffusion equation via radial basis functions. *Eur. Phys. J. Plus* **2015**, *130*, 47. [CrossRef]
94. Piret, C.; Hanert, E. A radial basis functions method for fractional diffusion equations. *J. Comput. Phys.* **2013**, *238*, 71–81. [CrossRef]
95. Li, C.P.; Cai, M. High-order approximation to Caputo derivatives and Caputo-type advection-diffusion equations: revisited. *Numer. Func. Anal. Optim.* **2017**, *38*, 861–890. [CrossRef]
96. Sousa, E.; Li, C. A weighted finite difference method for the fractional diffusion equation based on Riemann-Liouville derivative. *Appl. Numer. Math.* **2015**, *90*, 22–37. [CrossRef]
97. Sousa, E. Numerical approximations for fractional diffusion equations via splines. *Comput. Math. Appl.* **2011**, *62*, 938–944. [CrossRef]
98. Sousa, E. A second order explicit finite difference method for the fractional advection diffusion equations. *Comput. Math. Appl.* **2012**, *64*, 3141–3152. [CrossRef]
99. Li, C.P.; Chen, A. Numerical methods for fractional partial differential equations. *Int. J. Comput. Math.* **2018**, *95*, 1048–1099. [CrossRef]
100. Gao, G.H.; Sun, H.W.; Sun, Z.Z. Stability and convergence of finite difference schemes for a class of time-fractional sub-diffusion equations based on certain supeconvergence. *J. Comput. Phys.* **2015**, *280*, 510–528. [CrossRef]
101. Nasir, H.; Gunawardana, B. A second order finite difference approximation for the fractional diffusion equation. *Int. J. Appl. Phys. Math.* **2013**, *3*, 237–243. [CrossRef]
102. Zhao, L.J.; Deng, W.H. A series of high-order quasi-compact schemes for space fractional diffusion equations based on the superconvergent approximations for fractional derivatives. *Numer. Meth. Partial Diff. Equ.* **2015**, *31*, 1345–1381. [CrossRef]
103. Tian, W.Y.; Zhou, H.; Deng, W.H. A class of second order difference approximations for solving space fractional diffusion equations. *Math. Comput.* **2015**, *84*, 1703–1727. [CrossRef]
104. Zhou, H.; Tian, W.Y.; Deng, W.H. Quasi-compact finite difference schemes for space fractional diffusion equations. *J. Sci. Comput.* **2013**, *56*, 45–66. [CrossRef]
105. Chen, M.H.; Deng, W.H. Fourth order accurate scheme for the space fractional diffusion equations. *SIAM J. Numer. Anal.* **2014**, *52*, 1418–1438. [CrossRef]
106. Wu, R.F.; Ding, H.F.; Li, C.P. Determination of coefficients of high-order schemes for Riemann-Liouville derivative. *Sci. World J.* **2014**. [CrossRef] [PubMed]
107. Li, C.P.; Yi, Q. Modeling and computing of fractional convection equation. *Commun. Appl. Math. Comput.* **2019**, *1*, 565–595. [CrossRef]
108. Ding, H.F.; Li, C.P. High-order numerical algorithms for Riesz derivatives via constructing new generating functions. *J. Sci. Comput.* **2017**, *71*, 759–784. [CrossRef]
109. Ding, H.F.; Li, C.P. Fractional-compact numerical algorithms for Riesz spatial fractional reaction-dispersion equations. *Frac. Calc. Appl. Anal.* **2017**, *20*, 722–764. [CrossRef]
110. Ding, H.F.; Li, C.P. High-order numerical approximation formulas for Riemann-Liouville (Riesz) tempered fractional derivatives: Construction and application (II). *Appl. Math. Lett.* **2018**, *86*, 208–214. [CrossRef]
111. Tuan, V.K.; Gorenflo, R. Extrapolation to the limit for numerical fractional differentiation. *ZAMM J. Appl. Math. Mech.* **1995**, *75*, 646–648. [CrossRef]
112. Ding, H.F.; Li, C.P.; Chen, Y.Q. High-order algorithms for Riesz derivaive and their applications (I). *Abstr. Appl. Anal.* **2014**, *293*, 218–237.
113. Ding, H.F.; Li, C.P. High order algorithms for Riesz derivative and their applications (III). *Fract. Calc. Appl. Anal.* **2016**, *19*, 19–55. [CrossRef]
114. Çelik, C.; Duman, M. Crank-Nicolson method for the fractional diffusion equation with the Riesz fractional derivative. *J. Comput. Phys.* **2012**, *231*, 1743–1750. [CrossRef]
115. Li, C.P.; Yi, Q.; Kurths, J. Fractional convection. *J. Comput. Nonlinear Dyn.* **2017**, *13*. [CrossRef]
116. Ding, H.F.; Li, C.P.; Chen, Y.Q. High-order algorithms for Riesz derivative and their applications (II). *J. Comput. Phys.* **2015**, *293*, 218–237. [CrossRef]

117. Ding, H.F.; Li, C.P. High-order algorithms for Riesz derivative and their applications (V). *Numer. Meth. Partial Diff. Equ.* **2017**, *33*, 1754–1794. [CrossRef]
118. Hao, Z.P.; Sun, Z.Z.; Cao, W.R. A fourth-order approximation of fractional derivatives with its applications. *J. Comput. Phys.* **2015**, *281*, 787–805. [CrossRef]
119. Meerschaert, M.M.; Scheffler, H.P.; Tadjeran, C. Finite difference methods for two-dimensional fractional dispersion equation. *J. Comput. Phys.* **2006**, *211*, 249–261. [CrossRef]

© 2020 by the authors. Licensee MDPI, Basel, Switzerland. This article is an open access article distributed under the terms and conditions of the Creative Commons Attribution (CC BY) license (http://creativecommons.org/licenses/by/4.0/).

Article

Desiderata for Fractional Derivatives and Integrals

Rudolf Hilfer [1] and Yuri Luchko [2,*]

[1] ICP, Fakultät für Mathematik und Physik, Universität Stuttgart, Allmandring 3, 70569 Stuttgart, Germany; hilfer@icp.uni-stuttgart.de
[2] Fachbereich Mathematik-Physik-Chemie, Beuth Hochschule für Technik Berlin, Luxemburger Str. 10, 13353 Berlin, Germany
* Correspondence: luchko@beuth-hochschule.de

Received: 11 January 2019; Accepted: 25 January 2019; Published: 4 February 2019

Abstract: The purpose of this brief article is to initiate discussions in this special issue by proposing *desiderata* for calling an operator a fractional derivative or a fractional integral. Our *desiderata* are neither axioms nor do they define fractional derivatives or integrals uniquely. Instead they intend to stimulate the field by providing guidelines based on a small number of time honoured and well established criteria.

Keywords: fractional derivatives; fractional integrals; fractional calculus

MSC: 26A33; 34A08; 34K37; 35R11; 44A40

A list of six *desiderata*[1] is proposed that in our opinion would justify calling an operator D^α (or I^α) a fractional derivative (or a fractional integral) of non-integer order $\alpha \notin \mathbb{N}$. Derivatives and integrals of fractional order have a long history and, up until the recent proliferation of novel fractional derivatives, most definitions and interpretations of fractional operators seem to implicitly assume the *desiderata* of an operational calculus as formulated in this article.

Mathematical terms used in the formulation of our *desiderata* are defined in Appendix A. A family $\{D^\alpha, I^\alpha\}$ of operators with $\alpha \in \mathbb{Q}, \mathbb{R}$ or $\mathbb{C}$ is proposed to be called a *family of fractional derivatives D^α and integrals I^α of order α* (with $\operatorname{Re}\alpha \geq 0$)[2] if and only if it satisfies the following six *desiderata*:

(a) Integrals I^α and derivatives D^α of fractional order α should be linear operators on linear spaces[3].
(b) On some subset[4] $G_{(b)} \subset D(I^\alpha) \cap I^\beta[D(I^\beta)] \cap D(I^{\alpha+\beta}) \neq \emptyset$ the index law (semigroup property)

$$(I^\alpha \circ I^\beta)f = I^{\alpha+\beta} f \qquad (1)$$

holds true for $\operatorname{Re}\alpha \geq 0$ and $\operatorname{Re}\beta \geq 0$, where $D(I^\alpha)$ denotes the domain of I^α, and $\circ$ denotes composition of operators.

(c) Restricted to a suitable subset $G_{(c)} \subset D(I^\alpha)$ of the domain of I^α the fractional derivatives D^α of order α operate as left inverses

$$D^\alpha \circ I^\alpha = 1_{G_{(c)}} \qquad (2)$$

for all α with $\operatorname{Re}\alpha \geq 0$, where $1_{G_{(c)}}$ is the identity on $G_{(c)}$.

---

[1] properties to be desired.
[2] It is common to use only one of the symbols I or D in the sense that either $D^\alpha = I^{-\alpha}$ or $I^\alpha = D^{-\alpha}$. In this paper we keep the distinction between I and D by assuming $\operatorname{Re}\alpha \geq 0$ unless otherwise specified. This entails discussing the case $\operatorname{Re}\alpha = 0$ separately whenever necessary.
[3] Dependencies of I^α and D^α on other parameters are usually present, but notationally suppressed.
[4] Here the index (b) refers to *desideratum* (b). The same applies in *desiderata* (d)–(f) below.

(d) There is a subset $G_{(d)} \subset D(D^\alpha)$ of the domain of D^α such that the limits

$$g^1 = D^1 f = \lim_{\alpha \to 1} D^\alpha f, \qquad f \in G_{(d)}, \tag{3a}$$

$$g^0 = D^0 f = \lim_{\alpha \to 0} D^\alpha f, \qquad f \in G_{(d)}, \tag{3b}$$

exist in some sense and define linear maps $D^1 : G_{(d)} \to G_{(d)}$ resp. $D^0 : G_{(d)} \to G_{(d)}$.

(e) The limiting map $D^0 = 1_{G_{(d)}}$ is the identity on $G_{(d)}$, i.e., $g^0 = f$;

(f) The limiting map $D^1 = D$ is a derivation on $G_{(d)}$. This means it is possible to define a multiplication $\cdot : G_{(d)} \times G_{(d)} \to G_{(d)}$ on $G_{(d)}$ such that the Leibniz rule

$$D(f \cdot g) = g \cdot (Df) + f \cdot (Dg) \tag{4}$$

holds for all $f, g \in G_{(d)}$.

If the semigroup law (1) can be extended to all $\alpha \in \mathbb{R}$ or $\alpha \in \mathbb{C}$, we propose to speak of *fractional calculus*. Our *desiderata* are obviously inspired by operational calculus. Recall that an operational calculus is a continuous one-to-one mapping between an algebra of functions and an algebra of operators such that the neutral elements match and algebraic relations are preserved. Extending the algebra from polynomial functions to convergent power series suffices for an operational calculus. More singular functions, namely non-analytic power functions, are required for fractional calculus.

Desiderata differing substantially from those above have been formulated in [1] (p. 5) and [2] (p. 5). Envisaging exclusively analytic functions the criteria given in [1] (p. 5) are extremely restrictive. In theory and applications it is nowadays imperative to include more general functions, measures and also distributions into the purview.

Given the extreme restrictions in [1] (p. 5) a more recent proposal [2] went to opposite extremes. Little or no attention is given to a domain of definition for the fractional derivatives in [2]. Our *desiderata* for fractional derivatives again differ substantially from those in [2]. Rather than requiring some form of the generalized Leibniz rule ^{2}P5 in [2] for all α we desire the Leibniz rule only for $\alpha = 1$, and that differs not only from ^{2}P5, but also from ^{2}P3. In addition the identity rule ^{2}P2 in [2] does not restrict the admissible operators at all. As long as there is no continuity in α or a well defined limit, the identity rule can always be fulfilled, simply by setting $D^0 = 1$. More generally, Ref. [2] seems to neglect parameters other than α, or the topological and operator-theoretic implications of the limit in Equation (2) [2].

Our *desiderata* do not include non-locality of fractional derivatives. Fractional derivatives, that are local operators, were introduced in [3,4] and are discussed further in [5] and Section 7 of [6]. Contrary to ^{2}P3 in [2] we do not constrain the limits $\alpha \to n$ with $n \in \mathbb{N}$ for $n \geq 2$, because we wish to allow more generality.

To illustrate our *desiderata* in a simple case consider fractional operators of Riemann-Liouville type for complex valued functions $f : [a, b] \to \mathbb{C}$ on a compact interval $[a, b] \subset \mathbb{R}$ with $-\infty < a < b < \infty$ when the fractional order α is real and restricted to $0 \leq \alpha \leq 1$. The (right-sided) Riemann-Liouville fractional integrals I_{a+}^α of order $\alpha \geq 0$ with lower limit a are defined by setting $I_{a+}^0 f := f$ for $\alpha = 0$ and

$$(I_{a+}^\alpha f)(x) := \frac{1}{\Gamma(\alpha)} \int_a^x (x-y)^{\alpha-1} f(y) dy \tag{5}$$

for $\alpha > 0$ and $x \leq b$. The (right-sided) Riemann-Liouville-type fractional derivatives D_{a+}^α of order $0 \leq \alpha \leq 1$ are defined as [5] (p. 434)

$$(D_{a+}^{\alpha,\beta} f)(x) = \left(I_{a+}^{\beta(1-\alpha)} \frac{d}{dx} I_{a+}^{(1-\beta)(1-\alpha)} f \right)(x) \tag{6}$$

where the number $0 \leq \beta \leq 1$ parametrizes different types of fractional derivatives. The classical Riemann-Liouville derivative is of type $\beta = 0$, while the popular Liouville-Caputo derivative has type $\beta = 1$ [7] (p. 10).

Both operator families are linear and *desideratum* (a) can be fulfilled on numerous linear spaces due to the compactness of $[a, b]$. Examples are the Lebesgue spaces $L^p([a, b])$ with $1 \leq p \leq \infty$ or Hölder spaces $C^\gamma([a, b])$ with $\gamma > 0$. *Desideratum* (c) holds e.g. for $G_{(c)} = C^\gamma([a, b])$ with $\gamma > \alpha + \beta - \alpha\beta$ and $\alpha + \beta - \alpha\beta \neq 1$. The Riemann-Liouville fractional integrals (extended from $0 \leq \alpha \leq 1$ to $\alpha \geq 0$) are a strongly continuous semigroup of operators with respect to the parameter $\alpha \geq 0$, and obey the index law (1) in *desideratum* (b) for all $\alpha, \beta \geq 0$ on $G_{(b)} = L^p([a, b])$ or suitable subspaces. The *desiderata* (d), (e) and (f) can then be derived with the help of the semigroup property. For the Riemann-Liouville operators they hold e.g. for smooth (infinitely often differentiable) functions $f \in G_{(d)} = C^\infty([a, b])$.

For infinite intervals or for generalized functions the problem of domains may become more involved and our *desiderata* may become more restrictive. As an example consider the family of symmetric Riesz operators

$$(I^\alpha f)(x) := \frac{1}{2\Gamma(\alpha)\cos(\alpha\pi/2)} \int_{-\infty}^{\infty} |x-y|^{\alpha-1} f(y) dy \tag{7}$$

on the real line. In this case the limiting operator $D^1 = \lim_{\alpha \to -1} I^\alpha$ is again well defined, but does not fulfill *desideratum* (f). Instead, it fulfills Leibniz' formula for $D^n(fg)$ with $n = 2$, i.e.,

$$D^1[D^1(fg)] = g D^1(D^1 f) + 2(D^1 f)(D^1 g) + f D^1(D^1 g). \tag{8}$$

We propose to call such operators obeying Leibniz formula for $D^n(fg)$ with $n \geq 2$ pseudofractional derivatives or fractional pseudoderivatives.

Of course, our *desiderata* do not define fractional derivatives and integrals in a unique way. Still, they considerably restrict the set of admissible operators as seen above. In our opinion the above desiderata formulate crucial constraints for the development of a meaningful mathematical theory of fractional calculus and its reasonable applications.

Much work has been done on mathematical interpretations of fractional derivatives and integrals. The results are documented in numerous texts and treatises (see [6,8] for recent reviews). It seems however, that the connection (or not) of classical and recent fractional calculi with historical and contemporary forms of operational and functional calculi such as Heaviside-Mikusinski calculus, Dunford-Schwarz calculus, or Hille-Phillips calculus is a rich source of numerous open problems whose speedy solution would seem pertinent to advance and ultimately consolidate the field. We hope that the *desiderata* above are sufficiently restrictive to initiate a discussion of these pressing problems, and thereby stimulate readers and contributors to address some of these open problems in their areas of expertise and interest.

Appendix A

For the convenience of readers from non-mathematical disciplines we recall some definitions: A real (or complex) *linear space* (or *vector space*) over the field $\mathbb{R}$ (or $\mathbb{C}$) of real (or complex) numbers is a non-empty set X with two operations called addition and scalar multiplication fulfilling the usual rules of vector addition and multiplication of vectors with numbers[5].

---

[5] (a) for all $f, g \in X$ also $f + g \in X$, (b) $f + g = g + f$, (c) $f + (g + h) = (f + g) + g$, (d) there exists an element $0 \in X$ (called origin) such that $f + 0 = f$ for all $f \in X$, (e) for all $f \in X$ there is an element $-f \in X$ such that $f + (-f) = 0$ (f) for all $a \in \mathbb{R}$ (or $a \in \mathbb{C}$) and $f \in X$ an element $af \in X$ is defined, (g) for all $a \in \mathbb{R}$ (or $a \in \mathbb{C}$) and $f, g \in X$ one has $a(f + g) = af + ag$, (h) for all $a, b \in \mathbb{R}$ (or $a, b \in \mathbb{C}$) and $f \in X$ one has $(a + b)f = af + bf$, (i) for all $a, b \in \mathbb{R}$ (or $a, b \in \mathbb{C}$) and $f \in X$ one has $a(bf) = (ab)f$, and (j) $1f = f$ for all $f \in X$.

Let X, Y, Z be linear spaces (vector spaces). A *linear operator* $A : X \to Y$ is a linear subspace of the direct sum $X \oplus Y$, where

$$X \oplus Y := \{(f, g) : f \in X, g \in Y\} \tag{A1}$$

is the linear space of pairs (f, g) with $f \in X$, $g \in Y$ and addition defined as $(f_1, g_1) + (f_2, g_2) = (f_1 + f_2, g_1 + g_2)$ for all $f_i \in X$ and $g_i \in Y$. The *identity operator* $1_X : X \to X$ is defined as

$$1_X = 1 := \{(f, f) : f \in X\}. \tag{A2}$$

The *domain* $D(A)$ and *range* $R(A)$ of a linear operator $A : X \to Y$ are

$$D(A) := \{f \in X : \exists g \in Y \text{ s.t. } (f, g) \in A\} \tag{A3a}$$
$$R(A) := \{g \in Y : \exists f \in X \text{ s.t. } (f, g) \in A\}. \tag{A3b}$$

The *inverse* A^{-1} of A is defined as

$$A^{-1} := \{(g, f) \in Y \oplus X : (f, g) \in A\} \tag{A4}$$

with domain $D(A^{-1}) = R(A)$. For $A, B \in X \oplus Y$ their *sum* is defined as

$$A + B := \{(f, g + h) \in X \oplus Y : (f, g) \in A, (g, h) \in B\} \tag{A5}$$

with $D(A + B) = D(A) \cap D(B)$. For $A \in X \oplus Y, B \in Y \oplus Z$ the *composition* $B \circ A : X \to Z$ is the linear operator defined as

$$B \circ A := \{(f, h) \in X \oplus Z : \exists g \in Y \text{ s.t. } (f, g) \in A \text{ and } (g, h) \in B\} \tag{A6}$$

with $D(B \circ A) = \{f \in D(A) : \exists g \in D(B) \text{ s.t. } (f, g) \in A\}$.

Let $1 \le p < \infty$ be a fixed real number. The Lebesgue space $L^p([a, b])$ consists of those Lebesgue measurable functions $f : [a, b] \to \mathbb{C}$ on the intervall $[a, b] = [a, b] \subset \mathbb{R}$ for which the norm

$$\|f\|_p = \left(\int_{[a,b]} |f(x)|^p \, dx \right)^{1/p} \tag{A7}$$

is finite. For $p = \infty$ the space $L^\infty([a, b])$ is the set of all Lebesgue measurable functions such that $\|f\|_\infty = \operatorname{ess\,sup}_{x \in [a,b]} |f(x)|$ is finite where ess sup denotes supremum up to sets of Lebesgue measure zero (called essential supremum). The space $B([a, b])$ consists of all bounded functions on $[a, b]$. Its norm is $\|f\|_B = \sup_{x \in [a,b]} |f(x)|$. The space $C([a, b]) = C^0([a, b])$ consists of all continuous functions. Its norm is again $\|f\|_B = \sup_{x \in [a,b]} |f(x)|$ because continuous functions on a compact interval are also bounded. For $0 < \gamma \le 1$ and $f : [a, b] \to \mathbb{C}$ the number

$$\operatorname{Höl}_\gamma(f, [a, b]) := \sup \left\{ \frac{|f(x) - f(y)|}{|x - y|^\gamma}; x, y \in [a, b], x \ne y \right\} \in [0, \infty] \tag{A8}$$

is called *Hölder constant* of f on $[a, b]$ of Hölder order γ. The Hölder space $C^\gamma([a, b])$ is defined as

$$C^\gamma([a, b]) := \{f \in C([a, b]) : \operatorname{Höl}_\gamma(f, [a, b]) < \infty\} \tag{A9}$$

and is a Banach space for the norm

$$\|f\|_{C^\gamma} = \sup_{s\in[a,b]} |f(x)| + \text{Höl}_\gamma(f,[a,b]). \tag{A10}$$

For $\alpha = k+\gamma > 1$ with $k = 0,1,2,\ldots$ and $0 < \gamma \leq 1$ it is defined as

$$C^\alpha([a,b]) := \left\{ f \in C^k([a,b]) : \text{Höl}_\gamma(f^{(k)},[a,b]) < \infty \right\} \tag{A11}$$

with

$$\|f\|_{C^\alpha} = \|f\|_{C^k} + \|f^{(k)}\|_{C^\gamma} \tag{A12}$$

where $f^{(k)}$ is the k-th derivative of f.

A family $\{T(t)\}, t \geq 0$ of bounded linear operators $T(t) : X \to X$ on a Banach space X is called a *strongly continuous one-parameter semigroup* if it satisfies:

(a) $T(0) = 1_X$.
(b) $T(t)T(s) = T(t+s)$ for all $t,s \geq 0$.
(c) For every $x \in X$ the orbit maps $y_x : t \mapsto y_x(t) := T(t)x$ are continuous from $[0,\infty)$ into X.

References

1. Ross, B. A brief history and exposition of the fundamental theory of fractional calculus. In *Fractional Calculus and its Applications*; Ross, B., Ed.; Springer Verlag: Berlin, Germany, 1975; Volume 457, pp. 1–37.
2. Ortigueira, M.; Tenreiro-Machado, J. What is a fractional derivative? *J. Comput. Phys.* **2015**, *293*, 4–13. [CrossRef]
3. Hilfer, R. Thermodynamic Scaling Derived via Analytic Continuation from the Classification of Ehrenfest. *Phys. Scr.* **1991**, *44*, 321. [CrossRef]
4. Hilfer, R. Multiscaling and the Classification of Continuous Phase Transitions. *Phys. Rev. Lett.* **1992**, *68*, 190. [CrossRef] [PubMed]
5. Hilfer, R. Fractional Calculus and Regular Variation in Thermodynamics. In *Applications of Fractional Calculus in Physics*; Hilfer, R., Ed.; World Scientific: Singapore, 2000; p. 429.
6. Hilfer, R. Mathematical and physical interpretations of fractional derivatives and integrals. In *Handbook of Fractional Calculus and Applications, Volume 1: Basic Theory*; Kochubei, A., Luchko, Y., Eds.; De Gruyter: Berlin, Germany, 2019; p. 47.
7. Liouville, J. Mémoire sur quelques Questions de Geometrie et de Mecanique, et sur un nouveau genre de Calcul pour resoudre ces Questions. *J. l'Ecole Polytech.* **1832**, *XIII*, 1.
8. Kochubei, A.; Luchko, Y. Basic FC operators and their properties. In *Handbook of Fractional Calculus and Applications, Volume 1: Basic Theory*; Kochubei, A., Luchko, Y., Eds.; De Gruyter: Berlin, Germany, 2019; p. 23.

© 2019 by the authors. Licensee MDPI, Basel, Switzerland. This article is an open access article distributed under the terms and conditions of the Creative Commons Attribution (CC BY) license (http://creativecommons.org/licenses/by/4.0/).

Communication

Maximal Domains for Fractional Derivatives and Integrals

R. Hilfer * and T. Kleiner

ICP, Fakultät für Mathematik und Physik, Universität Stuttgart, Allmandring 3, 70569 Stuttgart, Germany
* Correspondence: hilfer@icp.uni-stuttgart.de

Received: 11 March 2020; Accepted: 6 June 2020; Published: 6 July 2020

Abstract: The purpose of this short communication is to announce the existence of fractional calculi on precisely specified domains of distributions. The calculi satisfy *desiderata* proposed above in *Mathematics* 7, 149 (2019). For the *desiderata* (a)–(c) the examples are optimal in the sense of having maximal domains with respect to convolvability of distributions. The examples suggest to modify *desideratum* (f) in the original list.

Keywords: fractional derivatives; fractional integrals; fractional calculus

MSC: 26A33; 34A08; 34K37; 35R11; 44A40

A list of six *desiderata* was recently proposed in [1] for calling families of operators $\{D^\alpha, I^\alpha\}$ with family index $\alpha \in \mathbb{I}$ from some index set $\mathbb{I} \subseteq \mathbb{C}$ fractional derivatives (D^α) and fractional integrals (I^α) of order $\alpha \notin \mathbb{N}$. Distributional domains for $\{D^\alpha, I^\alpha\}$ seem to require a minor modification of these *desiderata*.

Multiplication of distributions is ill-defined so that for distributions *desideratum* (f) (Leibniz rule) requires generalization. A slightly modified list of *desiderata* might read as follows:

(a) Integrals I^α and derivatives D^α of fractional order α should be linear operators on linear spaces.

(b) On some subset $G_{(b)} \subseteq D(I^\alpha) \cap I^\beta[D(I^\beta)] \cap D(I^{\alpha+\beta})$, $G_{(b)} \neq \emptyset$, $G_{(b)} \neq \{0\}$ the index law (semigroup property)

$$(I^\alpha \circ I^\beta) f = I^{\alpha+\beta} f \tag{1}$$

holds true for $\operatorname{Re} \alpha \geq 0$ and $\operatorname{Re} \beta \geq 0$, where $D(I^\alpha)$ denotes the domain of I^α.

(c) Restricted to a suitable subset $G_{(c)} \subseteq D(I^\alpha)$ of the domain of I^α the fractional derivatives D^α of order α operate as left inverses

$$D^\alpha \circ I^\alpha = 1_{G_{(c)}} \tag{2}$$

for all α with $\operatorname{Re} \alpha \geq 0$, where $\circ$ denotes composition of operators, and $1_{G_{(c)}}$ is the identity on $G_{(c)}$.

(d) Each of the two limits

$$g^1 = D^1 f = \lim_{\alpha \to 1} D^\alpha f, \qquad f \in G_{(d)}, \tag{3a}$$

$$g^0 = D^0 f = \lim_{\alpha \to 0} D^\alpha f, \qquad f \in G_{(d)}, \tag{3b}$$

should exist in some sense on some set $G_{(d)} \subseteq D(D^\alpha)$, $G_{(d)} \neq \emptyset$, $G_{(d)} \neq \{0\}$. Moreover, the limiting maps $D^1 : G_{(d)} \to G_{(d)}$ and $D^0 : G_{(d)} \to G_{(d)}$ should be linear.

(e) $D^0 = 1_{G_{(d)}}$ is the identity on $G_{(d)}$, i.e., $g^0 = f$ in Equation (3b).
(f) Endowed with a suitable multiplication $\odot : G_{(f)} \times G_{(d)} \to G_{(d)}$ the limiting map $D^1 = D$ obeys the Leibniz rule

$$D(f \odot g) = f \odot (Dg) + (Df) \odot g \quad (4)$$

for all $f \in G_{(f)}, g \in G_{(d)}$ with $G_{(f)} \neq \emptyset$, $G_{(f)} \neq \{0\}$. If $G_{(d)}$ consist of numerical functions, then $\odot$ is pointwise multiplication and $G_{(f)} = G_{(d)}$.

Given these modified *desiderata*, the objective in this short note is to introduce fractional calculi for distributions. Let us stress that the distributional domains $D(I^\alpha)$, $D(D^\alpha)$ given in Theorem 1 below are maximal in a precise mathematical sense. One cannot enlarge them without violating either the *desiderata* or the interpretation of fractional derivatives and integrals as convolution operators. Recall that numerous other mathematical interpretations exist [2], that may have different maximal domains. In this paper fractional operators are interpreted as convolutions with power law kernels (cf. [2], Equation (28)). A comprehensive analysis of convolutions with power law kernels on weighted spaces of continuous functions was recently given in [3].

Define the spaces of continuously differentiable functions, test functions, and smooth functions with bounded derivates

$$C^m(\mathbb{R}^d) := \{f : \mathbb{R}^d \to \mathbb{C} | f \text{ is } m\text{-times continuously differentiable}\} \quad (5a)$$

$$\mathcal{D}(\mathbb{R}^d) := \{f \in C^\infty(\mathbb{R}^d) | f \text{ has compact support}\} \quad (5b)$$

$$\mathcal{B}(\mathbb{R}^d) := \{f \in C^\infty(\mathbb{R}^d) | f \text{ has bounded derivatives}\} \quad (5c)$$

in the usual way [4]. The spaces $C^m, \mathcal{D}$ are endowed with the norm $\|f\|_\infty = \sup |f|$. The topology on $\mathcal{B}$ is induced by the seminorms $\|f\|_{N,g} = \sup\{\|g\partial^{n_1}...\partial^{n_d} f\|_\infty : n_i \in \mathbb{N}, \sum_i^d n_i \leq N\}$ with $N \in \mathbb{N}$ and $g \in C_v$, where C_v is the space of continuous functions vanishing at infinity.

The space of distributions $\mathcal{D}'$ is the topological dual of $\mathcal{D}$. The dual space $\mathcal{B}'$ is the space of integrable distributions. The pairing $\mathcal{D} \times \mathcal{D}' \to \mathbb{C}$ is denoted $\langle \cdot, \cdot \rangle$, the pairing $\mathcal{B} \times \mathcal{B}' \to \mathbb{C}$ as $\langle \cdot, \cdot \rangle_\mathcal{B}$.

Definition 1. *Two distributions $f_1, f_2 \in \mathcal{D}'(\mathbb{R}^d)$ are called convolvable iff $\varphi(f_1 \otimes f_2) \in \mathcal{B}'(\mathbb{R}^{2d})$ for all $\phi \in \mathcal{D}(\mathbb{R}^d)$, where $\varphi(x_1, x_2) = \phi(x_1 + x_2)$. Their convolution $f_1 * f_2$ is defined by requiring that*

$$\langle \phi, f_1 * f_2 \rangle = \langle 1, \varphi(f_1 \otimes f_2) \rangle_\mathcal{B} \quad (6)$$

holds for all $\phi \in \mathcal{D}(\mathbb{R}^d)$.

Let $\mathcal{D}'_+$ denote the space of causal distributions defined as elements $f \in \mathcal{D}'(\mathbb{R})$ whose support is bounded on the left.

Definition 2. *Fractional integrals I^α_+ and derivatives D^α_+ are defined for all $\alpha \in \mathbb{C}$ and all distributions $f \in \mathcal{D}'_+$ as convolution operators*

$$I^\alpha_+ f := K_\alpha * f \quad (7a)$$
$$D^\alpha_+ f := K_{-\alpha} * f \quad (7b)$$

with kernels

$$K_\alpha(x) = \begin{cases} \dfrac{x^{\alpha-1}}{\Gamma(\alpha)} & \text{for } x > 0 \\ 0 & \text{for } x \leq 0 \end{cases} \qquad \text{for Re } \alpha > 0 \qquad (8a)$$

$$K_\alpha(x) = \dfrac{d^m}{dx^m} K_{\alpha+m}(x) \qquad \text{for } -m < \text{Re } \alpha \leq 0, m \in \mathbb{N}. \qquad (8b)$$

The operators I^α_+ and D^α_+ are linear and continuous on $\mathcal{D}'_+$. The kernels $\{K_\alpha : \alpha \in \mathbb{C}\}$ form a convolution group

$$K_\alpha * K_\beta = K_{\alpha+\beta} \qquad (9)$$

for all $\alpha, \beta \in \mathbb{C}$. This entails the index law $I^\alpha_+(I^\beta_+ f) = I^{\alpha+\beta}_+ f$ for all $f \in \mathcal{D}'_+$ and $\alpha, \beta \in \mathbb{C}$. Clearly, all *desiderata* are fulfilled for $\{I^\alpha_+, D^\alpha_+\}$ with $D(I^\alpha_+) = D(D^\alpha_+) = G_{(b)} = G_{(c)} = G_{(d)} = \mathcal{D}'_+$ and $G_{(f)} = \mathcal{C}^\infty$.

The domain $\mathcal{D}'_+$ of causal distributions will now be enlarged using certain sets of lower semicontinuous functions as convolution weights. A function $f : \mathbb{R} \to \overline{\mathbb{R}}_+$, where $\overline{\mathbb{R}}_+ := [0, \infty]$, is called lower semicontinuous, if the set $\{f \leq a\}$ is closed for every $a \in \overline{\mathbb{R}}_+$. The set of all lower semicontinuous functions is denoted $\mathcal{I}$, the set of lower semicontinuous functions whose support is bounded on the left is denoted $\mathcal{I}_+$. For $(p,k) \in \mathbb{R} \times \mathbb{N}$ let

$$P^{p;k} := \left\{ f \in \mathcal{I} \mid \exists C > 0 \ \forall x \in \mathbb{R} : f(x) \leq C(1+|x|)^p [\log(e+|x|)]^k \right\} \qquad (10)$$

be the set of lower semicontinuous functions of power-logarithmic growth of order (p,k). Then

$$P_+ := \mathcal{I}_+ \cap \left(\bigcup_{q \in \mathbb{R}} P^{q;0} \right) \qquad (11a)$$

$$R_+ := \mathcal{I}_+ \cap \left(\bigcup_{k \in \mathbb{N}_0} P^{-1;k} \right) \qquad (11b)$$

are the sets of interest.

Definition 3. *Let $U \subseteq \mathcal{D}'$ and let $\mathcal{B}(\mathcal{D})$ denote the set of all bounded subsets of $\mathcal{D}$. Then*

$$(U)^*_{\mathcal{D}'} := \{ f \in \mathcal{D}' : (f,g) \text{ are convolvable for all } g \in U \} \qquad (12)$$

denotes the set of all distributions convolvable with the given set U. A locally convex topology $\mathcal{T}_U$ on $U \subseteq \mathcal{D}'$ is defined by the family of seminorms

$$\|f\|_{V,g} = (|f|_V * |g|_V)(0) = \int |f|_V(x)|g|_V(-x)dx \qquad (13)$$

with $V \subset \mathcal{D}, V \in \mathcal{B}(\mathcal{D})$ and $g \in (U)^*_{\mathcal{D}'}$. Here, the V-modulus of an element $f \in \mathcal{D}'$ is defined as

$$|f|_V(x) := \sup_{g \in V} |\langle f(\cdot), g(\cdot - x) \rangle| \qquad (14)$$

for all $x \in \mathbb{R}$.

Theorem 1. *The convolution group* $\{K_\alpha : \alpha \in \mathbb{C}\}$, *resp.* $\{K_\alpha : \alpha \in i\mathbb{R}\}$, *can be extended from* $(\mathcal{D}'_+, \mathcal{T}_{\mathcal{D}'_+})$ *to operate linearly, bijectively, and continuously on the space* $(\mathsf{U}, \mathcal{T}_\mathsf{U})$ *with* $\mathsf{U} = (P_+)^*_{\mathcal{D}'}$, *resp.* $\mathsf{U} = (R_+)^*_{\mathcal{D}'}$, *in such a way that compact sets of indices α map to equicontinuous sets of operators.*

Corollary 1. *The desiderata (a)–(e) are fulfilled for* $\{I^\alpha_+, D^\alpha_+\}_{\alpha \in \mathbb{C}}$ *with*

$$\mathsf{D}(I^\alpha_+) = \mathsf{D}(D^\alpha_+) = \mathsf{G}_{(b)} = \mathsf{G}_{(c)} = \mathsf{G}_{(d)} = (P_+)^*_{\mathcal{D}'} \tag{15a}$$

and for $\{I^\alpha_+, D^\alpha_+\}_{\alpha \in i\mathbb{R}}$ *with* $\mathsf{D}^1, \mathsf{G}_{(d)}$ *as in* (15a) *and*

$$\mathsf{D}(I^\alpha_+) = \mathsf{D}(D^\alpha_+) = \mathsf{G}_{(b)} = \mathsf{G}_{(c)} = (R_+)^*_{\mathcal{D}'}. \tag{15b}$$

In both cases it is possible to choose $\mathsf{G}_{(f)} = \mathcal{B}$.

The proof of Theorem 1 and its corollary will be published elsewhere, because it is lengthy and giving it here would distract attention from the main message. The domains $\mathsf{D}(I^\alpha_+), \mathsf{D}(D^\alpha_+), \mathsf{G}_{(b)}, \mathsf{G}_{(c)}$ are maximal with respect to convolvability in both cases. The second case $\{I^\alpha_+, D^\alpha_+\}_{\alpha \in i\mathbb{R}}$ yields a (purely imaginary) "fractional calculus of order zero" in the sense that Re $\alpha = 0$ for all operators in that subset.

References

1. Hilfer, R.; Luchko, Y. Desiderata for Fractional Derivatives and Integrals. *Mathematics* **2019**, *7*, 149. [CrossRef]
2. Hilfer, R. Mathematical and physical interpretations of fractional derivatives and integrals. In *Handbook of Fractional Calculus with Applications: Basic Theory*; Kochubei, A., Luchko, Y., Eds.; Walter de Gruyter GmbH: Berlin, Germany, 2019; Volume 1, pp. 47–86.
3. Kleiner, T.; Hilfer, R. Weyl Integrals on Weighted Spaces. *Fract. Calc. Appl. Anal.* **2019**, *22*, 1225–1248. [CrossRef]
4. Schwartz, L. *Theorie des Distributions*; Hermann: Paris, France, 1966.

© 2020 by the authors. Licensee MDPI, Basel, Switzerland. This article is an open access article distributed under the terms and conditions of the Creative Commons Attribution (CC BY) license (http://creativecommons.org/licenses/by/4.0/).

Article

Fractional Derivatives: The Perspective of System Theory

Manuel Duarte Ortigueira [1] and José Tenreiro Machado [2,*]

[1] CTS–UNINOVA and DEE of NOVA, School of Science and Technology, Nova University of Lisbon, Campus da FCT da UNL, Quinta da Torre, 2829–516 Caparica, Portugal; mdo@fct.unl.pt
[2] Department of Electrical Engineering, Institute of Engineering, Polytechnic of Porto, R. Dr. António Bernardino de Almeida, 431, 4249–015 Porto, Portugal
* Correspondence: jtm@isep.ipp.pt; Tel.: +351-22-8340500

Received: 31 December 2018; Accepted: 25 January 2019; Published: 5 February 2019

Abstract: This paper addresses the present day problem of multiple proposals for operators under the umbrella of "fractional derivatives". Several papers demonstrated that various of those "novel" definitions are incorrect. Here the classical system theory is applied to develop a unified framework to clarify this important topic in Fractional Calculus.

Keywords: fractional calculus; fractional derivatives; fractional anti-derivatives; fractional integrals; fractional operators

1. Introduction

The name *fractional derivative* (FD) is assigned to several mathematical operators, namely the Grünwald-Letnikov (GL), Liouville (L), Riemann-Liouville (RL), Caputo (C), Marchaud, Hadamard (H), Riesz, and others [1–5]. They are considered as generalisations of the classical derivative studied by Leibniz, Newton, Euler, and Lagrange, just to name a few of the most important mathematicians [1,2,6–10]. In previous papers we contributed to the on-going debate about the pros and cons of each formulation and we proposed possible approaches coherent with classic results in Science and Engineering [11,12]. In recent years, several operators were suggested claiming to be a "fractional derivative". This state of affairs motivated several papers showing the incorrectness of using the designation FD and even, in several cases, the absence of novelty [13–17]. We proposed a coherent framework for deciding if a given operator is a FD by formulating two criteria [11]. Nevertheless, the debate and the controversy remain. Some questions require suitable answers:

- What do we mean by "derivative"?
- What is the relation between derivative and integral?
- Frequently different notions of "integral" are mixed. Should not we use different notations or distinct names?
- When can we say that an operator is *fractional*?
- Should we consider a framework where integer and non integer orders co-exist and are mixed?
- What do we mean by *fractional calculus*? Should it be the calculus involving, at least, one "non integer order" derivative?
- Can we consider as "fractional operator" any expression involving a convolution of a function and a given kernel?
- Is it reasonable to choose a classic operator, to change its form by introducing a parameter and to call it FD?
- How can we call "fractional derivative" to an operator that is itself solution of a linear differential equation?

- The existence of a non integer parameter is reason for the use of the word "fractional"?

Several of these questions and others were discussed in round tables in the scope of several meetings held by the fractional calculus community [18,19]. Nevertheless, present day state of affairs reveals that they were not sufficient to stimulate all researchers for a systematic definition of the fundamental concepts. We observe the emergence of a plethora of assumed new operators that are named as novel or generalised fractional derivatives. Often it is also claimed that such operators fit better the experimental data. Obviously, from an application point of view, such lightly written words would need a systematic and solid testing with data from many distinct scientific areas, and the comparison with the results provided by classical derivatives (it is important to remark that this requires long observation intervals to capture long range memory effects). Furthermore, from a formal point of view, the good or bad fit into data, or the so-called "generalisation" by modifying some kernel, are not necessarily correct in mathematical terms when thinking on the properties of FD. Quoting Henri Poincaré *Mathematics has a threefold purpose. It must provide an instrument for the study of nature. But this is not all: it has a philosophical purpose, and, I dare say, an aesthetic purpose.* The main aim in this paper is to continue the discussion and try to establish a framework for avoiding misinterpretations and controversial or, even the incorrect, use of definitions.

Having these ideas in mind, this paper is organized as follows. Section 2 introduces the main terms and assumptions adopted in the follow-up. Section 3 recalls the definitions and fundamental properties of classic derivatives. Several aspects for the unification of concepts in the perspective of system theory are also discussed. Motivated by these ideas, Section 4 addresses the definition of fractional derivative. In these initial sections we assume that the independent variable t is continuous. However, computational systems are being increasingly important. Section 5 considers the case of the "discrete-time" operators. Finally, in Section 6 several conclusions and additional comments are presented.

2. Glossary and Assumptions

In Science it is important to define precisely the concepts that we are talking about. In fractional calculus there is a considerable confusion in the adopted terminology having, in some cases, different names for the same operator. Here we try to clarify the meaning for different terms in order to avoid such problem. Therefore, we start with some fundamental terms. Later, when necessary, we will introduce others, that are necessary in the rest of the paper.

- Anti-causal
  An anti-causal system is causal under reverse time flow. A system is *anti-causal* if the output at any instant depends only on values of the input at the present and future time.
- Anti-derivative
  The operator that is simultaneously the left and right inverse of the derivative will be called *anti-derivative*. It will be used to compute the definite integral through the Barrow formula [20]. This should be not confused with the negative order derivative, that needs not to be inverse of a derivative.
- Backward
  Reverse time flow—from future to past.
- Causal operator or system
  A system is *causal* if the output at any instant depends only on values of the input at the present and past time [21].
- Derivative
  Derivative (first order) of a function, $f(t)$ is the limit of the ratio of the change in such function to the corresponding change in its independent variable as the latter change approaches zero. It will be represented by $Df(t)$, $f'(t)$, or $\frac{df(t)}{dt}$.

- Forward
  Normal time flow—from past to future.
- Fractional
  Fractional will have the meaning of *non integer* real number.
- Integral
  In strict mathematical therms, there are several definitions of integral. However, the simplest is the Riemann integral that we can state as the numerical measure of the area under the graph of a given positive function, above the horizontal axis, and bounded on the sides by ordinates drawn at the endpoints of a specified interval. This is usually called *definite integral* and it is distinct from the *indefinite integral*, also called *primitive*.
- Primitive
  The operator that is only the right inverse of the derivative will be called *primitive*.

This paper tries to answer some of the initial questions. To clarify concepts, we adopt the designation "unified derivative" instead of "fractional derivative".

We assume that

- We work on $\mathbb{R}$. Nonetheless, this is not a limitation. If the function at hand is defined on any sub-interval in $\mathbb{R}$, we can extend the definition of the function to the whole real line with null values.
- We do not address the proof of existence of the operators.
- We use the two-sided Laplace transform (LT):

$$F(s) = \mathcal{L}\left[f(t)\right] = \int_{\mathbb{R}} f(t) e^{-st} dt, \tag{1}$$

where $f(t)$ is any function defined on $\mathbb{R}$ and $F(s)$ is its transform, provided that it has a non empty region of convergence

- The Fourier transform (FT), $\mathcal{F}\left[f(t)\right]$, is obtained from the LT through the substitution $s = i\omega$ with $\omega \in \mathbb{R}$ and $i = \sqrt{-1}$
- The functions and distributions have Laplace and/or Fourier transforms
- Current properties of the Dirac delta distribution, $\delta(\cdot)$, and its derivatives, $\delta'(\cdot), \delta''(\cdot) \cdots$, will be used
- The standard convolution operation will be adopted

$$f(t) * g(t) = \int_{\mathbb{R}} f(\tau) g(t-\tau) d\tau. \tag{2}$$

- The order of any fractional derivative, α, is any real number. We will not consider the complex order, since it gives non Hermitian derivatives.
- The multi-valued expressions s^α and $(-s)^\alpha$ will be used. To obtain functions from them we will fix for branch-cut lines the negative real half axis for the first and the positive real half axis for the second; for both the first Riemann surface is chosen.
- The Heaviside unit step will be represented by $\varepsilon(t)$ and the signum function by $sgn(t)$. These functions are related by $sgn(t) = 2\varepsilon(t) - 1$.
- We define the "floor" of a real number α as the integer $N = \lfloor \alpha \rfloor$ verifying $N \leq \alpha < N+1$.

3. The Classic Derivatives and Their Inverses

3.1. Elemental Derivatives

We find in the literature three standard definitions of (order 1) derivative [12] (the called quantum derivative will not be considered here [22]). These elemental derivatives can be considered as "seeds" for the notion of high level derivatives. Such derivatives are:

Definition 1.

- Forward or causal

$$D_f f(t) = \lim_{h \to 0} \frac{f(t) - f(t-h)}{h}, \qquad (3)$$

- Backward or anti-causal

$$D_b f(t) = \lim_{h \to 0} \frac{f(t+h) - f(t)}{h}. \qquad (4)$$

Remark 1. *Substituting $-h$ for $+h$ interchanges the definitions, meaning that we only have to consider $h > 0$.*

- Two-sided or centred

$$D_c f(t) = \lim_{h \to 0} \frac{f(t+h/2) - f(t-h/2)}{h}. \qquad (5)$$

Remark 2. *The expression (5) was used in [23,24] to obtain two different two-sided (centred) fractional derivatives. Later, in Section 3.4, we will recover the general formulation of these derivatives.*

Remark 3. *Most literature on Differential Calculus uses definition (4) only due to historical reasons.*

In terms of the Laplace transform we have

- Forward or causal

$$\mathcal{L}\left[D_f f(t)\right] = \lim_{h \to 0} \frac{1 - e^{-sh}}{h} F(s) = sF(s), \qquad (6)$$

- Backward or anti-causal

$$\mathcal{L}\left[D_b f(t)\right] = \lim_{h \to 0} \frac{e^{sh} - 1}{h} = sF(s), \qquad (7)$$

- Two-sided or acausal

$$\mathcal{F}\left[D_c f(t)\right] = \lim_{h \to 0} \frac{e^{sh/2} - e^{-sh/2}}{h} F(i\omega) = sF(s) \qquad (8)$$

Remark 4. *Note that, although different, the LT of the three derivatives is the same and valid in the whole complex plane.*

It is straightforward to invert the relations (3) and (4), and we obtain

$$D_f^{-1} f(t) = \lim_{h \to 0} \sum_{n=0}^{\infty} f(t - nh) \cdot h, \qquad (9)$$

$$D_b^{-1} f(t) = \lim_{h \to 0} \sum_{n=0}^{\infty} f(t + nh) \cdot h. \qquad (10)$$

Using the LT, we have

$$\mathcal{L}\left[D_f^{-1}f(t)\right] = \lim_{h\to 0} h \sum_{n=0}^{\infty} e^{-sh} F(s) = \frac{1}{s}F(s), \quad Re(s) > 0, \tag{11}$$

$$\mathcal{L}\left[D_b^{-1}f(t)\right] = \lim_{h\to 0} h \sum_{n=0}^{\infty} e^{sh} F(s) = \frac{1}{s}F(s), \quad Re(s) < 0. \tag{12}$$

Remark 5. *Note the appearance of the regions of convergence (ROC) subsets of $\mathbb{C}$. This important fact is tied with causality [21,25].*

3.2. First Unification

The repeated use of the above derivatives and anti-derivatives leads to closed formulae valid for any integer order, $N \in \mathbb{Z}$ [12], such that:

$$D_f^N f(t) = \lim_{h\to 0^+} \frac{\sum_{n=0}^{\infty} \frac{(-N)_n}{n!} f(t-nh)}{h^N}, \tag{13}$$

$$D_b^N f(t) = (-1)^N \lim_{h\to 0^+} \frac{\sum_{n=0}^{\infty} \frac{(-N)_n}{n!} f(t+nh)}{h^N}, \tag{14}$$

respectively, where $(a)_k = a(a+1)(a+2)\ldots(a+k-1)$ denotes the Pochammer symbol. Expressions (13) and (14) reflect, in a unified way, all integer order derivatives and anti-derivatives. Therefore, we can use only the word derivative independently of having positive or negative order.

The corresponding LT are given by:

$$\mathcal{L}\left[D_f^N f(t)\right] = s^N F(s), \quad Re(s) > 0, \tag{15}$$

$$\mathcal{L}\left[D_b^N f(t)\right] = s^N F(s), \quad Re(s) < 0, \tag{16}$$

respectively. From the point of view of system theory, expressions (15) and (16) tell us that the derivative operator represents a system with transfer function (TF) given by $H(s) = s^N$. In this perspective, we consider the system approach with the integer order derivatives formulated by means of the two-sided LT property $\mathcal{L}\left[f^{(n)}(t)\right] = s^n L[f(t)]$, where $n \in \mathbb{Z}$, and it becomes clear the meaning of the sequence

$$\ldots s^{-n} \ldots s^{-2} \; s^{-1} \; 1 \; s^1 \; s^2 \ldots s^n \ldots \tag{17}$$

in the Laplace domain. Indeed, the corresponding time sequence is

$$\ldots \pm \frac{t^{n-1}}{(n-1)!} u(\pm t) \cdots \pm \frac{t^2}{2!} u(\pm t) \; \pm \frac{t^1}{1!} u(\pm t) \pm u(\pm t) \; \delta(t) \; \delta'(t) \; \delta''(t) \; \ldots \delta^{(n)}(t) \ldots \tag{18}$$

that allows us to write, for the causal definition (the other case is similar),

$$D_f^N f(t) = \int_0^\infty f(t-\tau) \frac{\tau^{N-1}}{(N-1)!} d\tau, \tag{19}$$

where we assume that, if $N \leq 0$, then $\frac{\tau^{N-1}}{(N-1)!} = \delta^{(N)}(\tau)$.

3.3. Second Unification

It is straightforward to extend formulae (13) and (14) to any real order. In fact, with $\alpha \in \mathbb{R}$ we can write

$$D_f^\alpha f(t) = \lim_{h \to 0^+} \frac{\sum_{n=0}^{\infty} \frac{(-\alpha)_n}{n!} f(t - nh)}{h^\alpha}, \qquad (20)$$

$$D_b^\alpha f(t) = e^{-i\alpha\pi} \lim_{h \to 0^+} \frac{\sum_{n=0}^{\infty} \frac{(-\alpha)_n}{n!} f(t + nh)}{h^\alpha}, \qquad (21)$$

that have LT

$$\mathcal{L}\left[D_f^\alpha f(t)\right] = s^\alpha F(s), \quad \mathrm{Re}(s) > 0, \qquad (22)$$

$$\mathcal{L}\left[D_b^\alpha f(t)\right] = s^\alpha F(s), \quad \mathrm{Re}(s) < 0, \qquad (23)$$

respectively. These relations, allow us to fill in the gaps in middle the discrete sequence (17) to obtain, for example

$$\ldots s^{-n} \ldots s^{-\pi} \ldots s^{-2} \ldots s^{-3/2} \ldots s^{-1} \ldots s^{-1/3} \ldots 1 \; s^1 \ldots s^{3/2} \ldots s^2 \ldots s^n \ldots . \qquad (24)$$

giving a meaning for s^α, $\pm \mathrm{Re}(s) > 0$. The inverse LT of this transfer function is

$$\mathcal{L}\left[s^\alpha\right] = \pm \frac{t^{-\alpha-1}}{\Gamma(-\alpha)} \varepsilon(\pm t) \qquad (25)$$

that leads to

$$D_f^\alpha f(t) = \int_0^\infty f(t - \tau) \frac{\tau^{-\alpha-1}}{\Gamma(-\alpha)} d\tau, \qquad (26)$$

generalising the causal expression (19) to real orders. For the anti-causal case, we get the general expression:

$$D_b^\alpha f(t) = e^{-i\alpha\pi} \int_0^\infty f(t + \tau) \frac{\tau^{-\alpha-1}}{\Gamma(-\alpha)} d\tau. \qquad (27)$$

3.4. Third Unification

The factor $e^{-i\alpha\pi}$ in (27) was already included by Liouville [26] to guarantee that $\mathcal{L}\left[D_b^\alpha f(t)\right] = s^\alpha F(s)$, for $\mathrm{Re}(s) < 0$. It apeared also in the backward GL derivative (21) and particular integer order cases. However, this factor may be of no relevance in many applications, especially when the independent variable is space, not time. If this term is removed, then we can join pairs of formulae into only one. We change also the nomenclature, using *left* for forward and *right* for backward. Therefore, (20) and (21) lead to

$$D_{l,r}^\alpha f(t) = \lim_{h \to 0^+} \frac{\sum_{n=0}^{\infty} \frac{(-\alpha)_n}{n!} f(t \pm nh)}{h^\alpha}, \qquad (28)$$

where the signs − and + are used for the left and right derivatives, respectively. The corresponding Liouville integral formulations are expressed by

$$D_{l,r}^\alpha f(t) = \int_0^\infty f(t \pm \tau) \frac{\tau^{-\alpha-1}}{\Gamma(-\alpha)} d\tau. \qquad (29)$$

The LT of these derivatives are

$$\mathcal{L}\left[D_{l,r}^\alpha f(t)\right] = (\pm s)^\alpha F(s), \quad \operatorname{Re}(\pm s) > 0. \tag{30}$$

From these results we conclude that

1. We can combine two derivatives of any orders, α and β, to obtain a third derivative

$$s^\alpha s^\beta = s^{\alpha+\beta} \tag{31}$$

2. If $f(t) = e^{i\omega t}$, $\omega \in \mathbb{R}$, then

$$D^\alpha e^{i\omega t} = (\pm\omega)^\alpha e^{i\omega t} = |\omega|^\alpha e^{\pm i\alpha\frac{\pi}{2}\operatorname{sgn}(\omega)} e^{i\omega t} \tag{32}$$

where the + and − signs refer to the left and right cases in (28) and (29), respectively.

3. The corresponding frequency responses are given by

$$H(i\omega) = |\omega|^\alpha e^{\pm i\alpha\frac{\pi}{2}\operatorname{sgn}(\omega)} \tag{33}$$

This result is important, since it expresses very clearly the unification of the derivatives and motivates a further development as discussed in the next subsection.

3.5. Fourth Unification

In (31) it is written that the combination of two derivatives of the same type (e.g., left) gives rise to another derivative of the same type. Now, we consider the combination of one derivative of each type.

Definition 2. *Consider two derivatives, causal and anti-causal, with orders α and β, having frequency responses $|\omega|^\alpha e^{i\alpha\frac{\pi}{2}\operatorname{sgn}(\omega)}$ and $|\omega|^\beta e^{-i\beta\frac{\pi}{2}\operatorname{sgn}(\omega)}$, respectively.*
We define a new derivative with frequency response

$$\Psi_\theta^\gamma(i\omega) = |\omega|^\gamma e^{i\theta\frac{\pi}{2}\operatorname{sgn}(\omega)}, \tag{34}$$

where $\gamma = \alpha + \beta$ is the order of the derivative and $\theta = \alpha - \beta$ is the parameter of asymmetry (sometimes called skewness).

It can be shown [23,24] that, if $\gamma > -1$, then the frequency response (34) corresponds to a two-sided derivative given by:

$$D_c^\gamma f(t) := \lim_{h \to 0^+} h^{-\gamma} \sum_{n=-\infty}^{+\infty} (-1)^n \cdot \frac{\Gamma(\gamma+1)}{\Gamma\left(\frac{\gamma+\theta}{2} - n + 1\right)\Gamma\left(\frac{\gamma-\theta}{2} + n + 1\right)} f(t - nh). \tag{35}$$

Suitable choices of the parameters γ and θ allow us to recover the causal and anti-causal derivatives. The particular cases of $\alpha = \beta$ and $\alpha - \beta = \pm 1$ are interesting and correspond to well-known operators as we will see later at Section 4.3.

3.6. Bode Diagrams

Bode diagrams are useful tools for the analysis and design of linear systems [21,25], since they provide a direct insight into models adopted in engineering and natural systems. This tool is of relevance when applied to the unified derivatives above discussed in Sections 3.2–3.5.

Definition 3. *From formula (34) define two spectra (Figure 1):*

1. *Amplitude spectrum*

$$A(\omega) = |\omega|^{\gamma} \tag{36}$$

2. *Phase spectrum*

$$\Phi(\omega) = \theta \frac{\pi}{2} sgn(\omega) \tag{37}$$

For real-valued functions, the amplitude and the phase are even and odd functions, respectively [21,25]. For this reason, we only need to represent log plots for positive frequencies that are called *Bode diagrams*. For $A(\omega)$ it is usual to express the amplitude in deciBell (dB). Then, it results

$$A(\omega)|_{dB} = \gamma 20 \log \omega \tag{38}$$

that is represented by a straight line with slop 20γ dB per decade (dB/dec). The phase $\Phi(\omega)$ is expressed in radians or degrees and represented by horizontal straight lines at $\alpha \frac{\pi}{2}$.

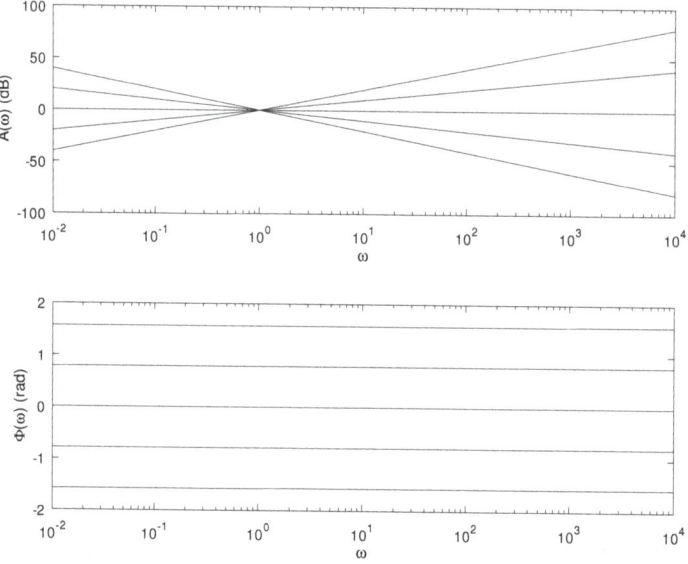

Figure 1. Bode plots for $\alpha = \{-1, -0.5, 0.5, 1\}$ with $\theta = \alpha$, corresponding to the amplitude and phase spectra, given in (36) and (37).

In Table 1, we consider some particular cases for the parameters γ and θ and we point out the name of the of resulting derivatives, assuming that $\alpha > 0$. We include also the Riesz and Feller potentials [1,2] that we will discuss in Section 4.

Table 1. Some of known derivatives obtained with particular values of γ and θ.

	γ	θ	Freq. Response	Name
1	α	α	$(i\omega)^\alpha$	causal derivative/Grünwald-Letnikov
2	$-\alpha$	$-\alpha$	$(i\omega)^{-\alpha}$	causal anti-derivative/Grünwald-Letnikov/Liouville
3	α	α	$(i\omega)^N (i\omega)^{\alpha-N}$	causal/Liouville
4	α	α	$(i\omega)^{\alpha-N} (i\omega)^N$	causal/Liouville-Caputo
5	α	$-\alpha$	$(-i\omega)^\alpha$	right derivative/Grünwald-Letnikov
6	$-\alpha$	$-\alpha$	$(-i\omega)^{-\alpha}$	right anti-derivative/Grünwald-Letnikov/Liouville
7	α	$-\alpha$	$(-i\omega)^N (-i\omega)^{\alpha-N}$	anti-causal/Liouville
8	α	$-\alpha$	$(-i\omega)^{\alpha-N} (-i\omega)^N$	right/Liouville-Caputo
9	α	0	$|\omega|^\alpha$	symmetric two-sided
10	$-\alpha$	0	$|\omega|^{-\alpha}$	Riesz potential
11	α	± 1	$i\,sgn(\omega)\,|\omega|^\alpha$	anti-symmetric two-sided
12	$-\alpha$	0	$i\,sgn(\omega)\,|\omega|^{-\alpha}$	Feller potential

This list includes the most relevant examples of application of our framework. Some operators such as, for example, the Erdélyi and Kober integrals fall outside this point of view and should not be considered derivatives.

Remark 6. *It can be shown that the derivatives defined by means of (34) verify the usual properties required for the FD to follow, namely the strict sense criterion proposed in [11].*

4. Derivative Definition Through Integral Formulations

4.1. Definition

The results from the previous section motivate the following definition of the unified derivative.

Definition 4. *We define the α-order "unified derivative" as the convolutional operator*

$$D_\theta^\alpha f(t) = \int_\mathbb{R} f(t-\tau)\psi_\theta^\alpha(\tau)d\tau, \tag{39}$$

where $\psi_\theta^\alpha(t)$, $t \in \mathbb{R}$, is the kernel of the derivative and $\theta \in \mathbb{R}$, is an asymmetry parameter that controls the characteristics of the derivative, namely the causality. The kernel $\psi_\theta^\alpha(t)$ is a function with Fourier transform $\Psi_\theta^\alpha(i\omega)$, $\omega \in \mathbb{R}$, such that the corresponding Bode diagram of amplitude is a straight line with slope 20α dB/dec and the phase is a horizontal straight line with value $\alpha\frac{\pi}{2}$.

4.2. A General Kernel

In (34) we obtained the frequency response of the unified derivative. If the inverse Fourier transform is $\psi_\theta^\alpha(t) = \mathcal{F}^{-1}\left[|\omega|^\alpha e^{i\frac{\pi}{2}\theta \cdot sgn(\omega)}\right]$, then it is known [23,24] that:

$$\psi_\theta^\alpha(t) = \frac{\sin\left[(\alpha + \theta \cdot sgn(t))\frac{\pi}{2}\right]}{2\sin(\alpha\pi)\,\Gamma(-\alpha)}|t|^{-\alpha-1}. \tag{40}$$

This is the general kernel that allows us to express the integral formulation of the unified derivative that can be written as:

$$D_\theta^\alpha f(t) = \frac{1}{2\sin(\alpha\pi)\,\Gamma(-\alpha)} \int_\mathbb{R} f(t-\tau) \sin\left[(\alpha + \theta \cdot sgn(\tau))\frac{\pi}{2}\right]|\tau|^{-\alpha-1}d\tau. \tag{41}$$

4.3. Some Particular Kernels

As seen above, the most interesting derivatives result from particular values of the parameters α and θ. Here, we analyse several cases as follows.

1. $\alpha = \theta = N \in \mathbb{N}_0$
   In this case, $\Psi_\theta^\gamma(i\omega) = (i\omega)^\alpha$ and $\psi_N^N(t) = \delta^{(N)}(t)$. It yields

$$D_l^N(t) = \int_0^\infty f(t-\tau)\delta^{(N)}(\tau)d\tau = \int_{-\infty}^\infty f(\tau)\delta^{(N)}(t-\tau)d\tau \quad (42)$$

   stating a well known property of the impulse distribution.

2. $\alpha = \theta \in \mathbb{R}^+$
   In this case, $\Psi_{-\alpha}^{-\alpha}(i\omega) = (i\omega)^{-\alpha}$ and $\psi_{-\alpha}^{-\alpha}(t) = \frac{t^{\alpha-1}}{\Gamma(\alpha)}\varepsilon(t)$. We obtain

$$D_l^{-\alpha}f(t) = \frac{1}{\Gamma(\alpha)}\int_0^\infty f(t-\tau)\tau^{\alpha-1}d\tau = \frac{1}{\Gamma(\alpha)}\int_{-\infty}^t f(\tau)(t-\tau)^{\alpha-1}d\tau \quad (43)$$

   corresponding to the causal Liouville anti-derivative (line 2 in Table 1).

3. $\alpha \in \mathbb{R}^+$ and $\theta = -\alpha$
   In this case, $\Psi_\alpha^{-\alpha}(i\omega) = (-i\omega)^{-\alpha}$ and $\psi_\alpha^{-\alpha}(t) = \frac{(-t)^{\alpha-1}}{\Gamma(\alpha)}\varepsilon(-t)$. Then, it results

$$D_r^{-\alpha}f(t) = \frac{1}{\Gamma(\alpha)}\int_0^\infty f(t+\tau)\tau^{\alpha-1}d\tau = \frac{1}{\Gamma(\alpha)}\int_t^\infty f(\tau)(\tau-t)^{\alpha-1}d\tau, \quad (44)$$

   that corresponds to line 6 in Table 1.

4. $\alpha \in \mathbb{R}^+$, $\theta = \alpha$
   Let $\Psi_\alpha^\alpha(i\omega) = (i\omega)^\alpha$. This is essentially the previous case 2 that leads to (43). However, the inverse FT produces a kernel that originates a singular integral. To avoid this problem we can use the properties of the *pseudo-functions* [27] that allow us to regularize (43) and use it for the derivative case. Let $N = \lfloor \alpha \rfloor + 1$. We can write (43) as [28]

$$D_l^\alpha f(t) = \frac{1}{\Gamma(-\alpha)}\int_0^\infty \tau^{-\alpha-1}\left[f(t-\tau) - \sum_0^{N-1}\frac{(-)^m f^{(m)}(t)}{m!}\tau^m\right]d\tau, \quad (45)$$

   that we will call regularized Liouville derivative. Similar regularised integrals can be obtained for (44).

Remark 7. *If $0 < \alpha < 1$ and $N = 1$, then we get*

$$D_l^\alpha f(t) = \frac{1}{\Gamma(-\alpha)}\int_0^\infty \tau^{-\alpha-1}\left[f(t-\tau) - f(t)\right]d\tau, \quad (46)$$

   that coincides with the Marchaud derivative [1]. Nonetheless, for $\alpha > 1$, the Marchaud operator is no longer a derivative.

5. $\alpha \in \mathbb{R}^+$, $\theta = \alpha$ and $N = \lfloor \alpha \rfloor + 1$
   In this case, we have two possibilities:

(a) $\Psi_\alpha^\alpha(i\omega) = \Psi_N^N(i\omega)\Psi_{\alpha-N}^{\alpha-N}(i\omega) = (i\omega)^N (i\omega)^{\alpha-N}$. This frequency response corresponds to a two-step derivative: integer order, N, derivative after a fractional anti-derivative of order $N - \alpha$. Instead of (41), we can write

$$D_l^\alpha f(t) = \frac{d^N}{dt^N} \frac{1}{\Gamma(-\alpha + N)} \int_{-\infty}^{t} (t-\tau)^{N-\alpha-1} f(\tau) d\tau, \qquad (47)$$

that is called Liouville derivative [1] (line 3 in Table 1).

(b) $\Psi_\alpha^\alpha(i\omega) = \Psi_{\alpha-N}^{\alpha-N}(i\omega)\Psi_N^N(i\omega) = (i\omega)^{\alpha-N} (i\omega)^N$. It is the reverse of the above: a fractional anti-derivative of order $N - \alpha$ after an integer order N derivative. Then, (41) assumes the form

$$D_l^\alpha f(t) = \frac{1}{\Gamma(-\alpha + N)} \int_{-\infty}^{t} (t-\tau)^{N-\alpha-1} \frac{d^N f(\tau)}{d\tau^N} d\tau. \qquad (48)$$

that is called Liouville-Caputo derivative [10] (line 4 in Table 1).

The corresponding right derivatives are easily obtained.

6. $\alpha \in \mathbb{R}^+$, $\theta = 0$
   In this case, $\Psi_0^{-\alpha}(i\omega) = |\omega|^{-\alpha}$ and the inverse FT of (40) is

$$\psi_0^\alpha(t) = \frac{1}{\cos\left(\alpha\frac{\pi}{2}\right)\Gamma(\alpha)} |t|^{\alpha-1},$$

that leads to the Riesz potential (line 10 in Table 1)

$$D_0^{-\alpha} f(t) = \frac{1}{\cos\left(\alpha\frac{\pi}{2}\right)\Gamma(\alpha)} \int_{\mathbb{R}} f(t-\tau) |\tau|^{\alpha-1} d\tau \qquad (49)$$

7. $\alpha \in \mathbb{R}^+$ and $\theta = 1$
   In this case, $\Psi_0^{-\alpha}(i\omega) = |\omega|^{-\alpha}$ and the inverse FT of (40) is

$$\psi_0^\alpha(t) = \frac{1}{\sin\left(\alpha\frac{\pi}{2}\right)\Gamma(\alpha)} |t|^{\alpha-1} \mathrm{sgn}(t).$$

that leads to the Riesz-Feller potential (line 12 in Table 1)

$$D_0^{-\alpha} f(t) = \frac{1}{\sin\left(\alpha\frac{\pi}{2}\right)\Gamma(\alpha)} \int_{\mathbb{R}} f(t-\tau) |\tau|^{\alpha-1} \mathrm{sgn}(\tau) d\tau \qquad (50)$$

8. $\alpha = 0$ and $\theta = 1$
   In this case, $\Psi_\theta^0(i\omega) = e^{i\frac{\pi}{2}\theta \mathrm{sgn}(\omega)} = i\,\mathrm{sgn}(\omega)$, and (40) leads to

$$D_1^0 f(t) = \frac{1}{\pi} \int_{-\infty}^{\infty} f(t-\tau) \frac{1}{\tau} d\tau \qquad (51)$$

which is the Hilbert transform of $f(t)$ [21,25].

Remark 8. *We note that the scheme we presented is a theretical base for supporting the development and aplications of the FD. Practical problems may require some kind of modification—see for example [29].*

4.4. Classic Riemann-Liouville, Caputo, and Hadamard derivatives

The classic formulations of Riemann-Liouville (RL) and Caputo (C) left derivatives ($\alpha > 0$) are obtained from the (47) and (48) assuming that $f(t)$ is defined on a given interval $[a, b]$ (we can set $b = \infty$). Therefore, for $t \in [a, b]$ the RL and C derivatives are given by

$$^{RL}D_l^\alpha f(t) = D_l^N \left[\frac{1}{\Gamma(-\alpha + N)} \int_a^t (t-\tau)^{N-\alpha-1} f(\tau) d\tau \right] \tag{52}$$

$$^{C}D_l^\alpha f(t) = \frac{1}{\Gamma(-\alpha + N)} \int_a^t (t-\tau)^{N-\alpha-1} f^{(N)}(\tau) d\tau, \tag{53}$$

respectively, where $N = \lfloor \alpha \rfloor + 1$.

Remark 9. *It is important to note that, although the function $f(t)$ has bounded support, both derivatives define non bonded support functions.*

Concerning the Hadamard derivative and anti-derivative cases and for $\alpha > 0$, we have [2]

$$^{H}D^\alpha[f(x)] = \left(x \frac{d}{dx}\right)^N \frac{1}{\Gamma(N-\alpha)} \int_a^x \left(\log \frac{x}{\xi}\right)^{N-\alpha+1} \frac{f(\xi) d\xi}{\xi} \tag{54}$$

and

$$^{H}D^{-\alpha}[f(x)] = \frac{1}{\Gamma(\alpha)} \int_a^x \left(\log \frac{x}{\xi}\right)^{\alpha-1} \frac{f(\xi) d\xi}{\xi}. \tag{55}$$

With the change of variable inside the integral, that is, with $\xi = e^\tau$ and $x = e^t$, we obtain a derivative of the RL type.

5. On the Discrete-Time Derivatives

There are several approaches into the discrete-time FD. The most interesting are

- The methodology based on time scales [30–32] that uses the nabla and delta derivatives;
- Infinite series based on the approaches by Tarasov [33,34].

The first approach has more similarities with the theory we presented here and consists of the framework presented in [32]. In fact, the discrete-time derivatives described there recover the GL forward and backward derivatives introduced in (20) and (21). However, it is not straightforward to introduce some tool similar to Bode diagrams because:

1. The frequency response $H(i\omega)$ is obtained from the TF given by $H(s)$, when s assumes values on the Hilger circle: $|s - \frac{1}{h}| = \frac{1}{h}$, where h is the sampling interval. Therefore, the domain is defined by $\omega \in (-\frac{\pi}{h}, \frac{\pi}{h}]$;
2. The eigenvalue of the nabla derivative corresponding to the eigenvector $e^{i\omega hn}$ is $s = \frac{1-e^{-i\omega h}}{h}$. If h is very small, then $s \approx i\omega$. Therefore, only for small values of h the derivative is represented by straight lines in log plots;
3. There are no studies for the two-sided derivatives recovering to the one described in Section 3.5. The one proposed in [34] has a different formulation and properties.

From these considerations we conclude that the topic of discrete-time derivatives, requires still further study for keeping the simplicity of Bode diagrams.

6. Conclusions

This paper discussed the problem of multiple attempts to have distinct operators under the umbrella of "fractional derivatives". One possible strategy is to discuss the validity of several operators recently proposed. Indeed, in previous papers it was demonstrated that such "novel" fractional derivatives are incorrect. Here we adopted an alternative strategy based on the classical system theory well known in applied sciences. Based on the tools of this theory we discussed a unified framework demystifying misleading, and often incorrect, formulations. Quoting again Henri Poincaré: *To doubt everything, or, to believe everything, are two equally convenient solutions; both dispense with the necessity of reflection.*

Author Contributions: These two authors contributed equally to this paper.

Funding: This work was funded by Portuguese National Funds through the FCT—Foundation for Science and Technology under the project PEst- UID/EEA/00066/2013.

Conflicts of Interest: The authors declare no conflict of interest.

Abbreviations

The following abbreviations are used in this manuscript:

FD	Fractional derivative
FI	Fractional integral
RL	Riemann-Liouville
L	Liouville
C	Caputo
GL	Grünwald-Letnikov
H	Hadamard

References

1. Samko, S.; Kilbas, A.; Marichev, O. *Fractional Integrals and Derivatives: Theory and Applications*; Gordon and Breach Science Publishers: Amsterdam, The Netherlands, 1993.
2. Kilbas, A.; Srivastava, H.; Trujillo, J. *Theory and Applications of Fractional Differential Equations*; North-Holland Mathematics Studies; Elsevier: Amsterdam, The Netherlands, 2006; Volume 204.
3. Dugowson, S. Les Différentielles Métaphysiques. Ph.D. Thesis, Université Paris Nord, Villetaneuse, France, 1994.
4. Kiryakova, V. A long standing conjecture failed? In *Transform Methods and Special Functions*; Institute of Mathematics and Informatics, Bulgarian Academy of Sciences: Sofia, Bulgaria, 1998; pp. 579–588.
5. Kiryakova, V. A brief story about the operators of the generalized Fractional Calculus. *Fract. Calc. Appl. Anal.* **2008**, *11*, 203–220.
6. Magin, R. *Fractional Calculus in Bioengineering*; Begell House Inc.: Redding, CT, USA, 2006.
7. Tarasov, V.E. *Fractional Dynamics: Applications of Fractional Calculus to Dynamics of Particles, Fields and Media*; Nonlinear Physical Science; Springer: Beijing, China, 2010.
8. Uchaikin, V.V. *Fractional Derivatives for Physicists and Engineers: Background and Theory*; Nonlinear Physical Science; Springer: Beijing, China, 2013.
9. Ortigueira, M.D. *Fractional Calculus for Scientists and Engineers*, 2nd ed.; Lecture Notes in Electrical Engineering; Springer: Berlin/Heidelberg, Germany, 2011.
10. Herrmann, R. *Fractional Calculus: An Introduction for Physicists*; World Scientific Publishing Co.: Singapore, 2011.
11. Ortigueira, M.D.; Machado, J.A.T. What is a fractional derivative? *J. Comput. Phys.* **2015**, *293*, 4–13. [CrossRef]
12. Ortigueira, M.; Machado, J.T. Which Derivative? *Fract. Fract.* **2017**, *1*, 3. [CrossRef]
13. Tarasov, V.E. No violation of the Leibniz rule. No fractional derivative. *Commun. Nonlinear Sci. Numer. Simul.* **2013**, *18*, 2945–2948. [CrossRef]
14. Ortigueira, M.D.; Machado, J.T. A critical analysis of the Caputo–Fabrizio operator. *Commun. Nonlinear Sci. Numer. Simul.* **2018**, *59*, 608–611. [CrossRef]
15. Giusti, A. A comment on some new definitions of fractional derivative. *Nonlinear Dyn.* **2018**, *93*, 1757–1763. [CrossRef]

16. Tarasov, V.E. No nonlocality: No fractional derivative. *Commun. Nonlinear Sci. Numer. Simul.* **2018**, *62*, 157–163. [CrossRef]
17. Abdelhakim, A.; Machado, J.A.T. A critical analysis of the conformable derivative. *Nonlinear Dyn.* **2019**, 1–11. [CrossRef]
18. Machado, J.T.; Mainardi, F.; Kiryakova, V. Fractional Calculus: Quo Vadimus? (Where are we Going?). *Fract. Calc. Appl. Anal.* **2015**, *18*, 495–526. [CrossRef]
19. Machado, J.T.; Mainardi, F.; Kiryakova, V.; Atanackovic, T. Fractional Calculus: D'où Venons-Nous? Que Sommes-Nous? Où Allons-Nous? (Contributions to Round Table Discussion held at ICFDA 2016). *Fract. Calc. Appl. Anal.* **2016**, *19*, 1074–1104. [CrossRef]
20. Ortigueira, M.; Machado, J.T. Fractional Definite Integral. *Fract. Fract.* **2017**, *1*, 2. [CrossRef]
21. Oppenheim, A.V.; Willsky, A.S.; Hamid, S. *Signals and Systems*, 2nd ed.; Prentice-Hall: Upper Saddle River, NJ, USA, 1997.
22. Ortigueira, M.D. The fractional quantum derivative and its integral representations. *Commun. Nonlinear Sci. Numer. Simul.* **2010**, *15*, 956–962. [CrossRef]
23. Ortigueira, M.D. Riesz potential operators and inverses via fractional centred derivatives. *Int. J. Math. Math. Sci.* **2006**, *2006*, 6. [CrossRef]
24. Ortigueira, M.D. Fractional central differences and derivatives. *J. Vib. Control* **2008**, *14*, 1255–1266. [CrossRef]
25. Roberts, M. *Signals and Systems: Analysis Using Transform Methods And Matlab*, 2nd ed.; McGraw-Hill: New York, NY, USA, 2003.
26. Liouville, J. Memóire sur le calcul des différentielles à indices quelconques. *J. l'École Polytech. Paris* **1832**, *13*, 71–162.
27. Gel'fand, I.M.; Shilov, G.E. *Generalized Functions. Volume I: Properties and Operations*; Academic Press: New York, NY, USA; London, UK, 1964.
28. Ortigueira, M.D.; Magin, R.L.; Trujillo, J.J.; Velasco, M.P. A real regularised fractional derivative. *Signal Image Video Process.* **2012**, *6*, 351–358. [CrossRef]
29. Wei, Y.; Chen, Y.; Cheng, S.; Wang, Y. A note on short memory principle of fractional calculus. *Fract. Calc. Appl. Anal.* **2017**, *20*, 1382–1404. [CrossRef]
30. Bastos, N.R.O. Fractional Calculus on Time Scales. Ph.D. Thesis, Aveiro University, Aveiro, Portugal, 2012.
31. Goodrich, C.; Peterson, A.C. *Discrete Fractional Calculus*; Springer: Cham, Switzerland, 2015.
32. Ortigueira, M.D.; Coito, F.J.; Trujillo, J.J. Discrete-time differential systems. *Signal Process.* **2015**, *107*, 198–217. [CrossRef]
33. Tarasov, V.E. Exact Discrete Analogs of Derivatives of Integer Orders: Differences as Infinite Series. *J. Math.* **2015**, *2015*, 134842. [CrossRef]
34. Tarasov, V.E. Exact discretization by Fourier transforms. *Commun. Nonlinear Sci. Numer. Simul.* **2016**, *37*, 31–61. [CrossRef]

© 2019 by the authors. Licensee MDPI, Basel, Switzerland. This article is an open access article distributed under the terms and conditions of the Creative Commons Attribution (CC BY) license (http://creativecommons.org/licenses/by/4.0/).

Article

Fractional Derivatives and Integrals: What Are They Needed For?

Vasily E. Tarasov [1,2],* and Svetlana S. Tarasova [2]

[1] Skobeltsyn Institute of Nuclear Physics, Lomonosov Moscow State University, Moscow 119991, Russia
[2] Faculty "Information Technologies and Applied Mathematics", Moscow Aviation Institute (National Research University), Moscow 125993, Russia; s.s.tarasova@bk.ru
* Correspondence: tarasov@theory.sinp.msu.ru; Tel.: +7-495-939-5989

Received: 19 December 2019; Accepted: 22 January 2020; Published: 25 January 2020

Abstract: The question raised in the title of the article is not philosophical. We do not expect general answers of the form "to describe the reality surrounding us". The question should actually be formulated as a mathematical problem of applied mathematics, a task for new research. This question should be answered in mathematically rigorous statements about the interrelations between the properties of the operator's kernels and the types of phenomena. This article is devoted to a discussion of the question of what is fractional operator from the point of view of not pure mathematics, but applied mathematics. The imposed restrictions on the kernel of the fractional operator should actually be divided by types of phenomena, in addition to the principles of self-consistency of mathematical theory. In applications of fractional calculus, we have a fundamental question about conditions of kernels of fractional operator of non-integer orders that allow us to describe a particular type of phenomenon. It is necessary to obtain exact correspondences between sets of properties of kernel and type of phenomena. In this paper, we discuss the properties of kernels of fractional operators to distinguish the following types of phenomena: fading memory (forgetting) and power-law frequency dispersion, spatial non-locality and power-law spatial dispersion, distributed lag (time delay), distributed scaling (dilation), depreciation, and aging.

Keywords: fractional calculus; fractional derivative; translation operator; distributed lag; time delay; scaling; dilation; memory; depreciation; probability distribution

MSC: 26A33 Fractional derivatives and integrals; 34A08 Fractional differential equations; 60E05 Distributions: general theory

---

1. Introduction

Why do we need fractional derivatives and integrals of non-integer order? We are not interested in the answer from the standpoint of philosophy or methodology of science. We are primarily interested in the answer from the point of view of applied mathematics, theoretical physics, economic theory, and other applied sciences. For application of fractional calculus [1–7], we want to have an answer in the form of exact mathematical statements that is formulated in precise and strict form. To get such an answer, it is required to formulate the question in mathematical form. The question should actually be formulated as a mathematical problem of applied mathematics, as a task for new research.

We also do not plan to delve into the "linguistic" question of which operators might be called fractional and which are not. The first author has already formulated their point of view on this issue in articles [8–12]. There are also many important contributions to this discussion (for example, see [13–16]). In this article, we do not plan to continue the discussion directly in this direction. We want to direct our discussion in a different direction. However, we will make an important remark for this paper. Please note that the proposed principle "No nonlocality. No fractional derivative" [11] cannot be turned into the principle "No memory. No fractional derivative". This is due to the fact that nonlocality in time cannot be reduced only to memory (about the concept of memory, see for example in articles [17,18]). It should also be noted here that the operators that describe the delay, lag, and scaling continuously distributed over time cannot be attributed to fractional operators if the distribution is described by probability density functions. These operators are integer order operators with distributed delay, lag, and scaling.

The fractional calculus, which is the theory of integrals and derivatives of fractional order, describes a wide variety of different types of operators with non-integer order [1–7]. Fractional calculus allows us to describe various phenomena and effects in natural and social sciences. For example, we should note the non-locality of power-law type, spatial dispersion of power type, fading memory, frequency dispersion of power type, intrinsic dissipation, the openness of systems (interaction with environment), fractional relaxation-oscillation, fractional viscoelasticity, fractional diffusion-waves, long-range interactions of power-law type, and many others [19,20].

In applied mathematics, it is important to have a tool that allows you to adequately select the type of fractional operators for the type of phenomena under consideration. It is necessary to have clear mathematical criteria for associating fractional operators of non-integer orders and those types of phenomena that they can describe. Differential and integral operators of non-integer orders are a powerful tool for modeling and description of processes that characterized by fading memory and spatial nonlocality. However, not all operators of non-integer orders can describe the effects of memory (or non-locality).

We should emphasize that not all fractional derivatives and integrals can be used for modeling the processes with memory. For example, the Kober and Erdelyi–Kober operators as well as the Caputo–Fabrizio integral and derivatives cannot be applied to describe phenomena with memory or spatial nonlocality. These operators can be applied only to describe processes with continuously distributed scaling (dilation) and lag (delay), respectively [21,22]. We also can state that these operators can be interpreted as derivatives and integrals of integer orders with scaling or lag, distributions of which are described by some probability density functions [21,22].

In application of the differential and integral operators with non-integer orders, a fundamental question arises about the correct subject interpretation of the different types of operators. Interpretation is not in the form of a description of one of the particular manifestations of real processes, but by one or another type of phenomena. We should clearly understand what type of effects and phenomena a given fractional operator of non-integer order can describe.

It is necessary to understand what types of fractional operators, what types of phenomena can be described in principle. The most important role in this description of phenomena must be understood by what types of fractional derivatives and integrals of non-integer order, in principle, what types (classes) of phenomena can describe.

In applications of fractional calculus, we can distinguish the following types (classes) of phenomena by some properties of kernels:

- fading memory (forgetting) and power-law frequency dispersion;
- spatial non-locality and power-law spatial dispersion;
- distributed lag (time delay);
- dictributed scaling (dilation);
- depreciation and aging.

These types of phenomena can be described by fractional operators of non-integer orders with some types of operator kernels. For these types of phenomena, we should have mathematical conditions on the operator kernels, which uniquely identify one of types of these phenomena.

Let us give some examples of the correspondence between the some fractional derivatives (or integrals) and the type of phenomena, which can be described by these operators in Table 1.

Examples of these type of phenomena in physics are described in Handbook of Fractional Calculus with Application [19,20].

Table 1. Examples of the correspondence between the some fractional derivatives (or integrals) and the type of phenomena.

№	Type of Phenomena:	Example of Fractional Operators:
1	Memory and Non-Locality in Time	Caputo and Riemann–Liouville
2	Spatial Non-Locality and Spatial Dispersion	Riess and Liouville
3	Distributed Time Delay and Lag	Caputo–Fabrizio
4	Distributed Dilation and Scaling	Kober and Erdelyi–Kober, Gorenflo–Luchko–Mainardi
5	Distributed Depreciation and Aging	Prabhakar and Kilbas–Saigo–Saxena

In this paper, we proposed the properties of operator kernels and corresponding types of phenomena. In fractional calculus, we do not have a list of correspondence between mathematical properties of the operators kernels and types of effects and phenomena. Mathematically rigorous conditions on the kernels of fractional differential and integral operators are necessary to distinguish between different types of phenomena and processes.

First of all, we must clearly distinguish between types of fractional operators and types of phenomena. This should not be just a list of examples of specific manifestations in the different sciences. In fractional calculus, we should have correspondence between the types of phenomena and the types of properties of operator kernels. In this article, we will explain in more detail the proposed approach to the interpretation of fractional derivatives and integrals.

2. Formulation of Mathematical Problem

This article does not claim to be a general consideration of fractional derivatives and integrals. To simplify the discussion of fractional operators, we will consider operators with respect to one variable t, which will be interpreted as time. A discussion of the problem of the relationship between the types of phenomena and the types of fractional operators will be constructed on the example of the following operator

$$\left(D_{(K)}f\right)(t) = \int_{t_0}^{t} K(t,\tau)\left(\mathcal{D}_{\tau}^{(n)}f(\tau)\right)d\tau, \tag{1}$$

where $\left(\mathcal{D}_{\tau}^{(n)}f\right)(\tau)$ is differential operator of the integer order n, where $n = 0, 1, 2, \ldots$, and $K(t,\tau)$ is a kernel of the operator. For example, we can consider the standard derivative of the integer order n, i.e.,

$$\left(\mathcal{D}_{\tau}^{(n)}f\right)(\tau) = f^{(n)}(\tau) = \frac{d^n f(\tau)}{d\tau^n}. \tag{2}$$

In general, the kernel $K(t,\tau)$ depends on the order n and initial point t_0, i.e., we should use $K_{n,t_0}(t,\tau)$. To simplify the notation, we will use $K(t,\tau)$, assuming that n and t_0 are already fixed. Expression (1) has a sense, if the integral (1) exists. In general, the function $f^{(n)}(\tau)$ does not have to be continuous function and the kernel $K(t,\tau)$ can have an integrable singularity of some kind.

Remark 1. *In general, we can consider other differential operators $\left(\mathcal{D}_\tau^{(n)} f\right)(\tau)$ of the integer order n instead of the standard derivative $f^{(n)}(\tau)$. For example, we can consider the operators (1), where differential operator $\left(\mathcal{D}_\tau^{(n)} f\right)(\tau)$ is defined in the form.*

$$\left(\mathcal{D}_\tau^{(n)} f\right)(\tau) = \prod_{k=0}^{n}\left(1 + \gamma + k + \beta^{-1}\tau\frac{d}{d\tau}\right), \tag{3}$$

which is used in the Gorenflo–Luchko–Mainardi (GLM) operator [23–25] with some parameters $\gamma \in \mathbb{R}$ and $\beta > 0$ of the kernel $K(t, \tau)$ (for details see Equations (1) and (12) in [24], and Equations (4) and (39) in [25]). This operator is also known as the left-sided Caputo-type modification of the Erdelyi–Kober fractional derivative (see Equation (12) in [24] (p. 362)). Please note that the GLM operator was introduced for the first time in [23] in connection with the scale-invariant solutions of the time-fractional diffusion-wave equation (see Equation (58) on [23] (p. 188)). The special form (3) is needed in order to make this operator a left-inverse operator to the Erdelyi–Kober integral operator (see Equation (13) in [24] (p. 362)). Emphasize that the main property of any generalized (fractional) derivative is to be a left-inverse operator to the corresponding generalized (fractional) integral operator. Please note that the Kober and Erdelyi–Kober operators [1,4], as well as the Caputo–Fabrizio operators [26–28], cannot be applied for modeling processes with fading memory or spatial nonlocality. These operators can be used only to describe continuously distributed scaling (dilation) and lag (delay), respectively (sections below). Therefore we also can state that these operators are interpreted as derivatives and integrals of integer orders with scaling or lag, distributions of which are described by probability density functions.

Remark 2. *We can also consider instead of $\left(\mathcal{D}_\tau^{(n)} f\right)(\tau)$ a fractional differential (or integral) operator $\left(\mathcal{D}_\tau^{(\alpha)} f\right)(\tau)$ of another type than the ones defined by the kernel $K(t, \tau)$. For example, we can use the Caputo fractional derivative, $\left(\mathcal{D}_\tau^{(\alpha)} f\right)(\tau) = \left(D_{C,0+}^\alpha f\right)(\tau)$, and the kernel is the probability density function of the gamma distriburion (for details, see the Section 7 of the article [29] and the papers [30–32]). Such a choice is necessary to describe the simultaneous presence of two such phenomena as distributed lag and fading memory.*

Let us give a formulation of a mathematical problem of applied mathematics, as task for new research in fractional calculus that will be illustrated in this paper below.

Mathematical problem of fractional calculus in application: What conditions must the kernel $K(t, \tau)$ of operator (1) have in order to describe one or another type of phenomena? It is necessary to obtain exact correspondences between sets of properties of kernel and type of phenomena.

In this paper, we describe the conditions on the kernel $K(t, \tau)$, which allow us to use operator of the form (1) to describe the following types of phenomena:

(Type I): Continuously Distributed Scaling (Dilation);
(Type II): Continuously Distributed Lag (Delay).

We also give some comments to the phenomena:

(Type III): Continuously Distributed Fading Memory;
(Type IV): Distributed Depreciation and Aging.

Let us give these conditions for phenomena of Types I and II in the form of the following statements. The conditions on the kernel $K(t, \tau)$ for phenomenon of Types III and IV are discussed in the separate sections of this paper.

Statement 1.

Let us assume that the kernel $K(t, \tau)$ of the operator (1) with $t_0 = 0$ satisfies the following conditions

$$K(\lambda t, \lambda \tau) = \lambda^{-1} K(t, \tau), \tag{4}$$

for all $\lambda > 0$, and the condition of non-negativity, and the normalization condition

$$K(1,x) \geq 0, \quad \int_0^1 K(1,x)\,dx = K < \infty \tag{5}$$

for all $x \in (0,1)$, where K is a finite positive constant. In this case, operator (1) can be represented (by using the change of variable $\tau \to x = \tau/t$) in the form

$$\left(D_{(K)}f\right)(t) = K \int_0^1 \rho_1(x)\, S_x\!\left(\mathcal{D}_\tau^{(n)} f(\tau)\right) dx \tag{6}$$

with a numerical factor K, where $\rho_1(x) = K(1,x)/K$ is the probability density function that satisfies the condition of non-negativity and the normalization condition

$$\rho_1(x) \geq 0, \quad \int_0^1 \rho_1(x)\,dx = 1, \tag{7}$$

and S_x is the scaling (dilation) operator

$$\begin{aligned}
S_x f(t) &= f(t\cdot x), \\
S_x\!\left(\mathcal{D}_z^{(n)} f(z)\right) &= \left(\mathcal{D}_z^{(n)} f(z)\right)_{z=t\cdot x}, \\
S_x f^{(n)}(t) &= \left(\tfrac{d^n f(z)}{dz^n}\right)_{z=t\cdot x}.
\end{aligned} \tag{8}$$

Then operator (1) describes the continuously distributed scaling (dilations). In physics and economics, the dilation is the change of scale of objects and processes.

Remark 3. *Please note that using property (4) also allows us to write the operator (1) as the Mellin-type convolution*

$$\left(D_{(K)}f\right)(t) = \int_0^t K\!\left(\tfrac{t}{\tau},1\right)\!\left(\mathcal{D}_\tau^{(n)} f(\tau)\right)\tfrac{d\tau}{\tau}, \tag{9}$$

which differs from the Mellin convolution by the upper limit of t instead of infinity. Using the kernel

$$K_H(x,1) = \begin{cases} K(x,1) & x > 1, \\ 0 & x \leq 1. \end{cases} \tag{10}$$

The operator (7) can be represented in the form

$$\left(D_{(K)}f\right)(t) = K_H *_M f^{(n)} = \int_0^\infty K_H\!\left(\tfrac{t}{\tau},1\right)\!\left(\mathcal{D}_\tau^{(n)} f(\tau)\right)\tfrac{d\tau}{\tau}, \tag{11}$$

where $*_M$ is the Mellin convolution [33,34]. This representation allows us to propose a generalization the operator (9) by using the of the Mellin convolution in the definition of these generalized operators [29].

Remark 4. *Operators (1) and (2) with kernel, which satisfies the conditions (4) and (5), cannot be considered to be fractional derivative of non-integer order for positive integer values of n. The correct interpretation of these operators is integer order derivatives with the continuously distributed scaling (dilation). Please note that as a basis for the definition of these operators, which actually are integer order operators, one can use expression (6) with conditions (7) instead of Equation (1) with conditions (4) and (5).*

To have fractional generalization of these operators there are two ways: (A) we can use a fractional differential (or integral) operator $\left(\mathcal{D}_\tau^{(\alpha)} f\right)(\tau)$ instead of $\left(\mathcal{D}_\tau^{(n)} f\right)(\tau)$; (B) we can also use the kernel $\rho_1(x)$, which is not satisfied the normalization condition (7). In the work [29], we proposed a fractional generalization of this type of operators by the way (A) to describe processes with fading memory and distributed scaling.

Remark 5. *In our opinion, the Kochubei's approach to general fractional calculus [35–37], which is based on the Laplace convolution, can be applied to formulate new general fractional calculus, which will be based on the Mellin convolution. Moreover, the general fractional operators (9) and (11) can be used to formulate a generalization of the Luchko operational calculus [24,38], where the Mellin convolution will be used instead of the Laplace convolution.*

Statement 2.

Let us assume that the kernel $K(t,\tau)$ of the operator (1) with $t_0 = -\infty$ satisfies the following condition

$$K(t,\tau) = K(t-\tau) \tag{12}$$

for all $t > \tau$, the condition of non-negativity and the normability (or the normalization) condition

$$K(x) \geq 0, \quad \int_0^\infty K(x)\,dx = K < \infty \tag{13}$$

for all $x \in (0,\infty)$, where K is a finite positive constant. In this case, operator (1) can be represented (by using the change of variable $\tau \to x = t - \tau$) in the form

$$\left(D_{(K)}f\right)(t) = K \int_0^\infty \rho_2(x) T_x\left(D_\tau^{(n)} f(\tau)\right) dx \tag{14}$$

with a finite positive constant K, where $\rho_2(x) = K(x)/K$ is the probability density function that satisfies the condition of non-negativity and the normalization condition

$$\rho_2(x) \geq 0, \quad \int_0^\infty \rho_2(x)\,dx = 1, \tag{15}$$

and T_x is the translation (shift, lag) operator

$$T_x f(t) = f(t-x), \quad T_x\left(D_\tau^{(n)} f(\tau)\right) = \left(D_z^{(n)} f(z)\right)_{z=t-x}. \tag{16}$$

Then operators (1) describe the continuously distributed lag (time delay).

Remark 6. *Given the above, we can state that the operator with kernel, which satisfies the conditions (12) and (13), cannot be interpreted as fractional derivative of non-integer order for positive integer values of n. The correct interpretation of this operator is integer order derivative with the continuously distributed lag [29]. As a basis for the definition of this operator, which is integer order operators, we can use expression (14) with conditions (15) instead of Equation (1) with conditions (12) and (13).*

To have a fractional generalization of this operator, there are two ways: (A) to use a fractional differential (or integral) operator $\left(D_\tau^{(\alpha)} f\right)(\tau)$ instead of $\left(D_\tau^{(n)} f\right)(\tau)$; (B) to use the kernel $K(t,\tau)$, for which the normalization condition (15) is violated. In the work [29], we proposed a fractional generalization of this type operators by the way (A) to describe processes with memory and distributed lag. The fractional derivatives and integrals of non-integer orders, in which lag (time delay) is described by continuous probability distributions, were proposed in [29] (pp. 148–154), and used in macroeconomic models [30–32]. An example of fractional operators with distributed lag is also suggested in the Section 7 of the paper [29] (pp. 148–154).

Remark 7. *Please note that general operators of type (1) with the kernel (12) and without the condition (8) were considered by Anatoly N. Kochubei in works [35–37]. These works suggested concept of a general fractional calculus by using the differential operator based on Laplace convolution. Kochubei proposed the mathematical conditions on kernel of general fractional derivative, which lead to the fact that this general operator has a right inverse operator (a kind of a general fractional integral).*

3. Continuously Distributed Scaling (Dilation): Erdelyi–Kober Operators

As a generalization of the Riemann–Liouville fractional integral was proposed by Herman Kober. The Kober fractional integral [4] (p. 106), of the order $\alpha > 0$ is defined as

$$\left(I^{\alpha}_{K;0+;\eta}f\right)(t) = \frac{t^{-\alpha-\eta}}{\Gamma(\alpha)} \int_0^t \tau^{\eta} (t-\tau)^{\alpha-1} f(\tau) d\tau, \tag{17}$$

where $\eta \in \mathbb{R}$. If function $f(t) \in L_p(\mathbb{R}_+)$, with $1 \leq p < \infty$, and $\eta > (1-p)/p$, the operator (17) is bounded [1] (p. 323). For $\eta = 0$, operator (17) can be expressed through the Riemann-Liouville integration by the expression

$$\left(I^{\alpha}_{K;0+;1}f\right)(t) = t^{-\alpha} \left(I^{\alpha}_{RL,0+}f\right)(t). \tag{18}$$

Changing the variable of integration by $\tau \to x = \tau/t$, the Kober operator (17) takes the form

$$\left(I^{\alpha}_{K;0+;\eta}f\right)(t) = \frac{1}{\Gamma(\alpha)} \int_0^1 x^{\eta}(1-x)^{\alpha-1} f(x\,t) dx. \tag{19}$$

Expression (19) allows us to use the probability density function (p.d.f.) of the beta distribution in the form

$$\rho_{\alpha;\beta}(x) = \frac{1}{B(\alpha,\beta)} x^{\alpha-1}(1-x)^{\beta-1} \text{ for } x \in [0,1], \tag{20}$$

and $\rho_{\alpha;\beta}(x) = 0$ if $x \notin [0,1]$, where $B(\alpha,\beta)$ is the beta function. Using (20), the Kober fractional integral is represented by the equation

$$\left(I^{\alpha}_{K;0+;\eta}f\right)(t) = K_{EK} \int_0^1 \rho_{\eta+1;\alpha}(x) f(x\cdot t) dx \tag{21}$$

with the constant

$$K_{EK} = \frac{\Gamma(\eta+\alpha+1)}{\Gamma(\eta+1)}. \tag{22}$$

We note that expression (21) contains $f(x\cdot t)$ instead of $f(x)$. Therefore the variable $x > 0$ can be interpreted as a random variable, which describes scaling (dilation) with the gamma distribution. Using the scaling operator S_x: $S_x f(t) = f(x\cdot t)$, the Kober fractional integral (17) is represent by the equation

$$\left(I^{\alpha}_{K;0+;\eta}f\right)(t) = K_{EK} \int_0^1 \rho_{\eta+1;\alpha}(x)(S_x f(t)) dx, \tag{23}$$

where K_{EK} is defined by Equation (22). Equation (23) leads to the interpretation of the Kober operator as an expected value, where $x > 0$ is a random variable that describes the scaling and has the beta distribution up to numerical factor (22).

As a result, expression (23) gives a possibility to state that the Kober operator (17) can be interpreted as a continuously distributed dilation operator, in which the scaling variable has the beta distribution up to a constant factor (22).

The proposed interpretation of the Kober operator (17) allows us to generalize this operator by using other the probability density function instead of the beta distribution (20) and other lower and upper limits of integral in Equation (23). For example, the generalized operator of continuously distributed scaling (dilation) is define [29] by the expression

$$\left(D_{(\rho;S)}f\right)(t) = \int_0^\infty \rho(x)\left(S_x\left(\mathcal{D}_t^{(n)}f\right)(t)\right)dx, \tag{24}$$

where $n = 0, 1, 2, \ldots$, and $\rho(x) \geq 0$ is the probability density function such that

$$\int_0^\infty \rho(x)dx = 1. \tag{25}$$

In Equation (24) it is assumed that the integral $\int_0^\infty \rho(x)\left|S_x\left(\mathcal{D}_t^{(n)}f\right)(t)\right|dx$ converges, where $\mathcal{D}_x^{(n)}f(x)$ and $\rho(x)$ are piecewise continuous or continuous functions on $\mathbb{R}$. Here we can consider $\mathcal{D}_t^{(n)}f(t) = f^{(n)}(t)$.

The Erdelyi–Kober type operator [4] (p. 105), is defined by the equation

$$\left(I_{EK;0+;\sigma,\eta}^\alpha f\right)(t) = \frac{\sigma\, t^{-\sigma(\alpha+\eta)}}{\Gamma(\alpha)} \int_0^t \tau^{\sigma(\eta+1)-1}\, (t^\sigma - \tau^\sigma)^{\alpha-1} f(\tau)d\tau, \tag{26}$$

where $\alpha > 0$ is the order of integration. To get the notation of the paper (see Equation (1) in p. 360, [24]), we should change the indexes: $\sigma \to \beta$, $\alpha \to \delta$, $\eta \to \gamma$. In the case $\sigma = 1$, operator (26) is represented in the form of the Kober operator (17). Operator (26) can be represented by the equation

$$\left(I_{EK;0+;\sigma,\eta}^\alpha f\right)(t) = K_{EK} \int_0^1 \rho_{EK}(x)(S_x f(t))dx \tag{27}$$

with the probability density function

$$\rho_{EK}(x) = \frac{\sigma}{B(\eta+1, \alpha)} x^{\sigma(\eta+1)-1}(1-x^\sigma)^{\alpha-1}, \tag{28}$$

and the constant factor K_{EK} defined by Equation (22). For $\sigma = 1$, the function (28) described beta distribution (20).

As a result, the Erdelyi–Kober and Kober operators are operators of integer orders with continuously distributed scaling (dilation). We should note that the fractional generalizations of these operators, which can be applied to describe simultaneously action of distributed scaling and fading memory, were proposed in [29].

As a result, we can state that the operators (1) with kernels (4) and (5), the operators (6) with different probability density functions (7), and operators (23), (24), (27) can be applied to describe continuously distributed scale phenomena in economics, physics, and other sciences.

4. Continuously Distributed Delay (Lag): Caputo–Fabrizio Operator

The Caputo–Fabrizio operator is proposed in [26–28]. The Caputo–Fabrizio operator $D_{CF}^{(\alpha)}$ of the non-integer order $\alpha \in (0,1)$ is defined (see Equation (2.2) of [26] (p. 74)) by the equation

$$\left(D_{CF}^{(\alpha)}f\right)(t) = \frac{m(\alpha)}{1-\alpha}\int_{t_0}^t \exp\left\{-\frac{\alpha}{1-\alpha}(t-\tau)\right\}f^{(1)}(\tau)d\tau, \tag{29}$$

where $f^{(1)}(\tau) = df(\tau)/d\tau$ is the standard derivative of first order, $m(\alpha)$ is a "normalization" function. For $n > 1$, the Caputo–Fabrizio operator of the order $\alpha + n \in (n, n+1)$ is defined (see, Equation (2.8) of [26] (p. 76)) by the expression

$$\left(D_{CF}^{(\alpha+n)} f\right)(t) = \left(D_{CF}^{(\alpha)} f^{(n)}\right)(t), \tag{30}$$

where $\alpha \in (0,1)$ and $f^{(n)}(\tau) = d^n f(\tau)/d\tau^n$ are the standard derivatives of integer order $n \in \mathbb{N}$. The Caputo–Fabrizio operator of the order $\alpha \in (n, n+1)$ is defined (see, Equation (2.8) of [26] (p.76)) by the expression

$$\left(D_{CF}^{(\alpha)} f\right)(t) = \frac{m(\alpha - n)}{n - \alpha + 1} \int_{t_0}^{t} \exp\left\{-\frac{\alpha - n}{n - \alpha + 1}(t - \tau)\right\} f^{(n+1)}(\tau) d\tau, \tag{31}$$

where $n = [\alpha]$. The Caputo–Fabrizio operators (31) of order $\alpha \in (n, n+1)$ with $t_0 = -\infty$ can be represented in the form

$$\left(D_{CF}^{(\alpha)} f\right)(t) = \frac{\lambda \, m(\alpha - n)}{\alpha - n} \int_{-\infty}^{t} \exp\{-\lambda (t - \tau)\} f^{(n+1)}(\tau) d\tau, \tag{32}$$

where

$$\lambda = \frac{\alpha - n}{n - \alpha + 1} \tag{33}$$

Changing the variable $\tau \to x = t - \tau$ of integration in (32), Equation (32) takes the form

$$\left(D_{CF}^{(\alpha)} f\right)(t) = \frac{\lambda \, m(\alpha - n)}{\alpha - n} \int_{0}^{\infty} \exp\{-\lambda x\} \, f^{(n+1)}(t - x) dx. \tag{34}$$

Equation (34) can be represented by expression (14) in the form

$$\left(D_{CF}^{(\alpha)} f\right)(t) = K_{CF} \int_{0}^{\infty} \rho(x) \left(T_x f^{(n+1)}(t)\right) dx, \tag{35}$$

where the positive constant K_{CF} is

$$K_{CF} = \frac{m(\alpha - n)}{\alpha - n}, \tag{36}$$

and $\rho(x)$ is the probability density function of the exponential distribution

$$\rho(x) = \lambda \exp(-\lambda x), \tag{37}$$

for $x > 0$ and $\rho(x) = 0$ for $x \leq 0$, where $\lambda > 0$ is the parameter that is often called the rate parameter or the speed of response [39] (p. 27). It is also used the parameter $T = 1/\lambda$ as time-constant of exponentially distributed lag. This parameter T is interpreted as the length of the time delay [39] (p.27). The kernel (37) is actively used in economics to describe processes with distributed lag [39] (p. 26). We should note that distribution (37) describes the time between events in a Poisson point process, which is the continuous analogue of the geometric distribution. It is well-known that this distribution has the key property of being memoryless.

In the work [22], it is proved that the Caputo–Fabrizio operator of the order $\beta = n - 1/(\lambda + 1)$, coincides with derivative of integer order with exponentially distributed lag, where λ is the rate parameter (33) of the distribution (37), and $n = [\beta] + 1$. Therefore, the Caputo–Fabrizio operator can be interpreted as an integer order derivative with the exponentially distributed time delay.

The existence of the time delay is based on the fact that the processes have a finite speed, and the change of the input does not lead to instant changes of output. In physical sciences it is well-known that the finite speed of the process does not mean that there is memory in the process. Therefore continuously distributed lag cannot be considered to be a dependence of the state of as process on its history. The time delay cannot be interpreted as a memory.

As a result, the Caputo–Fabrizio operators cannot be applied to modeling memory or spatial nonlocality in processes, but this operator describes continuously (exponentially) distributed time delay.

The proposed interpretation of the Caputo–Fabrizio operator (35) allows us to generalize this operator [29] by using other the probability density function instead of the exponential distribution (37). For example, the generalized operator of continuously distributed scaling (dilation) is define [29] by the expression

$$(D_{(\rho;T)}f)(t) = \int_0^\infty \rho(x)(T_x f^{(n)}(t))dx, \tag{38}$$

where $n = 0, 1, 2, \ldots$, and $\rho(x) \geq 0$ is the probability density function such that

$$\int_0^\infty \rho(x)dx = 1. \tag{39}$$

In Equation (38) it is assumed that the integral $\int_0^\infty \rho(x) \left| (T_x f^{(n)}(t)) \right| dx$ converges, where $f^{(n)}(x)$ and $\rho(x)$ are piecewise continuous or continuous functions on $\mathbb{R}$.

The fractional generalization of the Caputo–Fabrizio operator was proposed in [29] to take into account various distributions of delay time and power-law fading memory in one operator.

5. Continuously Distributed Fading Memory

To describe memory (the fading memory), we can use operators (1), for which the normability condition is not satisfied.

For example, the operator (1) with $t_0 = -\infty$ and the kernel

$$K(t, \tau) = \frac{1}{\Gamma(n-\alpha)}(t-\tau)^{n-\alpha-1} \tag{40}$$

is the left-sided Caputo fractional derivative of the order $\alpha \geq 0$ (see Equation (2.4.15) [4] (p. 92) for $a = -\infty$) that is defined by the equation

$$(D_{C+}^\alpha f)(t) = \frac{1}{\Gamma(n-\alpha)} \int_{-\infty}^t (t-\tau)^{n-\alpha-1} f^{(n)}(\tau)d\tau, \tag{41}$$

where $\Gamma(\alpha)$ is the gamma function, and $f^{(n)}(\tau)$ is the derivative of the integer order $n = [\alpha] + 1$ for non-integer values of α (and $n = \alpha$ for integer values of α). Changing the variable $\tau \to x = t - \tau$ operator (1) with the kernel (40) and $t_0 = -\infty$ can be represented in the form

$$(D_{(K)}f)(t) = \int_0^\infty K_c(x)(T_x f^{(n)}(t))dx, \tag{42}$$

where the kernel

$$K_c(x) = \frac{x^{n-\alpha-1}}{\Gamma(n-\alpha)} \tag{43}$$

cannot be interpreted as a probability density function since the normalization condition is violated

$$\int_0^\infty K_c(x)\,dx = \left(\frac{x^{n-\alpha}}{\Gamma(n-\alpha+1)}\right)_0^\infty = \infty \tag{44}$$

for non-integer values of α.

Let us describe some basic principles and properties of the kernel that should be taken into account to describe memory.

Principle of violation of normability. Processes with memory cannot be described by operators (1) if the operator kernel can be considered to be a probability density function. In other words, the memory function cannot be probability density function.

The requirement of violation of the normability conditions is not enough for a comprehensive description of fading memory. We should have conditions for the kernel of operator (1), which allow us to use this operator to described memory.

Principle of causality. The main condition that must be satisfied for all types of memory is the fulfillment of the causality principle. It is obvious that the operators that describe memory phenomena should satisfy the causality principle. In mathematical form, the causality principle can be realized by the Kramers–Kronig relations [18].

In addition to these relations, we can state that the right-sided fractional derivatives (for example the Riemann–Liouville, Liouville, and Caputo-type) cannot be used to processes with the memory. The right-sided fractional integrals and derivatives are defined for $\tau > t$, where t is the present time moment. Therefore these operators describe dependencies of processes on the future states. The left-sided fractional operators describe the past states of the process.

Principle of memory fading. The important property of memory is the memory fading. The principle of memory fading was first proposed by Ludwig Boltzmann, and then it was significantly developed by Vito Volterra. This principle states that the increasing of the time interval leads to a decrease in the contribution of impact to the response. The exact mathematical formulation of this principle is given in [40–44], it is more complicated than that required for us in this paper, which is restricted by the operators (1). Therefore we will use a simplified formulation of the principle of memory fading [17].

Let us consider two functions $f(\tau)$ and $y(t)$, which are interpreted as the impact and response variables respectively, and we will assume that these functions are connected by the equation

$$y(t) = \int_0^t K(t,\tau) \left(\mathcal{D}_\tau^{(n)} f(\tau) \right) d\tau. \tag{45}$$

Let us assume that $\mathcal{D}_\tau^{(n)} f(\tau)$ is different from zero on a finite time interval $\tau \in [0, T]$, and which is zero outside this interval ($\mathcal{D}_\tau^{(n)} f(\tau) = 0$ for $t > T$). This means that we consider $H(T-\tau)\mathcal{D}_\tau^{(n)} f(\tau)$ instead of $\mathcal{D}_\tau^{(n)} f(\tau)$ in Equation (45) with times $t \in [T, \infty)$. Then Equation (45) gives

$$y(t) = \int_0^T K(t,\tau) \left(\mathcal{D}_\tau^{(n)} f(\tau) \right) d\tau \text{ for } t < T. \tag{46}$$

We see that for $t > T$ there is no impact, but the response is different from zero ($y(t) \neq 0$ for $t > T$). This means that the memory about the impact, which acts on time interval $[0, T]$, is stored in the process. Therefore, we can state that this process saves the history of changes of the impact. Using the mean value theorem, there is a value $\xi \in [0, T]$ and Equation (46) can be written as

$$y(t) = K(t,\xi) \left(\mathcal{D}_\tau^{(n)} f(\tau) \right)_{\tau=\xi} T. \tag{47}$$

As a result, we can see that the behavior of the response $y(t)$ is determined by the behavior of the kernel $K(t,\tau)$ with fixed constant time $\tau = \xi$. The behavior of the kernel $K(t,\tau)$ at infinite increase of t ($t \to \infty$) and fixed τ determines the dynamics of the process with memory (See Table 2).

Table 2. Examples of the correspondence between the type of memory and type of operator kernels.

Type of Memory	Type of Kernel	Fading (Dissipation)
Memory of Insignificant Events	$\left\|\lim_{t\to\infty} K(t,\tau)\right\| = 0$	Fading Memory
Memory of Significant Events	$0 < \left\|\lim_{t\to\infty} K(t,\tau)\right\| < \infty$	Non-Fading Memory
Memory of Crises and Shocks	$\left\|\lim_{t\to\infty} K(t,\tau)\right\| = \infty$	Non-Fading Memory

Let us assume that there is the limit

$$\lim_{t\to\infty} K(t,\tau) = K_\infty(\tau) = K_\infty, \tag{48}$$

for all τ, when $\tau < t$. In this case, we can consider the three basic type of behavior of $K(t,\xi)$ at infinity $t \to \infty$.

First Type ($K_\infty = 0$): Memory of Insignificant Events (IE-memory). If the kernel tends to zero ($K(t,\tau) \to 0$) at $t \to \infty$, then the process completely forgets about the impact that acts in the past. Then the process that is described by Equation (47) is reversible (is repeated) in a sense. We can say that the memory effects did not lead to irreversible changes of the process, since the memory about the impact has not been preserved forever. Therefore this type of memory can be called "the memory with complete forgetting" (or the memory of insignificant events). As a result, the mathematical characteristic of processes with fading memory can be described by the operator kernels that satisfy the following Principle of Memory Fading memory: Memory, which is described by the operator (45), is fading if the kernel satisfies the condition

$$\lim_{t\to\infty} K(t,\tau) = 0 \tag{49}$$

for all fixed values of τ. The memory will be called the memory with power-law fading if there is a parameter $\alpha > 0$ such that the limit $\lim_{t\to\infty} t^{-\alpha} K(t,\tau)$ is a finite constant for fixed τ. For example, the kernel (40) of the left-sided Caputo fractional derivative describes the power-law memory fading.

Second Type ($0 < |K_\infty| < \infty$): Memory of Significant Events (SE-memory). If the kernel $K(t,\tau)$ tends to a finite limit at $t \to \infty$, the impact leads to the irreversible consequences in the sense that the memory of the impact is preserved forever. Therefore this type of memory can be called "the memory with remembering forever" (or memory of significant events).

Third Type ($K_\infty = \infty$): Memory of Crises and Shocks (CS-memory). Unbounded increase of the kernel $K(t,\tau)$ at $t \to \infty$ (with fixed τ) characterizes an unstable process with memory. This kernel cannot be used to describe stable processes. However, this type of kernels can be used in the various models, which take into account the processes with crises and shocks (for example in economy), when we can expect a manifestation of instability phenomena. The behavior of processes with memory at time t is determined by the behavior of the operator kernel (memory function) in the previous time instants $\tau < t$. Therefore, an unbounded increase in the memory function at infinity ($t \to \infty$) does not lead us to the rejection of consideration of such operator kernels. For example, in this type of memory one can assume that the operator kernel $K(t,\tau)$ is bounded for all $\tau < t$ for a fixed $t < \infty$. Therefore this type of memory can be called "the memory of crises and shocks".

Non-Monotony of Decrease. In general, the memory fading assumes a set of stronger restrictions on the operator kernels. For example, it is assumed that the fading memory is described by operator kernels, which tends to zero monotonically with increasing the time variable. This assumption means that it is less probable to expect of strengthening of the memory with respect to the more distant events. We should note that in economics the agents may remember sharp and significant changes of the variables despite the fact that these changes were more distant past compared to weaker changes in the near past. For this reason, in economics we can use operator kernels without property of monotonic decrease.

Principle of memory reversibility. In paper [17,18], we describe some general restrictions that can be imposed on the structure and properties of memory. For example we consider the principle of memory reversibility (the principle of memory recovery). The principle of memory reversibility is connected with the principle of duality of accelerator with memory and multiplier with memory, which is proposed in [45]. Mathematically this principle is based on the main property of any fractional derivative to be a left-inverse operator to the corresponding fractional integral operator.

Remark 8. *We should note that there is an addition restriction on the kernel of the operator (1). In general, to have a self-consistent mathematical theory of the operators (1), the general fractional derivative (1) with $n = 1, 2, 3, \ldots$ should be a left-inverse operator to the corresponding general fractional integral operator (1) with $n = 0$. This requirement leads us to a relationship between the type of the operator kernels $K(t, \tau)$ and the order (and type) of the operators $\left(D_\tau^{(n)} f(\tau)\right)$ of integer order $n = 1, 2, 3, \ldots$. For the kernel should depend on the order, i.e., $K(t, \tau) = K_n(t, \tau)$.*

Remark 9. *General fractional calculus was proposed by Anatoly N. Kochubei in [35–37] and based on the use of differential operators with Laplace convolution (the general Laplace-convolutional derivatives). The principle of memory reversibility means that the general operators should have right inverse (a kind of a fractional integral). We assume that the Kochubei approach to formulation of general fractional calculus, which is based on the Laplace convolution, can be applied to formulate new fractional calculus based on Mellin convolution. The general operators (the general Mellin -convolutional derivatives), which are based on Mellin convolution, and equations with these operators can be used to describe the scaling (dilation) phenomena in physics and economics.*

6. Properties of Kernels of Inverse Operators and Type of Phenomena

An addition restriction on the kernel of the operator (1) can be considered. The general operators (1) with $n = 1, 2, 3, \ldots$ can be considered to be the general fractional derivative. The general operators (1) with $n = 0$ can be considered to be general fractional integrals. In our opinion to have a self-consistent mathematical theory, the general fractional derivative (1) with $n = 1, 2, 3, \ldots$ should be a left-inverse operator to the corresponding general fractional integral operator (1) with $n = 0$. Therefore we proposed the following principle for fractional calculus: Any type of generalized (fractional) derivative should be a left-inverse operator to the corresponding type of generalized (fractional) integral operator. This principle can be considered to be a requirement of the existence of a generalization of the fundamental theorem of calculus, which is a theorem that links the concept of differentiating with the concept of integrating.

Obviously, this principle, this requirement lead us to a relationship between the type of the operator kernels $K_n(t, \tau)$ $n = 1, 2, 3, \ldots$, and the type of the kernel $K_0(t, \tau)$. Please note that this requirement also leads us to a relationship between the type of the operator kernels $K_n(t, \tau)$ and the order (and type) of the operators $\left(D_\tau^{(n)} f(\tau)\right)$ of integer order $n = 1, 2, 3, \ldots$ **First Question:** In connection with this principle, the natural question arises about the relationship between the properties of the kernels of fractional operators, considered to be the fractional integrals and as the fractional derivatives. In many cases, kernels belong to one type of functions. For example, the kernel of the left-sided Caputo fractional derivative (see Equation (2.4.15) in p. 92, [4]) has the form

$$K_n(t, \tau) = \frac{1}{\Gamma(n - \alpha)} (t - \tau)^{n - \alpha - 1} \tag{50}$$

for $= 1, 2, 3, \ldots$. This fractional derivative is the left-inverse operator for the left-sided Riemann–Liouville fractional integral. The kernel of this integral is described by Equation (50) with $n = 0$ and negative α (see Equation (2.1.1) in [4] (p. 69)). The same situations we have for the Erdelyi–Kober operator and other types of fractional operators. However, this is not true in the general case.

Remark 10. *In paper [17,18], we describe some general restrictions that can be imposed on the structure and properties of memory. These restrictions are proposed as the principle of memory reversibility (the principle of memory recovery). Mathematically this principle is based on the property of any fractional derivative to be a left-inverse operator to the corresponding fractional integral operator.*

Statement 3.

The generalized (fractional) derivative (1) with $n = 1, 2, 3, \ldots$ must be the left inverse operator to the corresponding generalized (fractional) integral operator (1) with $n = 0$. However, the kernels $K_n(t, \tau)$ with $n = 1, 2, 3, \ldots$ of operator (1) and the kernel $K_0(t, \tau)$ of the fractional integral operator (1) with $n = 0$ can belong to different types of functions.

To prove this statement, we give an example of fractional operators of distributed orders.

In general, the parameter α that is the order of the fractional derivative or integral and describes the memory fading, can be distributed on an interval with some probability density function (the weight function). In the simplest case, we can use the continuous uniform distribution (CUD). The fractional integrals and derivatives of the uniform distributed order can be expressed thought the continual fractional integrals and derivatives, which were suggested by Adam M. Nakhushev [46,47]. The operators of non-integer orders, which are left inverse to the continual fractional integrals and derivatives, are proposed by Arsen V. Pskhu in [48,49]. Using the continual fractional integrals and derivatives, which were suggested by Nakhushev, we can define the integral and derivatives of uniform distributed order. These operators will be called the Nakhushev fractional integrals and derivatives. The corresponding inverse operators are proposed by Pskhu and therefore operators, which are inverse to fractional CUD fractional operators, will be called the Pskhu fractional integrals and derivatives.

In works of Pskhu [48,49] the notations $D_{0+}^{[\alpha,\beta]}$ and $D_{0+}^{-[\alpha,\beta]}$ are used for positive ($0 < \alpha < \beta$) and negative ($\alpha < \beta \leq 0$) values of α and β. In our opinion, this leads to confusion and misunderstanding in applications. Therefore we will use new notations, which allow us to see explicitly the integration and differentiation of the fractional orders.

The Nakhushev fractional integral can be defined (see Equation (5.1.7) of [49] (p. 136) and [48]) defined in the form

$$I_N^{[\alpha,\beta]} X(t) = \frac{1}{\beta - \alpha} \int_\alpha^\beta I_{RL,a+}^\xi X(t) d\xi = \int_0^t W(\alpha, \beta, t - \tau) X(\tau) d\tau, \quad (51)$$

where we use the function

$$W(\alpha, \beta, t) = \frac{1}{(\beta - \alpha) t} \int_\alpha^\beta \frac{t^\xi d\xi}{\Gamma(\xi)}. \quad (52)$$

Using Equation (5.1.26) of [49] (p. 143), the Nakhushev fractional derivative can be written in the form

$$D_N^{[\alpha,\beta]} X(t) = \left(\frac{d}{dx}\right)^n \int_0^t W(n - \alpha, n - \beta, t - \tau) X(\tau) d\tau, \quad (53)$$

where $\beta > \alpha > 0$. Please note that the Nakhushev fractional derivatives cannot be considered to be inverse operators for the Nakhushev fractional integration. The Pskhu fractional derivatives are inverse to the Nakhushev fractional integration and the Pskhu fractional integrals are inverse to the Nakhushev fractional derivatives.

The Pskhu fractional integral can be defined (see Equation (5.1.7) of [49] (p. 136), and [48]) by the expression

$$I_P^{[\alpha,\beta]} X(t) = (\alpha - \beta) \int_0^t (t-\tau)^{\beta-1} E_{\beta-\alpha}\left[(t-\tau)^{\beta-\alpha}; \beta\right] X(\tau) d\tau, \qquad (54)$$

where $\beta > \alpha > 0$. where $E_\alpha[z; \beta]$ is the Mittag–Leffler function that is defined by the expression

$$E_\alpha[z; \beta] = \sum_{k=0}^{\infty} \frac{z^k}{\Gamma(\alpha k + \beta)}. \qquad (55)$$

Using Equation (5.1.7) of [49] (p. 136), we can define the Pskhu fractional derivative as

$$D_P^{[\alpha,\beta]} X(t) = (\alpha - \beta) \left(\frac{d}{dx}\right)^n \int_0^t (t-\tau)^{-\alpha} E_{\beta-\alpha}^{n-1}\left[(t-\tau)^{\beta-\alpha}; 1-\alpha\right] X(\tau) d\tau, \qquad (56)$$

where $\beta > \alpha > 0$ and the function $E_\alpha^\mu[z; \beta]$ is defined by the equation

$$E_\alpha^\mu[z; \beta] = \frac{\partial}{\partial \mu}(z^\mu E_\alpha[z; \beta + \mu])$$

As a result, we have that the Nakhushev fractional derivatives cannot be considered to be left-inverse operators for the Nakhushev fractional integrals [48,49]. Operators, which are left-inverse operator for the Nakhushev fractional derivatives and integrals, are the Pskhu fractional integrals and derivatives.

As a result, we proved that the kernels of the original and inverse operators can be of different types.

Second Question: If the kernels of generalized (fractional) derivative and the corresponding generalized (fractional) integral operator can be described by functions of different types, then the second natural question arises: Will these kernels describe the same types of phenomena? If the operator cores are different, then what is the difference in the phenomena described by these different types of cores? As a suggested answer on these questions, we can propose the following hypothesis.

Hypothesis of Duality: The kernels of the original and inverse operators of fractional calculus should describe dual types of phenomena.

This hypothesis is based on an attempt to answer the second question in the framework of economic interpretation, which is presented in the form of the principle of duality proposed in [45]. In this principle we describe duality of two basic economic concepts: the accelerator with memory and multiplier with memory (for details see [45]).

Remark 11. *We assume that the Kochubei approach to formulation of general fractional calculus, which is based on the Laplace convolution, can be applied to formulate new fractional calculus based on Mellin convolution. This allows us to describe duality of the economic concepts of the accelerator with scaling and multiplier with scaling.*

7. Memory with Lag: Distributed Lag Fractional Operators

In general, we can simultaneously take into account two different types of phenomena. For example, we can simultaneously take into account lagging and memory phenomena. For this, we proposed the distributed lag fractional calculus in [29]. Then this approach was applied to macroeconomic models.

To illustrate this approach, let us assume that the joint action of two phenomena: the lag with gamma distribution of delay time and the power-law fading memory. We will use the Caputo fractional derivatives to describe power-law memory. The continuously distributed delay time is described by the translation operator, where the delay time $\tau > 0$ is a random variable that is distributed on positive semiaxis. We can prove that the composition of these operators is represented as the Abel-type integral and integro-differential operators with the confluent hypergeometric Kummer function in the kernel.

The Caputo fractional derivative with gamma distributed lag is defined by the equation

$$\left(D_{T;C;0+}^{\lambda,a;\alpha} f\right)(t) = \int_0^t K_T^{\lambda,a}(\tau)\left(D_{C,0+}^{\alpha} f\right)(t-\tau)\,d\tau, \tag{57}$$

where the kernel $K_T^{\lambda,a}(\tau)$ is the probability density function of the gamma distribution

$$K_T^{\lambda,a}(\tau) = \begin{cases} \frac{\lambda^a \tau^{a-1}}{\Gamma(a)} \exp(-\lambda \tau) & \text{if } \tau > 0, \\ 0 & \text{if } \tau \leq 0, \end{cases} \tag{58}$$

with the shape parameter $a > 0$ and the rate parameter $\lambda > 0$. If $a = 1$, the function (58) describes the exponential distribution. Using the associative property of the Laplace convolution, the operators (57) can be represented [29] in the form

$$\left(D_{T;C;0+}^{\lambda,a;\alpha} f\right)(t) = \int_0^t K_{TRL}^{\lambda,a;n-\alpha}(\tau) f^{(n)}(t-\tau)\,d\tau, \tag{59}$$

where $n-1 < \alpha \leq n$, and the kernel $K_{TRL}^{\lambda,a;n-\alpha}(t)$ has the form

$$K_{TRL}^{\lambda,a;n-\alpha}(t) = \frac{\lambda^a \Gamma(a)}{\Gamma(a+n-\alpha)} t^{a+n-\alpha-1} F_{1,1}(a; a+n-\alpha; -\lambda t), \tag{60}$$

where $F_{1,1}(a;b;z)$ is the confluent hypergeometric Kummer function that is defined (see [4] (pp.29–30)) by the equation

$$F_{1,1}(a;c;z) = \frac{\Gamma(c)}{\Gamma(a)\Gamma(c-a)} \int_0^1 t^{a-1}(1-t)^{c-a-1} \exp(zt)\,dt = \sum_{k=0}^{\infty} \frac{\Gamma(a+k)\Gamma(c)}{\Gamma(a)\Gamma(c+k)} \frac{z^k}{k!}, \tag{61}$$

where $a, z \in \mathbb{C}$, $Re(c) > Re(a) > 0$ such that $c \neq 0, -1, -2, \ldots$ and series (61) is absolutely convergent for all $z \in \mathbb{C}$. It should be noted that the kernel (60) can be represented through the three parameter Mittag-Leffler function $E_{\alpha,\beta}^{\gamma}(z)$, which is also called the Prabhakar function, by using the equation $F_{1,1}(a;c;z) = \Gamma(c) E_{1,c}^{a}(z)$. The Laplace transform of fractional operator (59) has the form

$$\left(\mathcal{L}(D_{T;C;0+}^{\lambda,a;\alpha} f)(t)\right)(s) = \frac{\lambda^a}{(s+\lambda)^a}\left[s^{\alpha}(\mathcal{L}Y)(s) - \sum_{j=0}^{n-1} s^{\alpha-j-1} f^{(j)}(0)\right], \tag{62}$$

where $n-1 < \alpha \leq n$.

As a result, the kernel $K_{TRL}^{\lambda,a;n-\alpha}(\tau)$ of the proposed special kind of the Abel-type fractional derivative describes the joint phenomenon of the power-law fading memory and the continuously distributed lag. Using Theorem 6.5 in [29] (pp. 145–146), and results of [31,32], we can describe the solution of the fractional differential equation

$$\left(D_{T;C;0+}^{\lambda,a;\alpha} y\right)(t) = \omega y(t) + F(t), \tag{63}$$

where the operator $D_{T;C;0+}^{\lambda,a;\alpha}$ is defined by Equation (59), $\alpha > 0$ is the order of the operators, the parameters $a > 0$ and $\lambda > 0$ are the shape and rate parameters of the gamma distribution of delay time. The solution of Equation (63) can be represented in the form

$$y(t) = \sum_{j=0}^{n-1} S_{\alpha,a}^{\alpha-j-1}[\omega \lambda^{-a}, \lambda|t] y^{(j)}(0) + \frac{1}{\omega} F(t) - \frac{1}{\omega} \int_0^t S_{\alpha,a}^{\alpha}[\omega \lambda^{-a}, \lambda|t-\tau] F(\tau)\,d\tau, \tag{64}$$

with $n = [\alpha] + 1$, and $S_{\alpha,\delta}^{\gamma}[\mu, \lambda|t]$ is the special function that is defined by the expression

$$S_{\alpha,\delta}^{\gamma}[\mu, \lambda|t] = -\sum_{k=0}^{\infty} \frac{t^{\delta(k+1)-\alpha k-\gamma-1}}{\mu^{k+1}\Gamma(\delta(k+1) - \alpha k - \gamma)} F_{1,1}(\delta(k+1); \delta(k+1) - \alpha k - \gamma, -\lambda t), \quad (65)$$

where $F_{1,1}(a; b; z)$ is the confluent hypergeometric Kummer function (61).

In the connection with a possibility of composition of two or more kernels of operators that describe different phenomena, an important question arises about the following inverse mathematical problem. How we can identify and separate actions of two different type phenomena in it simultaneously action? In our opinion, the answer on this question is important to physics, mechanics, economics and other sciences.

8. Operator Kernel Behavior at Zero and Interpretation

In general, the type of behavior of the operator kernel (1) at $t \to 0$ can be important to different applications. We can assume the following type of behavior the kernel $K(t)$.

(1) The operator kernel tends to zero while the argument t tends to zero

$$\lim_{t \to 0+} K(t) = 0. \quad (66)$$

(2) The kernel $K(t)$ tends to finite nonzero constant while the argument t tends to zero

$$\lim_{t \to 0+} K(t) = K(0) = const. \quad (67)$$

(3) The kernel $K(t)$ tends to infinity as the argument t tends to zero

$$\lim_{t \to 0+} K(t) = \pm\infty.$$

A lot of kernels of the fractional integral and derivatives demonstrate only the third (or first) type of behavior at zero for non-integer orders. Let us describe some examples of the operator kernels that have this type of behavior.

The kernel of the Riemann–Liouville fractional integral has the form

$$K_{RLI}(t) = \frac{1}{\Gamma(\alpha)} t^{\alpha-1}, \quad (68)$$

where $\alpha > 0$ [4] (p.69). We see that

$$K_{RLI}(0) = \begin{cases} 0 & if \quad \alpha > 1 \\ 1 & if \quad \alpha = 1 \\ \infty & if \quad 0 < \alpha < 1 \end{cases}. \quad (69)$$

This means that kernel of the Riemann-Liouville fractional integral can demonstrate three type of behavior at zero ($t = 0$). However, the second type of behavior ($K_{PI}(0)$=const) cannot be realized for non-integer orders $\alpha > 0$.

The kernel of the Caputo and Riemann–Liouville fractional derivatives has the form

$$K_{CD}(t) = K_{RLD}(t) = \frac{1}{\Gamma(n-\alpha)} t^{n-\alpha-1}, \quad (70)$$

where $n = [\alpha] + 1$, and $n - 1 < \alpha < n$ for non-integer values of order α [4] (pp.70–91). We see that

$$K_{CD}(0) = \begin{cases} 0 & \text{if} \quad 0 < \alpha < n-1 \\ 1 & \text{if} \quad \alpha = n-1 \\ \infty & \text{if} \quad \alpha > n-1 \end{cases}. \tag{71}$$

This means that kernel of the Caputo and Riemann-Liouville fractional derivatives can demonstrate only one (singular) type of behavior at zero ($t = 0$) for non-integer orders. The other two cases ($K_{CD}(t) = 0$ and $K_{CD}(0) = 1$) are not implemented for the following reasons: (A) The case $\alpha = n - 1$ cannot be used for the Caputo derivative since we have $\alpha = n$ for integer values of α (see Equation (2.4.3) in [4] (p.91)). For this case, the Riemann–Liouville fractional derivative is standard derivative of integer order. (B) The case $0 < \alpha < n - 1$ cannot be used by definition the Caputo and Riemann-Liouville fractional derivatives that contains the condition $n - 1 < \alpha < n$ for non-integer values of order α. We have a similar situation for the Erdelyi–Kober and Kober operators.

As a result, we see that the power-law kernels of fractional derivatives have significantly less variability in the behavior properties at zero. Please note that the variety of properties of operator kernel at zero is important for applications of these operators in economics and physics, for example.

Let us note that some important phenomena are described only by the kernels with second type of behavior. For example, in economics this condition is used for the kernels that describe the depreciation of fixed assets (of capital), depreciation of equipment, obsolescence, aging, wear and tear [50] (p. 20). The kernel $K(t - \tau)$ characterizes the share of fixed assets put into operation at time τ and continuing to operate at time $t > \tau$. Obviously, in this case, the condition $K(0) = 1$ must be satisfied. For this, economics often use the exponential functions and the probability density function of the exponential distribution.

The kernels of the Riemann–Liouville, Caputo, Erdelyi–Kober fractional operators of non-integer order cannot be used to describe the depreciation or aging phenomena in economy. To describe these phenomena we can use the fractional operators with the Prabhakar function, the hypergeometric function, the Kummer (confluent hypergeometric) function in the kernels. In the framework of fractional calculus, these operators were proposed and described more than forty years ago in [51], (see also [52,53]) for the Prabhakar function, [54,55] the Kummer (confluent hypergeometric) function, and [56] (see also [1] (pp. 731–737)) for the hypergeometric function.

Please note that the operators with the Kummer (confluent hypergeometric) function in the kernels can be interpreted as the joint effect of two phenomena: the memory with power-law fading and the lag with gamma distribution of delay time. In the paper [29] (see Theorem 4.3 and Equation (4.48) p. 137; see also Equations (4.53) and (6.7)), we use the operators with the Kummer (confluent hypergeometric) function in the kernels that is Laplace convolution of the kernel of the Caputo fractional derivatives and probability density function of the gamma distribution that describes the distribution of the delay time $\tau > 0$.

The kernel of the Prabhakar fractional integral has the form

$$K_{PI}(t) = t^{\mu-1} E_{\rho,\mu}^{\gamma}[\omega t^{\rho}] = t^{\mu-1} \sum_{k=0}^{\infty} \frac{\Gamma(\gamma + k)}{\Gamma(\gamma)\Gamma(\rho k + \mu)} \frac{(\omega t^{\rho})^k}{k!}. \tag{72}$$

We can see that the kernel (72) can demonstrate three type of behavior at zero

$$K_{PI}(0) = \begin{cases} 0 & \text{if} \quad \mu > 1 \\ 1 & \text{if} \quad \mu = 1 \\ \infty & \text{if} \quad 0 < \mu < 1 \end{cases} \tag{73}$$

The kernel of the Kilbas–Saigo–Saxena fractional derivative [53] (that is also called the Prabhakar fractional derivative), which is proposed in [53] and it is left-inverse operator for the Prabhakar fractional integral, has the form

$$K_{PD}(t) = t^{n-\mu-1} E^{-\gamma}_{\rho,n-\mu}[\omega t^\rho], \qquad (74)$$

where $n \geq [Re(\mu)] + 1$ with $Re(\mu) > 0$. We should emphasize that in kernel (74), we can use all positive integer values $n \geq [Re(\mu)] + 1$, where $Re(\mu) > 0$ since n is defined as $n = [\mu + \nu] + 1$ with $Re(\mu), Re(\nu) > 0$ in Theorem 9 in [53] (p. 47)).

Using expression (74), we get the following properties of kernel (74) in the initial point

$$K_{PD}(0) = \begin{cases} 0 & if \quad 0 < \mu < n-1 \\ 1 & if \quad \mu = n-1 \\ \infty & if \quad \mu > n-1 \end{cases}. \qquad (75)$$

As a result, the kernel of the Kilbas–Saigo–Saxena fractional derivative can demonstrate three type of behavior at zero. Please note that this operator remains a fractional operator and under condition $\mu = n - 1$. This behavior significantly distinguishes this operator from other fractional derivatives, which usually have a singularity at zero.

Therefore to satisfy the initial conditions $K(0) = 1$ for the operator kernel, we can use the kernels with the Prabhakar function. These kernels allow us to use the fractional integrals and derivatives with the Prabhakar function in the kernel, which proposed in the works [51–53], to describe depreciation processes in economics. In addition, we can state that the kernel $K_{PI}(t)$ is the complete monotonic function for the case $\omega < 0, 0 < \rho, \mu \leq 1, 0 < \gamma \leq \mu/\rho$. The property of the complete monotonicity is important for the interpretation of operator kernels that describe standard depreciation phenomena. However, we can assume that the requirement of complete monotonicity for depreciation kernels is not necessary, when taking into account modernization of the equipment.

9. Conclusions

In this paper, we discussed an interpretation of fractional derivatives and integrals from the point of view of applied mathematics, theoretical physics, and economic theory. We state that it is important to connect all restrictions on the fractional operator kernels with types of phenomena, in addition to the self-consistency of mathematical theory. In applications of fractional calculus, we have a fundamental question about conditions of kernels of non-integer order operators that allow us to describe one or another type of phenomena. It is necessary to obtain exact correspondences between sets of properties of kernel and type of phenomena. In this paper, we describe some important properties of fractional operator kernels that can determine the characteristic features of certain types of phenomena. We consider the possible characteristic properties of kernels of fractional operators to distinguish the following types of phenomena: fading memory (forgetting) and power-law frequency dispersion; spatial non-locality and power-law spatial dispersion; distributed lag (time delay); distributed scaling (dilation); depreciation and aging.

Let us briefly describe possible directions for application of the proposed approach.

a) We should note the power-law kernels function can be used to consider an approximation of the generalized memory functions [57]. Using the generalized Taylor series in the Trujillo-Rivero-Bonilla form for the memory function, we proved [57] that the equations with memory functions can be represented through the Riemann–Liouville fractional integrals and the Caputo fractional derivatives of non-integer orders for wide class of the kernels. We can also note that the Abel-type fractional integral operator with Kummer function in the kernel (see Equation (37.1) in [1] (p. 731), and [32]) can be represented as an infinite series of the Riemann–Liouville fractional integrals.

b) We can have new types of phenomena in quantum theory, where we should take into account the intrinsic dissipation, the openness of systems, an interaction with environment [58–61].

c) We can expecte new types of phenomena in nonlinear, chaotic systems and for self-organization processes [62–65], where we should take into account the new types of attractors, patterns and effects.

At the same time, we emphasize that we have in mind not new regular applications of fractional calculus to the description of various particular phenomena in various science. We mean exact correspondence between the types of phenomena and the types of properties of fractional operator kernels.

Author Contributions: Contribution of V.E.T. is connected with fractional calculus. Contribution of S.S.T. is connected with probability theory. All authors have read and agreed to the published version of the manuscript.

Funding: This research received no external funding.

Conflicts of Interest: The authors declare no conflict of interest.

References

1. Samko, S.G.; Kilbas, A.A.; Marichev, O.I. *Fractional Integrals and Derivatives Theory and Applications*; Gordon and Breach: New York, NY, USA, 1993; p. 1006, ISBN 9782881248641.
2. Kiryakova, V. *Generalized Fractional Calculus and Applications*; Longman & J. Wiley: New York, NY, USA, 1994; p. 360, ISBN 9780582219779.
3. Podlubny, I. *Fractional Differential Equations*; Academic Press: San Diego, CA, USA, 1998; p. 340.
4. Kilbas, A.A.; Srivastava, H.M.; Trujillo, J.J. *Theory and Applications of Fractional Differential Equations*; Elsevier: Amsterdam, The Netherlands, 2006; p. 540, ISBN 9780444518323.
5. Diethelm, K. *The Analysis of Fractional Differential Equations: An Application-Oriented Exposition Using Differential Operators of Caputo Type*; Springer: Berlin, Germany, 2010; p. 247. [CrossRef]
6. Kochubei, A.N.; Luchko, Y. (Eds.) *Handbook of Fractional Calculus with Applications. Vol.1. Basic Theory*; De Gruyter: Berlin, Germany, 2019; ISBN 978-3-11-057081-6.
7. Kochubei, A.N.; Luchko, Y. (Eds.) *Handbook of Fractional Calculus with Applications. Vol.2. Fractional Differential Equations*; De Gruyter: Berlin, Germany, 2019; ISBN 978-3-11-057082-3.
8. Tarasov, V.E. No violation of the Leibniz rule. No fractional derivative. *Commun. Nonlinear Sci. Numer. Simul.* **2013**, *18*, 2945–2948. [CrossRef]
9. Tarasov, V.E. Leibniz rule and fractional derivatives of power functions. *J. Comput. Nonlinear Dyn.* **2016**, *11*. [CrossRef]
10. Tarasov, V.E. On chain rule for fractional derivatives. *Commun. Nonlinear Sci. Numer. Simul.* **2016**, *30*, 1–4. [CrossRef]
11. Tarasov, V.E. No nonlocality. No fractional derivative. *Commun. Nonlinear Sci. Numer. Simul.* **2018**, *62*, 157–163. [CrossRef]
12. Tarasov, V.E. Rules for fractional-dynamic generalizations: Difficulties of constructing fractional dynamic models. *Mathematics* **2019**, *7*. [CrossRef]
13. Ortigueira, M.D.; Tenreiro Machado, J.A. What is a fractional derivative? *J. Comput. Phys.* **2015**, *293*, 4–13. [CrossRef]
14. Stynes, M. Fractional-order derivatives defined by continuous kernels are too restrictive. *Appl. Math. Lett.* **2018**, *85*, 22–26. [CrossRef]
15. Hilfer, R.; Luchko, Y. Desiderata for fractional derivatives and integrals. *Mathematics* **2019**, *7*. [CrossRef]
16. Cresson, J.; Szafranska, A. Comments on various extensions of the Riemann–Liouville fractional derivatives: About the Leibniz and chain rule properties. *Commun. Nonlinear Sci. Numer. Simul.* **2020**, *82*. [CrossRef]
17. Tarasova, V.V.; Tarasov, V.E. Concept of dynamic memory in economics. *Commun. Nonlinear Sci. Numer. Simul.* **2018**, *55*, 127–145. [CrossRef]
18. Tarasov, V.E.; Tarasova, V.V. Criterion of existence of power-law memory for economic processes. *Entropy* **2018**, *20*. [CrossRef]
19. Tarasov, V.E. (Ed.) *Handbook of Fractional Calculus with Applications. Volumes 4. Application in Physics. Part A*; Walter de Gruyter GmbH: Berlin, Germany; Boston, MA, USA, 2019; ISBN 978-3-11-057170-7. [CrossRef]

20. Tarasov, V.E. (Ed.) *Handbook of Fractional Calculus with Applications. Volumes 5. Application in Physics. Part B*; Walter de Gruyter GmbH: Berlin, Germany; Boston, MA, USA, 2019; ISBN 978-3-11-057172-1. [CrossRef]
21. Tarasov, V.E.; Tarasova, S.S. Probabilistic interpretation of Kober fractional integral of non-integer order. *Prog. Fract. Differ. Appl.* **2019**, *5*, 1–5. [CrossRef]
22. Tarasov, V.E. Caputo-Fabrizio operator in terms of integer derivatives: Memory or distributed lag? *Comput. Appl. Math.* **2019**, *38*. [CrossRef]
23. Gorenflo, R.; Luchko, Y.; Mainardi, F. Wright functions as scale-invariant solutions of the diffusion-wave equation. *J. Comput. Appl. Math.* **2000**, *11*, 175–191. [CrossRef]
24. Hanna, L.A.-M.; Luchko, Y.F. Operational calculus for the Caputo-type fractional Erdélyi–Kober derivative and its applications. *Integral Transform. Spec. Funct.* **2014**, *25*, 359–373. [CrossRef]
25. Luchko, Y.; Trujillo, J.J. Caputo-type modification of the Erdelyi-Kober fractional derivative. *Fract. Calc. Appl. Anal.* **2007**, *10*, 249–268.
26. Caputo, M.; Fabrizio, M. A new definition of fractional derivative without singular kernel. *Prog. Fract. Differ. Appl.* **2015**, *1*, 73–85. [CrossRef]
27. Caputo, M.; Fabrizio, M. Applications of new time and spatial fractional derivatives with exponential kernels. *Prog. Fract. Differ. Appl.* **2016**, *2*, 1–11. [CrossRef]
28. Caputo, M.; Fabrizio, M. On the notion of fractional derivative and applications to the hysteresis phenomena. *Meccanica* **2017**, *52*, 3043–3052. [CrossRef]
29. Tarasov, V.E.; Tarasova, S.S. Fractional and integer derivatives with continuously distributed lag. *Commun. Nonlinear Sci. Numer. Simul.* **2019**, *70*, 125–169. [CrossRef]
30. Tarasov, V.E.; Tarasova, V.V. Phillips model with exponentially distributed lag and power-law memory. *Comput. Appl. Math.* **2019**, *38*. [CrossRef]
31. Tarasov, V.E.; Tarasova, V.V. Harrod-Domar growth model with memory and distributed lag. *Axioms* **2019**, *8*. [CrossRef]
32. Tarasov, V.E.; Tarasova, V.V. Dynamic Keynesian model of economic growth with memory and lag. *Mathematics* **2019**, *7*. [CrossRef]
33. Luchko, Y. Integral transforms of the Mellin convolution type and their generating operators. *Integral Transform. Spec. Funct.* **2008**, *19*, 809–851. [CrossRef]
34. Luchko, Y.; Kiryakova, V. The Mellin integral transform in fractional calculus. *Fract. Calc. Appl. Anal.* **2013**, *16*, 405–430. [CrossRef]
35. Kochubei, A.N. General fractional calculus, evolution equations, and renewal processes. *Integral Equ. Oper. Theory* **2011**, *71*, 583–600. [CrossRef]
36. Kochubei, A.N. General fractional calculus. Chapter 5. In *Handbook of Fractional Calculus with Applications*; Kochubei, A., Luchko, Y., Eds.; De Gruyter: Berlin, Germany; Boston, MA, USA, 2019; pp. 111–126. [CrossRef]
37. Kochubei, A.N.; Kondratiev, Y. Growth equation of the general fractional calculus. *Mathematics* **2019**, *7*. [CrossRef]
38. Luchko, Y. Operational method in fractional calculus. *Fract. Calc. Appl. Anal.* **1999**, *2*, 1–26.
39. Allen, R.G.D. *Mathematical Economics*, 2nd ed.; Macmillan: London, UK, 1959; p. 812. [CrossRef]
40. Wang, C.C. The principle of fading memory. *Arch. Ration. Mech. Anal.* **1965**, *18*, 343–366. [CrossRef]
41. Coleman, B.D.; Mizel, V.J. On the general theory of fading memory. *Arch. Ration. Mech. Anal.* **1968**, *29*, 18–31. [CrossRef]
42. Coleman, B.D.; Mizel, V.J. A general theory of dissipation in materials with memory. *Arch. Ration. Mech. Anal.* **1967**, *27*, 255–274. [CrossRef]
43. Coleman, B.D.; Mizel, V.J. Norms and semi-groups in the theory of fading memory. *Arch. Ration. Mech. Anal.* **1966**, *23*, 87–123. [CrossRef]
44. Saut, J.C.; Joseph, D.D. Fading memory. *Arch. Ration. Mech. Anal.* **1983**, *81*, 53–95. [CrossRef]
45. Tarasov, V.E.; Tarasova, V.V. Accelerator and multiplier for macroeconomic processes with memory. *IRA Int. J. Manag. Soc. Sci.* **2017**, *9*, 86–125. [CrossRef]
46. Nakhushev, A.M. On the positivity of continuous and discrete differentiation and integration operators that are very important in fractional calculus and in the theory of equations of mixed type. *Differ. Equ.* **1998**, *34*, 103–112.
47. Nakhushev, A.M. *Fractional Calculus and its Application*; Fizmatlit: Moscow, Russia, 2003; p. 272. (In Russian)

48. Pskhu, A.V. On the theory of the continual integro-differentiation operator. *Differ. Equ.* **2004**, *40*, 128–136. [CrossRef]
49. Pskhu, A.V. *Partial Differential Equations of Fractional Order*; Nauka: Moscow, Russia, 2005; p. 199. (In Russian)
50. Moiseev, N.N. *Simplest Mathematical Models of Economic Forecasting*; Znanie: Moscow, Russia, 1975.
51. Prabhakar, T.R. A singular integral equation with a generalized Mittag-Leffler function in the kernel. *Yokohama Math. J.* **1971**, *19*, 7–15.
52. Kilbas, A.A.; Saigo, M.; Saxena, R.K. Solution of Volterra integro-differential equations with generalized Mittag-Leffler function in the kernels. *J. Integral Equ. Appl.* **2002**, *14*, 377–396. [CrossRef]
53. Kilbas, A.A.; Saigo, M.; Saxena, R.K. Generalized Mittag-Leffler function and generalized fractional calculus operators. *Integral Transform. Spec. Funct.* **2004**, *15*, 31–49. [CrossRef]
54. Prabhakar, T.R. Some integral equations with Kummer's functions in the kernels. *Can. Math. Bull.* **1971**, *4*, 391–404. [CrossRef]
55. Prabhakar, T.R. Two singular integral equations involving confluent hypergeometric functions. *Math. Proc. Camb. Philos. Soc.* **1969**, *66*, 71–89. [CrossRef]
56. Prabhakar, T.R.; Kashyap, N.K. A new class of hypergeometric integral equations. *Indian J. Pure Appl. Math.* **1980**, *11*, 92–94.
57. Tarasov, V.E. Generalized memory: Fractional calculus approach. *Fractal Fract.* **2018**, *2*. [CrossRef]
58. Evans, M.W.; Grigolini, P.; Parravicini, G.P. (Eds.) *Memory Function Approaches to Stochastic Problems in Condensed Matter*; Intersicence/De Gruyter: New York, NY, USA, 1985; ISBN 978-0-470-14331-5.
59. Tarasov, V.E. Quantum dissipation from power-law memory. *Ann. Phys.* **2012**, *327*, 1719–1729. [CrossRef]
60. Tarasov, V.E.; Tarasova, V.V. Time-dependent fractional dynamics with memory in quantum and economic physics. *Ann. Phys.* **2017**, *383*, 579–599. [CrossRef]
61. Tarasov, V.E. Fractional quantum mechanics of open quantum systems. In *Handbook of Fractional Calculus with Applications. Volume 5: Applications in Physics, Part B*; Walter de Gruyter: Berlin, Germany, 2019; Chapter 11; pp. 257–277. [CrossRef]
62. Zaslavsky, G.M. *Hamiltonian Chaos and Fractional Dynamics*; Oxford University Press: Oxford, UK, 2008; ISBN 978-0199535484.
63. Tarasova, V.V.; Tarasov, V.E. Logistic map with memory from economic model. *Chaos Solitons Fractals* **2017**, *95*, 84–91. [CrossRef]
64. Tarasov, V.E. *Fractional Dynamics: Applications of Fractional Calculus to Dynamics of Particles, Fields and Media*; Springer: New York, NY, USA, 2010. [CrossRef]
65. Tarasov, V.E. Self-organization with memory. *Commun. Nonlinear Sci. Numer. Simul.* **2019**, *72*, 240–271. [CrossRef]

© 2020 by the authors. Licensee MDPI, Basel, Switzerland. This article is an open access article distributed under the terms and conditions of the Creative Commons Attribution (CC BY) license (http://creativecommons.org/licenses/by/4.0/).

Article

Good (and Not So Good) Practices in Computational Methods for Fractional Calculus

Kai Diethelm [1,2], Roberto Garrappa [3,4,*] and Martin Stynes [5]

1. Fakultät Angewandte Natur- und Geisteswissenschaften, University of Applied Sciences Würzburg-Schweinfurt, Ignaz-Schön-Str. 11, 97421 Schweinfurt, Germany; kai.diethelm@fhws.de
2. GNS mbH Gesellschaft für Numerische Simulation mbH, Am Gaußberg 2, 38114 Braunschweig, Germany
3. Department of Mathematics, University of Bari, Via E. Orabona 4, 70126 Bari, Italy
4. INdAM Research Group GNCS, Piazzale Aldo Moro 5, 00185 Rome, Italy
5. Applied and Computational Mathematics Division, Beijing Computational Science Research Center, Beijing 100193, China; m.stynes@csrc.ac.cn
* Correspondence: roberto.garrappa@uniba.it

Received: 26 January 2020; Accepted: 25 February 2020; Published: 2 March 2020

Abstract: The solution of fractional-order differential problems requires in the majority of cases the use of some computational approach. In general, the numerical treatment of fractional differential equations is much more difficult than in the integer-order case, and very often non-specialist researchers are unaware of the specific difficulties. As a consequence, numerical methods are often applied in an incorrect way or unreliable methods are devised and proposed in the literature. In this paper we try to identify some common pitfalls in the use of numerical methods in fractional calculus, to explain their nature and to list some good practices that should be followed in order to obtain correct results.

Keywords: fractional differential equations; numerical methods; smoothness assumptions; persistent memory

1. Introduction

The increasing interest in applications of fractional calculus, together with the difficulty of finding analytical solutions of fractional differential equations (FDEs), naturally forces researchers to study, devise and apply numerical methods to solve a large range of ordinary and partial differential equations with fractional derivatives.

The investigation of computational methods for fractional-order problems is therefore a very active research area in which, each year, a large number of research papers are published.

The task of finding efficient and reliable numerical methods for handling integrals and/or derivatives of fractional order is a challenge in its own right, with difficulties that differ in character but are no less severe than those associated with finding analytical solutions. The specific nature of these operators involves computational challenges which, if not properly addressed, may lead to unreliable or even wrong results.

Unfortunately, the scientific literature is rich with examples of methods that are inappropriate for fractional-order problems. In most cases these are just methods that were devised originally for standard integer-order operators then applied in a naive way to their fractional-order counterparts; without a proper knowledge of the specific features of fractional-order problems, researchers are often unable to understand why unexpected results are obtained.

The main aims of this paper are to identify a few major guidelines that should be followed when devising reliable computational methods for fractional-order problems, and to highlight the main peculiarities that make the solution of differential equations of fractional order a different—but surely more difficult and stimulating—task from the integer-order case. We do not intend merely to criticize weak or wrong methods, but try to explain why certain approaches are unreliable in fractional calculus and, where possible, point the reader towards more suitable approaches.

This paper is mainly addressed at young researchers or scientists without a particular background in the numerical analysis of fractional-order problems but who need to apply computational methods to solve problems of fractional order. We aim to offer in this way a kind of guide to avoid some of the most common mistakes which, unfortunately, are sometimes made in this field.

The paper is organized in the following way. After recalling in Section 2 some basic definitions and properties, we illustrate in Section 3 the most common ideas underlying the majority of the methods proposed in the literature: very often the basic ideas are not properly recognized and common methods are claimed to be new. In Section 4 we discuss why polynomial approximations can be only partially satisfactory for fractional-order problems and why they are unsuitable for devising high-order methods (as has often been proposed). The major problems related to the nonlocality of fractional operators are addressed in Sections 5 and 6 discusses some of the most powerful approaches for the efficient treatment of the memory term. Some remarks related to the numerical treatment of fractional partial differential equations are presented in Section 7 and some final comments are given in Section 8.

2. Basic Material and Notations

With the aim of fixing the notation and making available the most common definitions and properties for further reference, we recall here some basic notions concerning fractional calculus.

For $\alpha > 0$ and any $t_0 \in \mathbb{R}$, in the paper we will adopt the usual definitions for the fractional integral of Riemann–Liouville type

$$J_{t_0}^\alpha f(t) = \frac{1}{\Gamma(\alpha)} \int_{t_0}^t (t-\tau)^{\alpha-1} f(\tau) d\tau, \quad t > t_0, \tag{1}$$

for the fractional derivative of Riemann–Liouville type

$$^{RL}D_{t_0}^\alpha f(t) := D^m J_{t_0}^{m-\alpha} f(t) = \frac{1}{\Gamma(m-\alpha)} \frac{d^m}{dt^m} \int_{t_0}^t (t-\tau)^{m-\alpha-1} f(\tau) d\tau, \quad t > t_0 \tag{2}$$

and for the fractional derivative of Caputo type

$$^CD_{t_0}^\alpha f(t) := J_{t_0}^{m-\alpha} D^m f(t) = \frac{1}{\Gamma(m-\alpha)} \int_{t_0}^t (t-\tau)^{m-\alpha-1} f^{(m)}(\tau) d\tau, \quad t > t_0, \tag{3}$$

with $m = \lceil \alpha \rceil$ the smallest integer greater than or equal to α.

We refer to any of the many existing textbooks on this subject (e.g., [1–6]) for an exhaustive treatment of the conditions under which the above operators exist and for their main properties. We just recall here the relationship between $^{RL}D_{t_0}^\alpha$ and $^CD_{t_0}^\alpha$ expressed as

$$^CD_{t_0}^\alpha f(t) = {^{RL}D_{t_0}^\alpha} \left(f - T_{m-1}[f; t_0] \right)(t), \tag{4}$$

where $T_{m-1}[f;t_0]$ is the Taylor polynomial of degree $m-1$ for the function f about the point t_0,

$$T_{m-1}[f;t_0](t) = \sum_{k=0}^{m-1} \frac{(t-t_0)^k}{k!} f^{(k)}(t_0). \tag{5}$$

Moreover, we will almost exclusively consider initial value problems of Cauchy type for FDEs with the Caputo derivative, i.e.,

$$\begin{cases} D_{t_0}^\alpha y(t) = f(t,y(t)) \\ y(t_0) = y_0, \, y'(t_0) = y_0^{(1)}, \ldots, y^{(m-1)}(t_0) = y_0^{(m-1)}, \end{cases} \tag{6}$$

for some assigned initial values $y_0, y_0^{(1)}, \ldots, y_0^{(m-1)}$. A few general comments will also be made regarding problems associated with partial differential equations.

3. Novel or Well-Established Methods?

Quite frequently, one sees papers whose promising title claims the presentation of "new methods" or "a family of new methods" for some particular fractional-order operator. Papers of this type immediately capture the attention of readers eager for new and good ideas for numerically solving problems of this type.

But reading the first few pages of such papers can be a source of frustration, since what is claimed to be new is merely an old method applied to a particular (maybe new) problem. Now it is understandable that sometimes an old method is reinvented by a different author, maybe because it can be derived by some different approach or because the author is unaware of the previously published result (perhaps because it was published under an imprecise or misleading title). In fractional calculus, however, a different and quite strange phenomenon has taken hold: well-known and widely used methods are often claimed as "new" just because they are being applied to some specific problem. It seems that some authors are unaware that it is the development of new ideas and new approaches that leads to methods that can be described as new—not the application of known ideas to a particular problem. Even the application of well-established techniques to any of the new operators, obtained by simply replacing the kernel in the integral (1) with some other function, cannot be considered a truly novel method, especially when the extension to the new operator is straightforward.

Most of the papers announcing "new" methods are instead based on ideas and techniques that were proposed and studied decades ago, and sometimes proper references to the original sources are not even given.

In fact, there are a few basic and powerful methods that are suitable and extremely popular for fractional-order problems, and many proposed "new methods" are simply the application of the ideas behind them. It may therefore be useful to illustrate the main and more popular ideas that are most frequently (re)-proposed in fractional calculus, and to outline a short history of their origin and development.

3.1. Polynomial Interpolation and Product-Integration Rules

Solving differential equations by approximating their solution or their vector field by a polynomial interpolant is a very old and common idea. Some of the classical linear multistep methods for ordinary differential equations (ODEs), specifically those of Adams–Bashforth or Adams–Moulton type, are based on this approach.

In 1954 the British mathematician Andrew Young proposed [7,8] the application of polynomial interpolation to solve Volterra integral equations numerically. This approach turns out to be suitable for FDEs since (6) can be reformulated as the Volterra integral equation

$$y(t) = T_{m-1}[f;t_0](t) + \frac{1}{\Gamma(\alpha)} \int_{t_0}^{t} (t-u)^{\alpha-1} f(u,y(u)) du. \tag{7}$$

The approach proposed by Young is to define a grid $\{t_n\}$ on the solution interval $[t_0, T]$ (very often, but not necessarily, equispaced, namely $t_n = t_0 + hn$, $h = (T - t_0)/N$) and to rewrite (7) in a piecewise way as

$$y(t_n) = T_{m-1}[f;t_0](t_n) + \frac{1}{\Gamma(\alpha)} \sum_{j=0}^{n-1} \int_{t_j}^{t_{j+1}} (t_n - u)^{\alpha-1} f(u,y(u)) du, \tag{8}$$

then to replace, in each interval $[t_j, t_{j+1}]$, the vector field $f(u, y(u))$ by a polynomial that interpolates to f on the grid. This approach is particularly simple if one uses polynomials of degree 0 or 1 because then one can determine the approximation solely on the basis of the data at one of the subinterval's end points (degree 0; the *product rectangle method*) or at both end points (degree 1; the *product trapezoidal method*); thus, in these cases one need not introduce auxiliary points inside the interval or points outside the interval. Neither of these methods can yield a particularly high order of convergence, but as we shall demonstrate in Section 4, the analytic properties of typical solutions to fractional differential equations make it very difficult and cumbersome to achieve high-order accuracy irrespective of the technique used. Consequently, and because these techniques have been thoroughly investigated with respect to their convergence properties [9] and their stability [10] and are hence very well understood, the product rectangle and product trapezoidal methods are highly popular among users of fractional order models.

Higher-order methods have occasionally been proposed [11,12] but—as indicated above and discussed in more detail in Section 4—they tend to require rather uncommon properties of the exact solutions to the given problems and therefore are used only infrequently. We also have to notice that the effects of the lack of regularity on the convergence properties of product-integration rules have been studied since 1985 for Volterra integral equations [13] and since 2004 for the specific case of FDEs [14].

3.2. Approximation of Derivatives: L1 and L2 Schemes

A classical numerical technique for approximating the Caputo differential operator from (3) is the so-called *L1 scheme*. For $0 < \alpha < 1$, the definition of the Caputo operator becomes

$$\mathcal{D}_{t_0}^{\alpha} f(t) = \frac{1}{\Gamma(1-\alpha)} \int_{t_0}^{t} (t-\tau)^{-\alpha} f'(\tau) d\tau \text{ for } t > t_0.$$

The idea ([15], Equation (8.2.6)) is to introduce a completely arbitrary (i.e., not necessarily uniformly spaced) mesh $t_0 < t_1 < t_2 < \ldots < t_N$ and to replace the factor $f'(\tau)$ in the integrand by the approximation

$$f'(\tau) \approx \frac{f(t_{j+1}) - f(t_j)}{t_{j+1} - t_j} \quad \text{whenever } \tau \in (t_j, t_{j+1}).$$

This produces the approximation formula

$$\mathcal{D}_{t_0}^{\alpha} f(t_n) \approx \mathcal{D}_{t_0,L1}^{\alpha} f(t_n) = \frac{1}{\Gamma(2-\alpha)} \sum_{j=0}^{n-1} w_{n-j-1,n} (f(t_{n-j}) - f(t_{n-j-1}))$$

with
$$w_{\mu,n} = \frac{(t_n - t_\mu)^{1-\alpha} - (t_n - t_{\mu+1})^{1-\alpha}}{t_{n-\mu} - t_{n-\mu-1}}.$$

For smooth functions f (but only under this assumption!) and an equispaced mesh $t_j = t_0 + jh$, the convergence order of the L1 method is $\mathcal{O}(h^{2-\alpha})$.

By construction, the L1 method is restricted to the case $0 < \alpha < 1$. For $\alpha \in (1,2)$, the L2 method ([15], §8.2) provides a useful modification. In its construction, one starts from the representation

$$^C D_{t_0}^\alpha f(t) = \frac{1}{\Gamma(2-\alpha)} \int_{t_0}^t t^{1-\alpha} f''(t-\tau) d\tau,$$

which is valid for these values of α. Using now a uniform grid $t_j = t_0 + jh$, one replaces the second derivative of f in the integrand by its central difference approximation,

$$f''(t_n - \tau) \approx \frac{1}{h^2} \left(f(t_n - t_{k+1}) - 2f(t_n - t_k) + f(t_n - t_{k-1}) \right)$$

for $\tau \in [t_k, t_{k+1}]$, which yields

$$^C D_{t_0}^\alpha f(t_n) \approx {}^C D_{t_0,L2}^\alpha f(t_n) = \frac{h^{-\alpha}}{\Gamma(3-\alpha)} \sum_{k=-1}^n w_{k,n} f(t_{n-k}),$$

where now

$$w_{k,n} = \begin{cases} 1 & \text{for } k = -1, \\ 2^{2-\alpha} - 3 & \text{for } k = 0, \\ (k+2)^{2-\alpha} - 3(k+1)^{2-\alpha} + 3k^{2-\alpha} - (k-1)^{2-\alpha} & \text{for } 1 \leq k \leq n-2, \\ -2n^{2-\alpha} + 3(n-1)^{2-\alpha} - (n-2)^{2-\alpha} & \text{for } k = n-1, \\ n^{2-\alpha} - (n-1)^{2-\alpha} & \text{for } k = n. \end{cases}$$

A disadvantage of this method is that it requires the evaluation for f at the point $t_{n+1} = (n+1)h$ which is located outside the interval $[0, t_n]$.

The central difference used in the definition of the L2 method is symmetric with respect to one of the endpoints of the associated subinterval $[t_k, t_{k+1}]$, not with respect to its mid point. If this is not desired, one may instead use the alternative

$$f''(t_n - \tau) \approx \frac{1}{h^2} \left(f(t_{n-k-2}) - f(t_{n-k-1}) + f(t_{n-k+1}) - f(t_{n-k}) \right)$$

on this subinterval. This leads to the L2C method [16]

$$^C D_{t_0}^\alpha f(t_n) \approx {}^C D_{t_0,L2C}^\alpha f(t_n) = \frac{h^{-\alpha}}{2\Gamma(3-\alpha)} \sum_{k=-1}^{n+1} w_{k,n} f(t_{n-k})$$

with

$$w_{k,n} = \begin{cases} 1 & \text{for } k = -1, \\ 2^{2-\alpha} - 2 & \text{for } k = 0, \\ 3^{2-\alpha} - 2^{2-\alpha} & \text{for } k = 1, \\ (k+2)^{2-\alpha} - 2(k+1)^{2-\alpha} + 2(k-1)^{2-\alpha} - (k-2)^{2-\alpha} & \text{for } 2 \le k \le n-2, \\ -n^{2-\alpha} - (n-3)^{2-\alpha} + 2(n-2)^{2-\alpha} & \text{for } k = n-1, \\ -n^{2-\alpha} + 2(n-1)^{2-\alpha} - (n-2)^{2-\alpha} & \text{for } k = n, \\ n^{2-\alpha} - (n-1)^{2-\alpha} & \text{for } k = n+1. \end{cases}$$

Like the L2 method, the L2C method also requires the evaluation of f outside the interval $[0, t_n]$; one has to compute $f((n+1)h)$ and $f(-h)$. Both the L2 and the L2C method exhibit $\mathcal{O}(h^{3-\alpha})$ convergence behavior for $1 < \alpha < 2$ if f is sufficiently well behaved; the constants implicitly contained in the $\mathcal{O}$-terms seem to be smaller for the L2 method in the case $1 < \alpha < 1.5$ and for the L2C method if $1.5 < \alpha < 2$.

In the limit case $\alpha \to 1$, the L2 method reduces to first-order backward differencing, and the L2C method becomes the centered difference of first order; for $\alpha \to 2$ the L2 method corresponds to the classical second-order central difference.

3.3. Fractional Linear Multistep Methods

Fractional linear multistep methods (FLMMs) are less frequently used since their coefficients are, in general, not known explicitly but it is necessary to devise some algorithm for their (technically often difficult) computation. Nevertheless, since these methods allow us to overcome some of the issues associated with other approaches, it is worth giving a short presentation of their properties.

FLMMs were proposed by Lubich in 1986 [17] and studied in the successive works [18–20]. They extend to fractional-order integrals the quadrature rules obtained from standard linear multistep methods (LMMs) for ODEs.

Let us consider a classical k-step LMM of order $p > 0$ with first and second characteristic polynomials $\rho(z) = \rho_0 z^k + \rho_1 z^{k-1} + \cdots + \rho_k$ and $\sigma(z) = \sigma_0 z^k + \sigma_1 z^{k-1} + \cdots + \sigma_k$, namely

$$\sum_{j=0}^{k} \rho_j y_{n-j} = h \sum_{j=0}^{k} \sigma_j f(t_{n-j}), \quad \text{where } \delta(\xi) = \frac{\rho(1/\xi)}{\sigma(1/\xi)} \text{ is the generating function.} \tag{9}$$

FLMMs generalizing LMMs (9) for solving FDEs (7) are expressed as

$$y_n = T_{m-1}[f; t_0](t) + h^\alpha \sum_{j=0}^{\nu} w_{n,j} f(t_j, y_j) + h^\alpha \sum_{j=0}^{n} w_{n-j}^{(\alpha)} f(t_j, y_j), \tag{10}$$

where the convolution weights $w_n^{(\alpha)}$ are obtained from the power series expansion of $(\delta(\xi))^{-\alpha}$, namely

$$\sum_{n=0}^{\infty} w_n^{(\alpha)} \xi^n = \frac{1}{(\delta(\xi))^\alpha},$$

and the $w_{n,j}$ are some starting weights that are introduced to deal with the lack of regularity of the solution at the origin; they are obtained by solving, at each step n, the algebraic linear systems

$$\sum_{j=0}^{\nu} w_{n,j} j^\gamma = -\sum_{j=0}^{n} w_{n-j} j^\gamma + \frac{\Gamma(\gamma+1)}{\Gamma(1+\gamma+\alpha)} n^{\gamma+\alpha}, \quad \nu \in A_p, \tag{11}$$

with $\mathcal{A}_p = \{\gamma \in \mathbb{R} \mid \gamma = i + j\alpha, \, i,j \in \mathbb{N}, \, \gamma < p - 1\}$ and $\nu + 1$ the cardinality of $\mathcal{A}_p$.

The intriguing property of FLMMs is that, unlike product-integration rules, they are able to preserve the same convergence order p of the underlying LMMs if the LMM satisfies certain properties: it is required that $\delta(\xi)$ has no zeros in the closed unit disc $|\xi| \leq 1$ except for $\xi = 1$, and $|\arg \delta(\xi)| < \pi$ for $|\xi| < 1$. Thus, high-order FLMMs are possible without requiring the imposition of artificial smoothness assumptions as is required for methods based on polynomial interpolation.

But the price to be paid for this advantage may be not negligible: the convolution weights $\omega_n^{(\alpha)}$ are not known explicitly and must be computed by some (possibly sophisticated) method (a discussion for the general case is available in [17–20] while algorithms for FLMMs of trapezoidal type are presented in [21]). Moreover, high-order methods may require the solution of large or very large systems (11) depending on the equation order α and the convergence order p of the method; in some cases these systems are so ill-conditioned as to affect the accuracy of the method, a problem addressed in depth in [22].

One of the simplest methods in this family is obtained from the backward Euler method, whose generating function is $\delta(\xi) = (1 - \xi)$. Its convolution weights are hence the coefficients in the asymptotic expansion of $(1 - \xi)^{-\alpha}$, i.e., they are the coefficients in the binomial series

$$\omega_j^{(\alpha)} = (-1)^j \binom{-\alpha}{j} = \frac{\Gamma(-\alpha + 1)}{j! \Gamma(-\alpha - j + 1)}$$

and no starting weights are necessary since the convergence order is $p = 1$ and hence $\mathcal{A}_p$ is the empty set. One recognizes easily that the so-called Grünwald-Letnikov scheme is obtained in this case. Although this scheme was discovered in the nineteenth century in independent works of Grünwald and Letnikov, its interpretation as an FLMM may facilitate its analysis.

4. Classical Approximations Will Not Give High-Order Methods

Solutions of fractional-derivative problems typically exhibit weak singularities. This topic is discussed at length in the survey chapter [23] and it is known since earlier works on Volterra integral equations [24,25]. This singularity is a consequence of the weakly singular behavior of the kernels of integral and fractional derivatives and its importance, from a physical perspective, is related to the natural emergence of completely monotone (CM) relaxation functions in models whose dynamics is governed by these operators [26,27]; CM relaxation behaviors are indeed typical of viscoelastic systems with strongly dissipative energies [28].

In the present section we shall examine the effects of the singular behavior on numerical methods, in the context of initial value problems such as (6).

To grasp quickly the main ideas, we focus on a very simple particular case of (6): the problem

$$\mathcal{D}_0^\alpha y(t) = 1 \text{ for } t \in (0, T], \tag{12}$$

where $0 < \alpha < 1$ and, for the moment, we do not prescribe the initial condition at $t = 0$. The general solution of (12) is

$$y(t) = \frac{t^\alpha}{\Gamma(1 + \alpha)} + b, \text{ where } b \text{ is an arbitrary constant.} \tag{13}$$

This solution lies in $C[0, T] \cap C^1(0, T]$ but not in $C^1[0, T]$. This implies that standard techniques for integer-derivative problems, which require that $y \in C^1[0, T]$ (or a higher degree of regularity), cannot be used here without some modification. In particular one cannot perform a Taylor series expansion of the solution around $t = 0$ because $y'(0)$ does not exist.

What about the initial condition? If we prescribe a condition of the form $y(0) = y_0$ we get $b = y_0$ in (13), but the solution is still not in $C^1[0, T]$. One might hope that a Neumann-type condition of the form

$y'(0) = 0$ would control or eliminate the singularity in the solution, but a consideration of (13) shows that it is impossible to enforce such a condition; that is, the problem $\mathcal{D}_0^\alpha y(t) = 1$ on $(0, T]$ with $y'(0) = 0$ has no solution. This seems surprising until we recall a basic property of the Caputo derivative from ([1], Lemma 3.11): if $m - 1 < \beta < m$ for some positive integer m and $z \in C^m[0, T]$, then $\lim_{t \to 0} \mathcal{D}_0^\beta z(t) = 0$. Hence, if in (12) one has $y \in C^1[0, T]$, then taking the limit as $t \to 0$ in (12) we get $0 = 1$, which is impossible. That is, any solution y of (12) cannot lie in $C^1[0, T]$.

One can present this finding in another way: for the problem $\mathcal{D}_0^\alpha y(t) = f(t)$ on $(0, T]$ with $f \in C[0, T]$, if the solution $y \in C^1[0, T]$, then one must have $f(0) = 0$. This result is a special case of ([1], Theorem 6.26).

Remark 1. *For the problem $\mathcal{D}_0^\alpha y(t) = f(t)$ on $(0, T]$ with $0 < \alpha < 1$, if one wants more smoothness of the solution y on the closed interval $[0, T]$, then one must impose further conditions on the data: by ([1], Theorem 6.27), for each positive integer m, one has $y \in C^m[0, T]$ if and only if $0 = f(0) = f'(0) = \cdots = f^{(m-1)}(0)$.*

Conditions such as $f(0) = 0$ (and the even stronger conditions listed in Remark 1) impose an artificial restriction on the data f that should be avoided. Thus we continue by looking carefully at the consequence of dealing with a solution of limited smoothness.

Returning to (12) and imposing the initial condition $y(0) = b$, the unique solution of the problem is given by (13), where b is now fixed. Most numerical methods for integer-derivative initial value problems are based on the premise that on any small mesh interval $[t_i, t_{i+1}]$, the unknown solution can be approximated to a high degree of accuracy by a polynomial of suitable degree. But is this true of the function (13)? We now investigate this question.

Consider the interval $[0, h]$, where $h = t_1$. This is the mesh interval where the solution (13) is worst behaved.

Lemma 1. *Let $\alpha \in (0, 1)$. Consider the approximation of t^α by a linear polynomial $c_0 + c_1 t$ on the interval $[0, h]$. Suppose this approximation is uniformly $\mathcal{O}(h^\beta)$ accurate on $[0, h]$ for some fixed $\beta > 0$. Then one must have $\beta \leq \alpha$.*

Proof. Our hypothesis is that $|t^\alpha - (c_0 + c_1 t)| \leq C h^\beta$ for all $t \in [0, h]$ and some constant C that is independent of h and t. Consider the values $t = 0, t = h/2$ and $t = h$ in this inequality: we get

$$\begin{cases} 0 - (c_0 + 0) & = \mathcal{O}(h^\beta), \\ (h/2)^\alpha - (c_0 + c_1 h/2) & = \mathcal{O}(h^\beta), \\ h^\alpha - (c_0 + c_1 h) & = \mathcal{O}(h^\beta). \end{cases}$$

The first equation gives $c_0 = \mathcal{O}(h^\beta)$. Hence the other equations give $(h/2)^\alpha - c_1 h/2 = \mathcal{O}(h^\beta)$ and $h^\alpha - c_1 h = \mathcal{O}(h^\beta)$. Eliminate c_1 by multiplying the first equation by 2 then subtracting from the other equation; this yields $h^\alpha - 2(h/2)^\alpha = \mathcal{O}(h^\beta)$. But this cannot be true unless $\beta \leq \alpha$, since the left-hand side is simply a multiple of h^α because $\alpha \neq 1$. □

Lemma 1 says that the approximation of t^α on $[0, h]$ by *any* linear polynomial is at best $\mathcal{O}(h^\alpha)$. But the order of approximation $\mathcal{O}(h^\alpha)$ of t^α on $[0, h]$ is also achieved by the constant polynomial 0. That is: using a linear polynomial to approximate t^α on $[0, h]$ does not give an essentially better result than using a constant polynomial. In a similar way one can show that using polynomials of higher degree does not improve the situation: the order of approximation of t^α on $[0, h]$ is still only $\mathcal{O}(h^\alpha)$. This is a warning that when solving typical fractional-derivative problems, high-degree polynomials may be no better than low-degree polynomials, unlike the classical integer-derivative situation.

One can generalize Lemma 1 to any $\alpha > 0$ with α not an integer, obtaining the same result via the same argument. Furthermore, our investigation of the simple problem (12) can be readily generalised to the much more general problem (6); see ([1], Section 6.4).

Implications for the Construction of Difference Schemes

The discussion earlier in Section 4 implies that, to construct higher-order difference schemes for typical solutions of problems such as (12) and (6), one must use non-classical schemes, since the classical schemes are constructed under the assumption that approximations by higher-order polynomials gives greater accuracy. The same idea is developed at length in [29], one of whose results we now present.

Note: although [29] discusses only boundary value problems, an inspection reveals that its arguments and results are also valid (mutatis mutandis) for initial value problems such as (6) when $f = f(t)$, i.e., when the problem (6) is linear.

Let $\alpha > 0$ be fixed, with α not an integer. Consider the problem $D^\alpha y = f$ on $[0, T]$ with $y(0) = 0$. Assume that the mesh on $[0, T]$ is equispaced with diameter h, i.e., $x_i = ih$ for $i = 0, 1, \ldots, N$. Suppose that the difference scheme used to solve $D^\alpha y = f$ at each point x_i for $i > 0$ is $\sum_{j=0}^{i} a_{ij} y_j^N = f(t_i)$. It is reasonable to assume that $|a_{ij}| = \mathcal{O}(h^{-\alpha})$ for all i and j since we are approximating a derivative of order α (one can check that almost all schemes proposed for this problem have this property).

We have the following variant of ([29], Theorem 3.3).

Theorem 1. *Assume that our scheme achieves order of convergence p for some $p > \alpha$ when $f(t) = Ct^k$ for all $k \in \{0, 1, \ldots, \lceil p - \alpha - 1 \rceil\}$. Then for each fixed positive integer i, the coefficients of the scheme must satisfy the following relationship:*

$$\lim_{h \to 0} \left(h^\alpha \sum_{j=0}^{i} j^{k+\alpha} a_{ij} \right) = \frac{i^k \Gamma(\alpha + k + 1)}{\Gamma(k+1)} \quad \text{for } k = 0, 1, \ldots, \lceil p - \alpha - 1 \rceil. \tag{14}$$

Proof. Fix $k \in \{0, 1, \ldots, \lceil p - \alpha - 1 \rceil\}$. This implies that $k < p - \alpha$. Choose for simplicity

$$f(t) = \frac{\Gamma(k + \alpha + 1)}{\Gamma(k+1)} t^k.$$

Then the true solution of our initial value problem is $y(t) = t^{k+\alpha}$. Fix a positive integer i. Then

$$\sum_{j=0}^{i} a_{ij} y_j^N = f(t_i) = \frac{\Gamma(k + \alpha + 1)}{\Gamma(k+1)} (ih)^k.$$

Hence, using the hypothesis that our scheme achieves order of convergence p and $|a_{ij}| = \mathcal{O}(h^{-\alpha})$,

$$\lim_{h \to 0} \left(h^\alpha \sum_{j=0}^{i} j^{k+\alpha} a_{ij} \right) = \lim_{h \to 0} h^{-k} \sum_{j=0}^{i} a_{ij} y(t_j)$$

$$= \lim_{h \to 0} h^{-k} \left\{ \frac{\Gamma(k+\alpha+1)}{\Gamma(k+1)} (ih)^k + \sum_{j=0}^{i} a_{ij} \left[y(x_j) - y_j^N \right] \right\}$$

$$= \lim_{h \to 0} \left[\frac{\Gamma(k+\alpha+1)}{\Gamma(k+1)} i^k + \mathcal{O}(h^{p-\alpha-k}) \right]$$

$$= \frac{\Gamma(k+\alpha+1)}{\Gamma(k+1)} i^k,$$

since $k < p - \alpha$. □

Theorem 1 implies that schemes that fail to satisfy (14) cannot achieve an order of convergence greater than $\mathcal{O}(h^\alpha)$ at each mesh point. (This is consistent with the approximation theory result of Lemma 1.) For example, in the case $0 < \alpha < 1$, it follows from Theorem 1 that the well-known L1 scheme is at best $\mathcal{O}(h^\alpha)$ accurate.

Remark 2. *To avoid the consequences of results such as Theorem 1, one can impose data restrictions such as $f(0) = 0$. This is discussed in ([29], Section 5), where theoretical and experimental results show an improvement in the accuracy of standard difference schemes, but only for a restricted class of problems.*

5. Failed Approaches to Treat Non-Locality

Non-locality is one of the major features of fractional-order operators. Indeed, fractional integrals and derivatives are often introduced as a mathematical formalism with the primary purpose of encompassing hereditary effects in the modeling of real-life phenomena when theoretical or experimental observations suggest that the effects of external actions do not propagate instantaneously but depend on the history of the system.

On the one hand, non-locality is a very attractive feature that has driven most of the interest and success of the fractional calculus; on the other hand, non-locality introduces severe computational difficulties that researchers try to overcome in different ways.

Unfortunately, some attempts to treat non-locality are unreliable and lead to wrong results. This is the case of the naive implementation of the "finite memory principle" consisting in simply neglecting a large amount of the history solution; since on the basis of this technique it is however possible to devise more sophisticated and accurate approaches, we postpone its discussion to Section 6.

We have also to mention methods based on some kind of fractional Taylor expansion of the solution, such as

$$y(t) = \sum_{k=0}^{\infty} Y_k (t - t_0)^{k\alpha},$$

where the coefficients Y_k are determined by some suitable numerical technique.

When solving integer-order differential equations, it is possible to use Taylor expansions to approximate the solution at a given point t_1 and hence reformulate the same expansion by moving the origin to the new point t_1, thus generating a step-by-step method in which the approximation at t_{n+1} is evaluated on the basis of the approximation at t_n (or at additional previous points).

With fractional-order equations, instead, the above expansion holds only with respect to the point t_0 (the initial or starting point of the fractional differential operator) and it is not possible to generate a step-by-step method. Expansions of this type are therefore able to provide an accurate approximation only locally, i.e., very close to the starting point t_0; consequently, as discussed in [30], methods based on these expansions are usually unsuitable for FDEs.

Another failed approach is based on an attempt to exploit the difference between $y(t_{n+1})$ and $y(t_n)$ in the integral formulation (7): rewrite the solution at t_{n+1} as some increment of the solution at t_n, i.e.,

$$y(t_{n+1}) = y(t_n) + G_n(t, y(t)), \tag{15a}$$

then approximate the increment

$$G_n(t, y(t)) = \frac{1}{\Gamma(\alpha)} \int_{t_0}^{t_{n+1}} (t_{n+1} - u)^{\alpha - 1} f(u, y(u)) du - \frac{1}{\Gamma(\alpha)} \int_{t_0}^{t_n} (t_n - u)^{\alpha - 1} f(u, y(u)) du \tag{15b}$$

by replacing the vector field $f(t, y(t))$ in both integrals of (15b) by its (first-order) interpolating polynomial at the grid points t_{n-1} and t_n. Methods of this kind read as

$$y_{n+1} = y_n + P_n(y_{n-1}, y_n), \qquad (16)$$

with P_n a known function obtained by standard interpolation techniques. Approaches of this kind are called *two-step Adams–Bashforth methods* and attract researchers since they apparently transform the non-local problem into a local one (and thus, a difficult problem into a much easier one); in (15b) $G_n(t, y(t))$ is still a non-local term but these methods are strangely becoming quite popular despite the fact that, as discussed in [31], they are usually unreliable because in most cases they attempt to approximate the (implicitly) non-local contribution $G_n(t, y(t))$ by some purely local term.

Using interpolation at the points t_{n-1} and t_n to approximate $f(t, y(t))$ over the much larger intervals $[t_0, t_n]$ and $[t_0, t_{n+1}]$ is completely inappropriate. It is well known that polynomial interpolation may offer accurate approximations within the interval of the data points, in this case in $[t_{n-1}, t_n]$; but outside this interval (where an extrapolation is made instead of an interpolation), the approximation becomes more and more inaccurate as the integration intervals $[t_0, t_n]$ and $[t_0, t_{n+1}]$ in (15b) become larger and larger, i.e., as the integration proceeds and n increases.

The consequence is that completely untrustworthy results must be expected from methods based on this idea.

Note that the fundamental flaw of this approach is not the decomposition (15) but the local (and hence inappropriate) way (16) in which the history is handled. Indeed, it is possible to construct technically correct and efficient algorithms on the basis of (15), for example if one treats the increment term (15b) by a numerical method that is cheaper in computational cost than the method used for the local term [32].

6. Some Approaches for the Efficient, and Reliable, Treatment of the Memory Term

The non-locality of the fractional-order operator means that it is necessary to treat the memory term in an efficient way. This term is commonly identified to be the source of a computational complexity which, especially in problems of large size, requires adequate strategies in order to keep the computational cost at a reasonable level, and indeed this observation has led to many investigations of (more or less successful) approaches to reduce the computational cost. It should be noted however that the high number of arithmetic operations is not the only potential difficulty that the memory term introduces. There is another more fundamental issue, which seems to have attracted much less attention: the history of the process not only needs to be taken into account in the computation but, in order to be properly handled, also needs to be *stored* in the computer's memory. While the required amount of memory is usually easily available in algorithms for solving ordinary differential equations, the memory demand may be too high for efficient handling in the case of, e.g., time-fractional partial differential equations where finite element techniques are used to discretize the spatial derivatives.

Most finite-difference methods for FDEs require at each time step the evaluation of some convolution sum of the form

$$y_n = \phi_n + \sum_{j=0}^{n} c_j y_{n-j} \quad \text{or} \quad y_n = \phi_n + \sum_{j=0}^{n} c_j f(t_{n-j}, y_{n-j}), \quad n = 1, 2, \ldots, N, \qquad (17)$$

where ϕ_n is a term which mainly depends on the initial conditions or other known information.

A naive straightforward evaluation of (17) has a computational cost proportional to $\mathcal{O}(N^2)$ and, when integration with a small-step size or on a large integration interval is required, the value of N can be extremely large and leads to prohibitive computational costs.

For this reason different approaches for a fast, efficient and reliable treatment of the memory term in non-local problems have been devised. We provide here a short description of some of the most interesting methods of this type. The influence of these approaches on the memory requirements will be addressed as well.

6.1. Nested Mesh Techniques

Several different concepts can be subsumed under the heading of so-called *nested meshes*. The general idea is based on the observation that the convolution sum in Equation (17) stems from a discretization of a fractional integral or differential operator that uses all the previous grid points as nodes. One can then ask whether it is really neccessary to use all these nodes or whether one could save effort by including only a subset of them by using a second, less fine mesh—i.e., a mesh nested inside the original one.

6.1.1. The Finite Memory Principle

The simplest idea in this class is the *finite memory principle* ([5], §7.3). It is based on defining a constant $\tau > 0$, the so-called memory length, and replacing (for $t > t_0 + \tau$) the memory integral term that extends over the interval $[t_0, t]$ by the integral over $[t - \tau, t]$ with the same integrand function. Technically speaking, this amounts to "forgetting" the entire history of the process that is more than τ units of time in the past, so the memory has a finite and fixed length τ instead of the variable length $t - t_0$ that may, in a long running process, be very much longer. From an algorithmic point of view, the finite memory method truncates the convolution sum in Equation (17) to a sum where j runs from $n - \nu$ to n for some fixed ν. This has a number of significant advantages:

- The computational complexity of the nth time step is reduced from $\mathcal{O}(n)$ to $\mathcal{O}(1)$. Therefore, the combined total complexity of the overall method with N time steps is reduced from $\mathcal{O}(N^2)$ to $\mathcal{O}(N)$.
- At no point in time does one need to access the part of the process history that is more than ν time steps in the past. Therefore, all those previous time steps can be removed from the active memory, and the memory requirement also decreases from $\mathcal{O}(N)$ to $\mathcal{O}(1)$.

Unfortunately, this idea also has severe drawbacks. Specifically, it has been shown in [33] that the convergence order of the underlying discretization technique is lost completely. In other words, one cannot prove that the algorithm converges as the (maximal) step size goes to 0. Therefore, the method is not recommended for practical use.

6.1.2. Logarithmic Memory

To overcome the shortcomings of the finite memory principle, two related but not identical methods, both of which are also based on the nested mesh concept, have been developed in [33,34]. The common idea of both these approaches is the way in which the distant part of the memory is treated. Rather than ignoring it completely as the finite memory principle does, they do sample it, but on a coarser mesh; indeed the fundamental principle is to introduce not just one coarsening level, but to use, say, the step size h on the most recent part of the memory, step size wh (with some parameter $w > 1$) on the adjacent region, $w^2 h$ on the next region, etc. The main difference between the two approaches of [33,34] then lies in the way in which the transition points from one mesh size to the next are chosen.

Specifically, as indicated in Figure 1, the method of Ford and Simpson [33] starts at the current time and fills subintervals of prescribed lengths from right to left with appropriately speced mesh points. This will lead to a reduction of the computational cost to $\mathcal{O}(N \log N)$ while retaining the convergence order of the underlying scheme [33]. However, as indicated in Figure 1, it is common that the left end point of the leftmost coarsely subdivided interval does not match the initial point. In this case, one can either fill the remaining subinterval at the left end of the full interval with a fine mesh (which increases the computational cost but also reduces the error) or simply ignore the contribution from this subinterval (which reduces the computational complexity but slightly increases the error; however, since the memory length still grows with the number of steps, this does not imply the complete loss of accuracy observed in the finite memory principle). In either case, grid points from the fine mesh that are not currently used in the nested mesh may become active again in future steps. Therefore, all previous grid points need to be kept in memory, so the required amount of memory space remains at $\mathcal{O}(N)$.

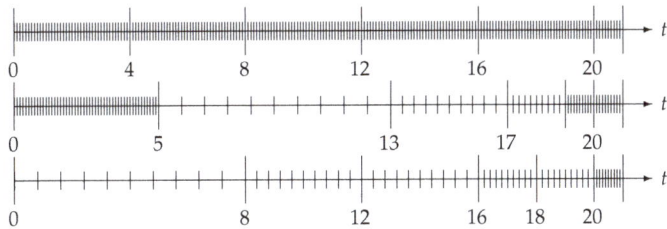

Figure 1. Full mesh (**top**) and nested meshes proposed in [33] (**center**) and in [34] (**bottom**). The meshes are shown for the time instant $t = 21$ and the basic step size $h = 1/10$.

In contrast, the approach of Diethelm and Freed [34] starts to fill the basic interval from left to right, i.e., it begins with the subinterval with the coarsest mesh and then moves to the finer-mesh regions. The final result is also a method with an $\mathcal{O}(N \log N)$ computational cost, and with the same convergence order as the Ford-Simpson method; but its selection strategy for grid points implies that points that are inactive in the current step will never become active again in future steps, and consequently the history data for these inactive points can be eliminated from the main memory. This reduces the memory requirements to only $\mathcal{O}(\log N)$.

6.2. A Method Based on the Fast Fourier Transform Algorithm

An effective approach for the fast evaluation of the convolution sums in (17) was proposed in [35,36]. The main idea is to split each of these sums in a way that enables the exploitation of the fast Fourier transform (FFT) algorithm. To provide a concise description, let us introduce the notations

$$T_p(n) = \sum_{j=p}^{n} c_{n-j} g_j, \quad S_{p,q}(n) = \sum_{j=p}^{q} c_{n-j} g_j, \quad n \geq p,$$

where $g_j = y_j$ or $g_j = f(t_j, y_j)$ according to the formula used in (17). Thus the numerical methods described by (17) can be recast as

$$y_n = \phi_n + T_0(n), \quad n = 1, 2, \ldots, N.$$

The algorithm described in [35,36] is based on splitting $T_0(n)$ into one or more partial sums of type $S_{p,q}(n)$ and just one final convolution sum $T_p(n)$ of a maximum (fixed) length r. Thus, the computation is simply initialized as

$$T_0(n) = \sum_{j=0}^{n} c_{n-j} g_j \qquad n \in \{1, 2, \ldots, r-1\}$$

and the following r values of $T_0(n)$ are split into the two terms

$$T_0(n) = S_{0,r-1}(n) + T_r(n) \qquad n \in \{r, r+1, \ldots, 2r-1\}.$$

Similarly, for the computation of the next $2r$ values, $T_0(n)$ is split according to

$$T_0(n) = \begin{cases} S_{0,2r-1}(n) + T_{2r}(n) & n \in \{2r, 2r+1, \ldots, 3r-1\} \\ S_{0,2r-1}(n) + S_{2r,3r-1}(n) + T_{3r}(n) & n \in \{3r, 3r+1, \ldots, 4r-1\} \end{cases}$$

and the further $4r$ summations are split according to

$$T_0(n) = \begin{cases} S_{0,4r-1}(n) + T_{4r}(n) & n \in \{4r, 4r+1, \ldots, 5r-1\} \\ S_{0,4r-1}(n) + S_{4r,5r-1}(n) + T_{5r}(n) & n \in \{5r, 5r+1, \ldots, 6r-1\} \\ S_{0,4r-1}(n) + S_{4r,6r-1}(n) + T_{7r}(n) & n \in \{6r, 6r+1, \ldots, 7r-1\} \\ S_{0,4r-1}(n) + S_{4r,6r-1}(n) + S_{6r,7r-1}(n) + T_{8r}(n) & n \in \{7r, 7r+1, \ldots, 8r-1\} \end{cases}$$

and this process is continued until all terms $T_0(n)$, for $n \leq N$, are evaluated.

Note that in the above splittings the length $\ell(p,q) = q - p + 1$ of each sum $S_{p,q}$ is always some multiple of r with a power of 2 as multiplying factor (i.e., the possible length of $S_{q,p}(n)$ is $r, 2r, 4r, 8r$ and so on).

For clarity, the diagram in Figure 2 illustrates the way in which the computation on the main triangle $T_0 = \{(n,j) : 0 \leq j \leq n \leq N\}$ is split into partial sums identified by the (red-labeled) squares $S_{p,q} = \{(n,j) : q+1 \leq n \leq q + \ell(p,q), p \leq j \leq q\}$ and final blocks denoted by the (blue-labeled) triangles $T_p = \{(n,j) : p \leq j \leq n \leq p+r-1\}$.

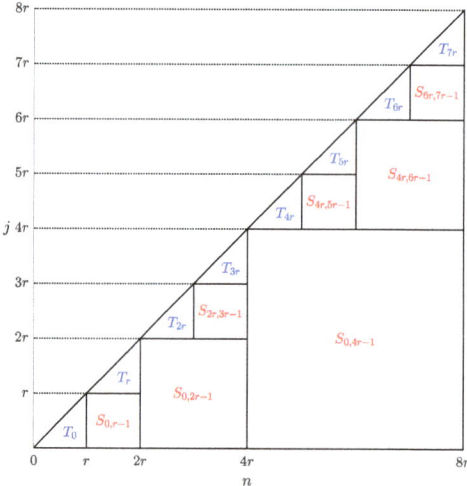

Figure 2. Splitting of the computation of $T_0(n)$ into partial sums $S_{p,q}$ (red-labeled squares) and final blocks T_p (blue-labeled triangles).

Each of the final blocks $T_{\ell r}(n)$, $n = \ell r, \ell r + 1, \ldots, (\ell+1)r - 1$, is computed by direct summation requiring $r(r+1)/2$ floating-point operations. The evaluation of the partial sums $S_{q,p}(n)$ can instead be performed by the FFT algorithm (see [37] for a comprehensive description) which requires a number of floating-point operations proportional to $2\ell \log_2 2\ell$, with $\ell = \ell(p,q)$ the length of each partial sum $S_{q,p}(n)$, since r is a power of 2.

In the optimal case in which both r and N are powers of 2, each partial sum $S_{p,q}$ that must be computed together with its length, number and computational cost is described in Table 1.

Table 1. Partial sums, their length, number and computational cost for the evaluation of $T_0(N)$.

Partial Sums	Len.	No.	Cost
$S_{0,\frac{N}{2}-1}$	$\frac{N}{2}$	1	$\mathcal{O}(N \log_2 N)$
$S_{0,\frac{N}{4}-1}, S_{\frac{N}{2},\frac{3N}{4}-1}$	$\frac{N}{4}$	2	$\mathcal{O}(\frac{N}{2} \log_2 \frac{N}{2})$
$S_{0,\frac{N}{8}-1}, S_{\frac{N}{4},\frac{3N}{8}-1}, S_{\frac{N}{2},\frac{5N}{8}-1}, S_{\frac{3N}{4},\frac{7N}{8}-1}$	$\frac{N}{8}$	4	$\mathcal{O}(\frac{N}{4} \log_2 \frac{N}{4})$
$S_{0,\frac{N}{16}-1}, S_{\frac{N}{8},\frac{3N}{16}-1}, S_{\frac{N}{4},\frac{5N}{16}-1}, S_{\frac{3N}{8},\frac{7N}{16}-1}, S_{\frac{N}{2},\frac{9N}{16}-1}, S_{\frac{5N}{8},\frac{11N}{16}-1}, S_{\frac{3N}{4},\frac{13N}{16}-1}, S_{\frac{7N}{8},\frac{15N}{16}-1}$	$\frac{N}{16}$	8	$\mathcal{O}(\frac{N}{8} \log_2 \frac{N}{8})$
$\vdots$	$\vdots$	$\vdots$	$\vdots$
$S_{0,r-1}, S_{2r,3r-1}, S_{4r,5r-1}, S_{6r,7r-1}, S_{8r,9r-1}, \ldots$	r	$s = \frac{N}{2r}$	$\mathcal{O}(\frac{N}{s} \log_2 \frac{N}{s})$

Furthermore, N/r final blocks $T_{\ell r}$, each of length r, are also computed in $r(r+1)/2$ floating-point operations and hence the total amount of floating point operations is proportional to

$$N \log_2 N + 2 \left(\frac{N}{2} \log_2 \frac{N}{2} \right) + 4 \left(\frac{N}{4} \log_2 \frac{N}{4} \right) + \cdots + s \left(\frac{N}{s} \log_2 \frac{N}{s} \right) + \frac{N}{r} \frac{r(r+1)}{2} =$$

$$= \sum_{j=0}^{\log_2 s} N \log_2 \frac{N}{2^j} + N \frac{r+1}{2} = \mathcal{O}(N (\log_2 N)^2), \quad s = \frac{N}{2r},$$

which turns out, for sufficiently large N, to be consistently significantly smaller than the number $\mathcal{O}(N^2)$ required by the direct summation of $T_0(N)$.

Although the whole procedure may appear complicated and requires some extra effort in coding, it turns out to be quite efficient since it can be applied to different methods of the form (17) and does not affect their accuracy. This preservation of accuracy is because the technique does take into account the entire history of the process in the same way as the straightforward approach mentioned above whose computational cost is $\mathcal{O}(N^2)$. Thus, one does need to keep the entire history data in active memory, but one avoids the requirement of using special meshes. All the Matlab codes for FDEs described in [10,21,38], and freely available on the Mathworks website [39], make use of this algorithm.

6.3. Kernel Compression Schemes

Although the terminology "kernel compression scheme" has been introduced only recently for a few specific works [40–42], we use it here to describe a collection of methods that were proposed at various times by various authors and are all based on essentially the same principle: approximation of the solution of a non-local FDE by means of (possibly several) local ODEs. We provide here just the main ideas underlying this approach and we will refer the reader to the literature for a more comprehensive coverage of the subject.

Actually, these are standalone methods (usually classified as nonclassical methods [43]) and not just algorithms improving the efficiency of the treatment of the memory term; for this reason they could have been discussed in Section 3 along with the other methods for FDEs. But since one of their main achievements (and the motivation for their introduction) is to handle memory and computational issues related to the long and persistent memory of fractional-order problems, we consider it appropriate to discuss them in the present section.

For ease of presentation we consider only $0 < \alpha < 1$ but the extension to any positive α is only a technical matter. The basic idea starts from some integral representation of the kernel of the RL integral (1), e.g.,

$$\frac{t^{\alpha-1}}{\Gamma(\alpha)} = \frac{\sin(\alpha\pi)}{\pi} \int_0^\infty e^{-rt} r^{-\alpha} dr, \qquad (18)$$

which, thanks to standard quadrature rules, can be approximated by exponential sums

$$\frac{t^{\alpha-1}}{\Gamma(\alpha)} = \sum_{k=1}^{K} w_k e^{-r_k t} + e_K(t), \qquad (19)$$

where the error $e_K(t)$ and the computational complexity related to the number K of nodes and weights depend on the choice among the many possible quadrature rules. When applying this approximation instead of the exact integral in the integral formulation (7), the solution of the FDE (6) is rewritten as

$$y(t) = y_0 + \sum_{k=1}^{K} w_k \int_{t_0}^{t} e^{-r_k(t-u)} f(u, y(u)) du + E_K(t). \qquad (20)$$

Each of the integrals in (20) is actually the solution of an initial value problem:

$$\begin{cases} y^{[k]}(t) = -r_k y^{[k]}(t) + f(t, y^{[k]}(t)) \\ y^{[k]}(t_0) = 0, \end{cases} \qquad (21)$$

which can be numerically approximated by standard ODE solvers, yielding approximations $y_n^{[k]}$ on some grid $\{t_n\}$. If the quadrature rule is chosen so as to make the error $E_K(t)$ so small that it can be neglected, an approximate solution of the original FDE (6) can be obtained step-by-step as

$$y_n = y_0 + \sum_{k=1}^{K} \tilde{w}_k y_n^{[k]},$$

where each $y_n^{[k]}$ depends only on $y_{n-1}^{[k]}$ or on a few other previous values, according to the selected ODE solver.

In practice, a non-local problem (the FDE) with non-vanishing memory is replaced by K local problems (the ODEs) each demanding a smaller computational effort and the memory storage is restricted to $\mathcal{O}(pK)$ if a p-step ODE solver is used for each of the ODEs (21).

Obviously, the idea sketched above requires several further technical details to work properly. First, an accurate error analysis is needed to ensure that the overall error is below the target accuracy. This is a very delicate task because it involves the investigation of the interaction between the quadrature rule used to approximate the integral in (20) and the ODE solver applied to the system (21), which can be a highly nontrivial matter. Moreover, some substantial additional problems must be addressed. For instance, A-stable methods should generally be preferred when solving the system (21) since some of the $r_k > 0$ can be very large and give rise to stiff problems.

A non-negligible issue is that it is not possible to find a quadrature rule approximating (18) in a uniform manner with respect to all relevant values of t, i.e. with the same accuracy for any $t \geq t_1$ where t_1 is the first mesh point to the right of the initial point t_0 or for all $t \geq t_0$ (in either case, the singularity at t_0 indeed makes the integral quite difficult to be approximated). To overcome this difficulty, several different approaches have been proposed.

In a series of pioneering works [44–46], where a complex contour integral

$$\frac{t^{\alpha-1}}{\Gamma(\alpha)} = \frac{1}{2\pi i} \int_{\mathcal{C}} e^{st} s^{-\alpha} ds$$

is chosen to approximate the kernel, the integration interval $[t_0, T]$ is divided into a sequence of subintervals of increasing lengths, and different quadrature rules (on different contours $\mathcal{C}$) are used in each of these intervals. While high accuracy can be obtained, this strategy is quite complicated and requires the use of more expensive complex arithmetic.

In [40–42] the integral in (7) is divided into local and history terms

$$y(t) = y_0 + \underbrace{\frac{1}{\Gamma(\alpha)} \int_{t_0}^{t-\delta t} (t-u)^{\alpha-1} f(u, y(u)) du}_{\text{History term}} + \underbrace{\frac{1}{\Gamma(\alpha)} \int_{t-\delta t}^{t} (t-u)^{\alpha-1} f(u, y(u)) du}_{\text{Local term}}$$

for a fixed $\delta t > 0$. This confines the singularity of the kernel to the local term, which can be approximated by standard methods for weakly singular integral equations (e.g., a product-integration rule) with a reduced computational cost and an insignificant memory requirement. The kernel in the history term no longer contains any singularity and can be safely approximated by (19) which applies now just for $t > \delta t$.

To obtain the highest possible accuracy, Gaussian quadrature rules are usually preferred. A rigorous and technical error analysis is necessary to tune parameters in an optimal way. Several implementations of approaches of this kind have been proposed (e.g., see [47–51]) but owing to their technical nature, a comparison to decide which method is in general the most convenient is difficult; we just refer to the interesting results presented in [52].

7. Some Remarks about Fractional Partial Differential Equations

Even though this paper is essentially devoted to the numerical solution of ordinary differential equations of fractional order and the computational treatment of the associated differential and integral operators, a few comments should be made regarding numerical methods for partial fractional differential equations (PDEs).

Remark 3. *The issues discussed in Section 4 are relevant to partial differential equations also. Indeed, it is shown in [53] that imposing excessive smoothness requirements on the solutions to a partial differential equation (e.g., for the sake of simplifying the error analysis or for obtaining a higher convergence order) has drastic implications regarding the class of admissible problems; in particular, the choice of the forcing function $f(x,t)$ in a linear initial-boundary value problem will then completely determine the initial condition in the problem.*

Our second remark regarding partial differential equations deals with a totally different aspect.

Remark 4. *Typical algorithms for time-fractional partial differential equations contain separate discretisation techniques with respect to the time variable and the space variable(s). A current trend is to employ a very high order method for the discretisation of the (non-fractional) differential operator with respect to the space variable. While this might seem an attractive approach at first sight, it has a number of disadvantages. Specifically, while this leads to a smaller discretization error in the space variable, it also increases the algorithm's overall complexity and makes the understanding of its properties more difficult. This complexity would be acceptable if the overall error could be reduced significantly. But since the overall error comprises not only the error from the space discretisation but* also the contribution from the time approximation, it follows that to reduce the overall error, one must force this latter component to be very small also. As indicated above, we cannot expect to achieve a high convergence order in this variable, so the only way to reach this goal is to choose the time step size very small (in comparison with the space mesh size). From Section 6 we conclude that a standard algorithm with a higher-than-linear complexity is likely to lead to prohibitive run times, and even if the time discretisation uses a method with a linear or almost linear complexity, this very small step size requirement will still imply a high overall cost. Therefore, the use of a high-order space discretisation in a time-fractional partial differential equation is usually inadvisable.

8. Concluding Remarks

In this paper we have tried to describe some issues related to the correct use of numerical methods for fractional-order problems. Unlike integer-order ODEs, numerical methods for FDEs are in general not taught in undergraduate courses and, very often, non-specialists are unaware of the peculiarities and major difficulties that arise in the numerical treatment of FDEs and fractional PDEs.

The availability of only a few well-organized textbooks and monographs in this field, together with the presence of many incorrect results in the literature, makes the situation even more difficult.

Some of the ideas collected in this paper were discussed in the lectures of the Training School on "Computational Methods for Fractional-Order Problems", held in Bari (Italy) during 22–26 July 2019, and promoted by the Cost Action CA15225—*Fractional-order systems: analysis, synthesis and their importance for future design*.

We believe that the scientific community should make an effort to raise the level of knowledge in this field by promoting specific academic courses at a basic level and/or by organizing training schools.

Author Contributions: Formal analysis, K.D., R.G. and M.S.; Investigation, K.D., R.G. and M.S.; Writing—original draft, K.D., R.G. and M.S.; Writing—review & editing, K.D., R.G. and M.S. All authors have read and agreed to the published version of the manuscript.

Funding: The cooperation which has lead to this article has been initiated and promoted within the COST Action CA15225, a network supported by COST (European Cooperation in Science and Technology). The work of Roberto Garrappa is also supported under a GNCS-INdAM 2019 Project. The work of Kai Diethelm was also supported by the German Federal Ministry of Education and Research (BMBF) under Grant No. 01IS17096A. The research of Martin Stynes is supported in part by the National Natural Science Foundation of China under grant NSAF U1930402.

Conflicts of Interest: The authors declare no conflict of interest.

Abbreviations

The following abbreviations are used in this manuscript:

CM	Complete monotonicity
FDE	Fractional differential equation
FLMM	Fractional linear multistep method
LMM	Linear multistep method
ODE	Ordinary differential equation
PDE	Partial differential equation

References

1. Diethelm, K. *The Analysis of Fractional Differential Equations*; Lecture Notes in Mathematics; Springer-Verlag: Berlin, Germany, 2010; Volume 2004, p. viii+247.
2. Kilbas, A.A.; Srivastava, H.M.; Trujillo, J.J. *Theory and Applications of Fractional Differential Equations*; North-Holland Mathematics Studies; Elsevier Science B.V.: Amsterdam, The Netherlands, 2006; Volume 204, p. xvi+523.

3. Mainardi, F. *Fractional Calculus and Waves in Linear Viscoelasticity*; Imperial College Press: London, UK, 2010; p. xx+347.
4. Miller, K.S.; Ross, B. *An Introduction to the Fractional Calculus and Fractional Differential Equations*; A Wiley-Interscience Publication; John Wiley & Sons, Inc.: New York, NY, USA, 1993; p. xvi+366.
5. Podlubny, I. *Fractional Differential Equations*; Mathematics in Science and Engineering; Academic Press Inc.: San Diego, CA, USA, 1999; Volume 198, p. xxiv+340.
6. Samko, S.G.; Kilbas, A.A.; Marichev, O.I. *Fractional Integrals and Derivatives*; Gordon and Breach Science Publishers: Yverdon, Switzerland, 1993; p. xxxvi+976.
7. Young, A. Approximate product-integration. *Proc. R. Soc. Lond. Ser. A.* **1954**, *224*, 552–561.
8. Young, A. The application of approximate product integration to the numerical solution of integral equations. *Proc. R. Soc. Lond. Ser. A.* **1954**, *224*, 561–573.
9. Diethelm, K.; Ford, N.J.; Freed, A.D. A predictor-corrector approach for the numerical solution of fractional differential equations. *Nonlinear Dyn.* **2002**, *29*, 3–22. [CrossRef]
10. Garrappa, R. On linear stability of predictor-corrector algorithms for fractional differential equations. *Int. J. Comput. Math.* **2010**, *87*, 2281–2290. [CrossRef]
11. Yan, Y.; Pal, K.; Ford, N.J. Higher order numerical methods for solving fractional differential equations. *BIT Numer. Math.* **2014**, *54*, 555–584. [CrossRef]
12. Li, Z.; Liang, Z.; Yan, Y. High-order numerical methods for solving time fractional partial differential equations. *J. Sci. Comput.* **2017**, *71*, 785–803. [CrossRef]
13. Dixon, J. On the order of the error in discretization methods for weakly singular second kind Volterra integral equations with nonsmooth solutions. *BIT* **1985**, *25*, 624–634. [CrossRef]
14. Diethelm, K.; Ford, N.J.; Freed, A.D. Detailed error analysis for a fractional Adams method. *Numer. Algorithms* **2004**, *36*, 31–52. [CrossRef]
15. Oldham, K.B.; Spanier, J. Theory and applications of differentiation and integration to arbitrary order. In *The Fractional Calculus*; Academic Press: New York, NY, USA; London, UK, 1974; p. xiii+234.
16. Lynch, V.E.; Carreras, B.A.; del Castillo-Negrete, D.; Ferreira-Mejias, K.M.; Hicks, H.R. Numerical methods for the solution of partial differential equations of fractional order. *J. Comput. Phys.* **2003**, *192*, 406–421. [CrossRef]
17. Lubich, C. Discretized fractional calculus. *SIAM J. Math. Anal.* **1986**, *17*, 704–719. [CrossRef]
18. Lubich, C. Convolution quadrature and discretized operational calculus. I. *Numer. Math.* **1988**, *52*, 129–145. [CrossRef]
19. Lubich, C. Convolution quadrature and discretized operational calculus. II. *Numer. Math.* **1988**, *52*, 413–425. [CrossRef]
20. Lubich, C. Convolution quadrature revisited. *BIT* **2004**, *44*, 503–514. [CrossRef]
21. Garrappa, R. Trapezoidal methods for fractional differential equations: theoretical and computational aspects. *Math. Comput. Simul.* **2015**, *110*, 96–112. [CrossRef]
22. Diethelm, K.; Ford, J.M.; Ford, N.J.; Weilbeer, M. Pitfalls in fast numerical solvers for fractional differential equations. *J. Comput. Appl. Math.* **2006**, *186*, 482–503. [CrossRef]
23. Stynes, M. Singularities. In *Handbook of Fractional Calculus With Applications*; De Gruyter: Berlin, Germany, 2019; Volume 3, pp. 287–305.
24. Miller, R.K.; Feldstein, A. Smoothness of solutions of Volterra integral equations with weakly singular kernels. *SIAM J. Math. Anal.* **1971**, *2*, 242–258. [CrossRef]
25. Lubich, C. Runge-Kutta theory for Volterra and Abel integral equations of the second kind. *Math. Comput.* **1983**, *41*, 87–102. [CrossRef]
26. Hanyga, A. A comment on a controversial issue: A generalized fractional derivative cannot have a regular kernel. *Fract. Calc. Appl. Anal.* **2020**, *23*, 211–223. [CrossRef]
27. Giusti, A. General fractional calculus and Prabhakar's theory. *Commun. Nonlinear Sci. Numer. Simul.* **2019**, *83*, 105114. [CrossRef]
28. Hanyga, A. Physically acceptable viscoelastic models. In *Trends in Applications of Mathematics to Mechanics*; Hutter, K., Wang, Y., Eds.; Shaker Verlag: Aachen, Germany, 2005; pp. 125–136.

29. Stynes, M.; O'Riordan, E.; Gracia, J.L. Necessary conditions for convergence of difference schemes for fractional-derivative two-point boundary value problems. *BIT* **2016**, *56*, 1455–1477. [CrossRef]
30. Sarv Ahrabi, S.; Momenzadeh, A. On failed methods of fractional differential equations: the case of multi-step generalized differential transform method. *Mediterr. J. Math.* **2018**, *15*, 149. [CrossRef]
31. Garrappa, R. Neglecting nonlocality leads to unreliable numerical methods for fractional differential equations. *Commun. Nonlinear Sci. Numer. Simul.* **2019**, *70*, 302–306. [CrossRef]
32. Deng, W.H. Short memory principle and a predictor-corrector approach for fractional differential equations. *J. Comput. Appl. Math.* **2007**, *206*, 174–188. [CrossRef]
33. Ford, N.J.; Simpson, A.C. The numerical solution of fractional differential equations: Speed versus accuracy. *Numer. Algorithms* **2001**, *26*, 333–346. [CrossRef]
34. Diethelm, K.; Freed, A.D. An Efficient Algorithm for the Evaluation of Convolution Integrals. *Comput. Math. Appl.* **2006**, *51*, 51–72. [CrossRef]
35. Hairer, E.; Lubich, C.; Schlichte, M. Fast numerical solution of nonlinear Volterra convolution equations. *SIAM J. Sci. Statist. Comput.* **1985**, *6*, 532–541. [CrossRef]
36. Hairer, E.; Lubich, C.; Schlichte, M. Fast numerical solution of weakly singular Volterra integral equations. *J. Comput. Appl. Math.* **1988**, *23*, 87–98. [CrossRef]
37. Henrici, P. Fast Fourier methods in computational complex analysis. *SIAM Rev.* **1979**, *21*, 481–527. [CrossRef]
38. Garrappa, R. Numerical Solution of Fractional Differential Equations: A Survey and a Software Tutorial. *Mathematics* **2018**, *6*, 16. [CrossRef]
39. Garrappa, R. Mathworks Author's Profile. Available online: https://www.mathworks.com/matlabcentral/profile/authors/2361481-roberto-garrappa (accessed on 26 January 2020).
40. Baffet, D. A Gauss-Jacobi kernel compression scheme for fractional differential equations. *J. Sci. Comput.* **2019**, *79*, 227–248. [CrossRef]
41. Baffet, D.; Hesthaven, J.S. A kernel compression scheme for fractional differential equations. *SIAM J. Numer. Anal.* **2017**, *55*, 496–520. [CrossRef]
42. Baffet, D.; Hesthaven, J.S. High-order accurate adaptive kernel compression time-stepping schemes for fractional differential equations. *J. Sci. Comput.* **2017**, *72*, 1169–1195. [CrossRef]
43. Diethelm, K. An investigation of some nonclassical methods for the numerical approximation of Caputo-type fractional derivatives. *Numer. Algorithms* **2008**, *47*, 361–390. [CrossRef]
44. López-Fernández, M.; Lubich, C.; Schädle, A. Adaptive, fast, and oblivious convolution in evolution equations with memory. *SIAM J. Sci. Comput.* **2008**, *30*, 1015–1037. [CrossRef]
45. Lubich, C.; Schädle, A. Fast convolution for nonreflecting boundary conditions. *SIAM J. Sci. Comput.* **2002**, *24*, 161–182. [CrossRef]
46. Schädle, A.; López-Fernández, M.; Lubich, C. Fast and oblivious convolution quadrature. *SIAM J. Sci. Comput.* **2006**, *28*, 421–438. [CrossRef]
47. Banjai, L.; López-Fernández, M. Efficient high order algorithms for fractional integrals and fractional differential equations. *Numer. Math.* **2019**, *141*, 289–317. [CrossRef]
48. Fischer, M. Fast and parallel Runge-Kutta approximation of fractional evolution equations. *SIAM J. Sci. Comput.* **2019**, *41*, A927–A947. [CrossRef]
49. Jiang, S.; Zhang, J.; Zhang, Q.; Zhang, Z. Fast evaluation of the Caputo fractional derivative and its applications to fractional diffusion equations. *Commun. Comput. Phys.* **2017**, *21*, 650–678. [CrossRef]
50. Li, J.R. A fast time stepping method for evaluating fractional integrals. *SIAM J. Sci. Comput.* **2010**, *31*, 4696–4714. [CrossRef]
51. Zeng, F.; Turner, I.; Burrage, K. A stable fast time-stepping method for fractional integral and derivative operators. *J. Sci. Comput.* **2018**, *77*, 283–307. [CrossRef]

52. Guo, L.; Zeng, F.; Turner, I.; Burrage, K.; Karniadakis, G.E.M. Efficient multistep methods for tempered fractional calculus: Algorithms and simulations. *SIAM J. Sci. Comput.* **2019**, *41*, 2510–2535. [CrossRef]
53. Stynes, M. Too much regularity may force too much uniqueness. *Fract. Calc. Appl. Anal.* **2016**, *19*, 1554–1562. [CrossRef]

© 2020 by the authors. Licensee MDPI, Basel, Switzerland. This article is an open access article distributed under the terms and conditions of the Creative Commons Attribution (CC BY) license (http://creativecommons.org/licenses/by/4.0/).

Article

On Fractional Operators and Their Classifications

Dumitru Baleanu [1,2,*] and Arran Fernandez [3]

1. Department of Mathematics, Cankaya University, Balgat 06530, Ankara, Turkey
2. Institute of Space Sciences, R76900 Magurele-Bucharest, Romania
3. Department of Mathematics, Faculty of Arts and Sciences, Eastern Mediterranean University, 99628 Famagusta, Northern Cyprus, via Mersin-10, Turkey; arran.fernandez@emu.edu.tr
* Correspondence: dumitru@cankaya.edu.tr

Received: 22 June 2019; Accepted: 6 September 2019; Published: 8 September 2019

Abstract: Fractional calculus dates its inception to a correspondence between Leibniz and L'Hopital in 1695, when Leibniz described "paradoxes" and predicted that "one day useful consequences will be drawn" from them. In today's world, the study of non-integer orders of differentiation has become a thriving field of research, not only in mathematics but also in other parts of science such as physics, biology, and engineering: many of the "useful consequences" predicted by Leibniz have been discovered. However, the field has grown so far that researchers cannot yet agree on what a "fractional derivative" can be. In this manuscript, we suggest and justify the idea of classification of fractional calculus into distinct classes of operators.

Keywords: fractional calculus; integral transforms; convergent series

MSC: 26A33

1. Background

Fractional calculus is a venerable branch of mathematics, first conceptualised in 1695 in a series of letters. L'Hopital posed the question to Leibniz of what would happen if the order of differentiation were taken to be $\frac{1}{2}$, and Leibniz replied [1]:

"It appears that one day useful consequences will be drawn from these paradoxes."

After these prophetic words, however, Leibniz did not propose a definition, leaving this task to the later scientists who followed him.

The concepts of fractional differentiation and fractional integration were examined further over the course of the 18th and 19th centuries. The topic attracted the attention of mathematical giants such as Riemann [2], Liouville [3], Abel [4], Laurent [5], and Hardy and Littlewood [6,7]. Detailed discussions of the history of fractional calculus may be found in [8–11]; here, we wish to focus on a few key points concerning the directions in which the field developed.

The "paradoxes" described by Leibniz were resolved by later authors, but this is not to say that the field of fractional calculus is now wholly free of open problems. One recurring issue through the centuries has been the existence of multiple conflicting definitions. In the mid-19th century, several different definitions of fractional calculus had already been proposed: Liouville had created one definition based on differentiating exponential functions and another based on an integral formula for inverse power functions, while Lacroix had created a different definition based on differentiating power functions. The definitions of Liouville and Lacroix are not equivalent, which led some critics to conclude that one must be "correct" and the other "wrong". De Morgan, however, wrote [12] that:

"Both these systems, then, may very possibly be parts of a more general system."

His words, like those of Leibniz 145 years earlier, were prophetic. Both Liouville's formula and Lacroix's are in fact special cases of what is now called the Riemann–Liouville definition of fractional calculus. This involves an arbitrary constant of integration c, which when set to zero yields Lacroix's formula and when set to $-\infty$ yields Liouville's.

This general Riemann–Liouville definition, for the fractional derivative and fractional integral of an arbitrary function, emerged in the late 19th century through a complex-analysis approach. Although the Riemann–Liouville formula is now used mostly in a real-analysis context, its original motivation came from generalising the Cauchy integral formula for repeated derivatives of a complex analytic function. Now, Riemann–Liouville is the most common way of defining fractional calculus. In this model, the fractional integral and fractional derivative of a function $f(x)$ are defined as follows:

$$^{RL}_cI^\nu_x f(x) = \frac{1}{\Gamma(\nu)} \int_c^x (x-t)^{\nu-1} f(t)\,dt, \qquad \mathrm{Re}(\nu) > 0;$$

$$^{RL}_cD^\nu_x f(x) = \frac{d^n}{dx^n}\,^{RL}_cI^{n-\nu}_x f(x), \quad n = \lfloor \mathrm{Re}(\nu) \rfloor + 1, \qquad \mathrm{Re}(\nu) \geq 0.$$

This definition is sufficiently general to cover the formulae both of Liouville and of Lacroix. However, it is still not the only proposed way of defining fractional calculus: multiple conflicting formulae persist to this day, confusing many newcomers to the field who expect to see a single definition of fractional derivatives just like there is a single definition of the first-order derivative. Fractional calculus may be called an "extension of meaning" [13], but there is more than one way to extend meaning. The Riemann–Liouville model can be used to describe processes with power-law behaviour, due to the power-function kernel in the definition of the integral transform, but there are many other types of behaviours that occur in nature and that cannot be described by simple power functions.

In the late 20th century, fractional calculus began to undergo a large increase in popularity and research output. The first international conference on fractional calculus was organised in 1974 in the USA; the same year also saw the publication of the first textbook [14] devoted to this field. Since then, fractional calculus has become a very active field of research, with several specialist journals on the topic. Applications have been discovered in many fields of science, as summarised in [15–18] and the references therein. In particular, the intermediate property of fractional-calculus operators is vital for the modelling of certain intermediate physical processes, e.g., in viscoelasticity [19,20]. Fractional calculus has also become a standard part of the graduate mathematics curriculum in some universities, with several textbooks [8,11,14,21–23] that can function as an introduction to the field for students and young researchers.

From the point of view of research, currently there are several differing perspectives and directions of exploration, which in some respects may be in opposition to each other. In the following section, we propose a possible way of resolving these issues.

2. The Question of Classification

In recent years, two trends have emerged in the consideration of fractional-calculus operators, motivated by a number of different considerations.

Firstly, there exists a desire to **explore and create new definitions and models** for fractional integral and differential operators. Dozens of definitions have been proposed in the 2010s alone, with a wide variety of types and properties [24–27]. One motivation here is the pure mathematician's desire to generalise: for example, to go beyond simple power functions and extend definitions to cover a whole host of different kernel functions. Another motivation is the applied scientist's need for models to describe accurately a wide variety of different systems: several definitions of fractional calculus have been inspired directly by real-world applications. The result of both types of research is to expand the field of fractional calculus. However, the question arises of how far the field can be stretched and still be called "fractional calculus", and the validity of some definitions has been debated.

Secondly, there exists a desire to **impose criteria and strict definitions** for what we call a "fractional derivative" or "fractional integral": which operators between functions should be named as such and which should not. The proposals range from strict requirements to mere suggestions, and multiple different criteria have been proposed [13,28–31]. The motivation here is to create a mathematical framework for fractional calculus, to know the boundaries of the field. Metric spaces and vector spaces, for example, have rigorous definitions and strict sets of criteria, so why not fractional integrals and fractional derivatives? The result of such a system would be to restrict the study of fractional calculus within certain boundaries. However, there is no consensus on where the boundaries should be drawn: opinions differ widely on what the criteria should be.

At first, these two ways of thinking seem very different. One seeks to expand the field without regard for boundaries, while the other seeks to restrict the field to within prescribed boundaries. However, as both points of view have some merit, we would like to seek a middle path, a way of satisfying both the desire for generalisation and diversification and the desire for rigorous classification.

The key lies in considering the valid motivations for both approaches. Mathematical structures have an aesthetic, intuitive logic, which guides our path to choosing appropriate criteria to define them and which often connects directly or indirectly with their physical applications. These real-world connections are of paramount importance: if one particular mathematical model emerges from some real data, then that model must be worth studying, and so we should not exclude it from consideration by imposing overly strict criteria.

The desire for generalisation and the desire for criteria, which seem opposed to each other, may both be satisfied by considering **broad classes of fractional-calculus operators**. We recall again the words of Augustus de Morgan, quoted above: if different definitions seem in contradiction, it is worth considering whether they may be unified as part of "a more general system". Ideally, such a system would be itself part of fractional calculus. Formally, then, we seek to define sets $\mathscr{A}, \mathscr{B}, \mathscr{C}$, etc. (we do not presume to know how many such sets will emerge), of operators between function spaces, such that each element of each of these sets may be interpreted as a "fractional operator" acting on functions and such that each set has some unifying properties which enable useful results to be proven for the entire class. We do not impose any requirements in general on how large or small these classes should be, or which function spaces they should act between, as we believe such a system should be able to cover many different families of operators.

Fractional calculus has been usefully interpreted in connection with many different branches of mathematics: for example, distributional calculus, functional calculus, spectral theory, Cauchy integrals, and Laplace transforms, as described and summarised in [32] (pp. 58–64). Our aim here is related but different: instead of embedding the whole of fractional calculus into other fields of analysis, we seek to create classifications within fractional calculus itself. Some recent studies [30,33–35] have proposed general classes of operators that are broad enough to cover many existing models of fractional calculus but still narrow enough to be rigorously analysed themselves. This approach is optimal for several reasons:

1. It satisfies the desire for generalisation. Any class of fractional-calculus operators will be more general than any one particular model, and the specific models can be studied as before within this framework or as special cases of the general class. If real-world applications give rise to a new model of fractional calculus, it may be able to fit into such a class, and then many of its properties would be known directly from general theorems about the class.
2. It also satisfies, to a certain degree, the desire for restrictions and criteria. Not all types of fractional calculus fall into one particular class, but each class can be studied in its own right; its defining attributes could be considered as "axioms" or criteria for that particular class. Thus, it is possible to study fractional calculus within the framework of certain prescribed conditions, without dismissing everything outside that framework as invalid.

In shaping the mathematical theory of fractional calculus, we should look beyond single specific formulae and create wider avenues of study. This will eliminate the need for many different research papers proving the same results in the same way for many different types of fractional calculus: instead, we can prove them just once for a whole class and then deduce the individual results as special cases. From the applications point of view, a particular collection of real data can be fitted to a particular model of fractional calculus which is already known as a special case of one of these broad classes.

At some point in the future, it may be possible to create a "most general" definition of fractional calculus by defining one single class $\mathscr{F}$ that covers all fractional operators and nothing else, with all the other classes of fractional derivatives and integrals as subsets. However, we believe that such a breakthrough is not imminent. We must wait to discover the full range of applications before we can decide where to draw the boundaries of the field, and at present, new applications of fractional calculus are still being discovered all the time. It would be hasty to restrict the field too far now and then discover after a few years that the restrictions exclude those fractional-calculus operators that are most useful in real-world modelling.

3. The Class of Analytic Kernels

To illustrate the ideas discussed in the previous section, we shall conduct a detailed analysis of one general class of fractional-calculus operators that was recently proposed in [35]. First we consider briefly some of the many models of fractional calculus that may be covered by this class.

- A model proposed by Atangana and Baleanu [25], which was defined more rigorously in [36] and whose applications have been discussed in [37–39], utilises an integral transform with a one-parameter Mittag-Leffler function ($E_\nu(z) = \sum_{n=0}^{\infty} \frac{z^n}{\Gamma(n\nu+1)}$ for $\mathrm{Re}(\nu) > 0$) in the kernel and an arbitrary normalisation function multiplier:

$$^{AB}_c I_x^\nu f(x) = \frac{1-\nu}{B(\nu)} f(x) + \frac{\nu}{B(\nu)} {}^{RL}_c I_x^\nu f(x);$$

$$^{ABRL}_c D_x^\nu f(x) = \frac{B(\nu)}{1-\nu} \cdot \frac{d}{dx} \int_c^x E_\nu\left(\tfrac{-\nu}{1-\nu}(x-t)^\nu\right) f(t)\, dt;$$

$$^{ABC}_c D_x^\nu f(x) = \frac{B(\nu)}{1-\nu} \int_c^x E_\nu\left(\tfrac{-\nu}{1-\nu}(x-t)^\nu\right) f'(t)\, dt.$$

- A model due to Prabhakar [40], which was formally connected to fractional calculus in [41] and whose applications have been discussed in [42,43], utilises an integral transform with a three-parameter Mittag-Leffler function ($E_{\mu,\nu}^\rho(z) = \sum_{n=0}^{\infty} \frac{(\rho)_n z^n}{\Gamma(n\mu+\nu)}$ for $\mathrm{Re}(\mu), \mathrm{Re}(\nu) > 0$) in the kernel:

$$^P_c I_x^{\mu,\nu,\rho,\omega} f(x) = \int_c^x (x-t)^{\nu-1} E_{\mu,\nu}^\rho\left(\omega(t-x)^\mu\right) f(t)\, dt;$$

$$^P_c D_x^{\mu,\nu,\rho,\omega} f(x) = \frac{d^n}{dx^n} {}^P_c I_x^{\mu,n-\nu,-\rho,\omega} f(x), \quad n = \lfloor \mathrm{Re}(\nu) \rfloor + 1.$$

- A model known as tempered fractional calculus [44,45], utilises an integral transform with the product of a power function and an exponential function in the kernel:

$$^T_c I_x^{(\alpha,\beta)} f(x) = \frac{1}{\Gamma(\alpha)} \int_c^x (x-t)^{\alpha-1} e^{-\beta(x-t)} f(t)\, dt;$$

$$^T_c D_x^{(\alpha,\beta)} f(x) = \left(\frac{d}{dx} + \beta\right)^n \left({}^T_c I_x^{(n-\alpha,\beta)} f(x)\right), \quad n = \lfloor \mathrm{Re}(\nu) \rfloor + 1.$$

- A model due to Srivastava et al. [26] utilises an integral transform with a Fox H-function in the kernel:

$$^{SH}_c I_x^{\omega;m,n,p,q;\alpha,\beta} f(x) = \int_c^x (x-t)^{\alpha-1} H_{p,q}^{m,n}\left(\omega(x-t)^\beta\right) f(t)\, dt,$$

where $H_{p,q}^{m,n}(z) = \frac{1}{2\pi i}\int_{\mathcal{L}}\Theta(s)z^s\,ds$ with $\mathcal{L}$ being a Mellin–Barnes contour from $-i\infty$ to $i\infty$ and $\Theta(s) = \frac{\prod_{j=1}^{m}\Gamma(c_j-d_js)\prod_{j=1}^{n}\Gamma(1-a_j+b_js)}{\prod_{j=n+1}^{p}\Gamma(a_j-b_js)\prod_{j=m+1}^{q}\Gamma(1-c_j+d_js)}$ with parameters satisfying the conditions stated in [26].

The definition presented in [35] is general enough to cover all of the above as special cases, while not so general as to lose its connection to fractional calculus. For this reason, we use it as an example of a broad class of fractional-calculus operators as discussed in the previous section. We may define a class $\mathcal{A}$ consisting of all operators given by the following general integral transform formula:

$$^A_c I_x^{\alpha,\beta} f(x) = \int_c^x (x-t)^{\alpha-1} A\left((x-t)^\beta\right) f(t)\,dt, \tag{1}$$

where c is a constant in the extended real line (often taken as zero or $-\infty$), α and β are complex parameters with positive real parts, and $A(z) = \sum_{k=0}^{\infty} a_k z^k$ is a general analytic function whose coefficients $a_k \in \mathbb{C}$ are permitted to depend on α and β. We may consider x as a real variable larger than c; function spaces for f are discussed below. Many properties of this newly-defined operator were already proved in [35]; here, as well as providing a brief summary of these, we shall extend the discussion by considering more properties and potential subclassifications.

Part of fractional calculus. The following series formula, proved in [35], expresses this integral transform directly in terms of the Riemann–Liouville fractional integral:

$$^A_c I_x^{\alpha,\beta} f(x) = \sum_{k=0}^{\infty} a_k \Gamma(\beta k + \alpha) \, ^{RL}_c I_x^{\beta k + \alpha} f(x). \tag{2}$$

Formally, we may write this series formula as a relation between functional operators:

$$^A_c I_x^{\alpha,\beta} = A_\Gamma\left(^{RL}_c I_x^\beta\right) \, ^{RL}_c I_x^\alpha, \tag{3}$$

where A_Γ is the transformed analytic function defined by:

$$A(z) = \sum_{k=0}^{\infty} a_k z^k \Rightarrow A_\Gamma(z) = \sum_{k=0}^{\infty} a_k \Gamma(\beta k + \alpha) z^k. \tag{4}$$

From the relation (2), it is clear that the general operator (1) can always be described using only the classical Riemann–Liouville fractional integral, which is indisputably part of fractional calculus. Thus, we contend that it makes sense to consider the general operator (1) as always a part of fractional calculus as well. It is already known [35] that the series formula (2) may be used to prove various useful properties, such as for example the product rule and chain rule [46,47], for the general operator (1) directly from the corresponding known result for Riemann–Liouville.

Generalisation of well-known models. It was verified in [35], or is clear from the definitions, that all four of the specific example models of fractional calculus mentioned above are special cases of the general definition (1). Of course, this class does not cover all possible types of fractional calculus: there are also many that are not special cases of (1). These include the Hadamard and Erdelyi–Kober definitions, and some definitions involving special functions applied to $1 - \frac{t}{x}$ instead of $x-t$, like [27,48].

Now we have confirmed that it makes sense to use (1) as the definition of a class of fractional-calculus operators: not all of fractional calculus, not just one specific model, but a general class that covers many cases and can be analysed in its own right. We continue with a further analysis of this class, its properties, and subclasses.

Historical connections and integral transform. The transformation between A and A_Γ defined by Equation (4) has some historical significance. In one of his "notes" working on what we now call the Mittag-Leffler function, Gösta Mittag-Leffler himself [49] considered the following transformation:

$$F(x) = \sum_{n=0}^{\infty} k_n x^n \Rightarrow F_\beta(x) = \sum_{n=0}^{\infty} \frac{k_n}{\Gamma(\beta n + 1)} x^n. \tag{5}$$

After relabelling notation, it is clear that Mittag-Leffler's transformation of F to F_β is precisely the inverse of the transformation (4) from A to A_Γ in the case where $\alpha = 1$. Mittag-Leffler noted that using $F(x) = \frac{1}{1-x}$ yields $F_\beta(x) = E_\beta(x)$, which we now call the Mittag-Leffler function. Thus, the study of the general class (1) intimately involved with the transformation (4) has some historical justification.

Furthermore, Mittag-Leffler [49] found the following relation between the functions in (5):

$$F(x) = \int_0^\infty e^{-\omega} F_\beta(\omega^\beta x) \, d\omega.$$

By a natural extension of this result to the case of general α, β, we obtain the following integral transform between A and A_Γ:

$$A_\Gamma(z) = \int_0^\infty e^{-\omega} \omega^{\alpha-1} A(\omega^\beta z) \, d\omega. \tag{6}$$

Going back to the classics is often a useful endeavour, and indeed, Mittag-Leffler's 1905 paper provided us with an elegant integral formula (6) for transforming between the functions A and A_Γ, which are important in the analysis of the class (1) of fractional models.

Local and non-local operators. In classical fractional calculus such as the Riemann–Liouville model, the operators are non-local. Like integrals, fractional derivatives depend not just on the behaviour of a function near a single point, but also on its behaviour in a wider region. This non-locality is often useful in modelling physical processes that have memory effects.

For our general class, the fractional integrals are always non-local since they are defined by an integral from c to x. The fractional derivatives as discussed in [35] are also non-local, except in the very special case when they reduce to the standard differentiation operations $\frac{d^n}{dx^n}$. This reminds us that our class does not cover the entirety of what has been called fractional calculus: any operators with locality properties are not contained in this class and must be classified using some other class.

Possession or lack of a semigroup property. One important property of any fractional-calculus operator is whether or not it has a semigroup property in one (or more) of the parameters associated with the operator. For example, in the Riemann–Liouville model, fractional integrals have a semigroup property while fractional derivatives do not. It is natural to ask, is the mth derivative/integral of the nth derivative/integral always equal to the $(m+n)$th derivative/integral?

For the general class (1), it was proved in [35] that a semigroup property in both α and β is impossible, but a semigroup property in the first parameter α can be obtained under the following condition on the coefficients a_k for the analytic function A:

$$\sum_{m+n=k} a_n(\alpha_1, \beta) a_m(\alpha_2, \beta) B(\alpha_1 + n\beta, \alpha_2 + m\beta) = a_k(\alpha_1 + \alpha_2, \beta) \qquad \forall k \in \mathbb{Z}_0^+. \tag{7}$$

It is easy to see that this class is general enough to cover both some fractional models with a semigroup property (such as Riemann–Liouville and Prabhakar) and some without a semigroup property (such as Atangana–Baleanu). However, Equation (7) gives us an explicit condition to know whether a given special case possesses a semigroup property or not.

We note that a semigroup property is not always required by physical motivations: fractional models either with or without such properties can be used to describe real-world problems [50].

Singular and non-singular operators. Another property that has been subject to much discussion is the singularity or non-singularity of fractional-calculus operators. The classical Riemann–Liouville model is defined by a singular integral, due to the power function $(x-t)^{\nu-1}$ in the integrand, but the singularity is integrable provided that $\text{Re}(\nu) > 0$. Some other models [24,25] have been promoted due to the non-singularity of their defining integrals.

Again, the class (1) is general enough to cover both some singular and some non-singular fractional-calculus operators. This time it is easy to find a condition for which is which. We write $v_0(A) \geq 0$ for the valency (multiplicity or ramification index) of the analytic function $A(z)$ at the point $z = 0$, so that $A(z) = z^{v_0(A)} B(z)$ for some function B that is analytic and nonzero in a neighbourhood of $z = 0$. Then, the general integral transform (1) is non-singular if:

$$\text{Re}\,(\alpha + \beta v_0(A)) \geq 1$$

(the most usual case is $\alpha = 1, v_0(A) = 0$), and it has an integrable singularity if:

$$0 < \text{Re}\,(\alpha + \beta v_0(A)) < 1.$$

(In the case where $\text{Re}\,(\alpha + \beta v_0(A)) \leq 0$, we have a non-integrable singularity, and the integral (1) is not defined since the function cannot be integrated near $t = x$.)

Again, neither singularity nor non-singularity is always required by physical motivations. Both singular and non-singular fractional-calculus operators have discovered many applications to real-world problems [51].

Dual operators. The definition (1) is, for a left-sided fractional integral operator, the integration being performed from c to x. We can equally well define a right-sided fractional integral operator, for x contained in some fixed interval $[c, d]$, namely:

$${}^A_x I^{\alpha,\beta}_d f(x) = \int_x^d (t-x)^{\alpha-1} A\left((t-x)^\beta\right) f(t)\,dt. \qquad (8)$$

This modified operator has the property that it is the dual of the original left-sided fractional integral operator:

$$\int_c^d \left({}^A_c I^{\alpha,\beta}_x f(x)\right) g(x)\,dx = \int_c^d f(x) \left({}^A_x I^{\alpha,\beta}_d g(x)\right) dx.$$

This can be quickly proved using Fubini's theorem, and it is an analogue of the integration by parts rule for standard integrals and Riemann–Liouville fractional integrals.

Functional bounds. The operator ${}^A_c I^{\alpha,\beta}_x$ defined by (1) defines a map between function spaces, and it may be useful to consider bounds and properties of this functional map.

In [35] it was proved that ${}^A_c I^{\alpha,\beta}_x$ is bounded on the space $L^1[c, c+R]$, with

$$\left\|{}^A_c I^{\alpha,\beta}_x f(x)\right\|_{L^1} \leq R^{\text{Re}(\alpha)} \sup_{|z| < R^{\text{Re}(\beta)}} |A(z)| \left\|f(x)\right\|_{L^1}.$$

Using Young's inequality for convolutions, we can prove that the same operator is also bounded on any L^p space, with

$$\left\|{}^A_c I^{\alpha,\beta}_x f(x)\right\|_{L^p} \leq R^{\text{Re}(\alpha)} \sup_{|z| < R^{\text{Re}(\beta)}} |A(z)| \left\|f(x)\right\|_{L^p},$$

for all $p \in [1, \infty]$. This functional space bound strengthens the pure mathematical foundation for the general class of operators, and it may be useful in the future study of fractional differential equations using operators in this class.

Fractionally-iterated operators. Some fractional operators in the literature have arisen by means of iteration. The process here is to start from some standard operator K between functions, write a formula for the iterated operator K^n, and then generalise that formula to non-integer values of n. This idea is of course what gave rise to fractional calculus in the first place, with the starting operator being simply $K = \frac{d}{dx}$, but it is also possible to apply the same process from a starting operator K which is already fractional.

However, doing so does not always yield a new fractional operator. In some cases, the process does give rise to new types of fractional calculus [52,53], but this relies on the semigroup property not being valid for the starting operator D. For example, if K is the Riemann–Liouville fractional integral ${}^{RL}_c I^\alpha$, then $K^n = {}^{RL}_c I^{n\alpha}$, and so, the fractionally-iterated operator $K^\nu = {}^{RL}_c I^{\nu\alpha}$ is also a Riemann–Liouville fractional integral, not a new type of operator.

Some of the issues around fractional iteration were also discussed in ([54], §5).

4. Conclusions

Fractional calculus is currently in a stage of rapid and continuous expansion and development. Right now, several different fractional-calculus operators are being proposed, with many different behaviours such as singular or non-singular, semigroup law or none, etc. On the other hand, several classifications of fractional-calculus operators have been suggested, proposing a variety of possible conditions that might be imposed. Some models of fractional calculus are subject to debate, being acceptable under one classification system, but not another.

There are many different points of view and approaches being taken in the study of fractional calculus. In terms of real-world problems, it is important to remember that not everything is known: some systems and behaviours are not yet understood using fractional calculus. In our opinion, going to the extremes—e.g., creating operators without regard for applications, or imposing hard conditions for all potential fractional-calculus operators—will not lead to significant progress in the understanding of the still hidden flavours of fractional calculus and their applications.

Instead of imposing criteria, we suggest organising fractional-calculus operators into classes having different types of properties. One large class of operators, presented in detail in this manuscript, is one example of a class with real-world applications where both singular and non-singular operators, both with and without semigroup properties, may live together in the same class. We think the words "true" and "false" are too simplistic to describe the complex process of debates that is occurring nowadays.

Author Contributions: Conceptualisation, D.B. and A.F.; investigation, D.B. and A.F.; writing—original draft preparation, D.B. and A.F.; writing—review and editing, D.B. and A.F.

Funding: This research received no external funding.

Conflicts of Interest: The authors declare no conflict of interest.

References

1. Leibniz, G.W. *Mathematische Schriften: aus den Handschriften der Königlichen Bibliothek zu Hannover. Briefwechsel zwischen Leibniz, Wallis, Varignon, Guido Grandi, Zendrini, Hermann und Freiherrn von Tschirnhaus*; Druck und Verlag von H. W. Schmidt: Halle, Germany, 1859; Volume 1.
2. Riemann, B. Versuch einer allgemeinen Auffassung der Integration und Differentiation. In *Gessamelte Mathematische Werke*; Dedekind, R., Weber, H., Eds.; Druck und Verlag: Leipzig, Germany, 1876.
3. Liouville, J. Mémoire Sur quelques Questions de Géometrie et de Mécanique, et sur un nouveau genre de Calcul pour résoudre ces Questions. *J. L'École Polytech.* **1832**, *13*, 1–69.
4. Abel, N.H. Solution de quelques problèmes á l'aide d'intégrales définies. In *Oeuvres Complètes de Niels Henrik Abel*; Sylow, L., Lie, S., Eds.; CUP: Cambridge, UK, 1881.
5. Laurent, H. Sur le calcul des dérivées à indices quelconques. *Nouv. Ann. MathÉmatiques J. Des Candidats Aux Écoles Polytech. Norm.* **1884**, *3*, 240–252.
6. Hardy, G.H.; Littlewood, J.E. Some properties of fractional integrals I. *Math. Z.* **1928**, *27*, 565–606. [CrossRef]
7. Hardy, G.H.; Littlewood, J.E. Some properties of fractional integrals II. *Math. Z.* **1932**, *34*, 403–439. [CrossRef]

8. Miller, K.S.; Ross, B. *An Introduction to the Fractional Calculus and Fractional Differential Equations*; Wiley: New York, NY, USA, 1993.
9. Dugowson, S. Les Différentielles Métaphysiques: Histoire et Philosophie de la Généralisation de l'ordre de Dérivation. Ph.D. Thesis, Université Paris Nord, Paris, France, 1994.
10. Hilfer, R. Threefold Introduction to Fractional Derivatives. In *Anomalous Transport: Foundations and Applications*; Klages, R., Radons, G., Sokolov, I.M., Eds.; John Wiley & Sons: Berlin, Germany, 2008.
11. Baleanu, D.; Diethelm, K.; Scalas, E.; Trujillo, J.J. *Fractional Calculus: Models and Numerical Methods*, 2nd ed.; World Scientific: New York, NY, USA, 2017.
12. De Morgan, A. *The Differential and Integral Calculus Combining Differentiation, Integration, Development, Differential Equations, Differences, Summation, Calculus of Variations with Applications to Algebra, Plane and Solid Geometry*; Baldwin and Craddock: London, UK, 1840.
13. Ross, B. A Brief History and Exposition of the Fundamental Theory of Fractional Calculus. In *Fractional Calculus and Its Applications*; Lecture Notes in Mathematics No. 457; Ross, B., Ed.; Springer: Heidelberg, Germany, 1975.
14. Oldham, K.B.; Spanier, J. *The Fractional Calculus*; Academic Press: San Diego, CA, USA, 1974.
15. Baleanu, D.; Lopes, A.M. (Eds.) *Handbook of Fractional Calculus with Applications, Volume 7: Applications in Engineering, Life and Social Sciences, Part A*; De Gruyter: Berlin, Germany, 2019.
16. Baleanu, D.; Lopes, A.M. (Eds.) *Handbook of Fractional Calculus with Applications, Volume 8: Applications in Engineering, Life and Social Sciences, Part B*; De Gruyter: Berlin, Germany, 2019.
17. Hilfer; R. (Ed.) *Applications of Fractional Calculus in Physics*; World Scientific: Singapore, 2000.
18. Sun, H.G.; Zhang, Y.; Baleanu, D.; Chen, W.; Chen, Y.Q. A new collection of real world applications of fractional calculus in science and engineering. *Commun. Nonlinear Sci. Numer. Simul.* **2018**, *64*, 213–231. [CrossRef]
19. El-Sayed, A.M.A.; Gaafar, F.M. Fractional calculus and some intermediate physical processes. *Appl. Math. Comput.* **2003**, *144*, 117–126. [CrossRef]
20. Bonfanti, A.; Fouchard, J.; Khalilgharibi, N.; Charras, G.; Kabla, A. A unified rheological model for cells and cellularised materials. preprint under review. [CrossRef]
21. Samko, S.G.; Kilbas, A.A.; Marichev, O.I. *Fractional Integrals and Derivatives: Theory and Applications*; Taylor & Francis: London, UK, 2002.
22. Kilbas, A.A.; Srivastava, H.M.; Trujillo, J.J. *Theory and Applications of Fractional Differential Equations*; Elsevier: Amsterdam, The Netherlands, 2006.
23. Podlubny, I. *Fractional Differential Equations*; Academic Press: San Diego, CA, USA, 1999.
24. Caputo, M.; Fabrizio, M. A new Definition of Fractional Derivative without Singular Kernel. *Prog. Fract. Differ. Appl.* **2015**, *1*, 73–85.
25. Atangana, A.; Baleanu, D. New fractional derivatives with nonlocal and non-singular kernel: Theory and application to heat transfer model. *Therm. Sci.* **2016**, *20*, 763–769. [CrossRef]
26. Srivastava, H.M.; Harjule, P.; Jain, R. A general fractional differential equation associated with an integral operator with the H-function in the kernel. *Russ. J. Math. Phys.* **2015**, *22*, 112–126. [CrossRef]
27. Çetinkaya, A.; Kiymaz, I.O.; Agarwal, P.; Agarwal, R. A comparative study on generating function relations for generalized hypergeometric functions via generalized fractional operators. *Adv. Differ. Equ.* **2018**, 156. [CrossRef]
28. Ortigueira, M.D.; Machado, J.A.T. What is a fractional derivative? *J. Comput. Phys.* **2015**, *293*, 4–13. [CrossRef]
29. Caputo, M.; Fabrizio, M. On the notion of fractional derivative and applications to the hysteresis phenomena. *Meccanica* **2017**, *52*, 3043–3052. [CrossRef]
30. Zhao, D.; Luo, M. Representations of acting processes and memory effects: general fractional derivative and its application to theory of heat conduction with finite wave speeds. *Appl. Math. Comput.* 2018, *346*, 531–544. [CrossRef]
31. Hilfer, R.; Luchko, Y. Desiderata for Fractional Derivatives and Integrals. *Mathematics* **2019**, *7*, 149. [CrossRef]
32. Hilfer, R. Mathematical and physical interpretations of fractional derivatives and integrals. In *Handbook of Fractional Calculus with Applications, Volume 1*; Kochubei, A., Luchko, Y., Eds.; de Gruyter: Berlin, Germany, 2019; pp. 47–85.
33. Atanacković, T.M.; Pilipović, S.; Zorica, D. Properties of the Caputo–Fabrizio fractional derivative and its distributional settings. *Fract. Calc. Appl. Anal.* **2018**, *21*, 29–44. [CrossRef]

34. Kochubei, A.N. General Fractional Calculus, Evolution Equations, and Renewal Processes. *Integr. Equ. Oper. Theory* **2011**, *71*, 83–600. [CrossRef]
35. Fernandez, A.; Özarslan, M.A.; Baleanu, D. On fractional calculus with general analytic kernels. *Appl. Math. Comput.* **2019**, *354*, 248–265. [CrossRef]
36. Baleanu, D.; Fernandez, A. On some new properties of fractional derivatives with Mittag-Leffler kernel. *Commun. Nonlinear Sci. Numer. Simul.* **2018**, *59*, 444–462. [CrossRef]
37. Baleanu, D.; Asad, J.H.; Jajarmi, A. The fractional model of spring pendulum: new features within different kernels. *Proc. Rom. Acad. Ser.* **2018**, *19*, 447–454.
38. Uçar, S.; Uçar, E.; Özdemir, N.; Hammouch, Z. Mathematical analysis and numerical simulation for a smoking model with Atangana–Baleanu derivative. *Chaos Solitons Fractals* **2019**, *118*, 300–306. [CrossRef]
39. Yusuf, A.; Qureshi, S.; Inc, M.; Aliyu, A.I.; Baleanu, D.; Shaikh, A.A. Two-strain epidemic model involving fractional derivative with Mittag-Leffler kernel. *Chaos* **2018**, *28*, 123121. [CrossRef]
40. Prabhakar, T.R. A singular integral equation with a generalized Mittag Leffler function in the kernel. *Yokohama Math. J.* **1971**, *19*, 7–15.
41. Kilbas, A.A.; Saigo, M.; Saxena, R.K. Generalized Mittag-Leffler function and generalized fractional calculus operators. *Integr. Transform. Spec. Funct.* **2004**, *15*, 31–49. [CrossRef]
42. Özarslan, M.A.; Kürt, C. Nonhomogeneous initial and boundary value problem for the Caputo-type fractional wave equation. *Adv. Differ. Equ.* **2019**, 199. [CrossRef]
43. Srivastava, H.M.; Fernandez, A.; Baleanu, D. Some new fractional calculus connections between Mittag–Leffler functions. *Mathematics* **2019**, *7*, 485. [CrossRef]
44. Li, C.; Deng, W.; Zhao, L. Well-posedness and numerical algorithm for the tempered fractional ordinary differential equations. *Discret. Contin. Dyn. Syst.* **2019**, *24*, 1989–2015.
45. Meerschaert, M.M.; Sabzikar, F.; Chen, J. Tempered fractional calculus. *J. Comput. Phys.* **2015**, *293*, 14–28.
46. Osler, T.J. Leibniz rule for fractional derivatives generalized and an application to infinite series. *Siam J. Appl. Math.* **1970**, *18*, 658–674. [CrossRef]
47. Osler, T.J. The fractional derivative of a composite function. *Siam J. Math. Anal.* **1970**, *1*, 288–293. [CrossRef]
48. Srivastava, H.M.; Saxena, R.K.; Parmar, R.K. Some Families of the Incomplete H-Functions and the Incomplete $\bar{H}$-Functions and Associated Integral Transforms and Operators of Fractional Calculus with Applications. *Russ. J. Math. Phys.* **2018**, *25*, 116–138. [CrossRef]
49. Mittag-Leffler, G. Sur la représentation analytique d'une branche uniforme "une fonction monogène": cinquième note. *Acta Math.* **1905**, *29*, 101–181. [CrossRef]
50. Atangana, A.; Gomez-Aguilar, J.F. Fractional derivatives with no-index law property: Application to chaos and statistics. *Chaos Solitons Fractals* **2018**, *114*, 516–535. [CrossRef]
51. Hristov, J. (Ed.) *The Craft of Fractional Modelling in Science and Engineering*; MDPI: Basel, Switzerland, 2018.
52. Jarad, F.; Uğurlu, E.; Abdeljawad, T.; Baleanu, D. On a new class of fractional operators. *Adv. Differ. Equ.* **2017**, 247. [CrossRef]
53. Fernandez, A.; Baleanu, D. On a new definition of fractional differintegrals with Mittag-Leffler kernel. *Filomat* **2019**, *33*, 245–254.
54. Fernandez, A.; Baleanu, D.; Srivastava, H.M. Series representations for models of fractional calculus involving generalized Mittag-Leffler functions. *Commun. Nonlinear Sci. Numer. Simul.* **2019**, *67*, 517–527. [CrossRef]

© 2019 by the authors. Licensee MDPI, Basel, Switzerland. This article is an open access article distributed under the terms and conditions of the Creative Commons Attribution (CC BY) license (http://creativecommons.org/licenses/by/4.0/).

Article

Some Alternative Solutions to Fractional Models for Modelling Power Law Type Long Memory Behaviours

Jocelyn Sabatier *, Christophe Farges and Vincent Tartaglione

IMS laboratory, Bordeaux University, UMR CNRS 5218, 351 Cours de la liberation, 33400 Talence, France; christophe.farges@u-bordeaux.fr (C.F.); vincent.tartaglione@u-bordeaux.fr (V.T.)
* Correspondence: Jocelyn.sabatier@u-bordeaux.fr

Received: 27 December 2019; Accepted: 22 January 2020; Published: 5 February 2020

Abstract: The paper first describes a process that exhibits a power law-type long memory behaviour: the dynamical behaviour of the heap top of falling granular matter such as sand. Fractional modelling is proposed for this process, and some drawbacks and difficulties associated to fractional models are reviewed and illustrated with the sand pile process. Alternative models that solve the drawbacks and difficulties mentioned while producing power law-type long memory behaviours are presented.

Keywords: fractional models; fractional differentiation; distributed time delay systems; Volterra equation; adsorption

1. Introduction

Research related to fractional differentiation has grown exponentially in recent years in many areas, including automatic control. In automatic control, many applications have been developed in dynamical system modelling using "fractional models". These models are mainly used to capture power law-type long memory input/output behaviours. In most of these applications, the models are described by differential equations that involve fractional derivatives or "fractional differential equations". For the multi-input, multi-output case, these models can be described by the equation:

$$\sum_{k=0}^{N_a} S_k \left(\frac{d}{dt}\right)^{v_{a_k}} y(t) = \sum_{k=0}^{N_b} T_k \left(\frac{d}{dt}\right)^{v_{b_k}} u(t) \quad N_a \in \mathbb{N}^*, \ N_b \in \mathbb{N}^* \tag{1}$$

in which $u(t) \in \mathbb{R}^m$ denotes the input vector, $y(t) \in \mathbb{R}^p$ denotes the output vector, $S_k \in \mathbb{R}^{p \times p}$, $T_k \in \mathbb{R}^{p \times m}$. $(d/dt)^{v_{a_k}}$ and $(d/dt)^{v_{b_k}}$ denote fractional differential operators of orders $v_{a_k} \in \mathbb{R}$ and $v_{b_k} \in \mathbb{R}$, respectively. These operators are defined in [1–4], and a detailed survey of the properties linked to these definitions can be found in [2].

If orders v_{a_k} and v_{b_k} in Relation (1) verify the relations $v_{a_{k_1}} = k_1/q$, $v_{b_{k_2}} = k_2/q$, $k_1 \in \mathbb{N}^*$ and $k_2 \in \mathbb{N}^*$, $q \in \mathbb{N}^*$, then the differentiation orders v_{a_k} and v_{b_k} are commensurate (multiple of the same rational number $v = 1/q$). Here, it is assumed that $N_a \geq N_b$. Using the order commensurability condition and for null initial conditions, the differential Equation (1) can be rewritten under the form:

$$\begin{cases} \frac{d^v}{dt^v} \zeta(t) = A\zeta(t) + Bu(t) \\ y(t) = C\zeta(t) + Du(t) \end{cases} \tag{2}$$

where $\zeta(t) \in \mathbb{R}^n$ is the pseudo-state vector, $v = 1/q$ is the fractional order of the model, and $A \in \mathbb{R}^{n \times n}$, $B \in \mathbb{R}^{n \times m}$, $C \in \mathbb{R}^{p \times n}$, and $D \in \mathbb{R}^{p \times m}$ are constant matrices. Model (2) is known in the literature under the name "fractional state space description", which was introduced for the first time in [5]. Alternatively, Models (1) and (2) can be described by transfer functions that involve non-integer powers of the Laplace variable s.

Although Models (1) and (2) are widely used in the literature, for modelling and beyond, several drawbacks associated with their use have been revealed in the last 10 years. Some of these problems result from too hasty "fractionalisations" of concepts dedicated to classical integer systems, without any physical justification. Thus, using an example, this paper aims:

- to illustrate these drawbacks and to show that alternative solutions exist for power law-type long memory behaviours modelling;
- to clarify the limits and benefits of fractional models.

In this paper, the first section defines the concept of "power law-type long memory behaviour" for linear time invariant (LTI) dynamical systems and gives some conditions in time and frequency domains for this class of systems to exhibit a power law-type behaviour. In Section 2, the dynamical behaviour of the heap top of falling granular matter such as sand is studied. This is an example of a process that exhibits a power law-type long memory behaviour. Then, fractional modelling is proposed for this process in Section 3, and some drawbacks and difficulties associated to fractional models are reviewed and illustrated with the sand pile process. Section 4 demonstrates that the power law-type behaviour of the sand pile process can be modelled by a non-linear model, thus demonstrating that other models than fractional models are possible for power law-type behaviours. Then, several alternative models that solve the drawbacks and difficulties mentioned while producing power law-type long memory behaviours are presented in Section 6.

2. Power Law-Type Long Memory Behaviours

In this paper, we intentionally use the expression "power law-type behaviours" and not "fractional behaviours", as the word fractional refers to fractional models, which are one of the means among others for modelling power law-type behaviours, and because the power can be other than a fractional number (a real number).

In the sequel, we will say that a system has a power law-like behaviour if its impulse response or if its frequency response exhibits a power law behaviour in a given time or frequency range. The term "power law" comes from the time series analysis field, as is recalled in the following subsection.

In the analysis of time series, long memory behaviours can be characterized in terms of their autocorrelation functions [6]. The autocorrelation highlights that the coupling between values of a signal at different times decreases slowly as the time difference increases. The decay of the autocorrelation function can be power-like and so is slower than exponential decay.

Thus, the concept of power law-type long memory is defined for signals in the time series field. The purpose of this section is to extend this concept to models that have output signals exhibiting power law-type long memory behaviour.

In Section 2.1, some properties of the spectral density of a system output signal and properties linking the autocorrelation functions of the input signal and the output signal are demonstrated in the general case of a linear time invariant (LTI) model. In Section 2.2, these properties are particularised to systems that have output signals exhibiting power law-type long memory behaviour, allowing to propose a general definition of a power law-type long memory model.

2.1. Spectral Density and Autocorrelation Functions of the Input Output Signals of an LTI System

Let $u(t)$ and $y(t)$ be respectively the input and the output of a dynamical LTI single input–single output model. Input $u(t)$ is assumed to be a white noise, and let $R_y(\xi)$ be the output autocorrelation defined by:

$$R_y(\xi) = \int_{-\infty}^{\infty} y(t+\xi)y(t)dt. \tag{3}$$

In addition, let $S_y(\omega)$ be the output power spectral density defined by:

$$S_y(\omega) = \int_{-\infty}^{\infty} R_y(\xi) e^{-j\omega\xi} d\xi. \tag{4}$$

The autocorrelation function $R_y(\xi)$ of the system output $y(t)$ is related to the autocorrelation function $R_u(\xi) = \int_{-\infty}^{\infty} u(t+\xi)u(t)dt$ of the system input $u(t)$ through the relation:

$$R_y(\xi) = \int_{-\infty}^{\infty} \int_{-\infty}^{\infty} u(t-p)h(p)dp \int_{-\infty}^{\infty} u(t+\xi-q)h(q)dqdt \tag{5}$$

or (if permutations of integrals are permitted)

$$R_y(\xi) = \int_{-\infty}^{\infty} \int_{-\infty}^{\infty} h(p)h(q)\left(\int_{-\infty}^{\infty} u(t-p)u(t+\xi-q)dt\right)dqdp. \tag{6}$$

Using the change of variable $t' = t - p$, Relation (6) becomes

$$R_y(\xi) = \int_{-\infty}^{\infty} \int_{-\infty}^{\infty} h(p)h(q)\left[\int_{-\infty}^{\infty} u(t')u(t'+\xi+p-q)dt'\right]dqdp \tag{7}$$

or

$$R_y(\xi) = \int_{-\infty}^{\infty} \int_{-\infty}^{\infty} h(p)h(q)R_u(\xi+p-q)dqdp. \tag{8}$$

If $u(t)$ is a white noise of variance σ, then $R_u(\xi) = \sigma\delta(\xi)$ where $\delta(.)$ is the Dirac function. Thus,

$$R_y(\xi) = \sigma \int_{-\infty}^{\infty} \int_{-\infty}^{\infty} h(p)h(q)\delta(\xi+p-q)dqdp. \tag{9}$$

Using Relation (4),

$$S_y(\omega) = \int_{-\infty}^{\infty} \int_{-\infty}^{\infty} \int_{-\infty}^{\infty} h(p)h(q)R_u(\xi+p-q)e^{-j\omega\xi}dqdpd\xi. \tag{10}$$

Using $\tau = \xi + p - q$, the previous relation becomes

$$S_y(\omega) = \int_{-\infty}^{\infty} h(p)e^{j\omega p}dp \int_{-\infty}^{\infty} h(q)e^{-j\omega q}dq \int_{-\infty}^{\infty} R_u(\tau)e^{-j\omega\tau}d\tau \tag{11}$$

and thus, if $H(j\omega)$ denotes the frequency response (and $H^*(j\omega)$ its conjugate) of the considered dynamical system:

$$S_y(\omega) = H(j\omega)H^*(j\omega)S_u(\omega) = \sigma|H(j\omega)|^2. \tag{12}$$

2.2. Power Law Concept Extended to LTI Systems

Let us now consider an LTI system whose impulse response is of the form

$$h(t) = \frac{K_t}{t^{1-\nu}} H_e(t) \quad \text{and} \quad H(j\omega) = \frac{K_\omega}{(j\omega)^\nu} \quad 0 < \nu < 2, \; K_t \in \mathbb{R}, \; K_\omega \in \mathbb{R}. \tag{13}$$

where $H_e(t)$ is the Heaviside function. According to Relation (12), the power spectral density of the system output to a white noise of variance σ is defined by

$$S_y(\omega) = \frac{\sigma K_\omega^2}{\omega^{2\nu}} \tag{14}$$

and exhibits a power law-type behaviour in the frequency domain. According to Relation (9) for a white noise input $u(t)$ of variance σ, the output autocorrelation is defined by

$$R_y(\xi) = \sigma \int_{p=0}^{\infty} \int_{q=0}^{\infty} h(p)h(q)\delta(\xi + p - q) dq dp \tag{15}$$

or as the integrated function is not equal to 0 only if $\xi + p = q$

$$R_y(\xi) = \sigma \int_{p=0}^{\infty} h(p)h(p+\xi) dp = \sigma \int_{p=0}^{\infty} \frac{K_t}{p^{1-\nu}} \frac{K_t}{(p+\xi)^{1-\nu}} dp \tag{16}$$

and thus if $\Gamma(.)$ denotes the Euler gamma function:

$$R_y(\xi) = \frac{\sigma K_t^2 4^{-\nu} \Gamma(\nu) \Gamma(\frac{1}{2} - \nu)}{\sqrt{\pi}} \xi^{2\nu - 1}. \tag{17}$$

Relation (17) demonstrates that the output signal autocorrelation exhibits a power law-type behaviour.

Definition 1 [Power law-type long memory system]. *A power law-type long memory system is an LTI system that has one of the following equivalent properties in a given time or frequency range:*

1. Its impulse response $h(t)$ slowly decays with respect to time according to:

$$h(t) = \frac{K_t}{t^{1-\nu}} H_e(t) \quad 0 < \nu < 2. \tag{18}$$

2. For a white noise input $u(t)$ of variance σ, its output autocorrelation function is:

$$R_y(\xi) = \frac{\sigma K_t^2 4^{-\nu} \Gamma(\nu) \Gamma(\frac{1}{2} - \nu)}{\sqrt{\pi}} \xi^{2\nu - 1}. \tag{19}$$

3. For a white noise input $u(t)$ of variance σ, its output power spectral density is:

$$S_y(\omega) = \frac{\sigma K_\omega^2}{\omega^{2\nu}}. \tag{20}$$

Definition 1 allows characterising the input output behaviour of the class of systems that is considered in this paper.

3. Sand Heap Growth: An Example of Power Law-Type Long Memory Behaviour

3.1. System Description

The dynamical behaviour of falling granular matter such as sand is studied (here, granulated sugar). As shown in Figure 1, it is assumed that the granular matter grows under a flow of sand $Q(t)$ and that the base of the cone created by the accumulation of matter can also grow with time. The experimental apparatus used to create the heap and to measure its height is also described in Figure 1. The sand falls from a conic tank and the height is measured using a webcam.

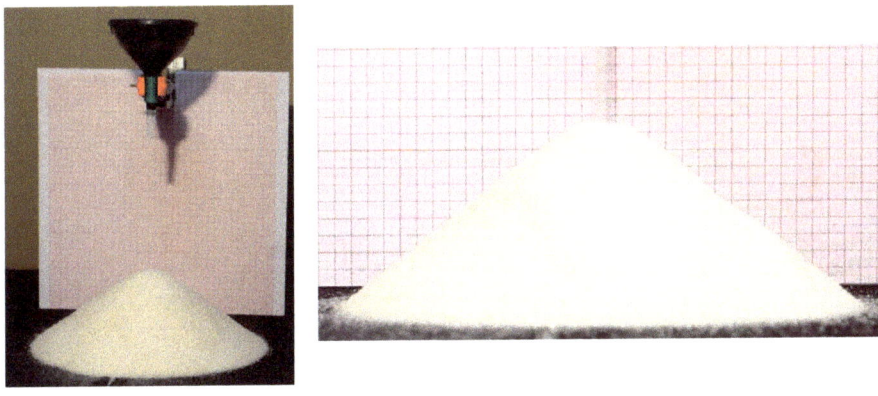

Figure 1. Illustration of sand pile growth (**right**) and description of the apparatus used to measure the heap height (**left**).

The time evolution of the sand heap top denoted as $h(t)$ is represented by Figure 2. The shape of the curve is similar to those represented in [7,8]. In order to show that this system has a power law-type long memory behaviour, the function

$$log[h(log(t))] \underset{\text{for large } t}{\sim} K_0 + vlog(t) \quad K_0 \in \mathbb{R} \tag{21}$$

is represented in Figure 3. For a large time duration, this figure shows that the curve behaves as a straight line:

$$K_0 + vlog(t) \tag{22}$$

thus highlighting that the considered system exhibits a power law-type behaviour. Indeed, if $h(t) = k_0 t^v$ then $log[h(t)] = log(k_0) + vlog(t) = K_0 + vlog(t)$. Thus, this system has Property 1 of Definition 1.

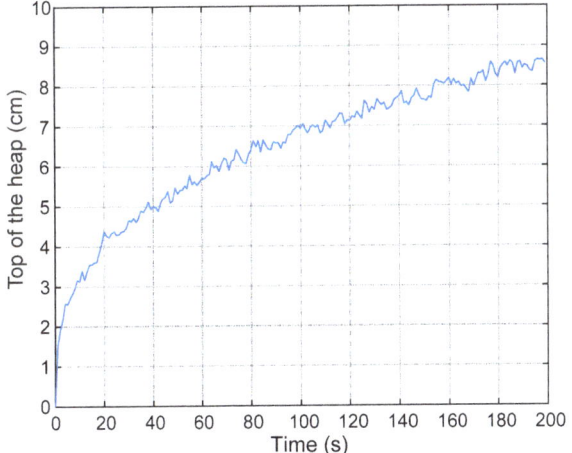

Figure 2. Heap top $h(t)$ variation.

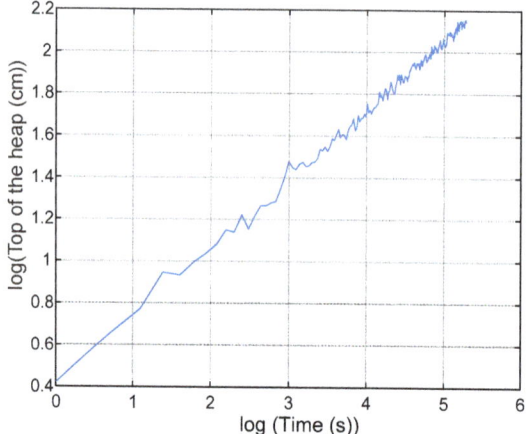

Figure 3. Function $\log[h(\log(t))]$.

3.2. Fractional Modelling of the Sand Pile Growth

As the system exhibits a power law-type behaviour, in a first approach, a fractional model was considered to model the system. The proposed model is defined by the transfer function

$$H(s) = \frac{K}{s^\nu} \quad K \in \mathbb{R}_+^*, \quad \nu \in \mathbb{R}, \quad 0 < \nu < 1. \tag{23}$$

Parameters K and ν are obtained through the minimisation of a quadratic criterion on the error between the measure and the model time response. The input of the model is assumed to be a Heaviside function of magnitude 1. The parameters obtained are:

$$K = 1.23 \quad \nu = 0.35. \tag{24}$$

Figure 4 shows a comparison of the measures and the model time response. This comparison reveals that the fractional model permits an accurate fitting of the measures thanks to a compact model involving only two parameters (K and ν). However, such a modelling approach comes with several drawbacks that are now described.

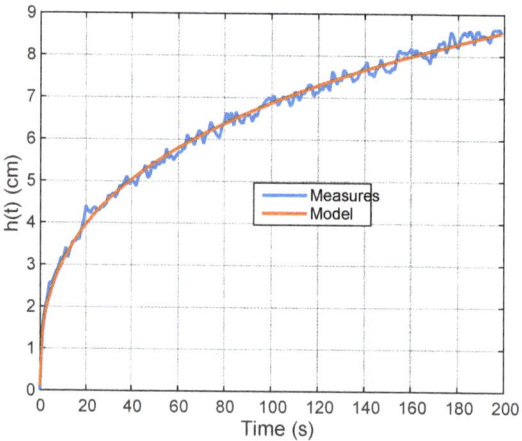

Figure 4. Comparison of the measures and the model response.

4. Drawbacks of Fractional Modelling

The drawbacks listed in the sequel hold for the fractional modelling approach done in the previous section and beyond. The fractional model obtained in the previous section is a particular form of the more general model

$$H(s) = \frac{T(s)}{R(s)} \qquad (25)$$

with $T(s) = \sum_{l=0}^{r} t_l s^{\beta_l}$ and $R(s) = \sum_{k=0}^{m} r_k s^{\alpha_k}$ where $r \in \mathbb{N}^*$, $m \in \mathbb{N}^*$, $t_l \in \mathbb{R}$, $r_k \in \mathbb{R}$, $\beta_{l+1} \geq \beta_l \geq 0$ and $\alpha_{k+1} \geq \alpha_k \geq 0$. The first drawback associated to this class of model is linked to its physical interpretation. The time constant distribution interpretation is often invoked [9] but does not reflect the internal behaviour of the modelled system, as for example for the case of the pile of sand. The other interpretations are not more satisfactory.

Drawback 1. *The physical interpretations proposed in the literature are not obtained based on the observation of a given phenomenon but result from purely mathematical discussions [9–17]. In the case of incommensurate orders, some interpretations can invalidate the obtained model [18].*

The impulse response of the Transfer Function (25), computed with the residue theorem using a Bromwich–Wagner path, can be written as [19]:

$$h(t) = h_p(t) + h_d(t) \qquad (26)$$

with

$$h_p(t) = \sum_{i=1}^{n} a_i e^{t p_i} \; n \in \mathbb{N}^*, \; a_i \in \mathbb{R}, \; p_i \in \mathbb{R}_-, \text{ and } h_d(t) = \int_0^\infty \mu(x) e^{-tx} dx. \qquad (27)$$

Function $h_p(t)$ is produced by the poles of the transfer functions $H(s)$ (residues of the Cauchy method). As explained in [20], the function $\mu(x)$ in $h_d(t)$ is defined by

$$\mu(x) = \frac{1}{2j\pi}\left[H\left((-x)^-\right) - H\left((-x)^+\right)\right] = \frac{1}{\pi} \frac{\sum_{k=1}^{m}\sum_{l=0}^{r} a_k q_l \sin(\pi(\alpha_k - \beta_l)) x^{\alpha_k - \beta_l}}{\sum_{k=0}^{m} a_k^2 x^{2\alpha_k} + \sum_{0 \leq k < l \leq m} 2 a_k a_l \cos(\pi(\alpha_k - \beta_l)) x^{\alpha_k - \beta_l}}. \qquad (28)$$

The Laplace transform of the function $h_d(t)$ is given by

$$h_d(s) = \int_0^\infty \frac{\mu(z)}{s+z} e^{-tz} dz. \qquad (29)$$

Such a relation shows that a fractional model exhibits poles distributed from 0 to $-\infty$, thus leading to the following drawback.

Drawback 2. *The memory of a fractional model is infinite and it exhibits infinitely slow and infinitely fast time constants (even if they are attenuated through the function $\mu()$, they exist), which excludes the possibility of linking the model internal variable to a physical variable.*

The infinite memory associated to fractional models can also be given by another interpretation. If an input $u(t)$ is applied to the submodel of the impulse response $h_d(t)$, the resulting output $y_d(t)$ is given by the relation [20,21]:

$$\begin{cases} \frac{\partial w(t,z)}{\partial t} = -zw(t,z) + u(t) \\ y_d(s) = \int_0^\infty \mu(z) w(t,z) dz \end{cases} \text{ with } z \in \mathbb{R}^+, \qquad (30)$$

which is known in the literature as diffusive representation [21]. The inverse spatial Fourier transform denoted by the symbol $\mathcal{F}^{-1}$ ($\mathcal{F}$ is for Fourier transform) applied to (30), leads to

$$\begin{cases} \frac{\partial \phi(t,\zeta)}{\partial t} = \frac{\partial^2 \phi(t,\zeta)}{\partial \zeta^2} + u(t)\delta(\zeta) \\ y_d(s) = \int_{-\infty}^{\infty} m(\zeta)\phi(t,\zeta)d\zeta \end{cases} \text{with } \zeta \in \mathbb{R} \qquad (31)$$

and

$$\phi(t,\zeta) = \mathcal{F}^{-1}\{w(t,4\pi^2z^2)\}, \quad m(\zeta) = \mathcal{F}^{-1}\{4\pi^2\zeta\mu(4\pi^2\zeta^2)\}.$$

Relation (31) allows us to claim that a fractional system can be associated to an infinite dimensional system described by a diffusion equation *on an infinite domain* ($\zeta \in \mathbb{R}$) [22]. It is this (double) infinite dimension requirement that creates the infinite memory mentioned above.

If Model (23) is used for the sand pile growth modelling, an infinite number of initial conditions is required (i.e., a state of infinite dimension is required). However, it is clear that the initial condition of the sand pile growth can be described using a single variable: the sand pile height $h(t)$ (a state for this system could be chosen as $h(t)$, making the sand pile growth model a first-order model).

This can also be illustrated in the thermal domain [23]. A fractional integrator is a solution of the heat equation (linking the thermal heat flux applied to the measured temperature) only if:

- the temperature measure is done at the point where the heat flux is applied; or
- an infinite dimension medium is considered.

Other spatial configurations can lead to power law-type behaviours but cannot be written under the form of Model (25) (exponential and hyperbolic functions are involved in the Laplace domain).

If orders β_l and α_k meet a commensurate condition in Relation (25), it can be rewritten as:

$$\begin{cases} \frac{d^\nu x(t)}{dt^\nu} = Ax(t) + Bu(t) \\ y(t) = Cx(t) + Du(t) \end{cases}. \qquad (32)$$

In this representation, an analysis of the units very quickly leads to doubts about the physical character of the coefficients in matrices A and B, leading to the following drawback.

Drawback 3. *The parameter units associated to description (parameters inside Matrices A and B) have no physical meaning (e.g., $sec^{-\nu}$ for parameters in Matrix A).*

Representation (32) is known in the literature as a "fractional state space description". However, this is an improper designation that results from a generalisation of concepts dedicated to integer systems without inquiring into the notion of state. This analysis is demonstrated in [24], and it leads to the following drawback.

Drawback 4. *Representation (32) is not a state space representation, as the variable x(t) does not have the properties of a state. That is why the terms "pseudo state" and "pseudo-state space description" were introduced [24].*

In Representation (32), as in the Transfer Function (25), the fractional differentiation operator $\frac{d^\nu}{dt^\nu}$ is not defined uniquely.

Drawback 5. *There are more than 30 definitions of the operator $\frac{d^\nu}{dt^\nu}$ [25].*

This multiplicity of definitions leads to developing results by choosing the most convenient definition to obtain them. This is why Caputo's definition became so popular, as it offers the possibility to take into account the initial conditions without taking into account all the past of the system. If from a mathematical point of view the definitions of Caputo, Riemann-Liouville, or others are in no way problematic, their use for the definition of fractional models is questionable. While fractional models are known to have a long and even infinite memory, the use of Caputo's derivative would make this memory disappear for a given time moment (initial time). This paradoxical situation led to several analyses that revealed the following drawback.

Drawback 6. *The initial conditions are not well taken into account in Representations (32) and (25) if the Caputo or Riemann–Liouville definitions are used [16,22,26,27].*

To solve these initialisation issues (and also the infinite memory issue), it was proposed in [28,29] to use a limited frequency band fractional integration operator in the definition of fractional models. Another consequence of the infinite memory of Model (32), and sometimes in contradiction with some results proposed in the literature, is the poor properties of the considered models.

Drawback 7. *Exact observability cannot be reached as all of the system's past must be known to predict its future [19].*

The analysis proposed in [19] could be extended to the analysis of controllability and flatness as model initialisation has an impact on these properties.

To avoid the multiplicity of definitions and the initial conditions problem, it was concluded in [24] that fractional integration is preferable in the definition of a fractional model and thus that Relation (32) should be rewritten under the form:

$$\begin{cases} x(t) = I_{t_0 \to -\infty}^{\nu}[Ax(t) + Bu(t)] \\ y(t) = Cx(t) + Du(t) \end{cases} \tag{33}$$

with:

$$I_{t_0 \to -\infty}^{\nu}[f(t)] = \frac{1}{\Gamma(\nu)} \int_{t_0}^{t} \frac{f(\tau)}{(t-\tau)^{1-\nu}} d\tau. \tag{34}$$

However, such a definition entails another drawback.

Drawback 8. *The fractional integration given by Relation (34) involves a singular kernel [30]; this leads to complications in the solution/simulation of the fractional order differential equations.*

Note that some non-singular kernels for modelling power law-type long memory behaviours have been proposed in [29].

In the case of the sand pile, the following section shows that all these drawbacks could have been avoided by using a different modelling approach while capturing accurately the power law behaviour.

5. Another Possible Model

Let $Q(t)$ be the flow of falling sand. If $V_c(t)$ denotes the sand heap cone volume with $V_c(t) = 1/3\pi r^2 h$, according to the notations introduced in Figure 5, the flow $Q(t)$ generates the volume variation of the cone:

$$\frac{dV_c(t)}{dt} = Q(t). \tag{35}$$

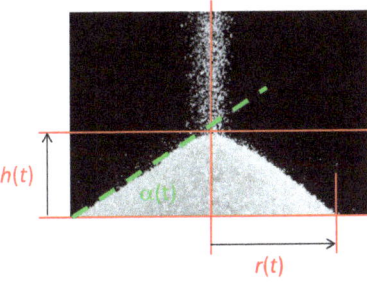

Figure 5. Notations for the characterisation of the sand heap growth.

As
$$\tan(\alpha) = \frac{h(t)}{r(t)} \quad \text{then} \quad V_c(t) = \frac{\pi h(t)^3}{3\tan^2(\alpha)} \quad (36)$$

and thus, under the hypothesis of a constant angle of repose α

$$\frac{dV_c(t)}{dt} = \frac{\pi h(t)^2}{\tan^2(\alpha)} \frac{dh(t)}{dt}. \quad (37)$$

Combining Relations (35) and (37), variation in the sand heap height is thus defined by the differential equation:

$$\frac{dh(t)}{dt} = \frac{\tan^2(\alpha)}{\pi h(t)^2} Q(t). \quad (38)$$

For $Q(t)$ constant, Model (35) can be rewritten as:

$$\frac{dh(t)}{dt} = \frac{a_0}{h(t)^2} H_e(t), \quad (39)$$

in which a_0 is a parameter and $H_e(t)$ is the Heaviside fonction. With the measures in Figure 2, parameter a_0 was computed with an optimisation algorithm aiming at minimising the error between the response of Model (38) and the measures. Parameter $a_0 = 1.07$ was obtained, and a comparison of the measures with the model response is shown by Figure 6.

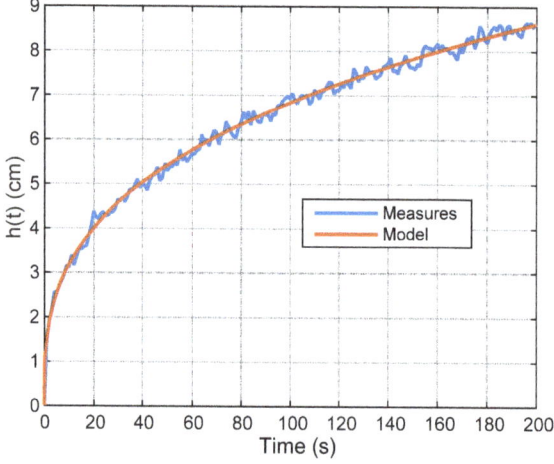

Figure 6. Comparison of the measures with the Model (35) response.

Similar to the fractional model, Model (39) also permits an accurate fitting of the system behaviour with a small number of parameters. However, Model (38) resolves most of the drawbacks mentioned in the following paragraph and in particular eliminates any questioning about the infinite space dimension and about initialization of the model.

Let us imagine that the experiment starts with a partially formed sand heap, as if the process had a past. Fractional modelling with Model (23) would impose the knowledge of all the system's past to restart the experiment, as if knowing the position of all the grains was necessary. However, in practice, this knowledge is not useful. It is not useful to know the position of all the grains of sand; it is only necessary to reconstitute a pile with similar geometric characteristics (angle of repose contained in parameter a_0 and heap height $h(t)$). This is exactly what Model (35) does. Only one state $h(t)$ and thus its initialisation is required. This example highlights the erroneous conclusions to which fractional modelling can lead. Admittedly, the temporal evolution fitting is very accurate, but the physical interpretation is not possible.

Due to the omnipresence of systems that exhibit power law-type behaviours, it appears important to develop new models that do not exhibit the above problems while being able to capture the corresponding dynamics. Some are proposed in the next section.

6. Beyond Fractional Models

6.1. Some Classes of Non-Linear Models

The previous section showed that models other than fractional models can be used to model power law-type long memory behaviours, in particular non-linear models. This is exactly what the authors did recently for the modelling of the adsorption process [31]. The adsorption process can be likened to the process of the random deposition of discs on a surface, which is denoted random sequential adsorption (RSA) and can be mathematically described as follows.

RSA Process: Let S be a square of size $L \times L$, $L \in \mathbb{R}_+^*$. Let $R \in \mathbb{R}_+^*$ with $R \ll L$ and $t = (t_k)_{k \in \mathbb{N}} \in \mathbb{R}^\mathbb{N}$ with $t_0 = 0$ and such that for all $k \in \mathbb{N}$, $t_{k+1} - t_k = \Delta t \in \mathbb{R}_+^*$. At $t = t_0$, the surface is empty. At each time t_k, a disk of radius R arrives on the surface S at a randomly chosen location. If the area corresponding to the disk is empty, the disk is placed at the location. If part of the corresponding area is covered by another disk, the disk goes back, and the configuration of S remains unchanged.

An example of the result produced by this process is shown in Figure 7.

Figure 7. A possible result for the random sequential adsorption (RSA) process.

If $\theta(t)$ denotes the density of the occupied area, it is explained in [32,33], and simulated in [31] that the covered surface can be described by a power law (see Figure 8).

$$\theta_\infty - \theta(t) \approx t^{-1/2}. \tag{40}$$

Given the power law behaviour of this process, a fractional model should be effective to describe the kinetic of the density $\theta(t)$. However, limitations on the ability of this kind of model to capture some properties of the RSA model were highlighted in [31] and are now summarised.

- With the RSA process (as for the sand pile process), if the flow is stopped, then the surface filling stops. If the flow restarts, the surface filling restarts from the same state. Such behaviour cannot be reproduced with a classical linear fractional model whose output relaxes for a null input.
- With a fractional modelling approach, an infinite dimensional model is obtained, requiring the entire model past knowledge for a proper initialisation. However, in practice, such knowledge is not required. Initialisation of the RSA process only requires the knowledge of the density $\theta(t)$ and a uniform distribution of the disks on the surface. Exact knowledge of the position of all the disks on the surface is not necessary, and thus not all the process history is required.

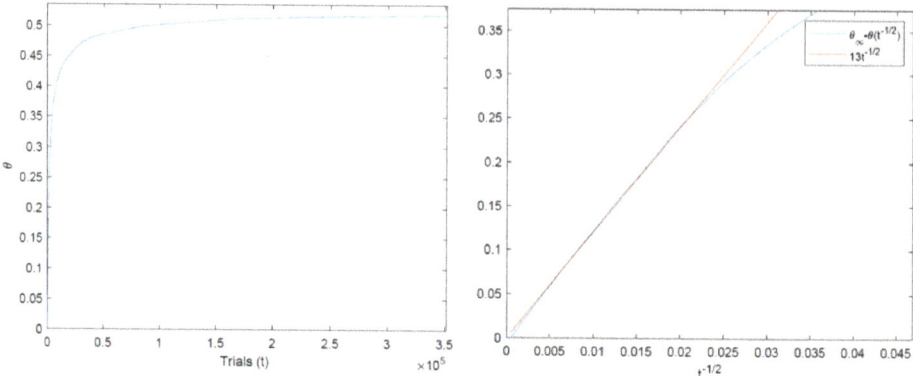

Figure 8. Density $\theta(t)$ of occupied area as a function of trials (**left**) and highlighting of the power law behaviour of $\theta_\infty - \theta(t)$ for large values of t (**right**).

To overcome these limitations, a model of the form

$$\dot{y}(t) = f(y)u(t) \tag{41}$$

was proposed in [31], in which $u(t)$ is the flow of disks that hit the surface, $y(t) = 1 - \theta(t)$ denotes the free surface density and

$$f(y) = (b_0 + b_1 y + b_2 y^2 + b_3 y^3 + b_4 y^4)\left[\frac{\tanh(100(y-0.5))}{2} - \frac{\tanh(60(y-0.5))}{2}\right] + (c_0 + c_1 y)\frac{\tanh(60(y-0.8))+1}{2}. \tag{42}$$

This model can be viewed as a serious alternative to fractional models as:

- It permits an accurate fitting of the RSA process kinetic in spite of its power law behaviour;
- It takes into account some non-linear behaviours in relation to the flow of incoming disks (or particles for the case of a real adsorption process);
- Its state is only of one dimension, and its initialisation only requires knowledge of the covered density;
- Its implementation does not require any approximation step.

6.2. Distributed Time Delay Models

Modelling of power law-type long memory behaviours is also possible using distributed time delay systems. This is exactly what is done in [34,35], in which the following class of time delay system is considered.

$$\frac{d}{dt}x(t) = A_0 x(t) + A_1 \int_0^{T_f} \eta(\tau) x(t-\tau) d\tau + Bu(t) \tag{43}$$

in which

$$\eta(t) = C_0 \left(\frac{\omega_l^\nu}{\Gamma(\nu)} t^{\nu-1} e^{-\omega_l t} - \frac{\omega_l^\nu}{\Gamma(\nu)} t^{\nu-1} e^{-\omega_m t} + \omega_l^\nu \omega_m^{1-\nu} e^{-\omega_m t} \right) \tag{44}$$

with

$$C_0 = \left| \frac{1}{\left(\frac{j}{\omega_l}+1\right)^\nu} - \left(\frac{\omega_l}{\omega_m}\right)^\nu \frac{1}{\left(\frac{j}{\omega_m}+1\right)^\nu} + \left(\frac{\omega_l}{\omega_m}\right)^\nu \frac{1}{\frac{j}{\omega_m}+1} \right|^{-1}. \tag{45}$$

As shown by Figure 9, the input/output frequency behaviour of such a model exhibits a power law behaviour in a frequency band that can be adjusted using coefficients A_0, A_1, and B.

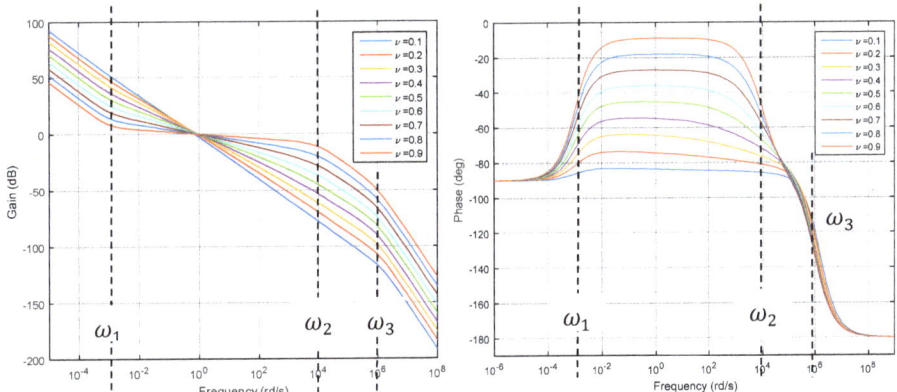

Figure 9. Gain (**left**) and phase (**right**) diagrams of $x(s)/u(s)$ for various values of v where corner frequencies $\omega_1 = 10^{-3}$ rd/s, $\omega_2 = 10^4$ rd/s, and $\omega_3 = 10^6$ rd/s depend on parameters A_0, A_1, and B [34].

In comparison with the Fractional Model (32), Model (43) has the following advantages:

- In Relation (43), the variable $x(t)$ can be viewed as a real state and a physical meaning can be associated to it;
- There is no longer any ambiguity in the operator used for the definition of Relation (43) (in Equation (32), Caputo's, Riemann–Liouville, or another can be chosen);
- Kernel $\eta(t)$ in Relation (43) is not singular, unlike the definition of fractional derivative in Equation (32);
- The memory of Model (43) is of finite length;
- Initialisation of Model (43) requires knowledge of its state on a finite length and is well defined.

6.3. First Kind Volterra Equations

It must be noted that Fractional Model (32), which is widely used in the literature, is a particular case of a Volterra equation of the first kind. According to [4]—p. 46 (if the fractional integral of order v of each component of vector $x(t)$ exists) and after first-order integration of both sides of the first equation in Relation (32), the following equations can be obtained:

$$\int_0^t \eta^{1-v}(t-\tau)x(\tau)d\tau = \int_0^t [Ax(\tau) + Bu(\tau)]d\tau \qquad (46)$$

where the kernel in Relation (46) is $\eta^{1-v}(t) = t^{-v}/\Gamma(1-v)$ and multiplies each component of vector $x(t)$. Thus, Representation (32) can be rewritten under the form of a Volterra equation of the first kind,

$$\int_0^t \left(\frac{t^{-v}}{\Gamma(1-v)}I_n - A\right)x(\tau)d\tau = v(t) \quad \text{with} \quad v(t) = \int_0^t Bu(\tau)d\tau, \quad y(t) = Cx(t), \qquad (47)$$

where I_n denotes an identity matrix with the same dimension as vector $x(t)$. Relation (47) demonstrates that a pseudo-state space description is a particular case of a Volterra equation of the first kind, as the kernel in Relation (32) has a fixed structure. Using a Volterra equation, the following class of model can be proposed

$$\int_0^t \eta(t-\tau)x(\tau)d\tau = v(t) \quad \text{with} \quad v(t) = \int_0^t u(\tau)d\tau, \quad y(t) = x(t) \qquad (48)$$

that generalises the pseudo-state space description (32) in two ways:

- Adapting the kernel $\eta(t)$ in Relation (48) (see also [29], it is possible to produce, with the same kind of equation, power law behaviours of various types (denoted explicit, implicit), but also many other long memory behaviours;
- In Relation (48), if $x(t) \in \mathbb{R}^n$, $\eta(t)$ is a matrix of kernels such that $\eta(t) = [\eta_{i,j}(t)]$, thus permitting great flexibility in the tuning of Relation (48). The case $\eta(t) = diag[\eta_i(t)]$ comes closer to the non-commensurate fractional pseudo-state space representation case, but it should be remembered that physical interpretations invalidate this kind of model [18].

Description (48) has another important advantage. Model memory can be limited by introducing a parameter T_f in the integral bounds such that:

$$\int_{t-T_f}^{t} \eta^v(t-\tau)x(\tau)d\tau = v(t). \tag{49}$$

Using the change of variable $\xi < t - \tau$, Relation (48) becomes:

$$\int_{0}^{T_f} \eta^v(\xi)x(t-\xi)d\xi = v(t) \quad \text{with} \quad v(t) = \int_{0}^{t} u(\tau)d\tau, \quad y(t) = x(t). \tag{50}$$

Relation (50) is close to Relation (43) and explicitly shows that knowledge of the model state $x(t)$ is required only on $\left[0, T_f\right]$ to compute its future.

7. Conclusions

This paper started from an illustrative example: sand pile growth under the effect of falling sand in the upper part of the heap. Using a simple experiment, it was shown that the pile growth exhibits a power law-type long memory behaviour. As fractional models also exhibit power law-type behaviours, they can be used to capture the input–output behaviour of such a system. However, several drawbacks are associated to this modelling, and were reviewed here. It is shown that a simple non-linear model permits a physical modelling of the considered system, thereby removing all the mentioned drawbacks. This leads to two conclusions:

- Even if fractional models permit an accurate fitting of power law-type input–output behaviours, they can give birth to disconnected issues of the system considered (initialisation, dimension, interpretations, ...)
- simpler more physical models can be obtained if we try to understand the physical origin of the behaviour.

This is what the authors did to model adsorption phenomena [31]. Yet again, a non-linear model proved to be more suitable than a fractional model for such a modelling problem. However, it is also shown in the rest of the present paper that other models such as distributed time delay models, or a Volterra equation of the first kind, also have the ability to produce power law behaviour without the drawbacks associated to fractional models.

Author Contributions: J.S. contributed all the paragraphs of the article. He contributed to the creation of the test bench giving the measurements of the growth of the sand heap (Section 3) and to the exploitation of these measures (Section 5). He also contributed to paragraph 2 that concerns the characterization of a power law type long memory phenomenon. The drawbacks listed in paragraph 4 result from a decade of reflection on non-whole models. He recently helped to find other models allowing modeling (Section 6). C.F. contributed to the characterization of a power law type long memory phenomenon of Section 2 and to the modelling approach of Section 6.1 for RSA phenomenon. He contributed to the reflections that permitted the writing of the Section 4. V.T. contributed to the modelling approach of Section 6.1 for RSA phenomenon. All authors have read and agreed to the published version of the manuscript.

Funding: This research received no external funding.

Conflicts of Interest: The authors declare no conflict of interest.

References

1. Miller, K.S.; Ross, B. *An Introduction to the Fractional Calculus and Fractional Differential Equations*; John Wiley & Sons: New York, NY, USA, 1993.
2. Oldham, K.B.; Spanier, J. *The Fractional Calculus; Theory and Applications of Differentiation and Integration to Arbitrary Order (Mathematics in Science and Engineering, V)*; Academic Press: Cambridge, MA, USA, 1974.
3. Podlubny, I. *Fractional Differential Equations. Mathematics in Sciences and Engineering*; Academic Press: Cambridge, MA, USA, 1999.
4. Samko, S.G.; Kilbas, A.A.; Marichev, O.I. *Fractional Integrals and Derivatives: Theory and Applications*; Gordon and Breach Science Publishers: London, UK, 1993.
5. Bagley, R.L.; Calico, R.A. Fractional order state equations for the control of viscoelastically damped structures. *J. Guid. Control Dyn.* **1991**, *14*, 304–311. [CrossRef]
6. Oppenheim, A.V.; Alan SWillsky, A.S.; Hamid, S. *Signals and Systems, Pearson New International Edition*; Pearson Education Limited: Harlow, UK, 1996.
7. Mandal, S.; Khakhar, D. Granular surface flow on an asymmetric conical heap. *J. Fluid Mech.* **2019**, *865*, 41–59. [CrossRef]
8. Pacheco-Vazquez, F.; Moreau, F.; Vandewalle, N.; Dorbolo, S. Sculpting sandcastles grain by grain: Self-assembled sand towers. *Phys. Rev. E* **2012**, *86*, 051303. [CrossRef] [PubMed]
9. Oustaloup, A. *Diversity and Non-Integer Differentiation for System Dynamics*; Wiley: Hoboken, NJ, USA, 2014.
10. Ben Adda, F. Geometric interpretation of the fractional derivative. *J. Fract. Calc.* **1997**, *11*, 21–52.
11. Podlubny, I. Geometric and physical interpretation of fractional integration and fractional differentiation. *J. Fract. Calc. Appl. Anal.* **2002**, *5*, 357–366.
12. Gorenflo, R. Afterthoughts on interpretation of fractional derivatives and integrals. In *Transform Methods and Special Functions, Varna'96, Institute of Mathematics and Informatics*; Rusev, P., Dimovski, I., Kiryakova, V., Eds.; Bulgarian Academy of Sciences: Bulgarian, Sofia, 1998.
13. Mainardi, F. Considerations on fractional calculus: Interpretations and applications. In *Transform Methods and Special Functions, Varna'96, Institute of Mathematics and Informatics*; Rusev, P., Dimovski, I., Kiryakova, V., Eds.; Bulgarian Academy of Sciences: Bulgarian, Sofia, 1998.
14. Nigmatullin, R.R. A fractional integral and its physical interpretation. *Theoret. and Math. Physics* **1992**, *90*, 242–251. [CrossRef]
15. Rutman, R.S. On physical interpretations of fractional integration and differentiation. *Theor. Math. Phys.* **1995**, *105*, 393–404. [CrossRef]
16. Sabatier, J.; Merveillaut, M.; Malti, R.; Oustaloup, A. On a Representation of Fractional Order Systems: Interests for the Initial Condition Problem. In Proceedings of the 3rd IFAC Workshop on "Fractional Differentiation and its Applications" (FDA'08), Ankara, Turkey, 5–7 November 2008.
17. Tenreiro Machado, J.A. A probabilistic Interpretation of the Fractional-Order differentiation. *J. Fract. Calc. Appl. Anal.* **2003**, *6*, 73–80.
18. Dokoumetzidis, A.; Magin, R.; Macheras, P. A commentary on fractionalization of multi-compartmental models. *Pharm. Pharm.* **2010**, *37*, 203–207. [CrossRef]
19. Sabatier, J.; Farges, C.; Merveillaut, M.; Feneteau, L. On observability and pseudo state estimation of fractional order systems. *Eur. J. Control* **2012**, *18*, 260–271. [CrossRef]
20. Matignon, D. Stability properties for generalized fractional differential systems. *ESAIM Proc.* **1998**, *5*, 145–158. [CrossRef]
21. Montseny, G. Diffusive representation of pseudo-differential time-operators. *ESAIM Proc.* **1998**, *5*, 159–175. [CrossRef]
22. Sabatier, J.; Merveillaut, M.; Malti, R.; Oustaloup, A. How to Impose Physically Coherent Initial Conditions to a Fractional System? *Commun. Nonlinear Sci. Numer. Simul.* **2010**, *15*, 1318–1326. [CrossRef]
23. Sabatier, J.; Nguyen, H.C.; Farges, C.; Deletage, J.Y.; Moreau, X.; Guillemard, F.; Bavoux, B. Fractional models for thermal modeling and temperature estimation of a transistor junction. *Adv. Differ. Equ.* **2011**. [CrossRef]
24. Sabatier, J.; Farges, C.; Trigeassou, J.C. Fractional systems state space description: Some wrong ideas and proposed solutions. *J. Vib. Control* **2014**, *20*, 1076–1084. [CrossRef]

25. De Oliveira, E.C.; Tenreiro Machado, J.A. A Review of Definitions for Fractional Derivatives and Integral. *Math. Probl. Eng.* **2019**, *2014*, 238459. [CrossRef]
26. Sabatier, J.; Farges, C. Comments on the description and initialization of fractional partial differential equations using Riemann-Liouville's and Caputo's definitions. *J. Comput. Appl. Math.* **2018**, *339*, 30–39. [CrossRef]
27. Ortigueira, M.D.; Coito, F.J. Initial Conditions: What Are We Talking about? In Proceedings of the Third IFAC Workshop on Fractional Differentiation, Ankara, Turkey, 5–7 November 2008.
28. Sabatier, J.; Rodriguez Cadavid, S.; Farges, C. Advantages of limited frequency band fractional integration operator in fractional models definition. In Proceedings of the Conference on Control, Decision and Information Technologies (CoDIT 2019), Paris, France, 23–26 April 2019.
29. Sabatier, J. Non-Singular Kernels for Modelling Power Law Type Long Memory Behaviours and Beyond. *Cybernetics* **2020**. accepted.
30. Caputo, M.; Fabrizio, M. A new definition of fractional derivative without singular kernel. *Prog. Fract. Differ. Appl.* **2015**, *1*, 73–85.
31. Tartaglione, V.; Farges, C.; Sabatier, J. Dynamical Modelling of Random Sequential Adsorption. In Proceedings of the European Control Conference ECC 2020, St-Petersburg, Russia, 12–15 May 2020.
32. Feder, J.; Giaever, I. Adsorption of ferritin. *J. Colloid Interface Sci.* **1980**, *78*, 144–154. [CrossRef]
33. Viot, P.; Tarjus, G.; Ricci, S.; Talbot, J. Random sequential adsorption of anisotropic particles. I. jamming limit and asymptotic behavior. *J. Chem. Phys.* **1992**, *97*, 5212–5218. [CrossRef]
34. Sabatier, J. Distributed time delay systems for power law long memory behaviors modelling. In Proceedings of the 58th Conference on Decision and Control (CDC 2019), Nice, France, 11–13 December 2019.
35. Sabatier, J. Power Law Type Long Memory Behaviors Modeled with Distributed Time Delay Systems. *Fract. Fractals* **2020**, *4*, 1. [CrossRef]

© 2020 by the authors. Licensee MDPI, Basel, Switzerland. This article is an open access article distributed under the terms and conditions of the Creative Commons Attribution (CC BY) license (http://creativecommons.org/licenses/by/4.0/).

Article

Fractional Integral Equations Tell Us How to Impose Initial Values in Fractional Differential Equations

Daniel Cao Labora

Department of Applied Mathematics I, School of Forest Engineering, Universidade de Vigo, Campus Universitario da Xunqueira, S/N, 36005 Pontevedra, Spain; daniel.cao.labora@uvigo.es

Received: 31 May 2020; Accepted: 2 July 2020; Published: 4 July 2020

Abstract: One major question in Fractional Calculus is to better understand the role of the initial values in fractional differential equations. In this sense, there is no consensus about what is the reasonable fractional abstraction of the idea of "initial value problem". This work provides an answer to this question. The techniques that are used involve known results concerning Volterra integral equations, and the spaces of summable fractional differentiability introduced by Samko et al. In a few words, we study the natural consequences in fractional differential equations of the already existing results involving existence and uniqueness for their integral analogues, in terms of the Riemann–Liouville fractional integral. In particular, we show that a fractional differential equation of a certain order with Riemann–Liouville derivatives demands, in principle, less initial values than the ceiling of the order to have a uniquely determined solution, in contrast to a widely extended opinion. We compute explicitly the amount of necessary initial values and the orders of differentiability where these conditions need to be imposed.

Keywords: fractional differential equations; initial values; existence; uniqueness

1. Introduction

One of the most typical trademarks involving Fractional Calculus is the wide range of opinions about the notions of what is a natural fractional version of some integer order concept and what is not. On the one hand, this plurality leads interesting debates and fosters a critical thinking about whether research is going "in the right direction" or not. On the other hand, it is difficult to handle such an amount of different notions and ideas in the extant literature. In particular, it is common to find lots of generalized fractional versions of a single integer order concept, some of them not very accurate or incoherent between them. These debates are still very alive nowadays [1].

In this frame, the task of this paper is to point out some relevant facts concerning the imposition of initial values for Riemann–Liouville fractional differential equations (FDE). Although the existence and study of FDE has been widely described, for instance in [2,3], in this paper, we provide strong reasons to reconsider the way of imposing initial values.

We have to highlight that our research has been conducted for the Riemann–Liouville fractional derivative, which is the most classical extension for the usual derivative. In addition, the results have been developed for the particular case of linear equations with constant coefficients. However, it seems natural that the ideas described here could be extended to much more general cases.

The main reason to study this issue for the Riemann–Liouville fractional derivative, and not for other fractional versions, is that it is the left inverse of the Riemann–Liouville fractional integral. In this sense, if we restrict the study of Fractional Calculus to functions defined on finite length intervals $[a, b]$, it is a big consensus that Riemann–Liouville fractional integral with base point a is the unique reasonable extension for the integral operator $\int_a^t$. The previous asseveration is not a simple opinion, since the Riemann–Liouville fractional integral can be characterized axiomatically in very reasonable terms.

Theorem 1 (Cartwright-McMullen, [4]). *Given a fixed $a \in \mathbb{R}$, there is only one family of operators $\left(I_{a+}^{\alpha}\right)_{\alpha>0}$ on $L^1[a,b]$ satisfying the following conditions:*

1. *The operator of order 1 is the usual integral with base point a. (Interpolation property)*
2. *The Index Law holds. That is, $I_{a+}^{\alpha} \circ I_{a+}^{\beta} = I_{a+}^{\alpha+\beta}$ for all $\alpha, \beta > 0$. (Index Law)*
3. *The family is continuous with respect to the parameter. That is, the following map $\mathrm{Ind}_a : \mathbb{R}^+ \longrightarrow \mathrm{End}_B\left(L^1[a,b]\right)$ given by $\mathrm{Ind}_a(\alpha) = I_{a+}^{\alpha}$ is continuous, where $\mathrm{End}_B\left(L^1[a,b]\right)$ denotes the Banach space of bounded linear endomorphisms on $L^1[0,b]$. (Continuity)*

This family is precisely given by the Riemann–Liouville fractional integrals, whose expression will be recalled during this paper.

Hence, it makes sense to study in detail fractional integral problems for the Riemann–Liouville fractional integral to derive consequences for the corresponding fractional equations afterwards. Finally, to draw the attention of curious readers, we mention again that one of the most interesting results that we have found out is that a FDE of order $\alpha > 0$ with Riemann–Liouville derivatives can demand, in principle, less initial values than $\lceil \alpha \rceil$ to have a uniquely determined solution. This result differs from a widely held opinion (see Theorem 1, Section 5.5, in [5]) which states that the necessary amount of initial values is $\lceil \alpha \rceil$. The reason for this discrepancy is that the question involving the "fractional smoothness" of the solutions is often neglected, since many results are derived after a not totally rigorous usage of certain mathematical concepts or results. In other words, it is important to build first the space where solutions lie in, to later seek solutions in that space.

A complete range of highlighted results with their implications can be consulted in Section 5, while the previous sections are devoted to the corresponding deductions.

Goal of the Work

The goal of this work is to study how should we impose initial values in fractional problems with a Riemann–Liouville derivative to ensure that they have a smooth and unique solution, where smooth simply means that the solution lies in a certain suitable space of fractional differentiability. To achieve this, we will depart from some known results involving the Riemann–Liouville fractional integral, since it arises as the natural generalization of the usual integral operator; recall Theorem 1.

First, we will recall some results that imply that fractional integral problems have always a unique solution. We also recall the fundamental notions concerning Fractional Calculus, and we pay special attention to the functional spaces where calculations are performed, and especially where fractional derivatives are well defined. Note that this point of "where are functions defined" is crucial to talk about existence or uniqueness of solution and is often neglected in the extant literature concerning Fractional Calculus. Indeed, to avoid this problem, much research has been conducted for Caputo derivatives instead of Riemann–Liouville, see, for instance, [6,7] or general comments in [8]. The ideas of this paragraph are developed in the second section, and most of them are available in the extant literature, except (to the best of the author's knowledge) Lemma 2, which plays a key role in the rest of the paper.

Second, we see how each FDE of order β is linked with a family of fractional integral problems, whose source term lives in a $\lceil \beta \rceil$ dimensional affine subspace of $L^1[0,b]$. This means that each solution to the FDE is a solution to one (and only one!) fractional integral problem of the $\lceil \beta \rceil$ dimensional family. Conversely, any solution to a fractional integral problem of the family solves the FDE, provided that the solution is smooth enough. In general, the set of source terms of the family of fractional integral problems that provide a smooth solution will consist in an affine subspace of $L^1[0,b]$ of a dimension lower than $\lceil \beta \rceil$. This is done in the third section.

Third, we characterize when a source term of the $\lceil \beta \rceil$ dimensional family induces a smooth solution, and thus a solution to the associated FDE. In particular, the affine space of such source terms is shown to have dimension $\lceil \beta - \beta_* \rceil$, where β_* is the highest order of differentiability in the FDE such that $\beta - \beta_* \notin \mathbb{N}$. This characterization induces a natural correspondence between each source term inducing a smooth solution for the integral problem and a vector of $\lceil \beta - \beta_* \rceil$ initial values fulfilled by the solution. More specifically, if we denote the solution to the FDE by $y(t)$, the initial values ensuring existence and uniqueness of solution are $D_{0+}^{\beta-k} y(0)$ for $k \in \{1, 2, \ldots, \lceil \beta - \beta_* \rceil\}$. The content of this paragraph is discussed in the fourth section.

Finally, we establish a section of conclusions to highlight the most relevant obtained results, and to point out to some relevant work that should be performed in the future to continue with this approach.

2. Fundamental Notions

In this section, we will introduce the fundamental notions of Fractional Calculus that we are going to use, together with their more relevant properties and some results of convolution theory that will be useful for our purposes. We assume that the reader is familiar with the basic theory of Banach spaces, Special Functions, and Integration Theory, especially the main facts involving the space of integrable functions over a finite length interval, denoted by $L^1[a, b]$, and the main properties of the Γ function.

2.1. The Riemann–Liouville Fractional Integral

We briefly introduce the Riemann–Liouville fractional integral, together with its most relevant properties. We make this introduction from the perspective of convolutions, since it will be relevant to notice that the Riemann–Liouville fractional integral is no more than a particular convolution operator, to apply later some adequate results of convolution theory.

Definition 1. *Given $f \in L^1[a, b]$, we defined its associated convolution operator as $C_a(f) : L^1[a, b] \longrightarrow L^1[a, b]$ defined as*

$$(f *_a g)(t) := (C_a(f) g)(t) := \int_a^t f(t - s + a) \cdot g(s) \, ds$$

for $g \in L^1[a, b]$ and $t \in [a, b]$. Under the previous notation, we say that f is the kernel of the convolution operator $C_a(f)$.

Definition 2. *We define the left Riemann–Liouville fractional integral of order $\alpha > 0$ of a function $f \in L^1[a, b]$ with base point a as*

$$I_{a+}^\alpha g(t) = \int_a^t \frac{(t - s)^{\alpha - 1}}{\Gamma(\alpha)} \cdot g(s) \, ds,$$

for almost every $t \in [a, b]$. In case that $\alpha = 0$, we just define

$$I_{a+}^0 g(t) = \operatorname{Id} g(t) = g(t).$$

Without loss of generality, we will assume that $a = 0$, since the results for a generic value of a can be achieved after developing them for $a = 0$ and applying a suitable translation. Moreover, when using the expression "Riemann–Liouville fractional integral", we will be referring to the left Riemann–Liouville fractional integral with base point $a = 0$.

Remark 1. *We observe that, for $\alpha > 0$, the Riemann–Liouville fractional integral operator I_{0+}^α can be written as a convolution operator $C_0(f)$, with kernel*

$$f(t) = \frac{t^{\alpha - 1}}{\Gamma(\alpha)}.$$

It is well known that the Riemann–Liouville fractional integral fulfills the following properties, see [9].

Proposition 1. *For every $\alpha, \beta \geq 0$:*

- I_{0+}^α is well defined, meaning that $I_{0+}^\alpha L^1[0,b] \subset L^1[0,b]$.
- I_{0+}^α is a continuous operator (equivalently, a bounded operator) from the Banach space $L^1[0,b]$ to itself.
- I_{0+}^α is an injective operator.
- I_{0+}^α preserves continuity, meaning that $I_{0+}^\alpha \mathcal{C}[0,b] \subset \mathcal{C}[0,b]$.
- We have the Index Law $I_{0+}^\beta \circ I_{0+}^\alpha = I_{0+}^{\alpha+\beta}$ for $\alpha, \beta \geq 0$. In particular, $I_{0+}^{\alpha+\beta} L^1[0,b] \subset I_{0+}^\beta L^1[0,b]$.
- Given $f \in L^1[0,b]$ and $\alpha \geq 1$, we have that $I_{0+}^\alpha f$ is absolutely continuous and, moreover, $I_{0+}^\alpha f(0) = 0$.

Furthermore, we will also use several times the following well known and straightforward remark, see [9].

Remark 2. *We have that, for $\beta > -1$ and $\alpha \geq 0$,*

$$I_{0+}^\alpha t^\beta = \frac{\Gamma(\beta+1)}{\Gamma(\alpha+\beta+1)} t^{\alpha+\beta}.$$

Indeed, $I_{0+}^\alpha t^\beta \in I_{0+}^\gamma L^1[0,b]$ if and only if $\alpha + \beta > \gamma - 1$.

2.2. The Riemann–Liouville Fractional Derivative

In this subsection, we will indicate the most relevant points when constructing the Riemann–Liouville fractional derivative. We will begin with a short introduction to absolutely continuous functions of order n, since the spaces where Riemann–Liouville differentiability is well defined can be understood as their natural generalization for the fractional case, see [9].

2.2.1. A Short Reminder Involving Absolutely Continuous Functions and the Fundamental Theorem of Calculus

It is widely known that absolutely continuous functions play a key role in several theories of Mathematical Analysis. These functions can be characterized via a "ε, δ" definition, but we only recall that absolutely continuous functions are, essentially, antiderivatives of functions in $L^1[0,b]$ up to addition with a constant. We will see later how this idea is highly relevant to construct the maximal spaces where Riemann–Liouville fractional derivatives are well defined.

Theorem 2 (Fundamental Theorem of Calculus). *Consider a real function f defined on an interval $[0,b] \subset \mathbb{R}$. Then, $f \in AC[0,b]$ if and only if there exists $\varphi \in L^1[0,b]$ such that*

$$f(t) = f(0) + \int_0^t \varphi(s)\, ds. \tag{1}$$

This last result allows us to define the derivative of an absolutely continuous function on $[0,b]$ as a certain function in $L^1[0,b]$. If $f \in AC[0,b]$, we define its derivative $D^1 f$ as the unique function $\varphi \in L^1[0,b]$ that makes (1) hold.

Remark 3. *It is relevant to have in mind that the previous definition makes sense because the antiderivative operator I_{0+}^1 is injective, recall Proposition 1. In particular,*

$$AC[0,b] = \langle\{1\}\rangle \oplus I_{0+}^1 L^1[0,b],$$

where "1" denotes the constant function with value 1.

Of course, it is possible to talk about absolutely functions of order n, for $n \in \mathbb{Z}^+$. In this case, for any $n \in \mathbb{Z}^+$, we say that $f \in AC^n[0,b]$ if and only if $f \in \mathcal{C}^{n-1}[0,b]$ and $D^{n-1}f \in AC[0,b]$.

Thus, $AC^n[0,b]$ consists of functions that can be differentiated n times, but the last derivative might be computable only in the weak sense of Fundamental Theorem of Calculus. Analogously to the previous remark, we have the following result, see page 3 of [9].

Remark 4. *We have that*
$$AC^n[0,b] = \left\langle \{1, t, \ldots, t^{n-1}\} \right\rangle \oplus I_{0+}^n L^1[0,b],$$
after applying n times the Fundamental Theorem of Calculus 2. Moreover, the sum is direct since the property $f \in I_{0+}^n L^1[0,b]$ implies that $f(0) = f'(0) = \cdots = f^{n-1}(0) = 0$, and the only polynomial of degree at most $n-1$ satisfying those conditions is the zero one.

The relevant observation is that the vector space of functions that can be differentiated n times, in the sense of Fundamental Theorem of Calculus 2, has two distinct parts that only share the zero function.

- The left part $\langle \{1, t, \ldots, t^{n-1}\} \rangle$ consists of polynomials of degree strictly lower than n. These functions describe, indeed, the kernel of the operator D^n and, thus,
$$\ker D^n = \left\langle \{1, t, \ldots, t^{n-1}\} \right\rangle.$$

- The right part $I_{0+}^n L^1[0,b]$ consists of functions that are obtained after integrating n times an element of $L^1[0,b]$, and hence it contains functions of trivial initial values until the derivative of order $n-1$.

At this point, we recall the following result, which is widely known and can be proved immediately with the previous definition.

Proposition 2. *If $n > m > 0$, we have that*
$$AC^n[0,b] \subset AC^m[0,b].$$

Although Proposition 2 seems irrelevant, it hides the key for a successful treatment of the fractional case. In the next part of the paper, we will reproduce a natural construction for the fractional analogue of the spaces $AC^n[0,b]$.

We will arrive to the same definition that was already presented in [9]. However, we will emphasize that the space of fractional differentiability of order α will never be contained in the space of order β, except if $\alpha - \beta \in \mathbb{N}$. This fact will cause FDEs to have less solutions than expected.

2.2.2. The Fractional Abstraction of the Spaces $AC^n[0,b]$

It is reasonable to define the Riemann–Liouville fractional derivative in such a way that it is the left inverse operator for the Riemann–Liouville fractional integral of the same order. After doing this, an easy analytical expression for its computation follows. Moreover, this explicit description can be extended to a bigger space, and it coincides with the definition available in the classical literature [9].

Property 1. *Consider $\alpha \geq 0$. The Riemann–Liouville fractional derivative of order α (and base point 0) fulfils that it is the left inverse of the Riemann–Liouville fractional integral of order α, meaning*
$$D_{0+}^\alpha I_{0+}^\alpha f = f,$$
for every $f \in L^1[0,b]$.

We should note that, due to the injectivity of the fractional integral, Property 1 defines the Riemann–Liouville fractional derivative on the space $I_{0+}^\alpha L^1[0,b]$. Moreover, it will be a surjective operator from $I_{0+}^\alpha L^1[0,b]$ to $L^1[0,b]$.

However, it is clear that we are ignoring something if we pretend that D_{0+}^α matches perfectly the usual derivative when α is an integer. In particular, we observe that, for an integer value of α, Property 1 only describes the behavior of D^α over the space $I_{0+}^\alpha L^1[0,b]$. Nevertheless, we are missing its definition over the supplementary part of $I_{0+}^\alpha L^1[0,b]$ in $AC^\alpha[0,b]$, which is ker D^α.

This problem is easily solved, since it is possible to describe Property 1 more explicitly. We just observe that the left inverse for I_{0+}^α is given by the expression $D_{0+}^\alpha = D_{0+}^{\lceil\alpha\rceil} I_{0+}^{\lceil\alpha\rceil-\alpha}$, due to the Fundamental Theorem of Calculus 2 and the Index Law in Proposition 1. Thus, one could define D_{0+}^α in a more general space than $I_{0+}^\alpha L^1[0,b]$, since the only necessary condition to define $D_{0+}^{\lceil\alpha\rceil} I_{0+}^{\lceil\alpha\rceil-\alpha} f$ is to ensure that $I_{0+}^{\lceil\alpha\rceil-\alpha} f \in AC^{\lceil\alpha\rceil}[0,b]$. Hence, the following definition is natural.

Definition 3. *For each $\alpha > 0$, we construct the following space*

$$\mathcal{X}_\alpha = \left(I_{0+}^{\lceil\alpha\rceil-\alpha}\right)^{-1}\left(AC^{\lceil\alpha\rceil}[0,b]\right),$$

which will be called the space of functions with summable fractional derivative of order α. If $\alpha = 0$, we define $\mathcal{X}_\alpha = L^1[0,b]$.

Remark 5. *Therefore, functions of $\mathcal{X}_\alpha$ are defined as the ones producing a function in $AC^{\lceil\alpha\rceil}[0,b]$ after being integrated $\lceil\alpha\rceil - \alpha$ times. This new function can be differentiated $\lceil\alpha\rceil$ times in the weak sense of Fundamental Theorem of Calculus 2.*

It is relevant to point out that the previous definition, although sometimes related, is different from other notions of "fractional spaces" available in the literature like, for instance, the Fractional Sobolev Spaces in Gagliardo's sense [10]. In our case, Definition 3 coincides with the one already presented in [9], and we can make it totally explicit.

Lemma 1. *For any $\alpha > 0$, we have that*

$$\mathcal{X}_\alpha = \left\langle \left\{ t^{\alpha-\lceil\alpha\rceil}, \ldots, t^{\alpha-2}, t^{\alpha-1} \right\} \right\rangle \oplus I_{0+}^\alpha L^1[0,b].$$

Proof. First, we check $\left\langle \left\{ t^{\alpha-\lceil\alpha\rceil}, \ldots, t^{\alpha-2}, t^{\alpha-1} \right\} \right\rangle \cap I_{0+}^\alpha L^1[0,b] = \{0\}$. If there is a function f in both summands, then $I_{0+}^{\lceil\alpha\rceil-\alpha} f$ will be simultaneously a polynomial of degree at most $\lceil\alpha\rceil - 1$, and a function in $I_{0+}^{\lceil\alpha\rceil} L^1[0,b]$. Therefore, $I_{0+}^{\lceil\alpha\rceil-\alpha} f$ has to be the zero function after repeating the argument in Remark 4 and, since fractional integrals are injective (Proposition 1), $f \equiv 0$.

It is clear that, applying $I_{0+}^{\lceil\alpha\rceil-\alpha}$ to the right-hand side, we will produce a function $AC^{\lceil\alpha\rceil}[0,b]$. Moreover, it is trivial that any function in $AC^{\lceil\alpha\rceil}[0,b]$ can be obtained in this way by virtue of Remark 2. Since the operator $I_{0+}^{\lceil\alpha\rceil-\alpha}$ is injective, the result follows. □

From the previous lemma, we get this immediate corollary.

Corollary 1. *Given $f \in L^1[0,b]$, we have that $f \in I_{0+}^\alpha L^1[0,b]$ if, and only if, $f \in \mathcal{X}_\alpha$ and also $D^s I_{0+}^{\lceil\alpha\rceil-\alpha} f(0) = 0$, for each $s \in \{0, \ldots, \lceil\alpha\rceil - 1\}$.*

Hence, we can use Property 1 and Corollary 1 to define the Riemann–Liouville fractional derivative, coinciding with Definition 2.4 in [9].

Definition 4. Consider $\alpha \geq 0$ and $f \in \mathcal{X}_\alpha$. We define the Riemann–Liouville fractional derivative of order α (and base point 0) as

$$D_{0+}^\alpha f := D_{0+}^{\lceil \alpha \rceil} \circ I_{0+}^{\lceil \alpha \rceil - \alpha} f,$$

where the last derivative might be only computed in the weak sense of the Fundamental Theorem of Calculus.

2.2.3. Properties of the Space $\mathcal{X}_\alpha$

We want to fully understand how D_{0+}^α works over $\mathcal{X}_\alpha$, and the most natural way is to split the problem into two parts, as suggested by Lemma 1. We already know that D_{0+}^α is the left inverse for I_{0+}^α, so we should study how D_{0+}^α behaves when applied to $\left\langle \left\{ t^{\alpha - \lceil \alpha \rceil}, \ldots, t^{\alpha - 2}, t^{\alpha - 1} \right\} \right\rangle$. It is a well known and straightforward computation that

$$D_{0+}^\alpha \left(\left\langle \left\{ t^{\alpha - \lceil \alpha \rceil}, \ldots, t^{\alpha - 2}, t^{\alpha - 1} \right\} \right\rangle \right) = \{0\}.$$

Hence, the kernel of D_{0+}^α has dimension $\lceil \alpha \rceil$ and is given by

$$\ker D_{0+}^\alpha = \left\langle \left\{ t^{\alpha - \lceil \alpha \rceil}, \ldots, t^{\alpha - 2}, t^{\alpha - 1} \right\} \right\rangle.$$

Moreover, we should note that, if $f(t) = a_0 t^{\alpha - \lceil \alpha \rceil} + \cdots + a_{\lceil \alpha \rceil - 1} t^{\alpha - 1}$ with $a_j \in \mathbb{R}$ for each $j \in \{0, 1, \ldots, \lceil \alpha \rceil - 1\}$, it is immediately necessary to do the following calculations from Remark 2, where $j \in \{1, \cdots, \lceil \alpha \rceil - 1\}$,

$$\begin{aligned}
\left(I_{0+}^{\lceil \alpha \rceil - \alpha} f \right)(0) &= a_0 \Gamma(\alpha - \lceil \alpha \rceil + 1), \\
\left(D_{0+}^{\alpha - \lceil \alpha \rceil + j} f \right)(0) &= a_j \Gamma(\alpha - \lceil \alpha \rceil + j + 1).
\end{aligned} \quad (2)$$

The previous formula generalizes the obtention of the Taylor coefficients for a fractional case and it can be used to codify functions in $\mathcal{X}_\alpha$ modulo $I_{0+}^\alpha L^1[0, b]$, since

$$\begin{aligned}
\left(I_{0+}^{\lceil \alpha \rceil - \alpha} g \right)(0) &= 0, \\
\left(D_{0+}^{\alpha - \lceil \alpha \rceil + j} g \right)(0) &= 0,
\end{aligned} \quad (3)$$

for $g \in I_{0+}^\alpha L^1[0, b]$, due to Proposition 1.

2.2.4. Intersection of Fractional Summable Spaces

In general, fractional differentiation presents some extra problems that do not exist when dealing with fractional integrals. One of the most famous ones is that there is no Index Law for fractional differentiation. One underlying reason for all these complications is the following one.

Remark 6. The condition $\alpha > \beta$ does not ensure $\mathcal{X}_\alpha \subset \mathcal{X}_\beta$, although the condition $\alpha - \beta \in \mathbb{N}$ trivially does. This makes Riemann–Liouville fractional derivatives somehow tricky, since the differentiability for a higher order does not imply, necessarily, the differentiability for a lower order with a different decimal part. In particular, this fact has critical implications when considering fractional differential equations, as we shall see in the paper, since the unknown function has to be differentiable for each order involved in the equation. These problems give an idea of why it can be a reasonable approach to work with fractional integrals instead, and try to inherit the obtained results for the case of fractional derivatives, instead of proving them for fractional derivatives directly.

Consequently, it is interesting to compute the exact structure of a finite intersection of such spaces of fractional differentiability of different orders. To the best of our knowledge, this result is not available in the extant literature.

Lemma 2. *Considering $\beta_n > \cdots > \beta_1 \geq 0$, we have that*

$$\bigcap_{j=1}^{n} \mathcal{X}_{\beta_j} = \left\langle \left\{ t^{\beta_n - \lceil \beta_n - \beta_* \rceil}, \ldots, t^{\beta_n - 1} \right\} \right\rangle \oplus I_{0^+}^{\beta_n} L^1[0,b],$$

where β_* is the maximum β_j such that $\beta_n - \beta_j \notin \mathbb{N}$. If such a β_j does not exist, the result still holds after defining $\beta_* = 0$.

In particular, $I_{0^+}^{\beta_n} L^1[0,b] \subset \bigcap_{j=1}^{n} \mathcal{X}_{\beta_j}$ and it has codimension $\lceil \beta_n - \beta_* \rceil$.

Proof. It is obvious that $\bigcap_{j=1}^{n} \mathcal{X}_{\beta_j} \subset \mathcal{X}_{\beta_n}$. Hence,

$$\bigcap_{j=1}^{n} \mathcal{X}_{\beta_j} \subset \mathcal{X}_{\beta_n} = \left\langle \left\{ t^{\beta_n - \lceil \beta_n \rceil}, \ldots, t^{\beta_n - 1} \right\} \right\rangle \oplus I_{0^+}^{\beta_n} L^1[0,b]. \tag{4}$$

It is clear that $I_{0^+}^{\beta_n} L^1[0,b]$ lies in $\bigcap_{j=1}^{n} \mathcal{X}_{\beta_j}$, for any $j \in \{1,\ldots,n\}$. This is simply because, due to the Index Law, $I_{0^+}^{\beta_n} L^1[0,b] \subset I_{0^+}^{\beta_j} L^1[0,b]$, since $\beta_n \geq \beta_j$.

Thus, the remaining question is to see when a linear combination of the $t^{\beta_n - k}$, where $k \in \{1,\ldots,\lceil \beta_n \rceil\}$, lies in $\bigcap_{j=1}^{n} \mathcal{X}_{\beta_j}$. The key remark is to realize that, for any finite set $\mathcal{F} \subset (-1,+\infty)$,

$$\sum_{\gamma \in \mathcal{F}} c_\gamma t^\gamma \in \mathcal{X}_{\beta_j}, \text{ where } c_\gamma \neq 0 \text{ for each } \gamma \in \mathcal{F},$$

if and only if $\gamma - \beta_j > -1$ or $\gamma - \beta_j \in \mathbb{Z}^-$, for every $\gamma \in \{1,\ldots,r\}$ with $c_\gamma \neq 0$. Consequently, it is enough study when $t^{\beta_n - k}$ lies in $\mathcal{X}_{\beta_j}$, and there are two options:

- If $\beta_n - \beta_j \in \mathbb{N}$, we know that $t^{\beta_n - k} \in \mathcal{X}_{\beta_j}$ always. This happens because either $\beta_n - k \in \{\beta_j - \lceil \beta_j \rceil, \ldots, \beta_j - 1\}$ or $\beta_n - k > \beta_j - 1$.
- In other case, β_n and β_j do not share decimal parts and we need to have $\beta_n - k > \beta_j - 1$ that can be rewritten as $k < \beta_n - \beta_j + 1$. If we want this to happen for every j such that $\beta_n - \beta_j \notin \mathbb{N}$, the condition is equivalent to $k < \beta_n - \beta_* + 1$, where β_* is the greatest β_j such that $\beta_n - \beta_j \notin \mathbb{N}$. Indeed, it can be rewritten as $1 \leq k \leq \lceil \beta_n - \beta_* \rceil$.

Therefore, the coefficients which are not necessarily null are the ones associated with $t^{\beta_n - k}$, where $k \in \{1,\ldots,\lceil \beta_n - \beta_* \rceil\}$. □

Remark 7. *Due to Lemma 2, any affine subspace of $\bigcap_{j=1}^{n} \mathcal{X}_{\beta_j}$ with dimension strictly higher than $\lceil \beta_n - \beta_* \rceil$ contains two distinct functions whose difference lies in $I_{0^+}^{\beta_n} L^1[0,b]$. Thus, in any vector subspace of $\bigcap_{j=1}^{n} \mathcal{X}_{\beta_j}$ with dimension strictly higher than $\lceil \beta_n - \beta_* \rceil$, there are infinitely many functions that lie in $I_{0^+}^{\beta_n} L^1[0,b]$.*

2.3. Fractional Integral Equations

Consider the fractional integral equation

$$\left(c_n I_{0^+}^{\gamma_n} + \cdots + c_1 I_{0^+}^{\gamma_1} + I_{0^+}^{\gamma_0} \right) x(t) = \tilde{f}(t),$$

where $f \in L^1[0,b]$, $\gamma_n > \cdots > \gamma_0 \geq 0$ and assume that it has a solution $x \in L^1[0,b]$. Since $I_{0^+}^{\gamma_n} L^1[0,b] \subset \cdots \subset I_{0^+}^{\gamma_0} L^1[0,b]$, and the left-hand side lies in $I_{0^+}^{\gamma_0} L^1[0,b]$, the condition $\tilde{f} \in I_{0^+}^{\gamma_0} L^1[0,b]$ is mandatory to ensure the existence of solution. In that case, we can apply the operator $D_{0^+}^{\gamma_0}$ to the previous equation and we obtain

$$\left(c_n I_{0^+}^{\alpha_n} + \cdots + c_1 I_{0^+}^{\alpha_1} + \text{Id} \right) x(t) = f(t), \tag{5}$$

where $D_{0^+}^{\gamma_0}\tilde{f} = f$ and $\alpha_j = \gamma_j - \gamma_0$ for $j \in \{1,\ldots,n\}$. If we use the notation

$$Y = c_n I_{0^+}^{\alpha_n} + \cdots + c_1 I_{0^+}^{\alpha_1},$$

Equation (5) can be rewritten as

$$(Y + \text{Id})\, x(t) = f(t). \tag{6}$$

Therefore, it is relevant to study the properties of the operator $Y + \text{Id}$ from $L^1[0,b]$ to itself, in order to understand Equation (6).

2.3.1. $Y + \text{Id}$ Is Bounded

This claim is a very well-known result, since each summand in Y is a bounded operator, and Id too, see [2]. It is also possible to prove this, just recalling that Y is a convolution operator with kernel in $L^1[0,b]$ and, thus, a bounded operator.

2.3.2. $Y + \text{Id}$ Is Injective

To prove that $Y + \text{Id}$ is injective, we will need a result concerning the annulation of a convolution. In a few words, we need to know what are the possibilities for the factors of a convolution, provided that the obtained result is the zero function. Roughly speaking, the classical result in this direction, known as the Titchmarsh Theorem, states that the integrand of the convolution from 0 to t is always zero, independently of t.

Theorem 3 (Titchmarsh, [11]). *Suppose that $f, g \in L^1[0,b]$ are such that $f *_0 g \equiv 0$. Then, there exist $\lambda, \mu \in \mathbb{R}^+$ such that the following three conditions hold:*

- $f \equiv 0$ in the interval $[0, \lambda]$,
- $g \equiv 0$ in the interval $[0, \mu]$,
- $\lambda + \mu \geq b$.

Remark 8. *In particular, the Titchmarsh Theorem states that the operator $C_0(f) : L^1[0,b] \longrightarrow L^1[0,b]$ is injective, provided that $f \in L^1[0,b]$ and that f is not null at any interval $[0, \lambda]$ for $\lambda > 0$.*

In particular, we need the following result.

Corollary 2. *The operator $Y + \text{Id}$ described in (6) is injective.*

Proof. Note that we can not apply Theorem 3 directly to $Y + \text{Id}$, since it is not a convolution operator due to the "Id" term. However, $I_{0^+}^1 \circ (T + \text{Id})$ is a convolution operator and we conclude, following Remark 8, that $I_{0^+}^1 \circ (Y + \text{Id})$ is injective. If the previous composition is injective, the right factor $(Y + \text{Id})$ has to be injective. □

2.3.3. $Y + \text{Id}$ Is Surjective

In this case, we will use the following result, concerning Volterra integral equations of the second kind. This result essentially states that some family of integral equations do always have a continuous solution, provided that the source term is continuous.

Theorem 4 (Rust, [12]). *Given $k \in L^1[0,b]$, the Volterra integral equation*

$$(C_0(k)\, v)(t) + v(t) := \int_0^t k(t-s) \cdot v(s)\, ds + v(t) = w(t)$$

has exactly one continuous solution $v \in C[0,b]$, provided that $w \in C[0,b]$ and that the following two conditions hold:

- If $h \in \mathcal{C}[0,b]$, then $C_a(k)\, h \in \mathcal{C}[0,b]$.
- If $n \in \mathbb{Z}^+$ is big enough, then $(C_0(k))^n = C_0(\tilde{k})$ for some $\tilde{k} \in \mathcal{C}[0,b]$,

Remark 9. *We know that the image of $Y + \mathrm{Id}$ will contain $\mathcal{C}[0,b]$, since fractional integrals map continuous functions into continuous functions (Proposition 1). Moreover, Y^n will be defined by a continuous kernel when $n \geq \alpha_1^{-1}$; recall that $\alpha_1 > 0$ was the least integral order in Y.*

We need to conclude that, indeed, the image of $Y + \mathrm{Id}$ is $L^1[0,b]$.

Corollary 3. *The operator $Y + \mathrm{Id}$ described in (6) is surjective.*

Proof. Consider $f \in L^1[0,b]$ and the equation

$$(Y + \mathrm{Id})\, x(t) = f(t).$$

We need to show that there is $x \in L^1[0,b]$ solving this problem. Observe that x solves the previous equation if and only if it solves

$$(Y + \mathrm{Id})\, (x(t) - f(t)) = -Y f(t),$$

but now the source term is in $I_{0^+}^{\alpha_1} L^1[0,b]$. If we repeat this idea inductively, we see that x solves the original equation if and only if

$$(Y + \mathrm{Id})\, (x(t) - (\mathrm{Id} - Y + \cdots + (-1)^n Y) f(t)) = (-1)^{n+1} Y^{n+1} f(t).$$

The right-hand side will be continuous for $n \geq \alpha_1^{-1}$ and, by Remark 9, it will have a solution. □

2.3.4. $Y + \mathrm{Id}$ Is a Bounded Automorphism in $L^1[0,b]$

We have already seen that $Y + \mathrm{Id}$ is bounded and bijective, and hence the inverse is also bounded due to the Bounded Inverse Theorem for Banach spaces. Therefore, we have the following result.

Theorem 5. *The operator $Y + \mathrm{Id}$, described in (6) is an invertible bounded linear map from the Banach space $L^1[0,b]$ to itself, whose inverse is also bounded.*

In particular, we get the following corollary

Theorem 6. *Given $f \in L^1[0,b]$, the equation*

$$(Y + \mathrm{Id})\, x(t) = f(t)$$

has exactly one solution $x \in L^1[0,b]$.

Although it is not the scope of this paper, we highlight that such an equation can be solved using classical techniques for integral equations or specific tools for the particular case of fractional integral equations, like the one exposed in [13].

3. Implications of Fractional Integral Equations in Fractional Differential Equations

It would be desirable to have a similar result to the previous one for the case of FDE, ensuring existence and uniqueness of solution. Of course, in order to ensure uniqueness of solution, it is necessary to add some "extra conditions" to the differential version of the equation. Namely, one possibility is to impose initial values. In particular, given an ODE of order $n \in \mathbb{Z}^+$ with unknown function u, we know that, after fixing the values $u(0), u'(0), \ldots, u^{(n-1)}(0)$, we can ensure uniqueness of solution under general hypotheses. The question is: what is the reasonable "fractional analogue" of the previous idea?

Before answering the previous question, we study the solutions of this general linear problem with constant coefficients

$$\left(c_1 D_{0+}^{\beta_1} + \cdots + c_{n-1} D_{0+}^{\beta_{n-1}} + D_{0+}^{\beta_n}\right) u(t) = w(t), \tag{7}$$

where $\beta_n > \cdots > \beta_1 \geq 0$ and $w \in L^1[0,b]$. Of course, the first point to answer is where should we look for the solution to (7). It is very relevant to clarify completely this issue, since there are classical references; for instance, see [5] (Theorem 1, Section 5.5), which state the following theorem or equivalent versions, which are not totally accurate as we shall comment on in the follow-up.

Theorem 7. *Consider a linear homogeneous fractional differential equation (for Riemann–Liouville derivatives) with constant coefficients and rational orders. If the highest order of differentiation is α, then the equation has $\lceil \alpha \rceil$ linearly independent solutions.*

It is important to note that several references are not clear enough about the notion of solution to a fractional differential equation. With the previous sentence, we mean that it is desirable to introduce a suitable space of differentiable functions first, to later discuss about the solvability of the fractional differential equation. We devote the rest of the paper to show that the previous theorem is only true in some weak sense. Indeed, after defining formally the notion of "strong solution", we will see that, in general, there are less than $\lceil \alpha \rceil$ linearly independent solutions. Indeed, only for those "strong" solutions it will be coherent to talk about initial values.

The inaccuracy involving Theorem 7, and similar results, relies in the fact that the notion of solution is not completely specified. Moreover, it is common to find "proofs" that use Laplace Transform techniques without enough mathematical rigor (the final step of inverting the transform is neglected), the order of infinite sums and linear operators is interchanged (without regarding if there are sufficient hypotheses that make it legit),...

If we go back to (7), we can make the following vital remark.

Remark 10. *We recall that, in the usual case of integer orders, we look for the solutions in $\mathcal{X}_{\beta_n}$. Although it is quite common to forget it, the underlying reason to do this is that $\bigcap_{j=1}^n \mathcal{X}_{\beta_j} = \mathcal{X}_{\beta_n}$ when every β_j is a non-negative integer. However, in general, this does not necessarily happen when the involved orders are non-integers. Thus, we may have $\bigcap_{j=1}^n \mathcal{X}_{\beta_j} \neq \mathcal{X}_{\beta_n}$ and, of course, a solution to Equation (7) has to lie in $\bigcap_{j=1}^n \mathcal{X}_{\beta_j}$.*

Consequently, it is convenient to know the structure of the set $\bigcap_{j=1}^n \mathcal{X}_{\beta_j}$, which has already been described in Lemma 2, to study existence and uniqueness of solution. Of course, to expect uniqueness of solution, some initial conditions have to be added to Equation (7), but this will be discussed in the next section. The fundamental remark is that solutions to Equation (7) fulfill

$$D_{0+}^{\beta_n}\left(c_1 I_{0+}^{\beta_n-\beta_1} + \cdots + c_{n-1} I_{0+}^{\beta_n-\beta_{n-1}} + \mathrm{Id}\right) u(t) = w(t). \tag{8}$$

Consequently, it is quite natural to make the following reflection. If $u(t)$ solves (7), it is because

$$\left(c_1 I_{0+}^{\beta_n-\beta_1} + \cdots + c_{n-1} I_{0+}^{\beta_n-\beta_{n-1}} + \mathrm{Id}\right) u(t) \in I_{0+}^{\beta_n} w(t) + \ker D_{0+}^{\beta_n}. \tag{9}$$

We will refer to the set of solutions to Equation (9), as the set of weak solutions. The previous terminology obeys the following reason: although a solution to (7) solves (9), the converse does not hold in general. Namely, a solution to (9) may not lie in $\bigcap_{j=1}^n \mathcal{X}_{\beta_j}$. Of course, if the weak solution lies in $\bigcap_{j=1}^n \mathcal{X}_{\beta_j}$, then it solves (7). The set of solutions to (7) will be called the set of strong solutions.

At this moment, we know two vital things:

- We have already described $\ker D_{0+}^{\beta_n} = \left\langle \left\{ t^{\beta_n-1}, \ldots, t^{\beta_n-\lceil \beta_n \rceil} \right\} \right\rangle$ that is a vector space of dimension $\lceil \beta_n \rceil$. Therefore, the set of weak solutions has dimension $\lceil \beta_n \rceil$ too, since it is the image of the affine space $I_{0+}^{\beta_n} w(t) + \ker D_{0+}^{\beta_n}$ via the automorphism $T^{-1} \in \text{Aut}_B(L^1[0,b])$.
- The dimension of the set of strong solutions is bounded from above by $\lceil \beta_n - \beta_* \rceil$. If the dimension were higher, due to Remark 7, we could find two different solutions to (7) whose difference would lie in $I_{0+}^{\beta_n} L^1[0,b]$. After writing their difference as $I_{0+}^{\beta_n} g$, where $g \neq 0$, it would trivially fulfill

$$\left(c_1 I_{0+}^{\beta_n - \beta_1} + \cdots + c_{n-1} I_{0+}^{\beta_n - \beta_{n-1}} + \text{Id} \right) g(t) = 0,$$

which is not possible since the linear operator on the left-hand side is injective.

From these remarks, there are some remaining points that need to be studied in detail. First, we prove that the bound $\lceil \beta_n - \beta_* \rceil$ is sharp by inspecting which of elements in $\ker D_{0+}^{\beta_n}$ guarantee that the weak solution associated with those elements is, indeed, a strong one.

Remark 11. We have $\bigcap_{j=1}^n \mathcal{X}_{\beta_j} \subset I_{0+}^{\beta_n - \lceil \beta_n - \beta_* \rceil + 1 - \varepsilon} L^1[0,b]$ for every increment $\varepsilon > 0$, but not for $\varepsilon = 0$. Moreover, if $f \in \ker D_{0+}^{\beta_n}$ is chosen as the right summand on the right-hand side in (9), we have that, for $\gamma \leq \beta_n$, $f \in I_{0+}^{\gamma} L^1[0,b]$ if and only if $u \in I_{0+}^{\gamma} L^1[0,b]$.

Therefore, after putting together the two previous ideas, we arrive to the following conclusion. In order to have a strong solution in (9), it is mandatory to select a source term $f \in \left\langle \left\{ t^{\beta_n-1}, \ldots, t^{\beta_n-\lceil \beta_n - \beta_* \rceil} \right\} \right\rangle$.

We see that the converse also holds, since $f \in \left\langle \left\{ t^{\beta_n-1}, \ldots, t^{\beta_n-\lceil \beta_n - \beta_* \rceil} \right\} \right\rangle$ can be shown to be sufficient, in order to have a strong solution.

Lemma 3. *If $u \in L^1[0,b]$ solves*

$$\left(c_1 I_{0+}^{\beta_n - \beta_1} + \cdots + c_{n-1} I_{0+}^{\beta_n - \beta_{n-1}} + \text{Id} \right) u(t) = I_{0+}^{\beta_n} w(t) + f(t) \tag{10}$$

for $f \in \left\langle \left\{ t^{\beta_n-1}, \ldots, t^{\beta_n-\lceil \beta_n - \beta_* \rceil} \right\} \right\rangle \subset \ker D_{0+}^{\beta_n}$, then $u \in \bigcap_{j=1}^n \mathcal{X}_{\beta_j}$.

Proof. If we use the notation $Y := c_1 I_{0+}^{\beta_n - \beta_1} + \cdots + c_{n-1} I_{0+}^{\beta_n - \beta_{n-1}}$, we deduce from Equation (10) that

$$(Y + \text{Id})(u(t) - f(t)) = I_{0+}^{\beta_n} w(t) - Y f(t).$$

Observe now that the summand $-Y f(t)$ can be decomposed into two parts, since two different situations can happen:

- If $\beta_n - \beta_j \notin \mathbb{N}$, we see that $I_{0+}^{\beta_n - \beta_j} t^{\beta_n - k}$ will be always in the space $I_{0+}^{\beta_n + (\beta_n - \beta_* + 1) - \lceil \beta_n - \beta_* \rceil - \varepsilon} L^1[0,b]$ for every $\varepsilon > 0$. This simply occurs because the worst choice is $\beta_j = \beta_*$ and $k = \lceil \beta_n - \beta_* \rceil$. Indeed, for ε small enough, the previous space is contained in $I_{0+}^{\beta_n} L^1[0,b]$, since $\beta_n - \beta_* + 1 - \lceil \beta_n - \beta_* \rceil$ is strictly positive.
- If $\beta_n - \beta_j \in \mathbb{N}$, there are two options:
  - If $\beta_n - \beta_j > \beta_n - \beta_*$, we have that $I_{0+}^{\beta_n - \beta_j} t^{\beta_n - k}$ lies again in $I_{0+}^{\beta_n} L^1[0,b]$, since the maximum value admitted for k is $\lceil \beta_n - \beta_* \rceil$.
  - If $\beta_n - \beta_j < \beta_n - \beta_*$, we have that $I_{0+}^{\beta_n - \beta_j} t^{\beta_n - k} \in \left\langle \left\{ t^{\beta_n - k'} \right\} \right\rangle$ for some $k' < k$.

Thus, we can write $I_{0+}^{\beta_n} w(t) - Y f(t) = I_{0+}^{\beta_n} w_1(t) + f_1(t)$, and arrive to the equation

$$\left(c_1 I_{0+}^{\beta_n - \beta_1} + \cdots + c_{n-1} I_{0+}^{\beta_n - \beta_{n-1}} + \text{Id}\right)(u(t) - f(t)) = I_{0+}^{\beta_n} w_1(t) + f_1(t).$$

Note that f lived in a $\lceil \beta_n - \beta_* \rceil$ dimensional vector space, but f_1 lives in a (at most) $\lceil \beta_n - \beta_* \rceil - 1$ dimensional vector space.

If we repeat the process, we obtain

$$\left(c_1 I_{0+}^{\beta_n - \beta_1} + \cdots + c_{n-1} I_{0+}^{\beta_n - \beta_{n-1}} + \text{Id}\right)(u(t) - f(t) - f_1(t)) = I_{0+}^{\beta_n} w_2(t) + f_2(t),$$

with f_2 lying in a (at most) $\lceil \beta_n - \beta_* \rceil - 2$ dimensional vector space. After enough iterations, the vector space has to be zero dimensional and we will have the situation

$$(Y + \text{Id})(u(t) - f(t) - \cdots - f_{r-1}(t)) = I_{0+}^{\beta_n} w_r(t) \in I_{0+}^{\beta_n} L^1[0, b].$$

Therefore, $u(t) - f(t) - \cdots - f_{r-1}(t) \in I_{0+}^{\beta_n} L^1[0, b]$. Finally, if we use that

$$f(t) + f_1(t) + \cdots + f_{r-1}(t) \in \left\langle \left\{ t^{\beta_n - 1}, \ldots, t^{\beta_n - \lceil \beta_n - \beta_* \rceil} \right\} \right\rangle,$$

it follows $u \in \left\langle \left\{ t^{\beta_n - 1}, \ldots, t^{\beta_n - \lceil \beta_n - \beta_* \rceil} \right\} \right\rangle \oplus I_{0+}^{\beta_n} L^1[0, b] = \bigcap_{j=1}^{n} X_{\beta_j}$. □

4. Smooth Solutions for Fractional Differential Equations

Until this point, we have checked that, in general, there are more weak solutions (a $\lceil \beta_n \rceil$ dimensional space) than strong solutions (a $\lceil \beta_n - \beta_* \rceil$ dimensional space). We have also seen how weak solutions are codified depending on the source term, more specifically depending on the element chosen in ker $D_{0+}^{\beta_n}$. Moreover, we know that, if the choice is made in a certain subspace of ker $D_{0+}^{\beta_n}$, then the obtained solution is a strong one. However, one could think about codifying strong solutions directly in the fractional differential equation via initial conditions, instead of using fractional integral problems and selecting a source term linked to a strong solution. Therefore, the last task should consist of relating the choices for ker $D_{0+}^{\beta_n}$ that give a strong solution with the corresponding initial conditions for the strong problem.

First, to simplify the notation, we reconsider Equation (7) with the additional hypotheses that each positive integer less then or equal to $\beta_n - \beta_1$ can be written as $\beta_n - \beta_j$ for some j.

$$D_{0+}^{\beta_n} \left(c_1 I_{0+}^{\beta_n - \beta_1} + \cdots + c_{n-1} I_{0+}^{\beta_n - \beta_{n-1}} + \text{Id}\right) u(t) = w(t)$$

This does not imply a loss of generality, since we can assume some $c_j = 0$, if needed. The only purpose of this assumption is to ease the notation in this proof, in the way that is described in the following paragraph.

If $\beta_* = \beta_{n-m}$, then $\beta_n - \beta_{n-m}$ is the least possible non-integer difference $\beta_n - \beta_j$. Thus, we can use the previous notational assumption to check that $\beta_n - \beta_{n-j} = j$ for $j < m$ and $\beta_n - \beta_{n-m} \in (m-1, m)$. Thus, $\lceil \beta_n - \beta_* \rceil = \lceil \beta_n - \beta_{n-m} \rceil = m$, and $c_{n-m+1}, \ldots, c_{n-1}$ are $m-1$ constants multiplying integrals of integer order in (10).

Now, we provide the main result of this section.

Lemma 4. *Under the previous notation, Equation (7) with given initial values $D_{0+}^{\beta_n-m}u(0),\ldots,D_{0+}^{\beta_n-1}u(0)$ has a unique solution in $\bigcap_{j=1}^{n}\mathcal{X}_{\beta_j}$. This solution coincides with the unique solution to (10), where the source term is the unique function $f \in \langle\{t^{\beta_n-1},\ldots,t^{\beta_n-m}\}\rangle$ fulfilling*

$$D_{0+}^{\beta_n-m}u(0) = D_{0+}^{\beta_n-m}f(0),$$
$$D_{0+}^{\beta_n-m+1}u(0) + c_{n-1}D_{0+}^{\beta_n-m}u(0) = D_{0+}^{\beta_n-m+1}f(0),$$
$$\cdots$$
$$D_{0+}^{\beta_n-1}u(0) + c_{n-1}D_{0+}^{\beta_n-2}u(0) + \cdots + c_{n-m+1}D_{0+}^{\beta_n-m}u(0) = D_{0+}^{\beta_n-1}f(0).$$

Proof. Consider again Equation (10)

$$\left(c_1 I_{0+}^{\beta_n-\beta_1} + \cdots + c_{n-1}I_{0+}^{\beta_n-\beta_{n-1}} + \mathrm{Id}\right)u(t) = I_{0+}^{\beta_n}w(t) + f(t),$$

Recall that we look for strong solutions to (10) that lie in the functional space $\langle\{t^{\beta_n-1},\ldots,t^{\beta_n-m}\}\rangle \oplus I_{0+}^{\beta_n}L^1[0,b]$, so we write

$$u(t) = d_1 t^{\beta_n-m} + \cdots + d_m t^{\beta_n-1} + I_{0+}^{\beta_n}\widetilde{u}(t).$$

Moreover, take into account that a strong choice for $f \in \ker D_{0+}^{\beta_n}$ allows us to write

$$f(t) = b_1 t^{\beta_n-m} + \cdots + b_m t^{\beta_n-1}.$$

Now, we will just derive the initial conditions after applying $D_{0+}^{\beta_n-k}$, for every $k \in \{1,\ldots,m\}$, and substituting $t = 0$ in (10).

On the right-hand side, this is easy, since $D_{0+}^{\beta_n-k}I_{0+}^{\beta_n}w(t) \in I_{0+}^{1}L^1[0,b]$ and, thus, the substitution at $t = 0$ gives zero. The function $D_{0+}^{\beta_n-k}f(t)$ can be computed trivially, due to the expression of f, obtaining

$$D_{0+}^{\beta_n-k}f(0) = \Gamma(\beta_n-k+1)\,b_k.$$

On the left-hand side, on the one hand, we have again a similar situation to the previous one, since $D_{0+}^{\beta_n-k}I_{0+}^{\beta_n-\beta_j}I_{0+}^{\beta_n}\widetilde{u}(t) \in I_{0+}^{k+(\beta_n-\beta_j)}L^1[0,b] \subset I_{0+}^{1}L^1[0,b]$ for any subindex $j \in \{1,\ldots,n\}$ and, thus, the substitution at $t = 0$ gives zero. On the other hand, $D_{0+}^{\beta_n-k}I_{0+}^{\beta_n-\beta_j}t^{\beta_n-l}$ has three possibilities:

- If $\beta_n - \beta_j > l - k$, then $D_{0+}^{\beta_n-k}I_{0+}^{\beta_n-\beta_j}t^{\beta_n-l}$ is a scalar multiple of a power of t with positive exponent. Thus, when we make the substitution at $t = 0$, we get 0.
- If $\beta_n - \beta_j = l - k$, then $D_{0+}^{\beta_n-k}I_{0+}^{\beta_n-\beta_j}t^{\beta_n-l} = \Gamma(\beta_n-l+1)$ is constant, and it is obviously defined for $t = 0$.
- If $\beta_n - \beta_j < l - k \leq m - 1$, then $\beta_n - \beta_j$ is an integer and the computation $D_{0+}^{\beta_n-k}I_{0+}^{\beta_n-\beta_j}t^{\beta_n-l}$ gives the zero function.

The interest of the previous trichotomy is that we never obtain some $t^{-\gamma}$ with $\gamma > 0$. In other cases, we would have huge trouble, since we could not evaluate the expression for $t = 0$. Fortunately, we can always apply $D_{0+}^{\beta_n-k}$ to Equation (10), for every value $k \in \{1,\ldots,m\}$, and substitute at $t = 0$. We arrive to the following linear system of equations:

$$D_{0+}^{\beta_n-m}u(0) = D_{0+}^{\beta_n-m}f(0),$$
$$D_{0+}^{\beta_n-m+1}u(0) + c_{n-1}D_{0+}^{\beta_n-m}u(0) = D_{0+}^{\beta_n-m+1}f(0),$$
$$\cdots$$
$$D_{0+}^{\beta_n-1}u(0) + c_{n-1}D_{0+}^{\beta_n-2}u(0) + \cdots + c_{n-m+1}D_{0+}^{\beta_n-m}u(0) = D_{0+}^{\beta_n-1}f(0).$$

Note that all the involved derivatives in the initial conditions have the same decimal part, since only coefficients $c_{n-1}, \cdots, c_{n-m+1}$ appear in the system. We also highlight that the system always has a unique solution, since it is triangular and it has no zero element in the diagonal. Therefore, a choice for f linked to a strong solution determines a vector of initial values $(D_{0+}^{\beta_n - m} u(0), \ldots, D_{0+}^{\beta_n - 1} u(0))$ and vice versa in a bijective way. □

We shall give two examples summarizing how to apply all the previous results.

Example 1. Consider the following fractional differential equation (strong problem):

$$\left(D_{0+}^{\frac{7}{3}} + 3 D_{0+}^{\frac{4}{3}} + 4 D_{0+}^{\frac{1}{3}} \right) u(t) = t^3$$

and define $\beta_1 = \frac{1}{3}, \beta_2 = \frac{4}{3}, \beta_3 = \frac{7}{3}$. In this case, note that $\beta_* = 0$, since all the differences $\beta_3 - \beta_j$ are integers. The strong solutions for the example will lie in $\bigcap_{j=1}^{3} \mathcal{X}_{\beta_j}$. The dimension of the affine space of strong solutions will be $\lceil \beta_3 \rceil = 3$, and the initial conditions that ensure existence and uniqueness of solution will be $D_{0+}^{\frac{4}{3}} u(0) = a_3$, $D_{0+}^{\frac{1}{3}} u(0) = a_2$ and $I_{0+}^{\frac{2}{3}} u(0) = a_1$.

Moreover, after left-factoring $D_{0+}^{\frac{7}{3}}$, we find that the associated family of weak problems is

$$\left(4 I_{0+}^2 + 3 I_{0+}^1 + \mathrm{Id} \right) u(t) = I_{0+}^{\frac{7}{3}} t^3 + f(t)$$

where $f(t) \in \left\langle \left\{ t^{\frac{4}{3}}, t^{\frac{1}{3}}, t^{-\frac{2}{3}} \right\} \right\rangle$, which lives in a three-dimensional space. The a priori weak, obtained solution is always strong since we have that $\lceil \beta_3 - \beta_* \rceil = \lceil \beta_3 \rceil$.

Finally, the relation between a choice for $f(t) = b_3 t^{\frac{4}{3}} + b_2 t^{\frac{1}{3}} + b_1 t^{-\frac{2}{3}}$ providing a strong solution and the initial conditions a_1, a_2 and a_3 is

$$a_1 = I_{0+}^{\frac{2}{3}} f(0) = b_1 \cdot \Gamma \left(1 - \tfrac{2}{3} \right),$$
$$a_2 + 3 a_1 = D_{0+}^{\frac{1}{3}} f(0) = b_2 \cdot \Gamma \left(1 + \tfrac{1}{3} \right),$$
$$a_3 + 3 a_2 + 4 a_1 = D_{0+}^{\frac{4}{3}} f(0) = b_3 \cdot \Gamma \left(1 + \tfrac{4}{3} \right).$$

Example 2. Consider the following fractional differential equation (strong problem):

$$\left(D_{0+}^{\frac{13}{4}} + 3 D_{0+}^{\frac{9}{4}} + D_{0+}^{2} + D_{0+}^{\frac{5}{4}} + D_{0+}^{\frac{1}{4}} \right) u(t) = t$$

and define $\beta_1 = 1, \beta_2 = \frac{5}{4}, \beta_3 = 2, \beta_4 = \frac{9}{4}, \beta_5 = \frac{13}{4}$. In this case, note that $\beta_* = \beta_3$, since it fulfills the property that $\beta_5 - \beta_*$ is the least possible non-integer difference $\beta_5 - \beta_j$. The strong solutions for the example will lie in $\bigcap_{j=1}^{5} \mathcal{X}_{\beta_j}$. The dimension of the affine space of strong solutions will be $\lceil \beta_5 - \beta_* \rceil = 2$ and the initial conditions that ensure existence and uniqueness of solution will be $D_{0+}^{\frac{9}{4}} u(0) = a_2$ and $D_{0+}^{\frac{5}{4}} u(0) = a_1$.

Moreover, after left-factoring $D_{0+}^{\frac{13}{4}}$, we find that the associated family of weak problems is

$$\left(I_{0+}^{\frac{9}{4}} + I_{0+}^{2} + I_{0+}^{\frac{5}{4}} + 3 I_{0+}^{1} + \mathrm{Id} \right) u(t) = I_{0+}^{\frac{13}{4}} t + f(t)$$

where $f(t) \in \left\langle \left\{ t^{\frac{9}{4}}, t^{\frac{5}{4}}, t^{\frac{1}{4}}, t^{-\frac{3}{4}} \right\} \right\rangle$, which lives in a four-dimensional space. The a priori weak, obtained solution will be strong if $f(t) \in \left\langle \left\{ t^{\frac{9}{4}}, t^{\frac{5}{4}} \right\} \right\rangle$.

Finally, the relation between a choice for $f(t) = b_2 t^{\frac{9}{4}} + b_1 t^{\frac{5}{4}}$ providing a strong solution and the initial conditions a_1 and a_2 is

$$a_1 = D_{0^+}^{\frac{5}{4}} u(0) = D_{0^+}^{\frac{5}{4}} f(0) = b_1 \cdot \Gamma\left(1 + \tfrac{5}{4}\right),$$
$$a_2 + 3 a_1 = D_{0^+}^{\frac{9}{4}} u(0) + 3 D_{0^+}^{\frac{5}{4}} u(0) = D_{0^+}^{\frac{9}{4}} f(0) = b_2 \cdot \Gamma\left(1 + \tfrac{9}{4}\right).$$

5. Conclusions

We summarize the conclusions obtained in this paper.

- We have recalled the main results involving existence and uniqueness of solution for linear fractional integral equations with constant coefficients.
- We have seen that, from each linear FDE with constant coefficients of order β_n, it is possible to derive a $\lceil \beta_n \rceil$ dimensional family of associated fractional integral equations, in a natural way. Moreover, each solution to the fractional differential equation fulfills exactly one of these fractional integral equations.
- We have shown that there exists a $\lceil \beta_n - \beta_* \rceil$ dimensional subfamily (of the $\lceil \beta_n \rceil$ dimensional family of associated fractional integral equations) such that each solution to a problem of the subfamily gives a solution to the original linear fractional differential equation of order β_n. This value β_* is obtained as the greatest differentiation order in the FDE such that $\beta_n - \beta_*$ is not an integer. If such a value does not exist, the same result holds after defining $\beta_* = 0$.
- We have seen how initial values at $t = 0$ for the derivatives of orders $\beta_n - \lceil \beta_n - \beta_* \rceil, \ldots, \beta_n - 1$ guarantee existence and uniqueness of solution to a linear fractional differential equation with constant coefficients of order β_n. We have described the correspondence between such initial values for the FDE and the selection of a source term in the $\lceil \beta_n - \beta_* \rceil$ dimensional subfamily of integral equations, in such a way that both problems have the same unique solution. If $\beta_n - \lceil \beta_n - \beta_* \rceil \in (-1, 0)$, this first initial value is imposed, indeed, for the fractional integral of order $\lceil \beta_n - \beta_* \rceil - \beta_n$.
- We expect that this idea can be extended to different types of fractional differential problems. It would be nice to amplify the scope of this work to a more general case than the one of constant coefficients. Furthermore, the same philosophy could be applied to other type of problems such as, for instance, periodic ones that have relevant applications [14].

Funding: This research received no external funding.

Conflicts of Interest: The author declares no conflict of interest.

References

1. Hilfer, R.; Luchko, Y. Desiderata for fractional derivatives and integrals. *Mathematics* **2019**, *7*, 149. [CrossRef]
2. Kilbas, A.; Srivastava, H.M.; Trujillo, J.J. *Theory and Applications of Fractional Differential Equations*; Elsevier: Amsterdam, The Netherlands, 2006.
3. Podlubny, I. *Fractional Differential Equations*; Academic Press: San Diego, CA, USA, 1999.
4. Cartwright, D.I.; McMullen, J.R. A note on the fractional calculus. *Proc. Edinb. Math. Soc.* **1978**, *21*, 79–80. [CrossRef]
5. Miller, K.S.; Ross, B. *An Introduction to the Fractional Calculus and Fractional Differential Equations*; John Wiley & Sons: Hoboken, NJ, USA, 1993.
6. Diethelm, K.; Ford, N.J. Numerical solution of the Bagley-Torvik equation. *BIT* **2002**, *42*, 490–507. [CrossRef]
7. Gorenflo, R.; Mainardi, F. Fractional Calculus: Integral and Differential Equations of Fractional Order. In *Fractals and Fractional Calculus in Continuum Mechanics*; Carpintieri, A., Mainardi, F., Eds.; Springer: Wien, Austria; New York, NY, USA, 1997; pp. 223–276.
8. Tenreiro Machado, J.A.; Mainardi, F.; Kiryakova, V.; Atanacković, T. Fractional Calculus: D'où venons-nous? Que sommes-nous? Où Allons-nous? *Fract. Calc. Appl. Anal.* **2016**, *19*, 1074–1104. [CrossRef]

9. Samko, S.; Kilbas, A.; Marichev, O. *Fractional Integrals and Derivatives. Theory and Applications*; Gordon and Breach: Yverdon, Switzerland, 1993.
10. Bergounioux, M.; Leaci, A.; Nardi, G.; Tomarelli, F. Fractional Sobolev spaces and functions of bounded variation of one variable. *Fract. Calc. Appl. Anal.* **2017**, *20*, 936–962. [CrossRef]
11. Titchmarsh, E.C. The zeros of certain integral functions. *Proc. Lond. Math. Soc.* **1926**, *25*, 283–302. [CrossRef]
12. Rust, W.M. A theorem on Volterra integral equations of the second kind with discontinuous kernels. *Am. Math. Mon.* **1934**, *41*, 346–350. [CrossRef]
13. Cao Labora, D.; Rodríguez-López, R. From fractional order equations to integer order equations. *Fract. Calc. Appl. Anal.* **2017**, *20*, 1405–1423. [CrossRef]
14. Staněk, S. Periodic problem for the generalized Basset fractional differential equation. *Fract. Calc. Appl. Anal.* **2015**, *18*, 1277–1290. [CrossRef]

© 2020 by the authors. Licensee MDPI, Basel, Switzerland. This article is an open access article distributed under the terms and conditions of the Creative Commons Attribution (CC BY) license (http://creativecommons.org/licenses/by/4.0/).

Article

Finite Difference Method for Two-Sided Two Dimensional Space Fractional Convection-Diffusion Problem with Source Term

Eyaya Fekadie Anley [1,2] and Zhoushun Zheng [1,*]

1. School of Mathematics and Statistics, Central South University, Changsha 410083, China; eyayafek@csu.edu.cn
2. Department of Mathematics, College of Natural and Computational Science, Arba-Minch University, Arba-Minch 21, Ethiopia
* Correspondence: zszheng@csu.edu.cn or 2009zhengzhoushun@163.com

Received: 17 August 2020; Accepted: 26 September 2020; Published: 29 October 2020

Abstract: In this paper, we have considered a numerical difference approximation for solving two-dimensional Riesz space fractional convection-diffusion problem with source term over a finite domain. The convection and diffusion equation can depend on both spatial and temporal variables. Crank-Nicolson scheme for time combined with weighted and shifted Grünwald-Letnikov difference operator for space are implemented to get second order convergence both in space and time. Unconditional stability and convergence order analysis of the scheme are explained theoretically and experimentally. The numerical tests are indicated that the Crank-Nicolson scheme with weighted shifted Grünwald-Letnikov approximations are effective numerical methods for two dimensional two-sided space fractional convection-diffusion equation.

Keywords: Crank–Nicolson scheme; weighted Shifted Grünwald–Letnikov approximation; space fractional convection-diffusion model; stability analysis; convergence order

MSC: 26A33; 35R11; 65M60

---

1. Introduction

Differential equation described fractional partial differential equations are appropriated to explain complex problems like viscoelasticity, electroanalytical chemistry, biology, fluid mechanics, engineering [1], physics [1,2], fractional operators [3] and flows in porous media [4–8]. Through the advection and dispersion processes, pollutants create a contaminant plume within an aquifer, the movement of which in an aquifer is described by transport model. One of the very rich transport model is advection–dispersion model, which is used to describe the transport phenomena in different fields of science. Solute transport is important to predict the solute concentration in aquifers, rivers, lakes and streams too.

Due to the fractional derivative property of differential operator of space fractional derivative, finding a numerical solution of fractional convection-diffusion equation is somehow difficult, specially for high dimensional case. Numerical methods for numerical approximations of one dimensional fractional convection-diffusion equations are the homotopy analysis transform method [9], the finite difference method [2,10–12], the collocation method [13–16], the Galerkin method [17–20] and the finite volume element method [21,22]. An improved matrix transform numerical method is proposed in Reference [23] to solve one dimensional space fractional advection–dispersion model and its analytical solution is found using padé approximation. Recently, space fractional convection-diffusion with variable coefficients are solved using shifted Grünwald-Letnikov difference operator for space and Crank-Nicolson scheme for time that produce second order convergence both in time and space with extrapolation was studied [24].

There are numerical schemes that used to solve two-dimension space fractional diffusion problems such as the alternating direction implicit (ADI) method [25–30], the Galerkin finite element method [31], the finite volume method [32] and the kronecker product splitting method [33]. ADI and CN-ADI spectral methods are used to solve two-dimensional Riesz space fractional diffusion equation with a non-linear reaction term with respect to their error estimates have been discussed (see References [34,35]). Reference [36] proposed a new group iterative scheme for the numerical solution of two dimensional time fractional advection-diffusion equation based on Caputo-type discretization of the fractional group scheme in combination with Crank-Nicolson scheme. The Crank-Nicolson Galerkin-fully discrete approximation method for two-dimensional space fractional advection–diffusion problem with optimal error estimation was investigated by Reference [37]. In Reference [38], comparative study of the finite element and difference method for two dimension space fractional advection–dispersion equation has been considered by modeling non-Fickian solute transport in groundwater. For the comparison they have used a backward-distance algorithm that used to extend the triangular elements to generic elements in the finite element analysis and a variable-step vector Grünwald-Letnikov formula to improve the solution accuracy of finite difference method. The stability and second order convergence are proved [39] by a novel finite volume method for the Riesz space distributed order advection-diffusion equation. Linear spline approximation for Riemann-Liouville fractional derivative and CNADI finite difference method for time discretization are applied for solving two-dimensional two-sided space factional convection-diffusion equation was explained (read the details in Reference [40]). Having the advantage of reduce multi-dimensional problems to one dimension and easy to implement, the ADI algorithm is the more selected technique for the discretization. Reference [41] has implemented unconditionally stable compact ADI method for two-dimensional Riesz space fractional diffusion problem with second order in time and fourth order accuracy in both spaces. Here, we need to construct weighted and shifted Grünwald-Letnikov difference operator (WSGD) with the Crank-Nicolson-ADI (CNADI)method for two-sided two dimension space fractional convection–diffusion problem to have second order both in time and space. The weighted and shifted Grünwald-Letnikov combined with CNADI also have been applied effectively for convection– dominance two-dimension two-sided space fractional convection–diffusion equation. It is suitable to apply the weighted combined with shifted Grünwald–Letnikov difference approximation for two-sided Riemann–Liouville fractional derivative to have second order accurate in space. Therefore, it is important to get a numerical scheme that leads to evaluate a two-sided two dimension space fractional convection–diffusion problem. Thus, this study has focused to have temporal and spatial second order convergence estimates for two dimensional two-sided space fractional convection–diffusion equations based on accurate finite difference method without extrapolation approach. The scheme has been judged using the Crank-Nicolson Peaceman Rachford alternating direction implicit (CNADI) method with the novel weighted Shifted Grünwald–Letnikov difference approximation (WSGD) and the algorithm has been supported with numerical simulation.

Consider the two-dimensional two-sided space fractional convection–diffusion problem with constant coefficients:

$$\frac{\partial u(x,y,t)}{\partial t} = c_x \frac{\partial^{\alpha_1} u(x,y,t)}{\partial |x|^{\alpha_1}} + c_y \frac{\partial^{\alpha_2} u(x,y,t)}{\partial |y|^{\alpha_2}} + d_x \frac{\partial^{\beta_1} u(x,y,t)}{\partial |x|^{\beta_1}} + d_y \frac{\partial^{\beta_2} u(x,y,t)}{\partial |y|^{\beta_2}} + p(x,y,t),$$

corresponding to initial condition:

$$u(x,y,0) = g(x,y), 0 \leq x \leq L_x, \ 0 \leq y \leq L_y, \qquad (1)$$

with the zero Dirichlet boundary conditions:

$$\begin{aligned} u(0,y,t) = 0; \quad u(L_x,y,t) = 0; \\ u(x,0,t) = 0; \quad u(x,L_y,t) = 0, \end{aligned} \quad (2)$$

where $0 < \alpha_1, \alpha_2 < 1, 1 < \beta_1, \beta_2 < 2, c_x, c_y \geq 0$ and $d_x, d_y > 0$ express the velocity parameter and positive diffusion coefficients, respectively.

Here, $u(x,y,t)$ is solute concentration expressed physically in References [42,43], and $p(x,y,t)$ is the source term so that the solute concentration transport is from left to right. For the case of integer order ($\alpha_1 = \alpha_2 = 1, \beta_1 = \beta_2 = 2$), Equation (6) gives to the two-dimension classical convection–diffusion equation (CDE). We have supposed that the two-dimensional space fractional convection–diffusion problem has sufficiently smooth and unique enough solutions.

The remain arrangement of this paper is organized as follows—in Section 2, we introduce some preliminary remarks, lemmas and definitions. We have shown the formulation of one dimensional Riesz space fractional convection–diffusion problem with Crank-Nicolson and weighted shifted Grünwald–Letnikov difference scheme in Section 3. In Section 4, we have described the formulation with discretization of two-dimensional Riesz space fractional convection–diffusion problem. In Section 5, unconditional stability and convergence order analysis of the scheme have done using CNADI-WSGD. In Section 6, numerical simulations are implemented to show the importance of our theoretical study and the conclusions are discussed in Section 7.

2. Preliminary Remarks

Definition 1. *The Riesz differential operator which is given by analytic continuation in the whole range $0 < \alpha \leq 2$ with $\alpha \neq 1$ as:*

$$\frac{\partial^\alpha u(x,t)}{\partial |x|^\alpha} = K\left[_{-\infty}D_x^\alpha + {_xD_\infty^\alpha} \right] u(x,t), \quad (3)$$

where the fractional derivative,

$$\begin{aligned} _{-\infty}D_x^\alpha u(x,t) &= \left(\frac{d}{dx}\right)^n \left[_{-\infty}I_x^{n-\alpha} u(x,t) \right] \\ _xD_\infty^\alpha u(x,t) &= \left(\frac{d}{dx}\right)^n \left[_xI_\infty^{n-\alpha} u(x,t) \right], \end{aligned} \quad (4)$$

with $n \in \mathbb{N}$ and coefficient $K = \frac{-1}{2\cos(\alpha\pi/2)}$, are the left and right Riemann-Liouville fractional derivatives. From this definition the fractional integral operators $_{-\infty}I_x^\alpha u(x,t)$ and $_xI_\infty^\alpha u(x,t)$ are the left and right Weyl fractional integrals as defined in Reference [44]:

$$\begin{cases} _{-\infty}I_x^\alpha u(x,t) = \frac{1}{\Gamma(\alpha)} \int_{-\infty}^x \frac{u(\eta,t)}{(x-\eta)^{1-\alpha}} d\eta, & \alpha > 0, \\ _xI_\infty^\alpha u(x,t) = \frac{1}{\Gamma(\alpha)} \int_x^\infty \frac{u(\eta,t)}{(\eta-x)^{1-\alpha}} d\eta, & \alpha > 0. \end{cases} \quad (5)$$

Lemma 1 ([44,45]). *Let $\alpha > 0$ and $\Gamma(.)$ represents gamma function, then the following are properties of binomial coefficients:*

1. $\binom{\alpha}{k} = \binom{\alpha-1}{k} + \binom{\alpha-1}{k-1}$.

2. $(-1)^k \binom{\alpha}{k} = (-1)^k \frac{\alpha(\alpha-1)(\alpha-2)\ldots(\alpha-k+1)}{k!}$.

3. $\frac{\Gamma(k-\alpha)}{\Gamma(-\alpha)\Gamma(k+1)} = (-1)^k \binom{\alpha}{k} = \binom{k-\alpha-1}{k}$.

4. $\lim_{m\to\infty}(-1)^{m-k}\binom{\alpha-k-1}{m-k}(m-k)^{\alpha-k}$

$= \lim_{m\to\infty}(-1)^{m-k}\dfrac{(-\alpha+k+1)(-\alpha+k+2)(-\alpha+k+3)\dots(-\alpha+m)}{(m-k)^{-\alpha+k}(m-k)!}$

$= \dfrac{1}{\Gamma(-\alpha+k+1)}.$

5. $\lim_{m\to\infty}\left(\dfrac{m}{m-k}\right)^{\alpha-k} = \lim\left(\dfrac{1}{1-\frac{k}{m}}\right)^{\alpha-k} = 1.$

Theorem 1. *Let $u(x)$ has $n-1$ continuous derivatives on the closed interval $[a,b]$ with the derivatives $u^{(n)}(x)$ are integrable for $x \geq a$ or $x \leq b$, then for each $\alpha\,(n-1 < \alpha \leq n)$, the left and right Riemann-Liouville fractional derivatives exist and coincide with the corresponding (left and right) Grünwald-Letnikov fractional derivatives.*

Proof. The left standard Grünwald-Letnikov fractional derivative is given by the limit expression on $[a,x]$,

$$_aD_x^\alpha u(x) = \lim_{h\to 0}\dfrac{1}{h^\alpha}\left(\sum_{k=0}^n (-1)^k \binom{\alpha}{k} u(x-kh)\right), \tag{6}$$

where $\dfrac{x-a}{n} = h = \dfrac{b-x}{n}, n-1 < \alpha \leq n$. Here our aim is to evaluate the limit described in Equation (6). For the evaluation of the limit, we are assuming the function $u(x)$ continuous on $[a,x]$ and for $\alpha > 0$, we have:

$$_aD_x^\alpha u(x) = \lim_{h\to 0}\dfrac{1}{h^\alpha}\left(\sum_{k=0}^n (-1)^k \binom{\alpha}{k} u(x-kh)\right) = \lim_{n\to\infty} U_h(x), \tag{7}$$

where $U_h(x) = \dfrac{1}{h^\alpha}\left(\sum_{k=0}^n (-1)^k \binom{\alpha}{k} u(x-kh)\right)$. We need to transform Equation (7) to the following form using the property 1 of Lemma 1.

$$\begin{aligned}
U_h(x) &= \dfrac{1}{h^\alpha}\left(\sum_{k=0}^n (-1)^k \binom{\alpha-1}{k} u(x-kh)\right) + \dfrac{1}{h^\alpha}\left(\sum_{k=0}^n (-1)^k \binom{\alpha-1}{k-1} u(x-kh)\right) \\
&= \dfrac{1}{h^\alpha}\left(\sum_{k=0}^n (-1)^k \binom{\alpha-1}{k} u(x-kh)\right) + \dfrac{1}{h^\alpha}\left(\sum_{k=0}^{n-1} (-1)^{k+1} \binom{\alpha-1}{k} u(x-(k+1)h)\right) \\
&= \dfrac{(-1)^n}{h^\alpha} \binom{\alpha-1}{n} u(a) + \dfrac{1}{h^\alpha}\left(\sum_{k=0}^{n-1} (-1)^k \binom{\alpha-1}{k} \Delta u(x-kh)\right),
\end{aligned} \tag{8}$$

where $\Delta u(x-kh) = u(x-kh) - u(x-(k+1)h)$ is the first order backward difference operator. Similarly, we have to apply property 1 of Lemma 1 repeatedly m times, after simplification we get:

$$\begin{aligned}
U_h(x) = &\sum_{p=0}^m (-1)^{n-p} \binom{\alpha-p-1}{n-p} h^{-\alpha} \Delta^p u(a+ph) \\
&+ \sum_{k=0}^{n-m-1} (-1)^k \binom{\alpha-m-1}{k} h^{-\alpha} \Delta^{m+1} u(x-kh).
\end{aligned} \tag{9}$$

Now, we need to evaluate the limit of Equation (9).

$$\lim_{n\to\infty} U_h(x) = \lim_{n\to\infty} U_{h_f}(x) + \lim_{n\to\infty} U_{h_s}(x),$$

where $U_{h_f}(x) = \sum_{k=0}^{m}(-1)^{n-p}\binom{\alpha-p-1}{n-p}h^{-\alpha}\Delta^p u(a+ph)$, which is the first sum and $U_{h_s}(x) = \frac{1}{h^\alpha}\left(\sum_{k=0}^{n-m-1}(-1)^k\binom{\alpha-m-1}{k}\Delta^{m+1}u(x-kh)\right)$, denote the second sum. Let us find the limit of pth-term of the first sum.

$$\begin{aligned}\lim_{n\to\infty} U_{h_f}(x) &= \lim_{n\to\infty}(-1)^{n-p}\binom{\alpha-p-1}{n-p}h^{-\alpha}\Delta^p u(a+ph) \\ &= \lim_{n\to\infty}(-1)^{n-p}\binom{\alpha-p-1}{n-p}(n-p)^{\alpha-p}\left(\frac{n}{n-p}\right)^{\alpha-p}(nh)^{-\alpha+p}\frac{\Delta^p u(a+ph)}{h^p} \\ &= (x-a)^{-\alpha+p}\lim_{n\to\infty}(-1)^{n-p}\binom{\alpha-p-1}{n-p}(n-p)^{\alpha-p}\times\lim_{n\to\infty}\left(\frac{n}{n-p}\right)^{\alpha-p} \qquad (10) \\ &\times\lim_{h\to 0}\frac{\Delta^p u(a+ph)}{h^p} = \frac{u^{(p)}(a)(x-a)^{-\alpha+p}}{\Gamma(-\alpha+p+1)}.\end{aligned}$$

In order to evaluate the limit of second sum U_{h_s}, we have to follow the property of binomial coefficients of Lemma 1.

$$\begin{aligned}\lim_{n\to\infty} U_{h_s} &= \lim_{k\to\infty}\left(\frac{1}{\Gamma(-\alpha+m+1)}\sum_{k=0}^{n-m-1}(-1)^k\Gamma(-\alpha+m+1)\binom{\alpha-m-1}{k}k^{-\alpha+m}\right) \\ &\times\lim_{h\to 0}h(hk)^{-\alpha+m}\frac{\Delta^{m+1}u(x-kh)}{h^{m+1}}.\end{aligned} \qquad (11)$$

From property 4 of Lemma 1, we have

$$\lim_{k\to\infty}(-1)^k\Gamma(-\alpha+m+1)\binom{\alpha-m-1}{k}k^{-\alpha+m} = 1. \qquad (12)$$

Moreover, if $m-\alpha > -1$, then

$$\lim_{h\to 0}\left(\sum_{k=0}^{n-m-1}h(hk)^{-\alpha+m}\frac{\Delta^{m+1}u(x-kh)}{h^{m+1}}\right) = \int_a^x(x-\eta)^{m-\alpha}u^{(m+1)}(\eta)d\eta. \qquad (13)$$

By considering Equations (12) and (13) we have that:

$$\lim_{h\to 0}\sum_{k=0}^{n-m-1}(-1)^k\binom{\alpha-m-1}{k}h^{-\alpha}\Delta^{m+1}u(x-kh) = \frac{1}{\Gamma(-\alpha+m+1)}\int_a^x(x-\eta)^{-\alpha+m}u^{(m+1)}(\eta). \qquad (14)$$

Now by combining Equations (10) and (14), we have finalized the general limit evaluation as:

$$_a D_x^\alpha u(x) = \lim_{n\to\infty}U_h(x) = \sum_{p=0}^m\frac{u^{(p)}(a)(x-a)^{-\alpha+p}}{\Gamma(-\alpha+p+1)} + \frac{1}{\Gamma(-\alpha+m+1)}\int_a^x(x-\eta)^{-\alpha+m}u^{(m+1)}(\eta)d\eta. \qquad (15)$$

By taking $n = m+1$ or $n-1 = m$ with $n-1 < \alpha \le n$, the left Grünwald-Letnikov fractional derivative over the closed interval $[a, x]$ is written as:

$$_a D_x^\alpha u(x) = \sum_{p=0}^{n-1}\frac{u^{(p)}(a)(x-a)^{-\alpha+p}}{\Gamma(-\alpha+p+1)} + \frac{1}{\Gamma(-\alpha+n)}\int_a^x(x-\eta)^{-\alpha+n-1}u^{(n)}(\eta)d\eta. \qquad (16)$$

Similarly, the right standard Grünwald-Letnikov fractional derivative on the closed interval $[x, b]$ is

$$_x D_b^\alpha u(x) = \sum_{p=0}^{n-1}\frac{u^{(p)}(b)(b-x)^{p-\alpha}}{\Gamma(p-\alpha+1)} + \frac{(-1)^n}{\Gamma(n-\alpha)}\int_x^b(\eta-x)^{n-\alpha-1}u^{(n)}(\eta)d\eta. \qquad (17)$$

Thus, for $a \to -\infty$, $u^{(p)}(x)$ approaches to zero and Equation (16) leads to have:

$$_{-\infty}D_x^\alpha u(x) = \frac{1}{\Gamma(n-\alpha)} \int_{-\infty}^x (x-\eta)^{n-\alpha-1} u^{(n)}(\eta) d\eta,$$
$$= \frac{1}{\Gamma(n-\alpha)} \frac{\partial^n}{\partial x^n} \int_{-\infty}^x (x-\eta)^{n-\alpha-1} u(\eta) d\eta, \qquad (18)$$

which gives the left Riemann-Liouville fractional derivative as we are expected and it is also exists for $n-1 < \alpha \leq n$. In a similar proof, we also have the right Riemann-Liouville fractional derivative as $b \to \infty$:

$$_xD_\infty^\alpha u(x) = \frac{(-1)^n}{\Gamma(n-\alpha)} \int_x^\infty (\eta-x)^{n-\alpha-1} u^{(n)}(\eta) d\eta,$$
$$= \frac{(-1)^n}{\Gamma(n-\alpha)} \frac{\partial^n}{\partial x^n} \int_x^\infty (\eta-x)^{n-\alpha-1} u(\eta) d\eta. \qquad (19)$$

□

Remark 1. *The left and right Riemann-Liouville fractional derivative of the function $u(x)$ with order α on a bounded domain $[0, L]$ are defined according to Theorem 1:*

Left Riemann-Liouville fractional derivative:

$$_0D_x^\alpha u(x) = \frac{1}{\Gamma(n-\alpha)} \frac{d^n}{dx^n} \int_0^x (x-\zeta)^{n-1-\alpha} u(\zeta) d\zeta. \qquad (20)$$

Right Riemann-Liouville fractional derivative:

$$_xD_L^\alpha u(x) = \frac{(-1)^n}{\Gamma(n-\alpha)} \frac{d^n}{dx^n} \int_x^L (\zeta-x)^{n-1-\alpha} u(\zeta) d\zeta. \qquad (21)$$

As it is discussed in Reference [46], the shifted Grünwald-Letnikov difference operator with first order

$$\Delta_{+x,p}^{(\alpha)} u(x) =_{-\infty} D_x^\alpha u(x) + O(h), \qquad (22)$$

which is defined as,

$$\Delta_{+x,p}^{(\alpha)} u(x) = \frac{1}{h^\alpha} \sum_{k=0}^\infty g_k^{(\alpha)} u(x-(k-p)h),$$
$$\Delta_{-x,p}^{(\alpha)} u(x) = \frac{1}{h^\alpha} \sum_{k=0}^\infty g_k^{(\alpha)} u(x-(k-p)h), \qquad (23)$$

approximates the left and right Riemann-Liouville fractional derivatives. Here p is an integer that shifts the approximation p-shift to the right and $g_k^\alpha = (-1)^k \binom{\alpha}{k}$ are the coefficients of the power series for the function $(1-z)^\alpha$,

$$(1-z)^\alpha = \sum_{k=0}^\infty (-1)^k \binom{\alpha}{k} z^k = \sum_{k=0}^\infty g_k^\alpha z^k \qquad (24)$$

for all $|z| \leq 1$, with:

$$g_0^{(\alpha)} = 1, \quad g_k^{(\alpha)} = \left(1 - \frac{\alpha+1}{k}\right) g_{k-1}^{(\alpha)}, \quad k = 1, 2, \dots. \qquad (25)$$

Lemma 2 ([47]). *The coefficients $g_k^{(\alpha)}$ satisfy the following properties for $0 < \alpha < 1$.*

$$\begin{cases} g_0^{(\alpha)} = 1, g_1^{(\alpha)} = -\alpha < 0, \\ g_2^{((\alpha)} < g_3^{(\alpha)} < \dots < 0, \\ \sum_{k=0}^\infty g_k^{(\alpha)} = 0, \sum_{k=0}^m g_k^{(\alpha)} > 0, m \geq 1. \end{cases} \qquad (26)$$

Lemma 3 ([47]). *The coefficients $g_k^{(\beta)}$ satisfy the following properties for the fractional order $1 < \beta < 2$.*

$$\begin{cases} g_0^{(\beta)} = 1, g_1^{(\alpha)} = -\beta < 0, \\ 1 \geq g_1^{((\alpha)} \geq g_2^{(\alpha)} \leq \ldots \geq 0, \\ \sum_{k=0}^{\infty} g_k^{(\alpha)} = 0, \sum_{k=0}^{m} g_k^{(\alpha)} < 0, m \geq 1. \end{cases} \tag{27}$$

Applying the above Theorem 1, and weighted shifted Grünwald-Letnikov fractional derivative derivation from Reference [46] for $0 < \alpha < 1, 1 < \beta \leq 2$, the left and right Riemann-Liouville fractional derivatives of $u(x)$ over a bounded interval at each point x can be formulated as:

$$_0D_x^\alpha u(x_m) = \frac{1}{h^\alpha} \sum_{k=0}^{m+1} w_k^{(\alpha)} u(x_{m-k+1}) + O(h^2)$$
$$_xD_L^\alpha u(x_m) = \frac{1}{h^\alpha} \sum_{k=0}^{N_x-m+1} w_k^{(\alpha)} u(x_{m+k-1}) + O(h^2) \tag{28}$$

and

$$_0D_x^\beta u(x_m) = \frac{1}{h^\beta} \sum_{k=0}^{m+1} w_k^{(\beta)} u(x_{m-k+1}) + O(h^2)$$
$$_xD_L^\beta u(x_m) = \frac{1}{h^\beta} \sum_{k=0}^{N_x-m+1} w_k^{(\beta)} u(x_{m+k-1}) + O(h^2), \tag{29}$$

where

$$w_0^{(\alpha)} = \frac{\alpha}{2} g_0^{(\alpha)}, \quad w_k^{(\alpha)} = \frac{\alpha}{2} g_k^{(\alpha)} + \frac{2-\alpha}{2} g_{k-1}^{(\alpha)}, k \geq 1$$
$$w_0^{(\beta)} = \frac{\beta}{2} g_0^{(\beta)}, \quad w_k^{(\beta)} = \frac{\beta}{2} g_k^{(\beta)} + \frac{2-\beta}{2} g_{k-1}^{(\beta)}, k \geq 1.$$

The properties of the weighted coefficients $w_k^{(\alpha)}$ and $w_k^{(\beta)}$ are discussed below.

Lemma 4 ([48]). *Assume that $0 < \alpha < 1$, then the coefficients $w_k^{(\alpha)}$ have the following properties:*

$$\begin{cases} w_0^{(\alpha)} = \frac{\alpha}{2} > 0, \ w_1^{(\alpha)} = \frac{2-\alpha-\alpha^2}{2} > 0, w_2^{(\alpha)} = \frac{\alpha(\alpha^2+\alpha-4)}{4} < 0, \\ w_2^{(\alpha)} < w_3^{(\alpha)} < w_4^{(\alpha)} < \ldots < 0, \\ \sum_{k=0}^{\infty} w_k^{(\alpha)} = 0, \ \sum_{k=0}^{m} w_k^{(\alpha)} > 0, m \geq 1. \end{cases} \tag{30}$$

Lemma 5 ([46]). *Assume that $1 < \beta \leq 2$, then the coefficients $w_k^{(\beta)}$ have the following properties:*

$$\begin{cases} w_0^{(\beta)} = \frac{\beta}{2} > 0, \ w_1^{(\beta)} = \frac{2-\beta-\beta^2}{2} > 0, w_2^{(\beta)} = \frac{\beta(\beta^2+\beta-4)}{4} < 0, \\ 1 \geq w_0^{(\beta)} \geq w_3^{(\beta)} \geq w_4^{(\beta)} \geq \ldots \geq 0, \\ \sum_{k=0}^{\infty} w_k^{(\beta)} = 0, \ \sum_{k=0}^{m} w_k^{(\beta)} < 0, m \geq 2. \end{cases} \tag{31}$$

3. Numerical Approximation for One Dimensional Two-Sided Convection-Diffusion Problem with Source Term

We have considered the one-dimensional two-sided space fractional convection–diffusion equation,

$$\frac{\partial u(x,t)}{\partial t} = c_x \frac{\partial^\alpha u(x,t)}{\partial |x|^\alpha} + d_x \frac{\partial^\beta u(x,t)}{\partial |x|^\beta} + p(x,t), (x,t) \in (0,L) \times (0,T) \tag{32}$$

with initial condition:
$$u(x,0) = g(x), \ 0 \leq x \leq L,$$
and with zero Dirichlet boundary conditions:
$$u(0,t) = 0, u(L,t) = 0, \ 0 < t \leq T,$$
where $0 < \alpha < 1, 1 < \beta < 2$.

The analytic solution for Riesz space fractional convection–diffusion equation is developed in Reference [49] using the spectral representation on a finite interval $[0, L]$. Reference [50] used Laplace transform and Fourier transform method for finding analytical solution of Riesz space fractional convection–diffusion problem with initial and zero Dirichlet boundary conditions. Here our discretization is based on the finite interval $[0, L]$ into a uniform mesh with the space step $h = L/N_x$ and the time step $\tau = T/N_t$, where N_x, N_t are positive integers and the set of grid points is denoted by $x_m = mh$ and $t_n = n\tau$ for $0 \leq m \leq N_x$ and $0 \leq n \leq N_t$. Let $t_{n+1/2} = (t_{n+1} + t_n)/2$ with $0 \leq n \leq N_t - 1$.

We have used the following notations for our formulation:
$$u_m^n = u(x_m, t_n), p_m^{n+1/2} = p(x_m, t_{n+1/2}), \delta_t u_m^n = \frac{u_m^{n+1} - u_m^n}{\tau}, c_x \geq 0, \ d_x > 0.$$

The Riesz space fractional convection–diffusion equation for $0 < \alpha < 1, 1 < \beta < 2$ can be written with following expression.

$$\frac{\partial^\alpha u(x,t)}{\partial |x|^\alpha} = -K_\alpha \left({}_0D_x^\alpha + {}_xD_L^\alpha \right) u(x,t), \quad \frac{\partial^\beta u(x,t)}{\partial |x|^\beta} = -K_\beta \left({}_0D_x^\beta + {}_xD_L^\beta \right) u(x,t) \quad (33)$$

Theorem 1 allows us to use the Riemann-Liouville fractional derivative definition for the formulation of the problem. The weighted shifted Grünwald-Letnikov derivative formula for approximating the two-sided fractional derivative is derived in References [46,48] for space fractional derivative and Crank-Nicolson scheme for time are used.

$$\frac{u_m^{n+1} - u_m^n}{\tau} = \frac{K_\alpha c_x}{h^\alpha} \left[\sum_{k=0}^{m+1} \omega_k^{(\alpha)} \frac{u_{m+1-k}^{n+1} + u_{m+1-k}^n}{2} + \sum_{k=0}^{N_x-m+1} \omega_k^{(\alpha)} \frac{u_{m-1+k}^{n+1} + u_{m-1+k}^n}{2} \right]$$
$$+ \frac{K_\beta d_x}{h^\beta} \left[\sum_{k=0}^{m+1} \omega_k^{(\beta)} \frac{u_{m+1-k}^{n+1} + u_{m+1-k}^n}{2} + \sum_{k=0}^{N_x-m+1} \omega_k^{(\beta)} \frac{u_{m-1+k}^{n+1} + u_{m-1+k}^n}{2} \right] + p_m^{n+1/2}, \quad (34)$$

where $K_\alpha = \frac{-1}{2\cos(\pi\alpha/2)}, K_\beta = \frac{-1}{2\cos(\pi\beta/2)}$. Then we have,

$$u_m^{n+1} - \frac{\bar{c}_x}{2} \left(\sum_{k=0}^{m+1} \omega_k^{(\alpha)} u_{m-k+1}^{n+1} + \sum_{k=0}^{N_x-m+1} \omega_k^{(\alpha)} u_{m+k-1}^{n+1} \right)$$
$$- \frac{\bar{d}_x}{2} \left(\sum_{k=0}^{m+1} \omega_k^{(\beta)} u_{m-k+1}^{n+1} + \sum_{k=0}^{N_x-m+1} \omega_k^{(\beta)} u_{m+k-1}^{n+1} \right)$$
$$= u_m^n + \frac{\bar{c}_x}{2} \left(\sum_{k=0}^{m+1} \omega_k^{(\alpha)} u_{m-k+1}^n + \sum_{k=0}^{N_x-m+1} \omega_k^{(\alpha)} u_{m+k-1}^n \right) \quad (35)$$
$$+ \frac{\bar{d}_x}{2} \left(\sum_{k=0}^{m+1} \omega_k^{(\beta)} u_{m-k+1}^n + \sum_{k=0}^{N_x-m+1} \omega_k^{(\beta)} u_{m+k-1}^n \right) + \tau p_m^{n+1/2},$$

where $\bar{c}_x = \frac{K_\alpha c_x \tau}{h^\alpha}, \bar{d}_x = \frac{K_\beta d_x \tau}{h^\beta}$.

Assume U_m^n be the numerical approximation of the solution u_m^n, then the CN-WSGD formulation for RSFCDEs become:

$$
\begin{aligned}
U_m^{n+1} &- \frac{\bar{c}_x}{2} \left(\sum_{k=0}^{m+1} \omega_k^{(\alpha)} U_{m-k+1}^{n+1} + \sum_{k=0}^{N_x-m+1} \omega_k^{(\alpha)} U_{m+k-1}^{n+1} \right) \\
&- \frac{\bar{d}_x}{2} \left(\sum_{k=0}^{m+1} \omega_k^{(\beta)} U_{m-k+1}^{n+1} + \sum_{k=0}^{N_x-m+1} \omega_k^{(\beta)} U_{m+k-1}^{n+1} \right) \\
&= U_m^n + \frac{\bar{c}_x}{2} \left(\sum_{k=0}^{m+1} \omega_k^{(\alpha)} U_{m-k+1}^n + \sum_{k=0}^{N_x-m+1} \omega_k^{(\alpha)} U_{m+k-1}^n \right) \\
&+ \frac{\bar{d}_x}{2} \left(\sum_{k=0}^{m+1} \omega_k^{(\beta)} U_{m-k+1}^n + \sum_{k=0}^{N_x-m+1} \omega_k^{(\beta)} U_{m+k-1}^n \right) + \tau p_m^{n+1/2}.
\end{aligned}
\quad (36)
$$

By denoting,

$$
a = \begin{pmatrix}
\omega_1^{(\alpha)} & \omega_0^{\alpha} & 0 & \cdots & 0 & 0 \\
\omega_2^{\alpha} & \omega_1^{(\alpha)} & \omega_0^{\alpha} & \cdots & 0 & 0 \\
\omega_3^{\alpha} & \omega_2^{(\alpha)} & \omega_1^{\alpha} & \cdots & 0 & 0 \\
\vdots & \vdots & \vdots & \ddots & \vdots & \vdots \\
\omega_{m-2}^{(\alpha)} & \omega_{m-3}^{(\alpha)} & \omega_{m-4}^{(\alpha)} & \cdots & \omega_1^{(\alpha)} & \omega_0^{(\alpha)} \\
\omega_{m-1}^{(\alpha)} & \omega_{m-2}^{(\alpha)} & \omega_{m-3}^{(\alpha)} & \cdots & \omega_2^{(\alpha)} & \omega_1^{(\alpha)}
\end{pmatrix},
$$

$$
b = \begin{pmatrix}
\omega_1^{(\beta)} & \omega_0^{\beta} & 0 & \cdots & 0 & 0 \\
\omega_2^{\beta} & \omega_1^{(\beta)} & \omega_0^{(\beta)} & \cdots & 0 & 0 \\
\omega_3^{(\beta)} & \omega_2^{(\beta)} & \omega_1^{(\beta)} & \cdots & 0 & 0 \\
\vdots & \vdots & \vdots & \ddots & \vdots & \vdots \\
\omega_{m-2}^{(\beta)} & \omega_{m-3}^{(\beta)} & \omega_{m-4}^{(\beta)} & \cdots & \omega_1^{(\beta)} & \omega_0^{(\beta)} \\
\omega_{m-1}^{(\beta)} & \omega_{m-2}^{(\beta)} & \omega_{m-3}^{(\beta)} & \cdots & \omega_2^{(\beta)} & \omega_1^{(\beta)}
\end{pmatrix},
$$

we have,

$$
A = \frac{\bar{c}_x}{2} \left(a + a^\top \right) + \frac{\bar{d}_x}{2} \left(b + b^\top \right). \quad (37)
$$

Therefore, the system of equations takes the form:

$$
(I - A) U^{n+1} = (I + A) U^n + \tau p^{n+\frac{1}{2}}, \quad (38)
$$

where I is the $(N_x - 1) \times (N_t - 1)$ identity matrix with $A_{m,j}$ as the matrix coefficients. These matrix coefficients for $m = 1, 2, 3, \ldots, N_x - 1, j = 1, 2, \ldots, N_x - 1$ are defined by:

$$
A_{m,j} = \begin{cases}
\frac{\bar{c}_x}{2} \left(\omega_0^{(\alpha)} + \omega_2^{(\alpha)} \right) + \frac{\bar{d}_x}{2} \left(\omega_0^{(\beta)} + \omega_2^{(\beta)} \right), & j = m - 1, \\
\frac{\bar{c}_x}{2} \left(\omega_0^{(\alpha)} + \omega_2^{(\alpha)} \right) + \frac{\bar{d}_x}{2} \left(\omega_0^{(\beta)} + \omega_2^{(\beta)} \right), & j = m + 1, \\
\bar{c}_x \omega_1^{(\alpha)} + \bar{d}_x \omega_1^{(\beta)}, & j = m, \\
\frac{\bar{c}_x}{2} \omega_{m-j+1}^{(\alpha)} + \frac{\bar{d}_x}{2} \omega_{m-j+1}^{(\beta)}, & j < m - 1, \\
\frac{\bar{c}_x}{2} \omega_{j-m+1}^{(\alpha_1)} + \frac{\bar{d}_x}{2} \omega_{j-m+1}^{(\beta)}, & j > m + 1.
\end{cases} \quad (39)
$$

For the convenience of implementation, using the matrix form of the grid functions,

$$u^n = [U_1^{n+1}, U_2^{n+1}, \ldots, U_{N_x-1}^n]^\top$$
$$p^{n+1/2} = [p_1^{n+1/2}, p_2^{n+1/2}, \ldots, p_{N_x-1}^{n+1/2}]^\top.$$

4. Formulation and Discretization of Two-Dimensional Riesz Space Fractional Convection Diffusion Equation with CNADI-WSGD Scheme

The analytic solution for two-dimensional Riesz space fractional anomalous diffusion equation is obtained by using the Fourier series expansion with homogeneous Dirichlet boundary condition. Let us take a bounded domain as $\Omega = [0, L_x] \times [0, L_y]$, $\Omega_t = [0, T]$ for our discretization of the problem. Here our aim is to find the full numerical approximation of the two-dimensional Riesz space fractional convection–diffusion problem with zero Dirichlet boundary condition over a finite domain $\Omega \times \Omega_t$.

Consider the two-dimensional two-sided space fractional convection-diffusion problem with constant coefficients as:

$$\begin{cases} \dfrac{\partial u(x,y,t)}{\partial t} = c_x \dfrac{\partial^{\alpha_1} u(x,y,t)}{\partial |x|^{\alpha_1}} + c_y \dfrac{\partial^{\alpha_2} u(x,y,t)}{\partial |y|^{\alpha_2}} + d_x \dfrac{\partial^{\beta_1} u(x,y,t)}{\partial |x|^{\beta_1}} \\ + d_y \dfrac{\partial^{\beta_2} u(x,y,t)}{\partial |y|^{\beta_2}} + p(x,y,t), \quad (x,y,t) \in \Omega \times \Omega_t, \\ u(x,y,0) = g(x,y), \quad (x,y) \in \Omega, \\ u(0,y,t) = 0, \quad u(L_x,y,t) = 0, \quad (y,t) \in [0, L_y] \times \Omega_t, \\ u(x,0,t) = 0, \quad u(x,L_y,t) = 0, \quad (x,t) \in [0, L_x] \times \Omega_t, \end{cases} \quad (40)$$

where $0 < \alpha_1, \alpha_2 < 1, 1 < \beta_1, \beta_2 < 2, c_x, c_y \geq 0$ and $d_x, d_y > 0$ express the velocity parameter and positive diffusion coefficients. Here the function $u(x,y,t)$ is specified as solute concentration under the groundwater. The Riesz space fractional-order derivative is defined as:

$$\dfrac{\partial^{\alpha_1} u(x,y,t)}{\partial |x|^{\alpha_1}} = K_{\alpha_1} \left[{}_0 D_x^{\alpha_1} + {}_x D_{L_x}^{\alpha_1} \right] u(x,y,t), \quad \dfrac{\partial^{\alpha_2} u(x,y,t)}{\partial |y|^{\alpha_2}} = K_{\alpha_2} \left[{}_0 D_y^{\alpha_2} + {}_y D_{L_y}^{\alpha_2} \right] u(x,y,t),$$

$$\dfrac{\partial^{\beta_1} u(x,y,t)}{\partial |x|^{\beta_1}} = K_{\beta_1} \left[{}_0 D_x^{\beta_1} + {}_x D_{L_x}^{\beta_1} \right] u(x,y,t), \quad \dfrac{\partial^{\beta_2} u(x,y,t)}{\partial |y|^{\beta_2}} = K_{\beta_2} \left[{}_0 D_y^{\beta_2} + {}_y D_{L_y}^{\beta_2} \right] u(x,y,t), \quad (41)$$

where

$$K_{\alpha_1} = \dfrac{-1}{2\cos(\pi\alpha_1/2)}, \quad K_{\alpha_2} = \dfrac{-1}{2\cos(\pi\alpha_2/2)}, \quad K_{\beta_1} = \dfrac{-1}{2\cos(\pi\beta_1/2)}, \quad K_{\beta_2} = \dfrac{-1}{2\cos(\pi\beta_2/2)}$$

and also from the coincides Theorem 1, we have the following left and right Riemann-Liouville fractional derivative definition for two dimension space fractional derivative.

$${}_0 D_x^{\alpha_1} u(x,y,t) = \dfrac{1}{\Gamma(1-\alpha_1)} \dfrac{\partial}{\partial x} \int_0^x (x-\eta)^{-\alpha_1} u(\eta, y, t) d\eta,$$

$${}_x D_{L_x}^{\alpha_1} u(x,y,t) = \dfrac{-1}{\Gamma(1-\alpha_1)} \dfrac{\partial}{\partial x} \int_x^{L_x} (\eta - x)^{-\alpha_1} u(\eta, y, t) d\eta,$$

$${}_0 D_x^{\beta_1} u(x,y,t) = \dfrac{1}{\Gamma(2-\beta_1)} \dfrac{\partial^2}{\partial x^2} \int_0^x (x-\eta)^{1-\beta_1} u(\eta, y, t) d\eta, \quad (42)$$

$${}_x D_{L_x}^{\beta_1} u(x,y,t) = \dfrac{1}{\Gamma(2-\beta_1)} \dfrac{\partial^2}{\partial x^2} \int_x^{L_x} (\eta - x)^{1-\beta_1} u(\eta, y, t) d\eta,$$

where $\Gamma(.)$ denotes the gamma function. In a similar way, we can express the Riesz space fractional operators $\dfrac{\partial^{\alpha_2} u(x,y,t)}{\partial |y|^{\alpha_2}}$ and $\dfrac{\partial^{\beta_2} u(x,y,t)}{\partial |y|^{\beta_2}}$ of orders $\alpha_2, \beta_2, (0 < \alpha_2 < 1, 1 < \beta_2 < 2)$ corresponding to y−direction. For time and space discretization, we use CNADI scheme and WSGD operator respectively. Let $u_{m,j}^n$ be the approximated solution of $u(x_m, y_j, t_n)$, $t_{n+1/2} = (t_n + t_{n+1})/2$, $p_{m,j}^{n+1/2} = p(x_m, y_j, t_{n+1/2})$, $h_x = \frac{L_x}{N_x}$, $h_y = \frac{L_y}{N_y}$, for the uniform space steps h_x, h_y and time-step $\tau = T/N_t$, $0 < m < N_x - 1$, $0 < j < N_y - 1$, $0 < n < N_t - 1$.

Therefore, the weighted and shifted-Grünwald-Letnikov difference operator with CN scheme for 2D-RSFCDE is expressed in the following formulation.

$$\frac{u_{m,j}^{n+1} - u_{m,j}^n}{\tau} = \frac{k_{\alpha_1} c_x}{h_1^{\alpha_1}} \left(\sum_{k=0}^{m+1} \omega_k^{(\alpha_1)} \frac{u_{m-k+1,j}^{n+1} + u_{m-k+1,j}^n}{2} + \sum_{k=0}^{N_x-m+1} \omega_k^{(\alpha_1)} \frac{u_{m+k-1,j}^{n+1} + u_{m+k-1,j}^n}{2} \right)$$
$$+ \frac{k_{\alpha_2} c_y}{h_2^{\alpha_2}} \left(\sum_{k=0}^{j+1} \omega_k^{(\alpha_2)} \frac{u_{m,j-k+1}^{n+1} + u_{m,j-k+1}^n}{2} + \sum_{k=0}^{N_y-j+1} \omega_k^{(\alpha_2)} \frac{u_{m,j+k-1}^{n+1} + u_{m,j+k-1}^n}{2} \right)$$
$$+ \frac{k_{\beta_1} d_x}{h_1^{\beta_1}} \left(\sum_{k=0}^{m+1} \omega_k^{(\beta_1)} \frac{u_{m-k+1,j}^{n+1} + u_{m-k+1,j}^n}{2} + \sum_{k=0}^{N_x-m+1} \omega_k^{(\beta_1)} \frac{u_{m+k-1,j}^{n+1} + u_{m+k-1,j}^n}{2} \right)$$
$$+ \frac{k_{\beta_2} d_y}{h_2^{\beta_2}} \left(\sum_{k=0}^{j+1} \omega_k^{(\alpha_2)} \frac{u_{m,j-k+1}^{n+1} + u_{m,j-k+1}^n}{2} + \sum_{k=0}^{N_y-j+1} \omega_k^{(\beta_2)} \frac{u_{m,j+k-1}^{n+1} + u_{m,j+k-1}^n}{2} \right) + p_{m,j}^{n+1/2}. \quad (43)$$

To simplify our formulation, it is possible to symbolize the following operator as:

$$\Delta_x^{(\alpha_1)} u_{m,j}^n = \frac{K_{\alpha_1} c_x}{h_1^{\alpha_1}} \left(\sum_{k=0}^{m+1} \omega_k^{(\alpha_1)} u_{m-k+1,j}^n + \sum_{k=0}^{N_x-m+1} \omega_k^{(\alpha_1)} u_{m+k-1,j}^n \right) + O(h_1^2)$$

$$\Delta_y^{(\alpha_2)} u_{m,j}^n = \frac{K_{\alpha_2} c_y}{h_2^{\alpha_2}} \left(\sum_{k=0}^{j+1} \omega_k^{(\alpha_2)} u_{m,j-k+1}^n + \sum_{k=0}^{N_y-j+1} \omega_k^{(\alpha_2)} u_{m,j+k-1}^n \right) + O(h_2^2) \quad (44)$$

$$\Delta_x^{(\beta_1)} u_{m,j}^n = \frac{K_{\beta_1} d_x}{h_1^{\beta_1}} \left(\sum_{k=0}^{m+1} \omega_k^{(\beta_1)} u_{m-k+1,j}^n + \sum_{k=0}^{N_x-m+1} \omega_k^{(\beta_1)} u_{m+k-1,j}^n \right) + O(h_1^2)$$

$$\Delta_y^{(\beta_2)} u_{m,j}^n = \frac{K_{\beta_2} d_y}{h_2^{\beta_2}} \left(\sum_{k=0}^{j+1} \omega_k^{(\beta_2)} u_{m,j-k+1}^n + \sum_{k=0}^{N_y-j+1} \omega_k^{(\beta_2)} u_{m,j+k-1}^n \right) + O(h_2^2).$$

By grouping like terms from Equations (43) and (44), we have:

$$\left[1 - \frac{\tau}{2} \left(\Delta_x^{(\alpha_1)} + \Delta_x^{(\beta_1)} \right) - \frac{\tau}{2} \left(\Delta_y^{(\alpha_2)} + \Delta_y^{(\beta_2)} \right) \right] u_{m,j}^{n+1}$$
$$= \left[1 + \frac{\tau}{2} \left(\Delta_x^{(\alpha_1)} + \Delta_x^{(\beta_1)} \right) + \frac{\tau}{2} \left(\Delta_y^{(\alpha_2)} + \Delta_y^{(\beta_2)} \right) \right] u_{m,j}^n + \frac{\tau}{2} p_{m,j}^{n+1/2} + \tau T_{m,j}^n, \quad (45)$$

where $T_{m,j}^n$ represent truncation error that can satisfy $\left| T_{m,j}^n \right| \leq \hat{k} \left(\tau^2 + h_1^2 + h_2^2 \right)$.

Let us define the operators:

$$\Delta_x^{(\alpha)} = \Delta_x^{(\alpha_1)} + \Delta_x^{(\beta_1)}$$
$$\Delta_y^{(\beta)} = \Delta_y^{(\alpha_2)} + \Delta_y^{(\beta_2)},$$

with these operator definitions, the CNADI-WSGD scheme for the 2D-RSFCDE with homogeneous Dirichlet boundary conditions can be defined as an operator form:

$$\left[1 - \frac{\tau}{2} \left(\Delta_x^{(\alpha)} + \Delta_x^{(\beta)} \right) \right] u_{m,j}^{n+1} = \left[1 + \frac{\tau}{2} \left(\Delta_x^{(\alpha)} + \Delta_y^{(\beta)} \right) \right] u_{m,j}^n + \tau p_{m,j}^{n+1/2}. \quad (46)$$

An alternating direction implicit Peacemann-Rachford is reduced a two-dimensional problem in to a one dimensional problem with a better computational efficient. For CNADI the operator can be expressed in the product form as:

$$\left(1 - \frac{\tau}{2}\Delta_x^{(\alpha)}\right)\left(1 - \frac{\tau}{2}\Delta_y^{(\beta)}\right) u_{m,j}^{n+1}$$
$$= \left(1 + \frac{\tau}{2}\Delta_x^{(\alpha)}\right)\left(1 + \frac{\tau}{2}\Delta_y^{(\beta)}\right) u_{m,j}^n + \frac{\tau}{2} p_{m,j}^{n+1/2}, \quad 1 \le m \le N_x - 1, \ 1 \le j \le N_y - 1, \quad (47)$$

which produce an additional perturbation error in the form of $\frac{\tau^2}{4}\Delta_x^\alpha \Delta_y^\beta \left(u_{m,j}^{n+1} - u_{m,j}^n\right)$ that has Taylor expansion as:

$$\frac{\tau^2}{4}\Delta_x^\alpha \Delta_y^\beta \left(u_{m,j}^{n+1} - u_{m,j}^n\right) = \frac{\tau^3}{4}\left(\left(\Delta_x^{(\alpha_1)} + \Delta_x^{(\beta_1)}\right)\left(\Delta_y^{(\alpha_2)} + \Delta_y^{(\beta_2)}\right) u_t\right)_{m,j}^{n+1/2}$$
$$+ \tau^3 O\left(\tau^2 + h_1^2 + h_2^2\right). \quad (48)$$

As compared to the approximation errors, the additional perturbation errors is insignificant and the scheme defined in Equation (45) has second order accuracy in both space and time which is $O\left(\tau^2 + h_1^2 + h_2^2\right)$.

The problem defined by Equation (47) can be simulated by the following efficient Peacemann-Rachford ADI approximation as it was presented in Reference [46] by considering $u_{m,j}^*$ as an intermediate solution to make a numerical solution $u_{m,j}^n$ at time t_n to the numerical solution $u_{m,j}^{n+1}$ at time t_{n+1}. The corresponding iterative algorithms are:

Algorithm 1: The first step is to solve the problem in the x-direction for each fixed y_j to find an intermediate solution $u_{m,j}^*$ in the form:

$$\left(1 - \frac{\tau}{2}\Delta_x^{(\alpha)}\right) u_{m,j}^* = \left(1 + \frac{\tau}{2}\Delta_y^{(\beta)}\right) u_{m,j}^n + \frac{\tau}{2} p_{m,j}^{n+1/2}. \quad (49)$$

Algorithm 2: The next step is to solve the problem in y-direction for each fixed x_m as:

$$\left(1 - \frac{\tau}{2}\Delta_x^{(\alpha)}\right) u_{m,j}^{n+1} = \left(1 + \frac{\tau}{2}\Delta_y^{(\beta)}\right) u_{m,j}^* + \frac{\tau}{2} p_{m,j}^{n+1/2}. \quad (50)$$

Algorithm 3: We need to apply the homogeneous Dirichlet boundary conditions:

$$u_{0,j}^n = u(0, y_j, t_n) = 0, u_{N_x,j}^n = u(L_x, y_j, t_n) = 0,$$
$$u_{m,0}^n = u(x_m, 0, t_n) = 0, u_{m,N_y}^n = u(x_m, L_y, t_n) = 0.$$

Therefore now compute the boundary condition for the intermediate solution $u_{m,j}^*$ which can be derived from subtracting Equation (50) from (49) to get:

$$u_{m,j}^* = \frac{1}{2}\left(1 - \frac{\tau}{2}\Delta_y^{(\beta)}\right) u_{m,j}^{n+1} + \frac{1}{2}\left(1 + \frac{\tau}{2}\Delta_y^{(\beta)}\right) u_{m,j}^n. \quad (51)$$

Therefore, the boundary conditions for $u_{m,j}^*$ needed to solve each set of equations.

$$u_{0,j}^* = \frac{1}{2}\left(1 - \frac{\tau}{2}\Delta_y^{(\beta)}\right) u_{0,j}^{n+1} + \frac{1}{2}\left(1 + \frac{\tau}{2}\Delta_y^{(\beta)}\right) u_{0,j}^n$$
$$u_{N_x,j}^* = \frac{1}{2}\left(1 - \frac{\tau}{2}\Delta_y^{(\beta)}\right) u_{N_x,j}^{n+1} + \frac{1}{2}\left(1 + \frac{\tau}{2}\Delta_y^{(\beta)}\right) u_{N_x,j}^n. \quad (52)$$

By setting $U_{m,j}^n$ be the numerical approximation to exact solution $u_{m,j}^n$, we get the finite difference approximation for Equation (47):

$$\left(1 - \frac{\tau}{2}\delta_x^{(\alpha)}\right)\left(1 - \frac{\tau}{2}\delta_y^{(\beta)}\right)U_{m,j}^{n+1} = \left(1 + \frac{\tau}{2}\delta_x^{(\alpha)}\right)\left(1 + \frac{\tau}{2}\delta_y^{(\beta)}\right)U_{m,j}^n + \tau p_{m,j}^{n+1/2} \tag{53}$$

$$U^n = \left[u_{1,1}^n, ..., u_{N_1,1}^n, u_{1,2}^n, ..., u_{N_1,2}^n, ..., u_{1,N_2}^n, ..., u_{N_1,N_2}^{n+1/2}\right]^T$$

$$p^{n+1/2} = \left[p_{1,1}^{n+1/2}, ..., p_{N_1,1}^{n+1/2}, p_{1,2}^{n+1/2}, ..., p_{N_1,2}^{n+1/2}, ..., p_{1,N_2}^{n+1/2}, ..., p_{N_1,N_2}^{n+1/2}\right]^T.$$

5. CNADI-WSGD Scheme for Theoretical Analysis of 2D-RSFCDE with Source Term

5.1. Stability and Convergence Analysis of CNADI-WSGD Scheme

For discussing the stability and convergence of the scheme, we need to write our problem in matrix form. Thus, the Equation (49) can be put as:

$$(I - A)u_l^* = (I + A)u_l^n + \frac{\tau}{2}p, 1 \leq l \leq N_y - 1 \tag{54}$$

with

$$u_l^n = \left(u_{1,l}^n, u_{2,l}^n, ..., u_{N_x-1,l}^n\right)^T,$$

$$u_l^* = \left(u_{1,l}^*, u_{2,l}^*, ..., u_{N_x-1,l}^*\right)^T,$$

$$p = (p(x_1, y_l, t_n), p(x_2, y_l, t_n), ..., p(x_{N_x-1}, y_l, t_n))^T,$$

and the coefficient of matrix $A = (a_{m,j})_{(N_x-1)\times(N_x-1)}$,

$$a_{m,j} = \begin{cases} \frac{\bar{c}_x}{2}\left(\omega_0^{(\alpha_1)} + \omega_2^{(\alpha_1)}\right) + \frac{\bar{d}_x}{2}\left(\omega_0^{(\beta_1)} + \omega_2^{(\beta_1)}\right), & j = m-1, \\ \frac{\bar{c}_x}{2}\left(\omega_0^{(\alpha_1)} + \omega_2^{(\alpha_1)}\right) + \frac{\bar{d}_x}{2}\left(\omega_0^{(\beta_1)} + \omega_2^{(\beta_1)}\right), & j = m+1, \\ \bar{c}_x\omega_1^{(\alpha_1)} + \bar{d}_x\omega_1^{(\beta_1)}, & j = m, \\ \frac{\bar{c}_x}{2}\omega_{m-j+1}^{(\alpha_1)} + \frac{\bar{d}_x}{2}\omega_{m-j+1}^{(\beta_1)}, & j < m-1, \\ \frac{\bar{c}_x}{2}\omega_{j-m+1}^{(\alpha_1)} + \frac{\bar{d}_x}{2}\omega_{j-m+1}^{(\beta_1)}, & j > m+1. \end{cases} \tag{55}$$

In a similar way, Equation (50) can be given in matrix form:

$$(I - B)\bar{u}_q^{n+1} = (I + B)\bar{u}_q^*, 1 \leq q \leq N_x - 1, \tag{56}$$

where,

$$\bar{u}_q^{n+1} = \left(u_{q,1}^{n+1}, u_{q,2}^{n+1}, ..., u_{q,N_y-1}^{n+1}\right)^T,$$

$$\bar{u}_q^* = \left(u_{q,1}^*, u_{q,2}^*, ..., u_{q,N_y-1}^*\right)^T,$$

and $B = (b_{m,j})_{(N_y-1) \times (N_y-1)}$,

$$b_{m,j} = \begin{cases} \dfrac{\bar{c}_y}{2}\left(\omega_0^{(\alpha_2)} + \omega_2^{(\alpha_2)}\right) + \dfrac{\bar{d}_y}{2}\left(\omega_0^{(\beta_2)} + \omega_2^{(\beta_2)}\right), & j = m-1, \\ \dfrac{\bar{c}_y}{2}\left(\omega_0^{(\alpha_2)} + \omega_2^{(\alpha_2)}\right) + \dfrac{\bar{d}_y}{2}\left(\omega_0^{(\beta_2)} + \omega_2^{(\beta_2)}\right), & j = m+1, \\ \bar{c}_y\omega_1^{(\alpha_2)} + \bar{d}_y\omega_1^{(\beta_2)}, & j = m, \\ \dfrac{\bar{c}_y}{2}\omega_{m-j+1}^{(\alpha_2)} + \dfrac{\bar{d}_y}{2}\omega_{m-j+1}^{(\beta_2)}, & j < m-1, \\ \dfrac{\bar{c}_y}{2}\omega_{j-m+1}^{(\alpha_2)} + \dfrac{\bar{d}_y}{2}\omega_{j-m+1}^{(\beta_2)}, & j > m+1 \end{cases} \qquad (57)$$

where, $\bar{c}_x = \dfrac{K_{\alpha_1} c_x \tau}{h_1^{\alpha_1}}$, $\bar{c}_y = \dfrac{K_{\alpha_2} c_y \tau}{h_2^{\alpha_2}}$, $\bar{d}_x = \dfrac{K_{\beta_1} d_x \tau}{h_1^{\beta_1}}$, $\bar{d}_y = \dfrac{K_{\beta_2} d_y \tau}{h_2^{\beta_2}}$.

Theorem 2. *Assume that $0 < \alpha_1, \alpha_2 < 1, 1 < \beta_1, \beta_2 \leq 2$, the coefficient matrices defined in Equations (55) and (57), then the diagonal matrix and coefficient matrix satisfy:*

$$|a_{m,m}| > \sum_{j=0, m \neq 1}^{N_x - 1} |a_{m,j}|, m = 1, 2, 3, ..., N_x - 1,$$

$$|b_{m,m}| > \sum_{j=0, m \neq 1}^{N_y - 1} |b_{m,j}|, m = 1, 2, 3, ..., N_y - 1, \qquad (58)$$

tells us that A and B which are defined in Equations (54) and (56) are strictly diagonally dominant.

Proof. First we will consider the diagonal dominance of the coefficient matrix $a_{m,j}$. Since $K_{\alpha_1} = \dfrac{1}{2\cos(\pi \alpha_1/2)} > 0$ and $K_{\beta_1} = \dfrac{1}{2\cos(\pi \beta_1/2)} < 0$ for $0 < \alpha_1 < 1, 1 < \beta_1 \leq 2$ implies that $\bar{c}_x = \dfrac{\tau K_{\alpha_1} c_x}{h_1^{\alpha_1}} > 0$ and $\bar{d}_x = \dfrac{\tau K_{\beta_1} d_x}{h_1^{\beta_1}} < 0$.

$$a_{m,m+1} = \dfrac{\bar{c}_x}{2}\left(\omega_0^{(\alpha_1)} + \omega_2^{(\alpha_1)}\right) + \dfrac{\bar{d}_x}{2}\left(\omega_0^{(\beta_1)} + \omega_2^{(\beta_1)}\right). \qquad (59)$$

From Lemmas 4 and 5, we have:

$$\omega_0^{(\alpha_1)} + \omega_2^{(\alpha_1)} = \dfrac{\alpha_1}{2} + \dfrac{\alpha_1(\alpha_1^2 + \alpha_1 - 4)}{4} < 0,$$

$$\omega_0^{(\beta_1)} + \omega_2^{(\beta_1)} = \dfrac{\beta_1}{2} + \dfrac{\beta_1(\beta_1^2 + \beta_1 - 4)}{4} > 0. \qquad (60)$$

Since $\bar{c}_x > 0$ and $\bar{d}_x < 0$, then we have:

$$a_{m,m+1} = \dfrac{\bar{c}_x}{2}\left(\omega_0^{(\alpha_1)} + \omega_2^{(\alpha_1)}\right) + \dfrac{\bar{d}_x}{2}\left(\omega_0^{(\beta_1)} + \omega_2^{(\beta_1)}\right) < 0,$$

$$a_{m,m-1} = \dfrac{\bar{c}_x}{2}\left(\omega_0^{(\alpha_1)} + \omega_2^{(\alpha_1)}\right) + \dfrac{\bar{d}_x}{2}\left(\omega_0^{(\beta_1)} + \omega_2^{(\beta_1)}\right) < 0. \qquad (61)$$

By looking Lemmas 4 and 5, we have seen that $\omega_1^{(\alpha_1)} > 0$ and $\omega_1^{(\beta_1)} < 0$, hence,

$$a_{m,m} = \bar{c}_x \omega_1^{(\alpha_1)} + \bar{d}_x \omega_1^{(\beta_1)} > 0. \qquad (62)$$

As we have shown from Lemma 4, when $k \geq 3$, $\omega_k^{(\alpha_1)} < 0$, then $\bar{c}_x \omega_k^{(\alpha_1)} < 0$. Similarly by seeing Lemma 5, when $k \geq 3$, $\omega_k^{(\beta_1)} > 0$, then $\bar{d}_x \omega_k^{(\beta_1)} < 0$. These indicates that the coefficient matrix $a_{m,j} < 0$ for $j > m+1, j < m-1$.

$$|a_{m,m}| > \sum_{j=0, m \neq 1}^{N_x - 1} |a_{m,j}|, m = 1, 2, 3, ..., N_x - 1,$$

which means matrix A defined by the coefficient matrix $a_{m,j}$, is strictly diagonally dominant. In the same way, the diagonally dominant result for matrix B can also be found as matrix A. □

5.2. Stability Analysis of the CNADI-WSGD Method

In order to study the stability and convergence analysis for the CNADI-WSGD scheme, we are focused on the following description.

Let $\chi_h = \left\{ v : v = \{v_{m,j}\} : \left\{ \{x_m = mh_1; y_j = jh_2\}_{m=0}^{N_x} \right\}_{j=0}^{N_y} \right\}$ be the mesh grid function. For any $v = v_{m,j} \in \chi_h$, we define our point-wise maximum norm as:

$$||v||_\infty = \max_{(m,j) \in \chi_h} |v_{m,j}|, \tag{63}$$

and the discrete L^2-norm

$$||v|| = \sqrt{h_1 h_2 \sum_{m=1}^{N_x - 1} \sum_{j=1}^{N_y - 1} v_{m,j}^2}. \tag{64}$$

Our next aim is to show the stability of CNADI-WSGD method which is defined as in the matrix form:

$$(I - A)((I - B)U^{n+1} = (I + A)((I + B)U^n + T^{n+1} \tag{65}$$

where the matrices A and B define the operator $\frac{\tau}{2}\Delta_x^{(\alpha)}$ and $\frac{\tau}{2}\Delta_y^{(\beta)}$, respectively. The vector T^{n+1} absorbs the source term $p_{m,j}^{n+1/2}$ and the Dirichlet boundary condition in the formulated problem.

Theorem 3. *Let $U_{m,j}^n$ be the numerical solution of the exact solution $u_{m,j}^n$, then CNADI-WSGD finite difference method (53) is unconditionally stable for $0 < \alpha_1, \alpha_2 < 1$ with $1 < \beta_1, \beta_2 \leq 2$.*

Proof. The matrices A and B are of size $(N_x - 1)(N_y - 1) \times (N_x - 1)(N_y - 1)$. The commutative property defined in Reference [51], allows us to obtain the unconditional stability of CNADI-WSGD method. The matrix A which is $(N_y - 1) \times (N_y - 1)$ block diagonal matrix whose blocks are $(N_x - 1 \times N_x - 1)$ square super triangular matrices which is expressed as $A = diag\left(A_1, A_2, ..., A_{N_y-1}\right)$. In the same way, the matrix B is a block matrix with $(N_x - 1) \times (N_x - 1)$ square diagonal matrices. The matrix B can be written as $B = [b_{m,j}]$, where each $b_{m,j}$ is an $(N_x - 1) \times (N_x - 1)$ matrix such that $b_{m,j}$ is a diagonal matrix $b_{m,j} = diag\left(b_{m,j}, b_{m,j}, ..., b_{m,j}\right)$ where $b_{m,j}$ is the $(m,j)^{th}$ entry of the matrix B defined above. As we have seen from Theorem 2, matrix A is diagonal dominant with entry $a_{m,m} > 0$. The sum of the absolute value of the off-diagonal entries on the row m of matrix A is:

$$\sum_{j=0, j\neq m}^{N_x-1} |a_{m,j}| = \sum_{j=0}^{m-2} |a_{m,j}| + \sum_{j=m+2}^{N_x-1} |a_{m,j}| + |a_{m,m+1}| + |a_{m,m-1}|,$$

$$= -\sum_{j=0}^{m-2} \left(\frac{\bar{c}_x}{2} \omega_{m-j+1}^{(\alpha_1)} + \frac{\bar{d}_x}{2} \omega_{m-j+1}^{(\beta_1)} \right) - \sum_{j=m+2}^{N_x-1} \left(\frac{\bar{c}_x}{2} \omega_{j-m+1}^{(\alpha_1)} + \frac{\bar{d}_x}{2} \omega_{j-m+1}^{(\beta_1)} \right),$$

$$- \bar{c}_x \left(\omega_0^{(\alpha_1)} + \omega_2^{(\alpha_1)} \right) - \bar{d}_x \left(\omega_0^{(\beta_1)} + \omega_2^{(\beta_1)} \right)$$

$$< \sum_{j=-\infty}^{m-2} \left(\frac{\bar{c}_x}{2} \omega_{m-j+1}^{(\alpha_1)} + \frac{\bar{d}_x}{2} \omega_{m-j+1}^{(\beta_1)} \right) - \sum_{j=m+2}^{\infty} \left(\frac{\bar{c}_x}{2} \omega_{j-m+1}^{(\alpha_1)} + \frac{\bar{d}_x}{2} \omega_{j-m+1}^{(\beta_1)} \right) \quad (66)$$

$$- \bar{c}_x \left(\omega_0^{(\alpha_1)} + \omega_2^{(\alpha_1)} \right) - \bar{d}_x \left(\omega_0^{(\beta_1)} + \omega_2^{(\beta_1)} \right),$$

$$= -\bar{c}_x \sum_{k=3}^{\infty} \omega_k^{(\alpha_1)} - \bar{d}_x \sum_{k=3}^{\infty} \omega_k^{(\beta_1)} - \bar{c}_x \left(\omega_0^{(\alpha_1)} + \omega_2^{(\alpha_1)} \right) - \bar{d}_x \left(\omega_0^{(\beta_1)} + \omega_2^{(\beta_1)} \right),$$

$$= \bar{c}_x \omega_1^{(\alpha_1)} + \bar{d}_x \omega_1^{(\beta_1)} - \bar{c}_x \sum_{k=0}^{\infty} \omega_k^{(\alpha_1)} - \bar{d}_x \sum_{k=0}^{\infty} \omega_k^{(\beta_1)},$$

$$= \bar{c}_x \omega_1^{(\alpha_1)} + \bar{d}_x \omega_k^{(\beta_1)} = |a_{m,m}|,$$

implies that,

$$\sum_{j=0, j\neq m}^{N_x-1} |a_{m,j}| < |a_{m,m}|.$$

Next we need to show that the eigenvalue of matrix A is negative real parts. For $0 < \alpha_1 < 1$, $1 < \beta_1 < 2$, we can see that,

$$\left| \lambda_1 - \bar{c}_x \omega_1^{(\alpha_1)} - \bar{d}_x \omega_1^{(\beta_1)} \right| \leq \frac{\bar{c}_x}{2} \left(\left| \sum_{k=0, k\neq 1}^{m+1} \omega_k^{(\alpha_1)} + \sum_{k=0, \neq 1}^{N_x-m+1} \omega_k^{(\alpha_1)} \right| \right) + \frac{\bar{d}_x}{2} \left(\left| \sum_{k=0, k\neq 1}^{m+1} \omega_k^{(\beta_1)} + \sum_{k=0, \neq 1}^{N_x-m+1} \omega_k^{(\beta_1)} \right| \right)$$

$$\leq \frac{\bar{c}_x}{2} \left(\sum_{k=0, k\neq 1}^{m+1} \left| \omega_k^{(\alpha_1)} \right| + \sum_{k=0, \neq 1}^{N_x-m+1} \left| \omega_k^{(\alpha_1)} \right| \right) + \frac{\bar{d}_x}{2} \left(\sum_{k=0, k\neq 1}^{m+1} \left| \omega_k^{(\beta_1)} \right| + \sum_{k=0, \neq 1}^{N_x-m+1} \left| \omega_k^{(\beta_1)} \right| \right). \quad (67)$$

We have noticed that,

$$\sum_{k=0}^{\infty} \omega_k^{(\alpha_1)} = 0, \quad \sum_{k=0}^{\infty} \omega_k^{(\beta_1)} = 0,$$

and

$$\sum_{k=0}^{N_x} \omega_k^{(\alpha_1)} + \sum_{k=0}^{N_x} \omega_k^{(\beta_1)} < - \left(\omega_1^{(\alpha_1)} + \omega_1^{(\beta_1)} \right).$$

Therefore,

$$\left| \lambda_1 - \bar{c}_x \omega_1^{(\alpha_1)} - \bar{d}_x \omega_1^{(\beta_1)} \right| \leq - \left(\bar{c}_x \omega_1^{(\alpha_1)} + \bar{d}_x \omega_1^{(\beta_1)} \right).$$

The eigenvalue λ_1 of matrix A satisfy,

$$- \left(\bar{c}_x \omega_1^{(\alpha_1)} + \bar{d}_x \omega_1^{(\beta_1)} \right) \leq \bar{c}_x \omega_1^{(\alpha_1)} + \bar{d}_x \omega_1^{(\beta_1)} \leq \lambda_1 \leq 0. \quad (68)$$

According to Greschgorin Theorem [52], the given eigenvalue of matrix A have non-positive real parts. Here we have noted that matrix A has an eigenvalue of λ_1 if and only if $(I - A)$ has an eigenvalue of $(1 - \lambda_1)$ if and only if $(I - A)^{-1}(I + A)$ has an eigenvalue of $(1 + \lambda_1)/(1 - \lambda_1)$. From the first part of this statement, we can concluded that all eigenvalues of the matrix $(I - A)$ have a spectral radius which is larger than unity indicates the matrix is invertible. Thus, every eigenvalue of the $(I - A)^{-1}(I + A)$ has a spectral radius which is less than 1. Similarly, we can show that matrix B also satisfy the same property as matrix A. From the scheme (53), we can express the error e^{n+1} in U^{n+1} at time t_{n+1} and the error e^n in U^n at time t_n as:

$$e^{n+1} = (I - A)^{-1} (I - B)^{-1} (I + A) (I + B) e^n, \tag{69}$$

where the identity matrix I is $(N_x - 1) \times (N_y - 1)$ square. Hence, Equation (53) can be put in the form:

$$e^n = \left((I - A)^{-1}(I + A)\right)^n \left((I - B)^{-1}(I + B)\right)^n e^0. \tag{70}$$

Letting λ_1 and λ_2 be an eigenvalue of matrices A and B respectively, then it results from Equation (69) that the real parts of λ_1 and λ_2 are both negative. The spectral radius of each matrix is less than unity, which has followed that $\left((I - A)^{-1}(I + A)\right)^n$ and $\left((I - A)^{-1}(I + A)\right)^n$ which converges to null matrix (see Reference [46]). Therefore, we have concluded the scheme defined in Equation (53), is unconditionally stable. □

5.3. Convergence Analysis of CNADI-WSGD Scheme

First of all we can express the truncation error of CNADI-WSGD difference method. So, it is easy to conclude that:

$$\frac{u(x_m, y_j, t_{n+1}) - u(x_m, y_j, t_n)}{\tau} = \left(\frac{\partial u(x, y, t)}{\partial t}\right)_{m,j}^{n+1/2} + O(\tau^2) \tag{71}$$

$$\begin{aligned}
&\left(c_x \frac{\partial^{\alpha_1} u(x, y, t)}{\partial |x|^{\alpha_1}} + c_y \frac{\partial^{\alpha_2} u(x, y, t)}{\partial |y|^{\alpha_2}}\right)_{m,j}^{n+1/2} \\
&= \frac{1}{2}\left(c_x \frac{\partial^{\alpha_1} u(x_m, y_j, t_{n+1})}{\partial |x|^{\alpha_1}} + c_y \frac{\partial^{\alpha_2} u(x_m, y_j, t_{n+1})}{\partial |y|^{\alpha_2}}\right) \\
&+ \frac{1}{2}\left(c_x \frac{\partial^{\alpha_1} u(x_m, y_j, t_n)}{\partial |x|^{\alpha_1}} + c_y \frac{\partial^{\alpha_2} u(x_m, y_j, t_n)}{\partial |y|^{\alpha_2}}\right) + O(\tau^2)
\end{aligned}$$

$$\begin{aligned}
&c_x \frac{\partial^{\alpha_1} u(x_m, y_j, t_n)}{\partial |x|^{\alpha_1}} + c_y \frac{\partial^{\alpha_2} u(x_m, y_j, t_n)}{\partial |y|^{\alpha_2}} \\
&= \bar{c}_x \left(\sum_{k=0}^{m+1} w_k^{(\alpha_1)} u_{m-k+1,j}^n + \sum_{k=0}^{N_x - m + 1} w_k^{(\alpha_1)} u_{m+k-1,j}^n\right) \\
&+ \bar{c}_y \left(\sum_{k=0}^{j+1} w_k^{(\alpha_2)} u_{m,j-k+1}^n + \sum_{k=0}^{N_y - m + 1} w_k^{(\alpha_2)} u_{m,j+k-1}^n\right) + O(h_1^2 + h_2^2),
\end{aligned} \tag{72}$$

where $0 < \alpha_1, \alpha_2 < 1$. It is the same to have a truncation error of $O(\tau^2)$ and $O\left(h_1^2 + h_2^2\right)$ for $1 < \beta_1, \beta_2 < 2$.

Therefore, the truncation error from Equation (43) is given by:

$$T_{m,j}^{n+1} = O(\tau^3 + \tau h_1^2 + \tau h_2^2).$$

Theorem 4. Assume $u_{m,j}^n$ be the analytic solution, and let $U_{m,j}^n$ be the approximation solution of the finite difference method (65), then for all $1 \leq n \leq N_t$, we have the estimate:

$$\|u_{m,j}^n - U_{m,j}^n\|_\infty \leq C(\tau^2 + h_1^2 + h_2^2), \tag{73}$$

where $\|u_{m,j}^n - U_{m,j}^n\|_\infty = \max_{1 \leq m \leq N_x, 1 \leq j \leq N_y} |u_{m,j}^n - U_{m,j}^n| = |e_{\hat{m},\hat{j}}^n|$, C is a positive constant independent of h_1, h_2 and τ with $\|.\|$ stands for the discrete L^2-norm.

Proof. Assume that $e_{m,j}^n$ be the error at grid points (x_m, y_j, t_n) can be defined as $e_{m,j}^n = u_{m,j}^n - U_{m,j}^n$ and denote $e^n = \left(e_{1,1}^n, e_{2,1}^n, ..., e_{N_x-1,1}^n, e_{1,2}^n, ..., e_{N_x-1,2}^n, ..., e_{1,N_y-1}^n, ..., e_{N_x-1,N_y-1}^n\right)^T$.

By looking to Equation (43), the error satisfies:

$$
\begin{aligned}
e_{m,j}^{n+1} &+ \frac{\bar{c}_x}{2}\left(\sum_{k=0}^{m+1} w_k^{(\alpha_1)} e_{m-k+1,j}^{n+1} + \sum_{k=0}^{N_x-m+1} w_k^{(\alpha_1)} e_{m+k-1,j}^{n+1}\right) + \frac{\bar{c}_y}{2}\left(\sum_{k=0}^{j+1} w_k^{(\alpha_2)} e_{m,j-k+1}^{n+1} + \sum_{k=0}^{N_y-j+1} w_k^{(\alpha_2)} e_{m,j+k-1}^{n+1}\right) \\
&+ \frac{\bar{d}_x}{2}\left(\sum_{k=0}^{m+1} w_k^{(\beta_1)} e_{m-k+1,j}^{n+1} + \sum_{k=0}^{N_x-m+1} w_k^{(\beta_1)} e_{m+k-1,j}^{n+1}\right) + \frac{\bar{d}_y}{2}\left(\sum_{k=0}^{j+1} w_k^{(\beta_2)} e_{m,j-k+1}^{n+1} + \sum_{k=0}^{N_y-j+1} w_k^{(\beta_2)} e_{m,j+k-1}^{n+1}\right) \\
&= e_{m,j}^n - \frac{\bar{c}_x}{2}\left(\sum_{k=0}^{m+1} w_k^{(\alpha_1)} e_{m-k+1,j}^n + \sum_{k=0}^{N_x-m+1} w_k^{(\alpha_1)} e_{m+k-1,j}^n\right) - \frac{\bar{c}_y}{2}\left(\sum_{k=0}^{j+1} w_k^{(\alpha_2)} e_{m,j-k+1}^n + \sum_{k=0}^{N_y-j+1} w_k^{(\alpha_2)} e_{m,j+k-1}^n\right) \\
&- \frac{\bar{d}_x}{2}\left(\sum_{k=0}^{m+1} w_k^{(\beta_1)} e_{m-k+1,j}^n + \sum_{k=0}^{N_x-m+1} w_k^{(\beta_1)} e_{m+k-1,j}^n\right) - \frac{\bar{d}_y}{2}\left(\sum_{k=0}^{j+1} w_k^{(\beta_2)} e_{m,j-k+1}^n + \sum_{k=0}^{N_y-j+1} w_k^{(\beta_2)} e_{m,j+k-1}^n\right) \\
&+ \tau O(\tau^2 + h_1^2 + h_2^2).
\end{aligned}
\tag{74}
$$

We have $e^0 = 0$, we have from Equations (43) and (74) if $n = 0$,

$$
\begin{aligned}
R_{m,j}^1 &= \frac{\bar{c}_x}{2}\left(\sum_{k=0}^{m+1} w_k^{(\alpha_1)} e_{m-k+1,j}^1 + \sum_{k=0}^{N_x-m+1} w_k^{(\alpha_1)} e_{m+k-1,j}^1\right) + \frac{\bar{c}_y}{2}\left(\sum_{k=0}^{j+1} w_k^{(\alpha_2)} e_{m,j-k+1}^1 + \sum_{k=0}^{N_y-j+1} w_k^{(\alpha_2)} e_{m,j+k-1}^1\right) \\
&+ \frac{\bar{d}_x}{2}\left(\sum_{k=0}^{m+1} w_k^{(\beta_1)} e_{m-k+1,j}^1 + \sum_{k=0}^{N_x-m+1} w_k^{(\beta_1)} e_{m+k-1,j}^1\right) + \frac{\bar{d}_y}{2}\left(\sum_{k=0}^{j+1} w_k^{(\beta_2)} e_{m,j-k+1}^1 + \sum_{k=0}^{N_y-j+1} w_k^{(\beta_2)} e_{m,j+k-1}^1\right)
\end{aligned}
\tag{75}
$$

if $n > 0$,

$$
\begin{aligned}
R_{m,j}^{n+1} &= \frac{\bar{c}_x}{2}\left(\sum_{k=0}^{m+1} w_k^{(\alpha_1)} e_{m-k+1,j}^{n+1} + \sum_{k=0}^{N_x-m+1} w_k^{(\alpha_1)} e_{m+k-1,j}^{n+1}\right) + \frac{\bar{c}_y}{2}\left(\sum_{k=0}^{j+1} w_k^{(\alpha_2)} e_{m,j-k+1}^{n+1} + \sum_{k=0}^{N_y-j+1} w_k^{(\alpha_2)} e_{m,j+k-1}^{n+1}\right) \\
&+ \frac{\bar{d}_x}{2}\left(\sum_{k=0}^{m+1} w_k^{(\beta_1)} e_{m-k+1,j}^{n+1} + \sum_{k=0}^{N_x-m+1} w_k^{(\beta_1)} e_{m+k-1,j}^{n+1}\right) + \frac{\bar{d}_y}{2}\left(\sum_{k=0}^{j+1} w_k^{(\beta_2)} e_{m,j-k+1}^{n+1} + \sum_{k=0}^{N_y-j+1} w_k^{(\beta_2)} e_{m,j+k-1}^{n+1}\right),
\end{aligned}
\tag{76}
$$

where $R_{m,j}^{n+1} \leq \tau c(\tau^2 + h_1^2 + h_2^2)$, $m = 1, 2, ..., N_x - 1$, $j = 1, 2, ..., N_y - 1$, $n = 1, 2, ..., N_t - 1$, c is positive constant independent of time step and space size. We have used the mathematical induction to prove our Theorem 4. Let $n = 1$ and assume $|e_{\hat{m},\hat{j}}| = \max_{1 \leq m \leq N_x-1, 1 \leq j \leq N_y-1} |e_{m,j}^1|$, we have the following expression.

$$\|e^1\|_\infty = |e^1_{\hat{m},\hat{j}}| \leq \frac{\bar{c}_x}{2}\left(\sum_{k=0}^{m+1}\omega_k^{(\alpha_1)}|e^1_{\hat{m}-k+1,\hat{j}}| + \sum_{k=0}^{N_x-m+1}\omega_k^{(\alpha_1)}|e^1_{\hat{m}+k-1,\hat{j}}|\right)$$

$$+ \frac{\bar{c}_y}{2}\left(\sum_{k=0}^{j+1}\omega_k^{(\alpha_2)}|e^1_{\hat{m},\hat{j}-k+1}| + \sum_{k=0}^{N_y-j+1}\omega_k^{(\alpha_2)}|e^1_{\hat{m},\hat{j}+k-1}|\right)$$

$$+ \frac{\bar{d}_x}{2}\left(\sum_{k=0}^{m+1}\omega_k^{(\beta_1)}|e^1_{\hat{m}-k+1,\hat{j}}| + \sum_{k=0}^{N_x-m+1}\omega_k^{(\alpha_1)}|e^1_{\hat{m}+k-1,\hat{j}}|\right)$$

$$+ \frac{\bar{d}_y}{2}\left(\sum_{k=0}^{j+1}\omega_k^{(\beta_2)}|e^1_{\hat{m},\hat{j}-k+1}| + \sum_{k=0}^{N_y-j+1}\omega_k^{(\beta_2)}|e^1_{\hat{m},\hat{j}+k-1}|\right)$$

$$\leq \left|\frac{\bar{c}_x}{2}\left(\sum_{k=0}^{m+1}\omega_k^{(\alpha_1)}e^1_{\hat{m}-k+1,\hat{j}} + \sum_{k=0}^{N_x-m+1}\omega_k^{(\alpha_1)}e^1_{\hat{m}+k-1,\hat{j}}\right)\right. \tag{77}$$

$$+ \frac{\bar{c}_y}{2}\left(\sum_{k=0}^{j+1}\omega_k^{(\alpha_2)}e^1_{\hat{m},\hat{j}-k+1} + \sum_{k=0}^{N_y-j+1}\omega_k^{(\alpha_2)}e^1_{\hat{m},\hat{j}+k-1}\right)$$

$$+ \frac{\bar{d}_x}{2}\left(\sum_{k=0}^{m+1}\omega_k^{(\beta_1)}e^1_{\hat{m}-k+1,\hat{j}} + \sum_{k=0}^{N_x-m+1}\omega_k^{(\alpha_1)}e^1_{\hat{m}+k-1,\hat{j}}\right)$$

$$+ \left.\frac{\bar{d}_y}{2}\left(\sum_{k=0}^{j+1}\omega_k^{(\beta_2)}e^1_{\hat{m},\hat{j}-k+1} + \sum_{k=0}^{N_y-j+1}\omega_k^{(\beta_2)}e^1_{\hat{m},\hat{j}+k-1}\right)\right|$$

$$= \left|R^1_{m,j}\right| \leq \tau C(\tau^2 + h_1^2 + h_2^2)$$

Assume that if $n \leq r$, $\|e^r\|_\infty \leq \tau C(\tau^2 + h_1^2 + h_2^2)$ hold and let $n = r+1$, let $|e^{r+1}_{\hat{m},\hat{j}}| = \max_{1 \leq m \leq N_x-1, 1 \leq j \leq N_y-1}|e^{r+1}_{m,j}|$. Thus,

$$\|e^{r+1}\|_\infty = |e^{r+1}_{\hat{m},\hat{j}}| \leq \frac{\bar{c}_x}{2}\left(\sum_{k=0}^{m+1}\omega_k^{(\alpha_1)}|e^{r+1}_{\hat{m}-k+1,\hat{j}}| + \sum_{k=0}^{N_x-m+1}\omega_k^{(\alpha_1)}|e^{r+1}_{\hat{m}+k-1,\hat{j}}|\right)$$

$$+ \frac{\bar{c}_y}{2}\left(\sum_{k=0}^{j+1}\omega_k^{(\alpha_2)}|e^{r+1}_{\hat{m},\hat{j}-k+1}| + \sum_{k=0}^{N_y-j+1}\omega_k^{(\alpha_2)}|e^{r+1}_{\hat{m},\hat{j}+k-1}|\right)$$

$$+ \frac{\bar{d}_x}{2}\left(\sum_{k=0}^{m+1}\omega_k^{(\beta_1)}|e^{r+1}_{\hat{m}-k+1,\hat{j}}| + \sum_{k=0}^{N_x-m+1}\omega_k^{(\alpha_1)}|e^{r+1}_{\hat{m}+k-1,\hat{j}}|\right)$$

$$+ \frac{\bar{d}_y}{2}\left(\sum_{k=0}^{j+1}\omega_k^{(\beta_2)}|e^{r+1}_{\hat{m},\hat{j}-k+1}| + \sum_{k=0}^{N_y-j+1}\omega_k^{(\beta_2)}|e^{r+1}_{\hat{m},\hat{j}+k-1}|\right)$$

$$\leq \left|\frac{\bar{c}_x}{2}\left(\sum_{k=0}^{m+1}\omega_k^{(\alpha_1)}e^{r+1}_{\hat{m}-k+1,\hat{j}} + \sum_{k=0}^{N_x-m+1}\omega_k^{(\alpha_1)}e^{r+1}_{\hat{m}+k-1,\hat{j}}\right)\right. \tag{78}$$

$$+ \frac{\bar{c}_y}{2}\left(\sum_{k=0}^{j+1}\omega_k^{(\alpha_2)}e^{r+1}_{\hat{m},\hat{j}-k+1} + \sum_{k=0}^{N_y-j+1}\omega_k^{(\alpha_2)}e^{r+1}_{\hat{m},\hat{j}+k-1}\right)$$

$$+ \frac{\bar{d}_x}{2}\left(\sum_{k=0}^{m+1}\omega_k^{(\beta_1)}e^{r+1}_{\hat{m}-k+1,\hat{j}} + \sum_{k=0}^{N_x-m+1}\omega_k^{(\alpha_1)}e^{r+1}_{\hat{m}+k-1,\hat{j}}\right)$$

$$+ \left.\frac{\bar{d}_y}{2}\left(\sum_{k=0}^{j+1}\omega_k^{(\beta_2)}e^{r+1}_{\hat{m},\hat{j}-k+1} + \sum_{k=0}^{N_y-j+1}\omega_k^{(\beta_2)}e^{r+1}_{\hat{m},\hat{j}+k-1}\right)\right|$$

$$= \left|R^{r+1}_{m,j}\right| \leq \tau C(\tau^2 + h_1^2 + h_2^2)$$

Therefore, there exists a positive constant c^* such that
$$\left|e_{m,j}^{r+1}\right|_\infty \le c^*(\tau^2 + h_1^2 + h_2^2),$$
which completes the proof. □

6. Numerical Simulations

1. Consider the one dimensional RSFCDEs over a bounded domain with initial and Dirichlet boundary conditions:
$$\begin{cases} \dfrac{\partial u(x,y,t)}{\partial t} = c_x \dfrac{\partial^\alpha u(x,t)}{\partial |x|^\alpha} + d_x \dfrac{\partial^\beta u(x,t)}{\partial |x|^\beta} + p(x,t), \\ u(x,0) = 0, 0 < x \le 1, \\ u(0,t) = u(1,t) = 0, 0 < t \le T, \end{cases}$$

with the source term:
$$\begin{aligned} p(x,t) =\ & t^{\beta-1} e^{\alpha t}(\beta + \alpha t) x^2 (1-x)^2 \\ & + \dfrac{c_x t^\beta e^{\alpha t}}{2\cos(\alpha\pi/2)} \Bigg[\dfrac{2}{\Gamma(3-\alpha)} \left(x^{2-\alpha} + (1-x)^{2-\alpha} \right) \\ & - \dfrac{12}{\Gamma(4-\alpha)} \left(x^{3-\alpha} + (1-x)^{3-\alpha} \right) + \dfrac{24}{\Gamma(5-\alpha)} \left(x^{4-\alpha} + (1-x)^{4-\alpha} \right) \Bigg] \\ & + \dfrac{d_x t^\beta e^{\alpha t}}{2\cos(\beta\pi/2)} \Bigg[\dfrac{2}{\Gamma(3-\beta)} \left(x^{2-\beta} + (1-x)^{2-\beta} \right) \\ & - \dfrac{12}{\Gamma(4-\beta)} \left(x^{3-\beta} + (1-x)^{3-\beta} \right) + \dfrac{24}{\Gamma(5-\beta)} \left(x^{4-\beta} + (1-x)^{4-\beta} \right) \Bigg]. \end{aligned}$$

The exact solution is
$$u(x,t) = t^\beta e^{\alpha t} x^2 (1-x)^2.$$

All the numerical simulations are done based on the finite space domain $\Omega \times \Omega_t$ where $\Omega = [0,1] \times [0,1]$ and $\Omega_t = [0,1]$. The order of convergence both in space and time are calculated using the formula:
$$\begin{aligned} Order_1 &= \dfrac{||E(h,\tau)||_\infty / ||E(h/2,\tau/2)||_\infty}{\log(2)}, \\ Order_2 &= \dfrac{||E(h_x,h_y,\tau)||_\infty / ||E(h_x/2,h_y/2,\tau/2)||_\infty}{\log(2)}, \end{aligned} \tag{79}$$

where $order_1$ is the rate of convergence for one-dimensional two-sided space fractional convection–diffusion equation and $order_2$ is rate convergence of two dimensional two-sided space fractional equation. $||E(h,\tau)||_\infty$ is the maximum error for one dimensional space fractional problem and $||E(h_x,h_y,\tau)||_\infty$ for two dimensional space fractional problem, denoted as $Max - Error$. As we have seen in Table 1, the second order convergence and the maximum error are confirmed at each grid size for convection-dominance (i.e., $c_x > d_x$) for one dimensional two-sided space fractional convection–diffusion equation with different space fractional order. As we have refined the grid size, the suitable maximum error is obtained. The convergence order and maximum error for a diffusion–dominance (i.e., $c_x < d_x$) one dimensional two-sided space fractional convection–diffusion problem are shown in Table 2. Figure 1 shows the good agreement of exact and numerical solution of one-dimensional convection–diffusion equation with the coefficients $c_x = 0.5, d_x = 1.5$ and with fractional orders $\alpha = 0.75, \beta = 1.85$ at $N_x = N_t = 100$ grid points.

Table 1. Convergence order and maximum error are produced with convection–dominance for example 1 at $T = 1, c_x = 2, d_x = 0.25$.

		$\alpha = 0.1$		$\alpha = 0.45$		$\alpha = 0.85$	
	$h = \tau$	Max − Error	$Order_1$	Max − Error	$Order_1$	Max − Error	$Order_1$
$\beta = 1.25$	1/10	8.3×10^{-3}	−	1.21×10^{-2}	−	1.70×10^{-2}	−
	1/20	2.2×10^{-3}	1.9156	3.2×10^{-3}	1.9189	4.4×10^{-3}	1.9500
	1/40	5.5354×10^{-4}	1.9907	8.0831×10^{-4}	1.9851	1.1×10^{-3}	2.0000
	1/80	1.3870×10^{-4}	2.0030	2.0291×10^{-4}	1.9941	2.7919×10^{-4}	1.9782
	1/160	3.4320×10^{-5}	2.0086	5.0701×10^{-5}	2.0008	6.9932×10^{-5}	1.9972
$\beta = 1.85$	1/10	8.8×10^{-3}	−	1.24×10^{-2}	−	1.81×10^{-2}	−
	1/20	2.4×10^{-3}	1.8745	3.3×10^{-3}	1.9098	4.8×10^{-3}	1.9149
	1/40	6.0999×10^{-4}	1.9762	8.5839×10^{-4}	1.9428	1.2×10^{-3}	2.0000
	1/80	1.5478×10^{-4}	1.9786	2.1846×10^{-4}	1.9743	3.1912×10^{-4}	1.9109
	1/160	3.8968×10^{-5}	1.9899	5.5120×10^{-5}	1.9867	8.0991×10^{-5}	1.9783

Table 2. Convergence order and maximum error produced with diffusion-dominance for example 1 at $T = 1, c_x = 0.25, d_x = 2$.

		$\alpha = 0.1$		$\alpha = 0.45$		$\alpha = 0.85$	
	$h = \tau$	Max − Error	$Order_1$	Max − Error	$Order_1$	Max − Error	$Order_1$
$\beta = 1.25$	1/10	8.3×10^{-3}	−	1.23×10^{-2}	−	1.75×10^{-2}	−
	1/20	2.1×10^{-3}	1.9827	3.2×10^{-3}	1.9425	4.5×10^{-3}	1.9594
	1/40	5.3416×10^{-4}	1.9750	7.9693×10^{-4}	2.0055	1.1×10^{-3}	2.0324
	1/80	1.3389×10^{-4}	1.9962	1.9979×10^{-4}	1.9960	2.8337×10^{-4}	1.9567
	1/160	3.3498×10^{-5}	1.9989	4.9986×10^{-5}	1.9989	7.0902×10^{-5}	1.9988
$\beta = 1.85$	1/10	9.5×10^{-3}	−	1.42×10^{-2}	−	2.01×10^{-2}	−
	1/20	2.5×10^{-3}	1.9260	3.7×10^{-3}	1.9403	5.3×10^{-3}	1.9231
	1/40	6.3804×10^{-4}	1.9702	9.5140×10^{-4}	1.9594	1.3×10^{-3}	2.0275
	1/80	1.6126×10^{-4}	1.9843	2.4052×10^{-4}	1.9839	3.4116×10^{-4}	1.9300
	1/160	4.0532×10^{-5}	1.9923	6.0459×10^{-5}	1.9921	8.5774×10^{-5}	1.9918

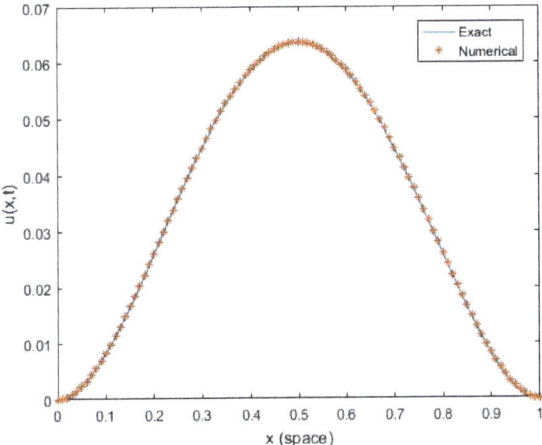

Figure 1. Comparison of exact and numerical solution for one-dimensional convection–diffusion equations (CDEs) at $\alpha = 0.75, \beta = 1.85$ for numerical example 1.

2. Consider two-dimensional diffusion problem.($c_x = c_y = 0$)

$$\begin{cases} \dfrac{\partial u(x,y,t)}{\partial t} = d_x \dfrac{\partial^{\beta_1} u(x,y,t)}{\partial |x|^{\beta_2}} + d_y \dfrac{\partial^{\beta_2} u(x,y,t)}{\partial |y|^{\beta_2}} + p(x,y,t) \\ u(x,y,0) = x^2(1-x)^2 y^2 (1-y)^2, 0 < x \le 1, 0 < y \le 1, \\ u(x,y,t)|_{\partial\Omega} = 0, 0 < t \le T, \ 1 < \beta_1, \beta_2 < 2, \end{cases}$$

with the source term:

$$\begin{aligned} p(x,y,t) &= \beta_1 (t+1)^{\beta_1-1} x^2 (1-x)^2 \beta_2 (t+1)^{\beta_2-1} y^2 (1-y)^2 \\ &+ \dfrac{d_x}{2\cos(\beta_1 \pi/2)} (t+1)^{\beta_1} \left[\dfrac{2}{\Gamma(3-\beta_1)} \left(x^{2-\beta_1} + (1-x)^{2-\beta_1} \right) \right. \\ &- \dfrac{12}{\Gamma(4-\beta_1)} \left(x^{3-\beta_1} + (1-x)^{3-\beta_1} \right) + \dfrac{24}{\Gamma(5-\beta_1)} \left(x^{4-\beta_1} + (1-x)^{4-\beta_1} \right) \Big] y^2 (1-y)^2 \quad (80) \\ &+ \dfrac{d_y}{2\cos(\beta_2 \pi/2)} (t+1)^{\beta_2} \left[\dfrac{2}{\Gamma(3-\beta_2)} \left(y^{2-\beta_2} + (1-y)^{2-\beta_2} \right) \right. \\ &- \dfrac{12}{\Gamma(4-\beta_2)} \left(y^{3-\beta_2} + (1-y)^{3-\beta_2} \right) + \dfrac{24}{\Gamma(5-\beta_2)} \left(y^{4-\beta_2} + (1-y)^{4-\beta_2} \right) \Big] x^2 (1-x)^2. \end{aligned}$$

Table 3 shows that the maximum error and order of convergence for two-dimensional two-sided space fractional diffusion equation with different space fractional orders by taking $c_x = 0 = c_y$. For this numerical simulation, we have used same step-size for space and time (i.e., $h_x = h_y = \tau$). The maximum time domain that used to obtain all the numerical results is $T = 1$ and the diffusion coefficients are $d_x = 2 = d_y$. The surface plot of $u(x,y,t)$ with the diffusion coefficients $d_x = 2.5$, $d_y = 1.5, \beta_1 = 1.25, \beta_2 = 1.85$ at the mesh points $h_1 = h_2 = \tau = 0.01$ is given in Figure 2.

Table 3. Convergence rate and maximum error produced for example 2 at $T = 1, d_x = 2 = d_y, h_x = h_y = \tau$.

		$\beta_1 = 1.25$		$\beta_1 = 1.5$		$\beta_1 = 1.95$	
	h_x, h_y, τ	Max − Error	Order$_2$	Max − Error	Order$_2$	Max − Error	Order$_2$
$\beta_2 = 1.85$	1/10	2.57×10^{-2}	−	2.75×10^{-2}	−	1.56×10^{-2}	−
	1/20	5.5×10^{-3}	2.2243	6.3×10^{-3}	2.0283	3.2×10^{-3}	2.2854
	1/40	7.4176×10^{-4}	2.4894	1.2×10^{-3}	2.3923	7.9692×10^{-4}	2.0056
	1/80	1.6630×10^{-4}	2.1565	2.4526×10^{-4}	2.2907	1.9977×10^{-4}	1.9961

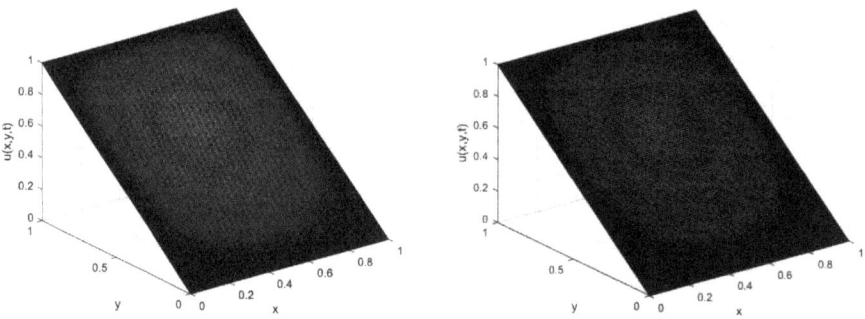

Figure 2. Surface of $u(x,y,t)$ for two-dimensional diffusion equation with $max - error = 1.7563 \times 10^{-4}$, $\beta_1 = 1.25, \beta_2 = 1.85$ for numerical example 2.

3. Let us consider the two-dimensional Riesz space fractional convection–diffusion problem with bounded domain:

$$\begin{cases} \dfrac{\partial u(x,y,t)}{\partial t} = c_x \dfrac{\partial^{\alpha_1} u(x,y,t)}{\partial |x|^{\alpha_1}} + c_y \dfrac{\partial^{\alpha_2} u(x,y,t)}{\partial |y|^{\alpha_2}} + d_x \dfrac{\partial^{\beta_1} u(x,y,t)}{\partial |x|^{\beta_1}} + d_y \dfrac{\partial^{\beta_2} u(x,y,t)}{\partial |y|^{\beta_2}} + p(x,y,t) \\ u(x,y,0) = 0, 0 < x \leq 1, 0 < y \leq 1, \\ u(x,y,t)|_{\partial\Omega} = 0, 0 \leq t \leq T, \end{cases}$$

with the source term:

$$\begin{aligned} p(x,y,t) &= t^{\beta_1-1} e^{\alpha_1 t}(\beta_1 + \alpha_1 t) x^2 (1-x)^2 y^2 (1-y)^2 \\ &+ \dfrac{c_x t^{\beta_1} e^{\alpha_1 t}}{2\cos(\alpha_1 \pi/2)} \left[\dfrac{2}{\Gamma(2-\alpha_1)} \left(x^{2-\alpha_1} + (1-x)^{2-\alpha_1} \right) \right. \\ &\left. - \dfrac{12}{\Gamma(4-\alpha_1)} \left(x^{3-\alpha_1} + (1-x)^{3-\alpha_1} \right) + \dfrac{24}{\Gamma(5-\alpha_1)} \left(x^{4-\alpha_1} + (1-x)^{4-\alpha_1} \right) \right] y^2 (1-y)^2 \\ &+ \dfrac{c_y t^{\beta_2} e^{\alpha_2 t}}{2\cos(\alpha_2 \pi/2)} \left[\dfrac{2}{\Gamma(2-\alpha_2)} \left(y^{2-\alpha_2} + (1-y)^{2-\alpha_2} \right) \right. \\ &\left. - \dfrac{12}{\Gamma(4-\alpha_2)} \left(y^{3-\alpha_2} + (1-y)^{3-\alpha_2} \right) + \dfrac{24}{\Gamma(5-\alpha_2)} \left(y^{4-\alpha_2} + (1-y)^{4-\alpha_2} \right) \right] x^2 (1-x)^2 \\ &+ \dfrac{d_x t^{\beta_1} e^{\alpha_1 t}}{2\cos(\beta_1 \pi/2)} \left[\dfrac{2}{\Gamma(2-\beta_1)} \left(x^{2-\beta_1} + (1-x)^{2-\beta_1} \right) \right. \\ &\left. - \dfrac{12}{\Gamma(4-\beta_1)} \left(x^{3-\beta_1} + (1-x)^{3-\beta_1} \right) + \dfrac{24}{\Gamma(5-\beta_1)} \left(x^{4-\beta_1} + (1-x)^{4-\beta_1} \right) \right] y^2 (1-y)^2 \\ &+ \dfrac{d_y t^{\beta_2} e^{\alpha_2 t}}{2\cos(\beta_2 \pi/2)} \left[\dfrac{2}{\Gamma(2-\beta_2)} \left(y^{2-\beta_2} + (1-y)^{2-\beta_2} \right) \right. \\ &\left. - \dfrac{12}{\Gamma(4-\beta_2)} \left(y^{3-\beta_1} + (1-y)^{3-\beta_2} \right) + \dfrac{24}{\Gamma(5-\beta_2)} \left(y^{4-\beta_2} + (1-y)^{4-\beta_2} \right) \right] x^2 (1-x)^2 \end{aligned}$$

The exact solution is,

$$u(x,y,t) = t^\beta e^{\alpha t} x^2 (1-x)^2 y^2 (1-y)^2.$$

In Table 4, we have found a numerical results that produce second order convergence rate and maximum error for two sided two dimensional space fractional convection–diffusion equation with diffusion–dominance ($c_x = 0.25 = c_y, d_x = d_y = 2$) phenomena. For this simulation we have taken a fixed value for $\beta_2(\beta_2 = 1.75)$ and for $\alpha_2(\alpha_2 = 0.5)$ with different values for α_1, β_1. Similarly in Table 5, we have considered the convection–dominance ($c_x = 2 = c_y, d_x = d_y = 0.25$) two-sided two-dimensional space fractional convection–diffusion problem with fixed $\beta_2(\beta_2 = 1.75)$ and for fixed $\alpha_2(\alpha_2 = 0.5)$. The order of convergence and maximum errors are calculated using the formula expressed in Equation (79). In Figure 3 the surface plot of exact and numerical solutions for two-dimensional convection–diffusion equations are investigated by considering the coefficients $c_x = c_y = 2.5, d_x = d_y = 1.5$ with orders $\alpha_1 = 0.75, \alpha_2 = 0.75, \beta_1 = 1.85, \beta_2 = 1.85$ at $N_x = N_y = N_t = 100$ mesh grid points.

Table 4. Convergence order produced with diffusion-dominance for example 3 at $T = 1, c_x = 0.25 = c_y, d_x = 2 = d_y, h_x = h_y = \tau$.

	h_x, h_y, τ	$\alpha_1 = 0.5$		$\alpha_1 = 0.75$		$\alpha_1 = 0.95$	
		Max − Error	Order$_2$	Max − Error	Order$_2$	Max − Error	Order$_2$
$\beta_1 = 1.25$	1/10	2.00×10^{-2}	−	2.09×10^{-2}	−	2.6×10^{-3}	−
	1/20	5.7×10^{-3}	1.8110	6.1×10^{-3}	1.7766	6.4607×10^{-4}	2.0087
	1/40	1.7×10^{-3}	1.7454	1.9×10^{-3}	1.6828	1.4898×10^{-4}	2.1166
	1/80	2.8802×10^{-4}	1.9962	6.6947×10^{-4}	1.7467	2.9810×10^{-5}	2.3213
$\beta_1 = 1.85$	1/10	1.52×10^{-2}	−	2.06×10^{-2}	−	2.34×10^{-2}	−
	1/20	3.6×10^{-3}	2.0780	5.00×10^{-3}	2.0426	5.5×10^{-3}	2.0890
	1/40	7.6214×10^{-4}	2.2399	1.1×10^{-3}	2.1844	1.0001×10^{-3}	2.4595
	1/80	1.0496×10^{-4}	2.4602	1.7907×10^{-4}	2.6189	1.8006×10^{-4}	2.4735
	1/160	1.9729×10^{-5}	2.4114	2.7175×10^{-5}	2.5202	3.5515×10^{-5}	2.3420

Table 5. Convergence order produced with convection-dominance for example 3 at $T = 1, c_x = 2 = c_y, d_x = 0.25 = d_y, h_x = h_y = \tau$.

	h_x, h_y, τ	$\alpha_1 = 0.5$		$\alpha_1 = 0.75$		$\alpha_1 = 0.95$	
		Max − Error	Order$_2$	Max − Error	Order$_2$	Max − Error	Order$_2$
$\beta_1 = 1.25$	1/10	1.51×10^{-2}	−	1.09×10^{-2}	−	4.1×10^{-3}	−
	1/20	2.00×10^{-3}	2.3163	1.9×10^{-3}	2.5203	1.4×10^{-3}	1.6502
	1/40	5.8273×10^{-4}	1.7791	4.8275×10^{-4}	1.9767	2.7307×10^{-4}	2.3581
	1/80	1.8883×10^{-4}	1.6257	1.5338×10^{-4}	1.6542	5.5304×10^{-5}	2.3038
$\beta_1 = 1.85$	1/10	1.80×10^{-2}	−	1.75×10^{-2}	−	1.39×10^{-2}	−
	1/20	4.3×10^{-3}	2.0656	4.1×10^{-3}	2.0937	2.4×10^{-3}	2.5340
	1/40	8.7170×10^{-4}	2.3024	7.6626×10^{-4}	2.4197	5.5154×10^{-4}	2.1215
	1/80	1.7428×10^{-4}	2.3224	1.3016×10^{-4}	2.5575	1.29899×10^{-4}	2.0861

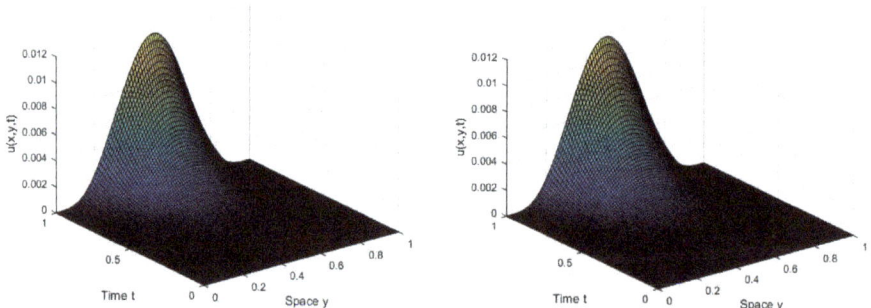

Figure 3. The surface of u(x,y,t) for $\alpha_1 = 0.75, \alpha_2 = 0.75, \beta_1 = 1.85, \beta_2 = 1.85$ for numerical example 3.

7. Conclusions

In our study, we have developed an algorithm for two-dimensional two-sided space fractional convection–diffusion problem using the CNADI difference method for time discretization combined with WSGD scheme for the approximation of space fractional derivative. We have used a shifted category of standard Grünwald-Letnikov difference method and weighted version of the shifted Grünwald-Letnikov difference approximation with CNADI scheme to have unconditionally stable and second order convergence both in space and time without extrapolation. Moreover, unconditional stability and second order convergence is justified for convection-dominance two-sided two dimension space fractional convection–diffusion equation. Our theoretical study and analysis has been confirmed by our numerical simulation in Section 6. We will consider the space fractional reaction convection–diffusion equation in our near further research.

Author Contributions: The authors contributed equally to the writing and approved the final manuscript of this paper. All authors have read and agreed to the final version of the manuscript.

Funding: This research was financially supported by the National Key Research (Grant No. 2017YFB0305601) and Development Program of China (Grant No. 2017YFB0701700).

Acknowledgments: We would like to express our thank to the editor and to the anonymous reviewers for their helpful comments for improving the article.

Conflicts of Interest: The authors confirmed that no conflict of interest.

Nomenclature

CN	Crank-Nicolson scheme.
ADI	Alternating direction implicit method.
CNADI	Crank-Nicolson alternating direction implicit method.
WSGD	Weighted shifted Grünwald-Letnikov difference operator.
RSFCDE	Riesz space fractional convection–diffusion equation.
2D-RSFCDE	Two-dimensional Riesz space fractional convection–diffusion equation.

References

1. Podlubny, I. *Fractional Differential Equations: An Introduction to Fractional Derivatives, Fractional Differential Equations, to Methods of Their Solution and Some of Their Applications*; Elsevier: New York, NY, USA, 1998; Volume 198.
2. Li, C.; Zeng, F. Finite difference methods for fractional differential equations. *Int. J. Bifurcat. Chaos* **2012**, *22*, 1230014. [CrossRef]
3. Diethelm, K. *The Analysis of Fractional Differential Equations*; Springer Science & Business Media: Berlin/Heidelberg, Germany, 2010.
4. Fomin, S.; Chugunov, V.; Hashida, T. Application of fractional differential equations for modeling the anomalous diffusion of contaminant from fracture into porous rock matrix with bordering alteration zone. *Trans. Porous Med.* **2010**, *81*, 187–205. [CrossRef]
5. Salman, W.; Gavriilidis, A.; Angeli, P. A model for predicting axial mixing during gas–liquid Taylor flow in microchannels at low Bodenstein number. *Chem. Eng. J.* **2004**, *101*, 391–396. [CrossRef]
6. Berkowitz, B.; Cortis, A.; Dror, I.; Scher, H. Laboratory experiments on dispersive transport across interfaces. *Water Resour. Res.* **2009**, *45*. [CrossRef]
7. Cortis, A.; Berkowitz, B. Computing "anomalous" contaminant transport in porous media. *Ground Water* **2005**, *43*, 947–950. [CrossRef]
8. Wang, K.; Wang, H. A fast characteristic finite difference method for fractional advection–diffusion equations. *Adv. Water. Resour.* **2011**, *34*, 810–816. [CrossRef]
9. Singh, J.; Swroop, R.; Kumar, D. A computational approach for fractional convection–diffusion equation via integral transforms. *Ain Shams Eng. J.* **2016**, *9*, 1019–1028. [CrossRef]
10. Wang, Y.M. A compact finite difference method for a class of time fractional convection–diffusion-wave equations with variable coefficients. *Numer. Algorithms* **2015**, *70*, 625–651. [CrossRef]
11. Gu, X.M.; Huang, T.Z.; Ji, C.C.; Carpentieri, B.; Alikhanov, A.A. Fast iterative method with a second-order implicit difference scheme for time-space fractional convection–diffusion equation. *J. Sci. Comput.* **2017**, *72*, 957–985. [CrossRef]
12. Cui, M. Combined compact difference scheme for the time fractional convection–diffusion equation with variable coefficients. *Appl. Math. Comput.* **2014**, *246*, 464–473. [CrossRef]
13. Tian, W.; Deng, W.; Wu, Y. Polynomial spectral collocation method for space fractional advection–diffusion equation. *Numer. Methods Partial Differ. Equ.* **2014**, *30*, 514–535. [CrossRef]
14. Bhrawy, A.H.; Baleanu, D. A spectral Legendre–Gauss–Lobatto collocation method for a space-fractional advection–diffusion equations with variable coefficients. *Rep. Math. Phys.* **2013**, *72*, 219–233. [CrossRef]

15. Saadatmandi, A.; Dehghan, M.; Azizi, M.R. The Sinc-Legendre collocation method for a class of fractional convection–diffusion equations with variable coefficients. *Commun. Nonlinear Sci. Numer. Simul.* **2012**, *17*, 4125–4136. [CrossRef]
16. Parvizi, M.; Eslahchi, M.R.; Dehghan, M. Numerical solution of fractional advection-diffusion equation with a nonlinear source term. *Numer. Algorithms* **2015**, *68*, 601–629. [CrossRef]
17. Li, C.; Zeng, F. *Numerical Methods for Fractional Calculus*; Chapman and Hall/CRC: Boca Raton, FL, USA, 2015.
18. Jin, B.; Lazarov, R.; Zhou, ; Z. A Petrov–Galerkin finite element method for fractional convection–diffusion equations. *SIAM J. Numer. Anal.* **2016**, *54*, 481–503. [CrossRef]
19. Aboelenen, T. A direct discontinuous Galerkin method for fractional convection–diffusion and Schrödinger-type equations. *Eur. Phys. J. Plus* **2018**, *133*, 316. [CrossRef]
20. Xu, Q.; Hesthaven, J.S. Discontinuous Galerkin method for fractional convection-diffusion equations. *SIAM J. Numer. Anal.* **2014**, *52*, 405–423. [CrossRef]
21. Gao, F.; Yuan, Y.; Du, N. An upwind finite volume element method for nonlinear convection–diffusion problem. *AJCM* **2011**, *1*, 264. [CrossRef]
22. Badr, M.; Yazdani, A.; Jafari, H. Stability of a finite volume element method for the time-fractional advection-diffusion equation. *Numer. Methods Partial Differ. Equ.* **2018**, *34*, 1459–1471. [CrossRef]
23. Ding, H.F.; Zhang, Y.X. New numerical methods for the Riesz space fractional partial differential equations, *Comput. Math. Appl.* **2012**, *63*, 1135–1146.
24. Anley, E.F.; Zheng, Z. Finite Difference Approximation Method for a Space Fractional Convection–Diffusion Equation with Variable Coefficients. *Symmetry* **2020**, *12*, 485. [CrossRef]
25. Zhang, Y.N.; Sun, Z.Z. Alternating direction implicit schemes for the two-dimensional fractional sub-diffusion equation. *J. Comput. Phys.* **2011**, *230*, 8713–8728. [CrossRef]
26. Tadjeran, C.; Meerschaert, M.M. A second-order accurate numerical method for the two-dimensional fractional diffusion equation. *J. Comput. Phys.* **2007**, *220*, 813–823. [CrossRef]
27. Meerschaert, M.M.; Scheffler, H.P.; Tadjeran, C. Finite difference methods for two-dimensional fractional dispersion equation. *J. Comput. Phys.* **2006**, *211*, 249–261. [CrossRef]
28. Valizadeh, S.; Borhanifar, A.; Malek, A. Compact ADI method for solving two-dimensional Riesz space fractional diffusion equation. *arXiv* **2018**, arXiv:1802.02015.
29. Wang, H.; Basu, T.S. A fast finite difference method for two-dimensional space-fractional diffusion equations. *SIAM J. Sci. Comput.* **2012**, *34*, A2444–A2458. [CrossRef]
30. Yin, X.; Fang, S.; Guo, C. Alternating-direction implicit finite difference methods for a new two-dimensional two-sided space-fractional diffusion equation. *Adv. Diff. Equ.* **2018**, *2018*, 389. [CrossRef]
31. Bu, W.; Tang, Y.; Yang, J. Galerkin finite element method for two-dimensional Riesz space fractional diffusion equations. *J. Comput. Phys.* **2014**, *276*, 26–38. [CrossRef]
32. Li, J.; Liu, F.; Feng, L.; Turner, I. A novel finite volume method for the Riesz space distributed-order diffusion equation. *Comput. Appl.* **2017**, *74*, 772–783. [CrossRef]
33. Chen, H.; Lv, W.; Zhang, T. A Kronecker product splitting preconditioner for two-dimensional space-fractional diffusion equations. *J. Comput. Phys.* **2018**, *360*, 1–14. [CrossRef]
34. Liu, F.; Chen, S.; Turner, I.; Burrage, K.; Anh, V. Numerical simulation for two-dimensional Riesz space fractional diffusion equations with a nonlinear reaction term. *Cent. Eur. J. Phys.* **2013**, *11*, 1221–1232. [CrossRef]
35. Zeng, F.; Liu, F.; Li, C.; Burrage, K.; Turner, I.; Anh, V. A Crank–Nicolson ADI spectral method for a two-dimensional Riesz space fractional nonlinear reaction-diffusion equation. *SIAM J. Numer. Anal.* **2014**, *52* 2599–2622. [CrossRef]
36. Balasim, A.T.; Ali, N.H.M. New group iterative schemes in the numerical solution of the two-dimensional time fractional advection-diffusion equation. *Cogent Math.* **2017**, *4*, 1412241. [CrossRef]
37. Zhao, Y.; Bu, W.; Huang, J.; Liu, D.Y.; Tang, Y. Finite element method for two-dimensional space-fractional advection–dispersion equations. *Appl. Math. Comput.* **2015**, *257*, 553–565. [CrossRef]
38. Pang, G.; Chen, W.; Sze, K.Y. A comparative study of finite element and finite difference methods for two-dimensional space-fractional advection–dispersion equation. *Adv. Appl. Math. Mech.* **2016**, *8*, 166–186. [CrossRef]
39. Li, J.; Liu, F.; Feng, L.; Turner, I. A novel finite volume method for the Riesz space distributed-order advection–diffusion equation. *Appl. Math. Model.* **2017**, *46*, 536–553. [CrossRef]

40. Chen, M.; Deng, W. A second-order numerical method for two-dimensional two-sided space fractional convection- diffusion equation. *Appl. Math. Model.* **2014**, *38*, 3244–3259. [CrossRef]
41. Zhang, Y.N.; Sun, Z.Z.; Zhao, X. Compact alternating direction implicit scheme for the two-dimensional fractional diffusion-wave equation. *SIAM J. Numer. Anal.* **2012**, *50*, 1535–1555. [CrossRef]
42. Benson, D.A.; Wheatcraft, S.; Meerschaert, M.M. Application of a fractional advection–dispersion equation. *Water Resour. Res.* **2000**, *36*, 1403–1412. [CrossRef]
43. Zhang, Y.; Meerschaert, M.M.; Neupauer, R.M. Backward fractional advection dispersion model for contaminant source prediction. *Water Resour. Res.* **2016**, *52*, 2462–2473. [CrossRef]
44. Samko, S.G.; Kilbas, A.A.; Marichev, O.I. *Fractional Integrals and Derivatives*; Gordon and Breach Science Publisher: Yverdon, Switzerland, 1993; Volume 1.
45. Podlubny, I. *Fractional Differential Equations Academic*; Academic Press: New York, NY, USA, 1999.
46. Tian, W.; Zhou, H.; Deng, W. A class of second order difference approximations for solving space fractional diffusion equations. *Math. Comp.* **2015**, *84*, 1703–1727. [CrossRef]
47. Liu, F.; Zhuang, P.; Anh, V.; Turner, I.; Burrage, K. Stability and convergence of the difference methods for the space–time fractional advection–diffusion equation. *Appl. Math. Comput.* **2007**, *191*, 12–20. [CrossRef]
48. Feng, L.; Zhuang, P.; Liu, F.; Turner, I.; Li, J. High-order numerical methods for the Riesz space fractional advection-dispersion equations. *arXiv* **2020**, arXiv:2003.13923.
49. Yang, Q.; Liu, F.; Turner, I. Numerical methods for fractional partial differential equations with Riesz space fractional derivatives. *Appl. Math. Model.* **2010**, *34*, 200–218. [CrossRef]
50. Shen, S.; Liu, F.; Anh, V.; Turner, I. The fundamental solution and numerical solution of the Riesz fractional advection–dispersion equation. *IMA J. Appl. Math.* **2008**, *73*, 850–872. [CrossRef]
51. Laub, A.J. *Matrix Analysis for Scientists and Engineers*; SIAM: Philadelphia, PA, USA, 2005; Volume 91.
52. Isaacson, E.; Keller, H.B. *Analysis of Numerical Methods*; Courier Corporation: Chelmsford, MA, USA, 2012.

Publisher's Note: MDPI stays neutral with regard to jurisdictional claims in published maps and institutional affiliations.

© 2020 by the authors. Licensee MDPI, Basel, Switzerland. This article is an open access article distributed under the terms and conditions of the Creative Commons Attribution (CC BY) license (http://creativecommons.org/licenses/by/4.0/).

Article

Intrinsic Discontinuities in Solutions of Evolution Equations Involving Fractional Caputo–Fabrizio and Atangana–Baleanu Operators

Christopher Nicholas Angstmann [1,*], Byron Alexander Jacobs [2] and Bruce Ian Henry [1] and Zhuang Xu [3]

1. School of Mathematics and Statistics, University of New South Wales, Sydney, NSW 2052, Australia; b.henry@unsw.edu.au
2. Department of Mathematics and Applied Mathematics, University of Johannesburg, Johannesburg 2006, South Africa; byronj@uj.ac.za
3. School of Physics, University of New South Wales, Sydney, NSW 2052, Australia; zhuang.xu@unsw.edu.au
* Correspondence: c.angstmann@unsw.edu.au

Received: 1 October 2020; Accepted: 11 November 2020; Published: 13 November 2020

Abstract: There has been considerable recent interest in certain integral transform operators with non-singular kernels and their ability to be considered as fractional derivatives. Two such operators are the Caputo–Fabrizio operator and the Atangana–Baleanu operator. Here we present solutions to simple initial value problems involving these two operators and show that, apart from some special cases, the solutions have an intrinsic discontinuity at the origin. The intrinsic nature of the discontinuity in the solution raises concerns about using such operators in modelling. Solutions to initial value problems involving the traditional Caputo operator, which has a singularity in its kernel, do not have these intrinsic discontinuities.

Keywords: Caputo–Fabrizio operator; Atangana–Baleanu operator; fractional falculus

MSC: 23A33

---

1. Introduction

In mathematical models based on evolution equations it is standard to restrict consideration to models whose solutions exist in a space of continuous functions. This is also true in cases of evolution equations with fractional order differential operators, such as the Riemann–Liouville operator [1] and the Caputo operator [2]. Both of these operators are based on integrals with power law kernels that are singular at the origin. In recent years, there has been a great deal of attention focussed on fractional differential operators based on integrals with non-singular kernels. Included in this are the Caputo–Fabrizio (CF) operator [3] and the Atangana–Baleanu in the sense of a Caputo (ABC) operator [4].

The CF and ABC operators have since been employed in numerous modelling applications, including applications to phase transitions [5], fluid and ground water flow [6], cancer treatment [7], and epidemiology [8], among others [9–12]. Many of the works in this field introduce the model evolution equations by the ad hoc "fractionalisation" of simply replacing integer order derivatives in traditional models with CF or ABC derivatives, without further phenomenological consideration.

On the theoretical side, there has been considerable effort devoted to understanding the interpretations of these differential operators. In a sequence of studies, Tarasov [13–15], Ortigueira and Machado [16], and Giusti [17] have shown that the CF differential operator should not be regarded as a fractional order operator and the ABC operator does not extend beyond the Caputo operator. Nevertheless, modelling applications still persist [8], as do numerical studies [18–20] and algebraic methods of solution [21]. We also note that, as presented by Hilfer and Luchko [22], there is no absolute agreement on a well defined set of properties to which a fractional derivative must adhere and it is not the aim of this work to interrogate such properties. A greater concern, expressed in the present work, is that CF and ABC operators are not generally suitable for modelling, whenever a solution is sought in a space of continuous functions. In particular, we show that we can construct well formed solutions to initial values problems (IVPs) with CF operators but the solutions have discontinuities at the origin. Many of the works discussed above make extensive use of integral transforms, and more specifically, the Laplace transform. A key point of the present contribution is highlighting that the Laplace transform is not bijective for the class of functions which solve IVPs under the CF operator and careful treatment of the solution near the initial condition is paramount.

The discontinuities in the solution of the IVPs necessitate the more general consideration of the derivatives in the definitions of both the CF and ABC operators. Such generalisations to distributional derivatives are well defined [23]. There is a large body of work concerned with the analysis of such derivatives, both for the case of the CF operator [24] and for the Riemann–Liouville, Caputo, and other fractional derivatives [25–27].

The remainder of this paper is organised as follows. We first construct solutions of CF IVPs and show that such solutions must, in general, feature a discontinuity at the origin. We then briefly discuss the impacts of these results on numerical methods for the solution of IVPs involving CF operators and point out a seemingly overlooked simple approach to the numerical evaluation of such equations. Next we repeat this treatment for the ABC operator and again show that solutions, in general, will have a discontinuity at the origin. Finally we consider a more traditional fractional derivative, the Caputo derivative, and show that solutions to Caputo IVPs can not feature such discontinuities.

2. Caputo–Fabrizio Operator

Definition 1 (The Caputo–Fabrizio operator). *The Caputo–Fabrizio (CF) operator is defined as [3]*,

$$^{CF}_{0}\mathcal{D}^{\alpha}_{t} u(t) = \frac{M(\alpha)}{1-\alpha} \int_0^t u'(\tau) \exp\left(-\frac{\alpha(t-\tau)}{1-\alpha}\right) d\tau, \tag{1}$$

for $0 \leq \alpha < 1$. Here $M(\alpha)$ is a weighting function such that $M(0) = M(1) = 1$.

It is typical and sufficient to take $M(\alpha) = 1$. The CF operator has been purported to be a fractional derivative when $0 < \alpha < 1$ which limits to an integer order derivative as $\alpha \to 1^-$ [3,28]. It should also be noted that the derivative in the integral may be considered in the distributional sense; for more information, see [24].

We will consider CF equations in the form of an IVP with

Definition 2 (A Caputo–Fabrizio Initial Value Problem). *A CF IVP is given by both a CF equation of the form*

$$^{CF}_{0}\mathcal{D}^{\alpha}_{t} u(t) = F(t), \tag{2}$$

with $F(t)$ a continuous function for $t \geq 0$ and an initial value,

$$u(0) = u_0, \tag{3}$$

with $u_0 \in \mathbb{R}$.

For now, it will suffice to say that we will consider solutions, $u(t)$, defined over the interval $t \in [0, \infty)$ without being overly concerned with the smoothness of such solutions other than asking for the derivative, at least in the distributional sense, to be well defined.

It should be noted that, in general, continuous solutions of this IVP do not exist [29]. Here we will consider a more general form of a solution by taking an ansatz such that

$$u'(t) = u_c'(t) + a\delta(t - 0^+), \tag{4}$$

where u_c is a continuous function, δ is a Dirac delta, and a is an unknown constant. Note that this form of a solution still permits a purely continuous form with $a = 0$. The form of this Dirac delta is chosen so that we have

$$\int_0^t \delta(\tau - 0^+)d\tau = H(t) = \begin{cases} 0 & t \leq 0, \\ 1 & t > 0. \end{cases} \tag{5}$$

With the ansatz in Equation (4) it follows that the solution will be of the form

$$u(t) = u(0) + u_c(t) - u_c(0) + aH(t). \tag{6}$$

We will further simplify this by taking $u_c(0) = u(0)$, to give

$$u(t) = u_c(t) + aH(t). \tag{7}$$

Care has to be taken here due to the fact that the Laplace transform is not bijective, i.e.,

$$\mathcal{L}^{-1}\{\mathcal{L}\{u(t)\}\} = u_c(t) + a \neq u(t). \tag{8}$$

As such, we can not rely on Laplace transform techniques to find solutions of this form.

Theorem 1. *For a CF IVP (Definition 2) assume that a solution in the form of the ansatz (Equation (7)) exists. Then such a solution is given by*

$$u(t) = u(0) + \frac{1-\alpha}{M(\alpha)}(F(t) - F(0)) + \frac{\alpha}{M(\alpha)}\int_0^t F(\tau)d\tau + \frac{(1-\alpha)F(0)}{M(\alpha)}H(t). \tag{9}$$

Proof. The solution is found by assuming that a solution in the form of the ansatz exists, substituting it into the IVP, and showing that the result is then consistent. To find the value of the unknown constant a from the ansatz we first substitute the Equation (4) into Equation (2), which gives

$$\frac{M(\alpha)}{1-\alpha}\int_0^t u_c'(\tau)\exp\left(-\frac{\alpha(t-\tau)}{1-\alpha}\right)d\tau + \frac{aM(\alpha)}{1-\alpha}\exp\left(-\frac{\alpha t}{1-\alpha}\right) = F(t) \tag{10}$$

for $t > 0$. Next we take the limit as t approaches 0 from above to give

$$\lim_{t\to 0^+}\frac{M(\alpha)}{1-\alpha}\int_0^t u_c'(\tau)\exp\left(-\frac{\alpha(t-\tau)}{1-\alpha}\right)d\tau + \lim_{t\to 0^+}\frac{aM(\alpha)}{1-\alpha}\exp\left(-\frac{\alpha t}{1-\alpha}\right) = \lim_{t\to 0^+}F(t) \tag{11}$$

and rearrange to find

$$a = \frac{(1-\alpha)F(0^+)}{M(\alpha)} = \frac{(1-\alpha)F(0)}{M(\alpha)} \tag{12}$$

as F is continuous.

To find the continuous part of this solution, we may differentiate the IVP, Equation (10), with respect to t:

$$\frac{M(\alpha)}{1-\alpha}u_c'(t) - \frac{\alpha M(\alpha)}{(1-\alpha)^2}\int_0^t u_c'(\tau)\exp\left(-\frac{\alpha(t-\tau)}{1-\alpha}\right)d\tau - \frac{a\alpha M(\alpha)}{(1-\alpha)^2}\exp\left(-\frac{\alpha t}{1-\alpha}\right) = F'(t). \tag{13}$$

From Equation (10) we also have

$$\int_0^t u_c'(\tau) \exp\left(-\frac{\alpha(t-\tau)}{1-\alpha}\right) d\tau = \frac{1-\alpha}{M(\alpha)} F(t) - a \exp\left(-\frac{\alpha t}{1-\alpha}\right). \tag{14}$$

Combining these two expressions gives an integer order differential equation for u_c,

$$u_c'(t) = \frac{\alpha}{M(\alpha)} F(t) + \frac{1-\alpha}{M(\alpha)} F'(t). \tag{15}$$

The integral form of this equation is

$$u_c(t) = u_c(0) + \frac{1-\alpha}{M(\alpha)} (F(t) - F(0)) + \frac{\alpha}{M(\alpha)} \int_0^t F(\tau) d\tau. \tag{16}$$

Hence the general form for the solution of the IVP is

$$u(t) = u(0) + \frac{1-\alpha}{M(\alpha)} (F(t) - F(0)) + \frac{\alpha}{M(\alpha)} \int_0^t F(\tau) d\tau + \frac{(1-\alpha)F(0)}{M(\alpha)} H(t). \tag{17}$$

□

From Equation (12) we see that the CF IVP only permits a continuous solution in the case where $F(0) = 0$. This requirement on the existence of continuous solutions has been noted in [29], where the result was obtained via Laplace transforms, and is the case considered in [30]. In all other cases the solution will involve both a continuous component and a step discontinuity at the origin.

This solution of the CF IVP can easily be alternatively verified by taking the CF operator of the solution to recover the original IVP,

$$\frac{M(\alpha)}{1-\alpha} \int_0^t u'(\tau) \exp\left(-\frac{\alpha(t-\tau)}{1-\alpha}\right) d\tau$$
$$= \int_0^t F'(\tau) \exp\left(-\frac{\alpha(t-\tau)}{1-\alpha}\right) d\tau$$
$$+ \frac{\alpha}{1-\alpha} \int_0^t F(\tau) \exp\left(-\frac{\alpha(t-\tau)}{1-\alpha}\right) d\tau + F(0) \exp\left(-\frac{\alpha t}{1-\alpha}\right) \tag{18}$$
$$= F(t).$$

Here we have simply applied integration by parts.

This general solution may alternatively be written as

$$u(t) = \begin{cases} u(0) & t = 0, \\ u(0) + \frac{1-\alpha}{M(\alpha)} F(t) + \frac{\alpha}{M(\alpha)} \int_0^t F(\tau) d\tau & t > 0. \end{cases} \tag{19}$$

It should be noted that this solution differs from the solution given in other papers, such as [28,31], although it is in agreement with the solution given in [17] for $t > 0$.

2.1. Weakening Continuity Requirements for $F(t)$

In the above we assumed that $F(t)$ was a continuous function. This is a little restrictive, as it excludes most of the interesting cases where $F(t)$ depends on the function $u(t)$, as $u(t)$ is not a continuous function. To accommodate a discontinuity at the origin we may remove the requirement that F is continuous and assume that we can write

$$F(t) = F_c(t) + bH(t), \tag{20}$$

where F_c is a continuous function and $b \in \mathbb{R}$. By following the same methodology as above, we have

$$\frac{M(\alpha)}{1-\alpha} \int_0^t u'_c(\tau) \exp\left(-\frac{\alpha(t-\tau)}{1-\alpha}\right) d\tau + \frac{aM(\alpha)}{1-\alpha} \exp\left(-\frac{\alpha t}{1-\alpha}\right) = F_c(t) + b \qquad (21)$$

for $t > 0$. By taking the limit from above as t tends to 0, we can find the unknown coefficient a.

$$a = \frac{(1-\alpha)(F_c(0)+b)}{M(\alpha)}. \qquad (22)$$

The continuous part of the solution will again be reduced to the solution of an ODE.

$$u'_c(t) = \frac{\alpha(F_c(t)+b)}{M(\alpha)} + \frac{1-\alpha}{M(\alpha)} F'_c(t). \qquad (23)$$

The general solution with a discontinuity at $t = 0$ for $F(t)$ is thus

$$u(t) = u(0) + \frac{1-\alpha}{M(\alpha)}(F_c(t) - F_c(0)) + \frac{\alpha}{M(\alpha)} \int_0^t (F_c(\tau)+b)d\tau + \frac{(1-\alpha)(F_c(0)+b)}{M(\alpha)} H(t). \qquad (24)$$

This solution is completely equivalent to the general solution given above in Equation (17). From this we see that the weakening of the continuity requirement did not effect the solution and we can attempt to solve IVPs of the form

$${}^{CF}_0 D^\alpha_t u(t) = F(u(t), t), \qquad (25)$$

with $u(0) = u_0$. In the case of IVPs of this form the given solution may only exist in the case that a, found via Equation (22), is real valued. For example with $F(u(t), t) = \frac{M(\alpha)}{1-\alpha} u(t)$, the relation in Equation (22) does not hold with $u_0 \neq 0$, and hence there is no solution of the ansatz form. Furthermore, for a non-linear equation, the resulting ODE for the continuous part of the solution, Equation (23), may not have solutions. Hence we must deal with each non-linear case individually.

2.2. Example Solutions for CF Initial Value Problems

We will construct some solutions of simple IVPs to illustrate the forms given above. In each case the validity of the solution can easily be seen by a direct substitution into the original equation.

2.2.1. Example CF IVP with $F(t) = 1$

Consider the CF IVP with

$${}^{CF}_0 D^\alpha_t u(t) = 1, \qquad (26)$$

and

$$u(0) = u_0. \qquad (27)$$

The solution of this IVP can be found directly via Equation (17) and is

$$u(t) = u_0 + \frac{\alpha}{M(\alpha)} t + \frac{1-\alpha}{M(\alpha)} H(t). \qquad (28)$$

Notice that the definition of $H(t)$ ensures that $u(0) = u_0$, but the solution has a step at $t = 0$. This is an illustrative example as it is simple to check against the definition of the CF operator.

2.2.2. Example CF IVP with $F(u(t), t) = -u_c(t)$

It is instructive to consider the differences induced by a discontinuity in F at $t = 0$. Here we present an example where F is taken to be the continuous part of the solution whilst in the next example we will show the case for F being the full solution.

Consider the CF IVP with

$$^{CF}_0\mathcal{D}^\alpha_t u(t) = -u_c(t), \qquad (29)$$

and

$$u(0) = u_0. \qquad (30)$$

This solution can be found first by solving the ODE for u_c, Equation (15), which can be rearranged to obtain,

$$u'_c(t) = -\frac{\alpha}{M(\alpha) + 1 - \alpha} u_c(t), \qquad (31)$$

subject to the initial condition $u_c(0) = u_0$. The continuous part of the solution is thus

$$u_c(t) = u_0 e^{-\frac{\alpha}{M(\alpha)+1-\alpha}t}. \qquad (32)$$

The discontinuous part of the solution is readily found from Equation (12) and combining the two will give the above solution. Thus the CF IVP has a solution of,

$$u(t) = u_0 e^{-\frac{\alpha}{M(\alpha)+1-\alpha}t} - \frac{(1-\alpha)u_0}{M(\alpha)} H(t). \qquad (33)$$

We can see that as $t \to \infty$ this solution changes sign and asymptotes to $-\frac{(1-\alpha)u_0}{M(\alpha)}$.

2.2.3. Example CF IVP with $F(u(t), t) = -u(t)$

Using the weakened form of the continuity requirement we can consider the IVP of the form

$$^{CF}_0\mathcal{D}^\alpha_t u(t) = -u(t), \qquad (34)$$

and

$$u(0) = u_0. \qquad (35)$$

As the right-hand side of this equation is dependent on the solution, we will first find the unknown coefficient from the ansatz via the relation given in Equation (22), with $b = a$ and $F_c(0) = u_c(0) = u_0$. This gives,

$$a = -\frac{(1-\alpha)u_0}{M(\alpha) + 1 - \alpha}. \qquad (36)$$

Following the same procedure as the previous example, we obtain the following ODE for the continuous part of the solution:

$$u'_c(t) = -\frac{\alpha}{M(\alpha) + 1 - \alpha}(u_c(t) + a). \qquad (37)$$

This can be solved with the initial condition $u_c(0) = u_0$ to give

$$u_c(t) = \frac{M(\alpha)u_0}{M(\alpha) + 1 - \alpha} e^{-\frac{\alpha}{M(\alpha)+1-\alpha}t} + \frac{(1-\alpha)u_0}{M(\alpha) + 1 - \alpha}. \qquad (38)$$

Again, combining the continuous and discontinuous parts of the solution will give the full solution,

$$u(t) = \frac{M(\alpha)u_0}{M(\alpha)+1-\alpha}e^{-\frac{\alpha}{M(\alpha)+1-\alpha}t} + \frac{(1-\alpha)u_0}{M(\alpha)+1-\alpha}(1-H(t)). \tag{39}$$

In contrast to the second example, this solution will remain positive, for $u_0 > 0$, and will asymptote to 0 as $t \to \infty$.

2.2.4. Example CF IVP with $F(u(t),t) = -\left(\frac{(1-\alpha)u_0^2}{M(\alpha)}\right)^2 - \frac{2(1-\alpha)u_0^2 u(t)}{M(\alpha)} - u(t)^2$

Again, using the weakened form of the continuity requirement, we can consider the IVP of the form

$${}_{0}^{CF}\mathcal{D}_t^\alpha u(t) = -\left(\frac{(1-\alpha)u_0^2}{M(\alpha)}\right)^2 - \frac{2(1-\alpha)u_0^2 u(t)}{M(\alpha)} - u(t)^2, \tag{40}$$

and

$$u(0) = u_0, \text{ with } u_0 < 0. \tag{41}$$

In this case we see that

$$F_c(t) + bH(t) = -\left(\frac{(1-\alpha)u_0^2}{M(\alpha)}\right)^2 - \frac{2(1-\alpha)u_0^2 u(t)}{M(\alpha)} - u(t)^2. \tag{42}$$

From the ansatz we also have

$$u(t) = u_c(t) + aH(t), \tag{43}$$

hence,

$$F_c(t) = -\left(\frac{(1-\alpha)u_0^2}{M(\alpha)}\right)^2 - \frac{2(1-\alpha)u_0^2 u_c(t)}{M(\alpha)} - u_c(t)^2 - 2a(u_c(t) - u_0), \tag{44}$$

and

$$b = -\frac{2(1-\alpha)u_0^2 a}{M(\alpha)} - a^2 - 2au_0. \tag{45}$$

From Equation (22) we then have the relation

$$a = \frac{(1-\alpha)}{M(\alpha)}\left(-\left(\frac{(1-\alpha)u_0^2}{M(\alpha)}\right)^2 - \frac{2(1-\alpha)u_0^3}{M(\alpha)} - u_0^2 - \frac{2(1-\alpha)u_0^2 a}{M(\alpha)} - a^2 - 2au_0\right). \tag{46}$$

This has solutions

$$a = -\frac{(1-\alpha)u_0^2}{M(\alpha)}. \tag{47}$$

From Equation (23) we have an ODE for the continuous part of the solution,

$$u_c'(t) = -\frac{\alpha(u_c(t)^2)}{M(\alpha)} + \frac{2(1-\alpha)}{M(\alpha)}u_c(t)u_c'(t). \tag{48}$$

This ODE has a solution,

$$u_c(t) = -\frac{M(\alpha)}{(1-\alpha)W_0\left(-\frac{M(\alpha)\exp\left(\frac{\alpha}{1-\alpha}t-\frac{M(\alpha)}{(1-\alpha)u_0}\right)}{(1-\alpha)u_0}\right)}, \tag{49}$$

where W_0 is a Lambert W function [32]. The full solution to the IVP is thus

$$u(t) = -\frac{M(\alpha)}{(1-\alpha)W_0\left(-\frac{M(\alpha)\exp\left(\frac{\alpha}{1-\alpha}t - \frac{M(\alpha)}{(1-\alpha)u_0}\right)}{(1-\alpha)u_0}\right)} - \frac{(1-\alpha)u_0^2}{M(\alpha)}H(t). \tag{50}$$

2.3. Numerical Considerations for Equations Involving the CF Operator

Equation (1) purports a memory effect by way of a convolution through time. Discretising the CF operator in this form leads to the unnecessary computation of memory terms. As is shown above, the solution to the IVP (2) may be obtained through the solution of the auxiliary ordinary differential Equation (15) (or in integral form Equation (17)).

The numerics contained within the recent literature are largely restricted to low order numerical methods. From the formulation presented in this work, we suggest that any numerical method appropriate for ODEs may be used to accurately solve IVPs with CF operators, and as such many efficient, highly accurate methods are available to these equations. To the best of the authors' knowledge no preceding work has proposed a numerical method which recovers discontinuous solutions to CF equations. As shown above these equations do not exhibit nontrivial continuous solutions, and as such require numerical methods tailored to recover the discontinuous dynamics.

3. The Atangana–Baleanu Operator

In a similar manner to the CF operator we will consider another non-singular kernel operator, the Atangana–Baleanu, in the sense of Caputo, (ABC) operator [4].

Definition 3 (The Atangana–Baleanu, in the sense of Caputo, Operator). *The ABC operator for $0 \leq \alpha < 1$ is defined as*

$$^{ABC}_0\mathcal{D}_t^\alpha u(t) = \frac{B(\alpha)}{1-\alpha}\int_0^t u'(\tau) E_\alpha\left(-\frac{\alpha(t-\tau)^\alpha}{1-\alpha}\right) d\tau, \tag{51}$$

where E_α is the Mittag-Leffler function, defined by:

$$E_\alpha(x) = \sum_{n=0}^\infty \frac{x^n}{\Gamma(\alpha n + 1)}. \tag{52}$$

Here $B(\alpha)$ is a normalisation constant, that must obey $B(0) = B(1) = 1$.

It is sufficient to take $B(\alpha) = 1$. The ABC operator is again often seen to be a fractional derivative as we recover an integer order derivative in the case $\alpha \to 1^-$. The use of the ABC operator as a fractional derivative is less contentious than the CF operator as it is non-local in time. As we are considering discontinuous solutions it is again necessary to interpret the derivative in the definition of the ABC operator in a distributional sense.

Again we will consider simple IVPs arising from this operator.

Definition 4 (An ABC Initial Value Problem). *An ABC IVP is given by both an ABC equation,*

$$^{ABC}_0\mathcal{D}_t^\alpha u(t) = F(t) \tag{53}$$

with $F(t)$ a continuous function in t, and an initial condition $u(0) = u_0$ for some $u_0 \in \mathbb{R}$.

To find a solution we consider the ansatz

$$u'(t) = u'_c(t) + a\delta(t - 0^+) \tag{54}$$

with the integral form of the solution

$$u(t) = u(0) + u_c(t) - u_c(0) + aH(t). \tag{55}$$

We will further simplify this by taking $u_c(0) = u(0)$, to give

$$u(t) = u_c(t) + aH(t). \tag{56}$$

Theorem 2. *For an ABC IVP (Definition 4) assume that a solution in the form of the ansatz (Equation (56)) exists. Then such a solution is given by*

$$u(t) = u(0) + \frac{1-\alpha}{B(\alpha)}(F(t) - F(0)) + \frac{\alpha}{B(\alpha)\Gamma(\alpha)} \int_0^t F(\tau)(t-\tau)^{\alpha-1} d\tau + \frac{(1-\alpha)F(0)}{B(\alpha)} H(t). \tag{57}$$

Proof. Combining Equation (53) and Equation (54) gives

$$\frac{B(\alpha)}{1-\alpha} \int_0^t u_c'(\tau) E_\alpha \left(-\frac{\alpha(t-\tau)^\alpha}{1-\alpha} \right) d\tau + \frac{aB(\alpha)}{1-\alpha} E_\alpha \left(-\frac{\alpha t^\alpha}{1-\alpha} \right) = F(t) \tag{58}$$

for $t > 0$. Taking the limit as t approaches 0 from above, we obtain

$$\lim_{t \to 0^+} \frac{B(\alpha)}{1-\alpha} \int_0^t u_c'(\tau) E_\alpha \left(-\frac{\alpha(t-\tau)^\alpha}{1-\alpha} \right) d\tau + \lim_{t \to 0^+} \frac{aB(\alpha)}{1-\alpha} E_\alpha \left(-\frac{\alpha t^\alpha}{1-\alpha} \right) = \lim_{t \to 0^+} F(t). \tag{59}$$

Thus the coefficient a is given by

$$a = \frac{(1-\alpha)F(0^+)}{B(\alpha)} = \frac{(1-\alpha)F(0)}{B(\alpha)}, \tag{60}$$

provided $F(x)$ is continuous.

As $u_c(t)$ is a continuous function and the Laplace transform is bijective over the space of continuous functions, Laplace transform techniques can be applied. Therefore we take the Laplace transform of both sides of Equation (58) from t to s domain to obtain

$$\frac{B(\alpha)}{1-\alpha} \frac{s^{\alpha-1}(s\mathcal{L}\{u_c(t)\} - u_c(0))}{s^\alpha + \frac{\alpha}{1-\alpha}} + \frac{aB(\alpha)}{1-\alpha} \frac{s^{\alpha-1}}{s^\alpha + \frac{\alpha}{1-\alpha}} = \mathcal{L}\{F(t)\}, \tag{61}$$

having used the result

$$\mathcal{L}\{t^{\alpha k+\beta-1} E_{\alpha,\beta}^{(k)}(yt^\alpha)\} = \frac{k! s^{\alpha-\beta}}{(s^\alpha - y)^{k+1}}, \quad \Re(s) > |y|^{1/\alpha}, \tag{62}$$

from [1]. By rearranging the equation above, we see that

$$s\mathcal{L}\{u_c(t)\} - u_c(0) = \frac{1-\alpha}{B(\alpha)}(s\mathcal{L}\{F(t)\} - F(0)) + \frac{\alpha}{B(\alpha)} s^{1-\alpha} \mathcal{L}\{F(t)\}. \tag{63}$$

To deal with the $s^{1-\alpha}\mathcal{L}\{F(t)\}$ on the RHS of Equation (63) we will utilise some results from fractional calculus. The Riemann–Liouville fractional derivative [1] of order $1 - \alpha$ with $0 \leq \alpha \leq 1$ is defined by

$$^{RL}_0\mathcal{D}^{1-\alpha}_t f(t) = \frac{1}{\Gamma(\alpha)} \frac{d}{dt} \int_0^t f(\tau)(t-\tau)^{\alpha-1} d\tau. \tag{64}$$

The Laplace transform of the Riemann–Liouville derivative is given by [33],

$$\mathcal{L}\{^{RL}_0\mathcal{D}^{1-\alpha}_t f(t)\} = s^{1-\alpha} \mathcal{L}\{f(t)\} + \lim_{t \to 0^+} {}^{RL}_0\mathcal{D}^{-\alpha}_t f(t), \tag{65}$$

where $^{RL}_0\mathcal{D}_t^{-\alpha}f(t)$ is a Riemann–Liouville fractional integral of order α. Furthermore, provided the limits exist, we also have [33,34],

$$\lim_{t \to 0^+} {}^{RL}_0\mathcal{D}_t^{-\alpha} f(t) = \lim_{t \to 0^+} \Gamma(1-\alpha) t^\alpha f(t), \tag{66}$$

and hence provided that $\lim_{t \to 0^+} F(t)$ exists we have

$$s^{1-\alpha}\mathcal{L}\{F(t)\} = \mathcal{L}\{{}^{RL}_0\mathcal{D}_t^{1-\alpha} F(t)\}. \tag{67}$$

The inverse Laplace transform of Equation (63) then gives the following ordinary integro-differential equation for $u_c(t)$:

$$\frac{d}{dt}u_c(t) = \frac{1-\alpha}{B(\alpha)}\frac{d}{dt}F(t) + \frac{\alpha}{B(\alpha)\Gamma(\alpha)}\frac{d}{dt}\int_0^t F(\tau)(t-\tau)^{\alpha-1}d\tau. \tag{68}$$

The integral form of the solution is obtained immediately from the equation above, the result is

$$u_c(t) = u_c(0) + \frac{1-\alpha}{B(\alpha)}(F(t) - F(0)) + \frac{\alpha}{B(\alpha)\Gamma(\alpha)}\int_0^t F(\tau)(t-\tau)^{\alpha-1}d\tau. \tag{69}$$

Hence the general form for the solution of the IVP is

$$u(t) = u(0) + \frac{1-\alpha}{B(\alpha)}(F(t) - F(0)) + \frac{\alpha}{B(\alpha)\Gamma(\alpha)}\int_0^t F(\tau)(t-\tau)^{\alpha-1}d\tau + \frac{(1-\alpha)F(0)}{B(\alpha)}H(t). \tag{70}$$

□

Again, we can alternatively verify this this solution by substituting the solution back to the ABC operator,

$$\frac{B(\alpha)}{1-\alpha}\int_0^t u'(\tau) E_\alpha\left(-\frac{\alpha(t-\tau)^\alpha}{1-\alpha}\right) d\tau$$
$$= F(0) E_\alpha\left(-\frac{\alpha t^\alpha}{1-\alpha}\right) + \int_0^t F'(\tau) E_\alpha\left(-\frac{\alpha(t-\tau)^\alpha}{1-\alpha}\right) d\tau \tag{71}$$
$$+ \frac{\alpha}{1-\alpha}\int_0^t \left({}^{RL}_0\mathcal{D}_t^{1-\alpha} F(t)\right) E_\alpha\left(-\frac{\alpha(t-\tau)^\alpha}{1-\alpha}\right) d\tau.$$

Note that both sides of Equation (71) are continuous; thus, the corresponding equation in Laplace space reads

$$\mathcal{L}\left\{\frac{B(\alpha)}{1-\alpha}\int_0^t u'(\tau) E_\alpha\left(-\frac{\alpha(t-\tau)^\alpha}{1-\alpha}\right) d\tau\right\}$$
$$= \frac{s^{\alpha-1}F(0)}{s^\alpha + \frac{\alpha}{1-\alpha}} + \frac{s^{\alpha-1}(s\mathcal{L}\{F(t)\} - F(0))}{s^\alpha + \frac{\alpha}{1-\alpha}} + \frac{\alpha}{1-\alpha}\frac{s^{\alpha-1}}{s^\alpha + \frac{\alpha}{1-\alpha}}\left(s^{1-\alpha}\mathcal{L}\{F(t)\}\right) \tag{72}$$
$$= \mathcal{L}\{F(t)\}.$$

We see that inverse Laplace transform of the equation above recovers the IVP:

$${}^{ABC}_0\mathcal{D}_t^\alpha u(t) = F(t). \tag{73}$$

We can note that the continuity requirement on $F(t)$ can be eased in the same manner as the CF operator by considering a discontinuous $F(t)$ such that $F(t) = F_c(t) + bH(t)$, where $F_c(t)$ is a continuous function. As such, we can attempt to consider IVPs of the form

$$^{ABC}_0\mathcal{D}^\alpha_t u(t) = F(u(t), t), \tag{74}$$

with $u(0) = u_0$. Again a solution of the ansatz form will only exist if a real valued constant a can be found such that the following relation holds,

$$a = \frac{(1-\alpha)(F_c(u_0, 0) + b)}{B(\alpha)}, \tag{75}$$

where F_c, a continuous function, and b, a real valued constant, are found from $F(u(t), t) = F_c(u(t), t) + bH(t)$.

3.1. Example ABC Initial Value Problems

3.1.1. Example ABC IVP with $F(t) = 1$

Consider the ABC IVP with

$$^{ABC}_0\mathcal{D}^\alpha_t u(t) = 1, \tag{76}$$

and

$$u(0) = u_0. \tag{77}$$

The solution of this IVP follows immediately from the Equation (70) and is

$$u(t) = u_0 + \frac{1}{B(\alpha)\Gamma(\alpha)} t^\alpha + \frac{1-\alpha}{B(\alpha)} H(t). \tag{78}$$

3.1.2. Example ABC IVP with $F(t) = u_c(t)$

Consider the ABC IVP with

$$^{ABC}_0\mathcal{D}^\alpha_t u(t) = u_c(t), \tag{79}$$

and

$$u(0) = u_0. \tag{80}$$

The solution can be found by independently calculating the continuous and discontinuous parts of the solution. The continuous part of the solution is found by first considering the equation for u_c in Laplace space, Equation (63), which can be rearranged to obtain

$$\mathcal{L}\{u_c(t)\} = \frac{s^{\alpha-1}}{s^\alpha - \frac{\alpha}{B(\alpha)-1+\alpha}} u_c(0). \tag{81}$$

The inverse Laplace transform from s to t domain then gives

$$u_c(t) = u_0 E_\alpha\left(\frac{\alpha}{B(\alpha)-1+\alpha} t^\alpha\right) \tag{82}$$

subject to the initial condition $u_c(0) = u_0$. The discontinuous part of the solution is found by calculating the coefficient a from Equation (60), giving

$$a = \frac{(1-\alpha)u_0}{B(\alpha)}. \tag{83}$$

The solution of the IVP is then as follows:

$$u(t) = u_0 E_\alpha\left(\frac{\alpha}{B(\alpha)-1+\alpha} t^\alpha\right) + \frac{(1-\alpha)u_0}{B(\alpha)} H(t). \tag{84}$$

3.1.3. Example ABC IVP with $F(u(t), t) = u(t)$

Using the weakened form of the continuity requirement we can consider the IVP of the form

$$^{ABC}_{0}D^\alpha_t u(t) = u(t), \qquad (85)$$

and

$$u(0) = u_0. \qquad (86)$$

Again the solution is found by considering the continuous and discontinuous parts separately. From Equation (60), we obtain

$$a = \frac{u_0(1-\alpha)}{B(\alpha) + \alpha - 1}. \qquad (87)$$

Subject to the initial condition $u_c(0) = u_0$. The continuous part of the solution can be found via its Laplace transform. Substituting the value for a into Equation (63) gives

$$\mathcal{L}(u_c(t)) = \frac{u_0 B(\alpha)}{B(\alpha) + \alpha - 1} \frac{s^{\alpha-1}}{s^\alpha - \frac{\alpha}{B(\alpha) + \alpha - 1}} - \frac{u_0(1-\alpha)}{B(\alpha) + \alpha - 1} s^{-1}. \qquad (88)$$

Inverting the Laplace transform then gives the continuous part of the solution,

$$u_c(t) = \frac{u_0 B(\alpha)}{B(\alpha) + \alpha - 1} E_\alpha \left(\frac{\alpha}{B(\alpha) - 1 + \alpha} t^\alpha \right) - \frac{u_0(1-\alpha)}{B(\alpha) + \alpha - 1}. \qquad (89)$$

Combining the continuous and discontinuous parts will then give the full solution,

$$u(t) = \frac{u_0 B(\alpha)}{B(\alpha) + \alpha - 1} E_\alpha \left(\frac{\alpha}{B(\alpha) - 1 + \alpha} t^\alpha \right) + \frac{u_0(1-\alpha)}{B(\alpha) + \alpha - 1} (H(t) - 1). \qquad (90)$$

4. Singular Kernel Operator Example: Caputo Derivative

Here we attempt to apply the same technique to a fractional derivative with a singular kernel, the Caputo derivative.

Definition 5 (Caputo Derivative). *The Caputo derivative of order α with $0 < \alpha < 1$, is defined as [2],*

$$^{C}_{0}D^\alpha_t u(t) = \frac{1}{\Gamma(1-\alpha)} \int_0^t u'(\tau)(t-\tau)^{-\alpha} d\tau. \qquad (91)$$

In order to explore the possibility of discontinuous solutions we will need to allow for the derivative in the Caputo definition to be interpreted in a distributional sense; see [25] for a more detailed exposition of the use of distributional derivatives in Caputo derivatives. We again consider a simple IVP.

Definition 6 (Caputo IVP). *A Caputo IVP is given by a Caputo equation,*

$$^{C}_{0}D^\alpha_t u(t) = F(t), \qquad (92)$$

with an initial condition, $u(0) = u_0$. We will assume that $F(t)$ a continuous function in t, and $u_0 \in \mathbb{R}$.

In a similar manner as above we consider an ansatz and look for solutions of the form

$$u(t) = u_c(t) + aH(t), \qquad (93)$$

with $u_c(t)$ a continuous function, a a constant, and $H(t)$ as defined in the previous sections.

Theorem 3. *For an Caputo IVP (Definition 6) assume that a solution in the form of the ansatz (Equation (93)) exists. Then such a solution does not possess a step discontinuity at $t = 0$.*

Proof. The proof follows from assuming a solution exists in the form of the ansatz and then showing that the only permitted value of the parameter a is zero. From the ansatz we have

$$u'(t) = u'_c(t) + a\delta(t - 0^+) \tag{94}$$

with $u'_c(t)$ being the derivative of a continuous function, δ a Dirac delta, and a some unknown constant. To obtain the value of the unknown constant we will substitute this into the IVP to give

$$\frac{1}{\Gamma(1-\alpha)} \int_0^t u'_c(\tau)(t-\tau)^{-\alpha} d\tau + \frac{a}{\Gamma(1-\alpha)} t^{-\alpha} = F(t). \tag{95}$$

Next we take the limit as $t \to 0^+$,

$$\lim_{t \to 0^+} \frac{1}{\Gamma(1-\alpha)} \int_0^t u'_c(\tau)(t-\tau)^{-\alpha} d\tau + \lim_{t \to 0^+} \frac{a}{\Gamma(1-\alpha)} t^{-\alpha} = F(0^+). \tag{96}$$

The left-hand side of this expression only exists if $a = 0$, whilst the right-hand side is well defined. As such solutions with a step discontinuity at the origin do not exist for the Caputo derivative. □

5. Conclusions

We have shown that the solutions of initial value problems using both the CF and ABC operators feature, in almost all cases, a discontinuity at the origin. The occurrence of the discontinuity is problematic for the application of the CF and ABC operators in modelling. Very few physical processes are well described by discontinuous functions, and fewer still with the discontinuity at the origin.

This discontinuity also raises issues with the use of these operators as fractional derivatives. Whilst both operators are generalisations of derivatives, in the sense that as $\alpha \to 1^-$ we recover an integer order derivative, the lack of smooth solutions is problematic. Many proponents of the use of these operators claim that the non-singular kernel is desirable, but as we have shown here in order for the solution of the IVP to exist the derivative of the solution must be singular, and thus the integrand is still singular.

In the Appendix A we have considered a more generalized ansatz for the solution of CF operator equations. This generalization does not give solutions to IVPs, but we can find solutions that diverge at the origin.

Traditional numerical methods are based on approximating the solution, and derivatives thereof, by their respective discrete counterparts. In the case where the solution exhibits a discontinuity, these approaches attempt to capture an infinite gradient in the same manner as the gradient of a smooth function. As such, resulting schemes are inadequate for capturing the dynamics of solutions admitted by the present IVPs. By decomposing the solution into discontinuous and continuous parts, traditional methods can be modified to approximate the continuous dynamics only.

In the case of the CF operator, the continuous part of the solution follows from a relatively simple integer order differential equation and the vast literature of methods are available to find efficient high order solutions. The numerical solution of the ABC operator equations is more complicated, as the operator does involve a history dependence.

In this work we have concentrated on the often explored case of $0 < \alpha < 1$, although extensions to the case $\alpha > 1$ are possible. Generalisations of the CF and ABC operators for larger values of α exist. To investigate the occurrence of discontinuities in such systems alternate forms of our ansatz would need to be taken.

We have also shown that this type of discontinuity at the origin can not occur in Caputo derivatives. The ansatz approach that we use is applicable to cases where we have derivatives appearing in integrands, such as the CF, ABC, and Caputo operators. This approach would need further modifications to be applicable to Riemann–Liouville type operators where the derivative occurs outside of the integral.

Author Contributions: Conceptualization, C.N.A., B.A.J., B.I.H. and Z.X.; formal analysis, C.N.A., B.A.J., B.I.H. and Z.X.; funding acquisition, C.N.A. and B.I.H.; investigation, C.N.A., B.A.J., B.I.H. and Z.X.; methodology, C.N.A., B.A.J. and B.I.H.; project administration, B.I.H.; supervision, C.N.A., B.A.J. and B.I.H.; writing—original draft, C.N.A., B.A.J., B.I.H. and Z.X.; writing—review editing, C.N.A., B.A.J., B.I.H. and Z.X. All authors have read and agreed to the published version of the manuscript.

Funding: This research was funded by the Australian Commonwealth Government ARC DP200100345. B.A.J. acknowledges support from the National Research Foundation of South Africa under grant number 129119.

Conflicts of Interest: The authors declare no conflict of interest.

Appendix A. Higher Order Singularities in the the Solution

We could consider a more generalised ansatz, with a higher order singularities,

$$u'(t) = u'_c(t) + \sum_{i=0}^{\infty} a_i \frac{d^i}{dt^i} \delta(t - 0^+), \tag{A1}$$

The solution in this case will then be of the form

$$u(t) = u_c(t) + a_0 H(t) + \sum_{i=0}^{\infty} a_{i+1} \frac{d^i}{dt^i} \delta(t - 0^+), \tag{A2}$$

where $\frac{d^i}{dt^i}\delta(t)$ denotes i-th derivative of the Dirac delta $\delta(t)$ with corresponding unknown constant a_i. Inserting this generalised ansatz into Equation (2), one finds

$$\frac{M(\alpha)}{1-\alpha} \int_0^t u'_c(\tau) \exp\left(-\frac{\alpha(t-\tau)}{1-\alpha}\right) d\tau + \frac{M(\alpha)}{1-\alpha} \sum_{i=0}^{\infty} a_i \left(\frac{\alpha}{1-\alpha}\right)^i \exp\left(-\frac{\alpha t}{1-\alpha}\right) = F_c(t) + b \tag{A3}$$

for $t > 0$. In the limit as $t \to 0$ from above, we have

$$\sum_{i=0}^{\infty} a_i \left(\frac{\alpha}{1-\alpha}\right)^i = \frac{(1-\alpha)(F_c(0^+) + b)}{M(\alpha)} = \frac{(1-\alpha)(F_c(0) + b)}{M(\alpha)}, \tag{A4}$$

since F_c is continous. Again, we make use of Leibniz's rule and differentiate Equation (A3) with respect to t; this gives

$$\frac{M(\alpha)}{1-\alpha} u'_c(t) - \frac{\alpha M(\alpha)}{(1-\alpha)^2} \int_0^t u'_c(\tau) \exp\left(-\frac{\alpha(t-\tau)}{1-\alpha}\right) d\tau - \frac{\alpha M(\alpha)}{(1-\alpha)^2} \sum_{i=0}^{\infty} a_i \left(\frac{\alpha}{1-\alpha}\right)^i \exp\left(-\frac{\alpha t}{1-\alpha}\right) = F'_c(t). \tag{A5}$$

From Equation (A3), one finds

$$\int_0^t u'_c(\tau) \exp\left(-\frac{\alpha(t-\tau)}{1-\alpha}\right) d\tau = \frac{(1-\alpha)(F_c(t) + b)}{M(\alpha)} - \sum_{i=0}^{\infty} a_i \left(\frac{\alpha}{1-\alpha}\right)^i \exp\left(-\frac{\alpha t}{1-\alpha}\right). \tag{A6}$$

Replacing the integral in Equation (A5) with RHS of Equation (A6) and rearranging yields

$$u'_c(t) = \frac{\alpha(F_c(t) + b)}{M(\alpha)} + \frac{1-\alpha}{M(\alpha)} F'_c(t). \tag{A7}$$

The integral solution of Equation (A7) is given as

$$u_c(t) = u_c(0) + \frac{1-\alpha}{M(\alpha)} (F_c(t) - F_c(0)) + \frac{\alpha}{M(\alpha)} \int_0^t (F_c(\tau) + b) d\tau. \tag{A8}$$

Hence the general form of the unbounded solution takes the following form

$$u(t) = u_c(0) + \frac{1-\alpha}{M(\alpha)}(F_c(t) - F_c(0)) + \frac{\alpha}{M(\alpha)}\int_0^t (F_c(\tau) + b)d\tau + a_0 H(t) + \sum_{i=0}^{\infty} a_{i+1}\frac{d^i}{dt^i}\delta(t - 0^+). \quad (A9)$$

Therefore, we see that the unbounded solution is non-unique with the only condition for the coefficients $a_i's$ given by Equation (A4). Note that generally the initial value $u(0) = u_0$ can not be imposed for solution of this form, as it involves a delta function and its distributional derivatives at the origin, unless we force the unknown coefficients $a_i = 0$ for $i \in \mathbb{Z}^+$. If we set $\{a_1, a_2, ...\} = 0$, from Equation (A4), it is then required that

$$a_0 = \frac{(1-\alpha)(F_c(0) + b)}{M(\alpha)}. \quad (A10)$$

This allows us to impose the initial condition $u(0) = u_0$, and the solution for the IVP in this case reads

$$u(t) = u(0) + \frac{1-\alpha}{M(\alpha)}(F_c(t) - F_c(0)) + \frac{\alpha}{M(\alpha)}\int_0^t (F_c(\tau) + b)d\tau + \frac{(1-\alpha)(F_c(0) + b)}{M(\alpha)}H(t). \quad (A11)$$

From this we see that higher order discontinuities at the origin can still produce solutions to the equation, but do not provide solutions for an IVP.

References

1. Podlubny, I. Fractional Differential Equations: An Introduction to Fractional Derivatives, Fractional Differential Equations, to Methods of Their Solution and Some of Their Applications. In *Mathematics in Science and Engineering*; Elsevier: Amsterdam, The Netherlands, 1999; Volume 198.
2. Caputo, M. Linear Models of Dissipation whose Q is almost Frequency Independent—II. *Geophys. J. Int.* **1967**, *13*, 529–539. [CrossRef]
3. Caputo, M.; Fabrizio, M. A new Definition of Fractional Derivative without Singular Kernel. *Prog. Fract. Differ. Appl.* **2015**, *1*, 73–85. [CrossRef]
4. Atangana, A.; Baleanu, D. New fractional derivatives with nonlocal and non-singular kernel: Theory and application to heat transfer model. *Therm. Sci.* **2016**, *20*, 763–769. [CrossRef]
5. Algahtani, O.J.J. Comparing the Atangana-Baleanu and Caputo–Fabrizio derivative with fractional order: Allen Cahn model. *Chaos Solitons Fractals* **2016**, *89*, 552–559. [CrossRef]
6. Atangana, A.; Baleanu, D. Caputo-Fabrizio derivative applied to groundwater flow within confined aquifer. *J. Eng. Mech.* **2017**, *143*, D4016005. [CrossRef]
7. Dokuyucu, M.A.; Celik, E.; Bulut, H.; Baskonus, H.M. Cancer treatment model with the Caputo-Fabrizio fractional derivative. *Eur. Phys. J. Plus* **2018**, *133*, 1–6. [CrossRef]
8. Baleanu, D.; Jajarmi, A.; Mohammadi, H.; Rezapour, S. A new study on the mathematical modelling of human liver with Caputo-Fabrizio fractional derivative. *Chaos Solitons Fractals* **2020**, *134*, 109705. [CrossRef]
9. Atangana, A.; Alkahtani, B.S.T. New model of groundwater flowing within a confine aquifer: Application of Caputo-Fabrizio derivative. *Arab. J. Geosci.* **2016**, *9*, 8. [CrossRef]
10. Ali, F.; Saqib, M.; Khan, I.; Sheikh, N.A. Application of Caputo-Fabrizio derivatives to MHD free convection flow of generalized Walters'-B fluid model. *Eur. Phys. J. Plus* **2016**, *131*, 377. [CrossRef]
11. Goufo, E.F.D. Application of the Caputo-Fabrizio Fractional Derivative without Singular Kernel to Korteweg-de Vries-Burgers Equation. *Math. Model. Anal.* **2016**, *21*, 188–198. [CrossRef]
12. Qureshi, S.; Yusuf, A. Modeling chickenpox disease with fractional derivatives: From caputo to atangana-baleanu. *Chaos Solitons Fractals* **2019**, *122*, 111–118. [CrossRef]
13. Tarasov, V.E. No violation of the Leibniz rule. No fractional derivative. *Commun. Nonlinear Sci. Numer. Simul.* **2013**, *18*, 2945–2948. [CrossRef]
14. Tarasov, V.E. On chain rule for fractional derivatives. *Commun. Nonlinear Sci. Numer. Simul.* **2016**, *30*, 1–4. [CrossRef]

15. Tarasov, V.E. No nonlocality. No fractional derivative. *Commun. Nonlinear Sci. Numer. Simul.* **2018**, *62*, 157–163. [CrossRef]
16. Ortigueira, M.D.; Machado, J.T. A critical analysis of the Caputo-Fabrizio operator. *Commun. Nonlinear Sci. Numer. Simul.* **2018**, *59*, 608–611. [CrossRef]
17. Giusti, A. A comment on some new definitions of fractional derivative. *Nonlinear Dyn.* **2018**, *93*, 1757–1763. [CrossRef]
18. Owolabi, K.M.; Atangana, A. Analysis and application of new fractional Adams-Bashforth scheme with Caputo-Fabrizio derivative. *Chaos Solitons Fractals* **2017**, *105*, 111–119. [CrossRef]
19. Qureshi, S.; Rangaig, N.A.; Baleanu, D. New numerical aspects of Caputo-Fabrizio fractional derivative operator. *Mathematics* **2019**, *7*, 374. [CrossRef]
20. Gao, W.; Ghanbari, B.; Baskonus, H.M. New numerical simulations for some real world problems with Atangana-Baleanu fractional derivative. *Chaos Solitons Fractals* **2019**, *128*, 34–43. [CrossRef]
21. Baleanu, D.; Mohammadi, H.; Rezapour, S. A fractional differential equation model for the COVID-19 transmission by using the Caputo-Fabrizio derivative. *Adv. Differ. Equ.* **2020**, *2020*, 299. [CrossRef]
22. Hilfer, R.; Luchko, Y. Desiderata for fractional derivatives and integrals. *Mathematics* **2019**, *7*, 149. [CrossRef]
23. Gel'fand, I.M.; Shilov, G.E. *Generalized Functions, Volume 1: Properties and Operations*; AMS Chelsea Publishing: New York, NY, USA, 1964.
24. Atanacković, T.M.; Pilipović, S.; Zorica, D. Properties of the Caputo-Fabrizio fractional derivative and its distributional settings. *Fract. Calc. Appl. Anal.* **2018**, *21*, 29–44. [CrossRef]
25. Li, C. Several Results of Fractional Derivatives in $D'(R_+)$. *Fract. Calc. Appl. Anal.* **2015**, *18*, 192–207. doi:10.1515/fca-2015-0013. [CrossRef]
26. Li, C.; Li, C.; Clarkson, K. Several Results of Fractional Differential and Integral Equations in Distribution. *Mathematics* **2018**, *6*, 97. [CrossRef]
27. Morales, M.G.; Došlá, Z.; Mendoza, F.J. Riemann-Liouville derivative over the space of integrable distributions. *Electron. Res. Arch.* **2020**, *28*, 567. [CrossRef]
28. Losada, J.; Nieto, J.J. Properties of a New Fractional Derivative without Singular Kernel. *Prog. Fract. Differ. Appl.* **2015**, *1*, 87–92. [CrossRef]
29. Capelas de Oliveira, E.; Jarosz, S.; Vaz, J., Jr. On the mistake in defining fractional derivative using a non-singular kernel. *arXiv* **2020**, arXiv:1912.04422v3.
30. Caputo, M.; Fabrizio, M. Applications of New Time and Spatial Fractional Derivatives with Exponential Kernels. *Prog. Fract. Differ. Appl.* **2015**, *2*, 1–11. [CrossRef]
31. Shaikh, A.; Tassaddiq, A.; Nisar, K.S.; Baleanu, D. Analysis of differential equations involving Caputo-Fabrizio fractional operator and its applications to reaction-diffusion equations. *Adv. Differ. Equ.* **2019**, *2019*, 178. [CrossRef]
32. Corless, R.M.; Gonnet, G.H.; Hare, D.E.G.; Jeffrey, D.J.; Knuth, D.E. On the Lambert W function. *Adv. Comput. Math.* **1996**, *5*, 329–359. [CrossRef]
33. LI, C.; Qian, D.; Chen, Y. On Riemann-Liouville and Caputo Derivatives. *Discret. Dyn. Nat. Soc.* **2011**, *2011*, 562494. [CrossRef]
34. Zhang, S. Monotone iterative method for initial value problem involving Riemann–Liouville fractional derivatives. *Nonlinear Anal. Theory Methods Appl.* **2009**, *71*, 2087–2093. [CrossRef]

Publisher's Note: MDPI stays neutral with regard to jurisdictional claims in published maps and institutional affiliations.

© 2020 by the authors. Licensee MDPI, Basel, Switzerland. This article is an open access article distributed under the terms and conditions of the Creative Commons Attribution (CC BY) license (http://creativecommons.org/licenses/by/4.0/).

MDPI
St. Alban-Anlage 66
4052 Basel
Switzerland
Tel. +41 61 683 77 34
Fax +41 61 302 89 18
www.mdpi.com

Mathematics Editorial Office
E-mail: mathematics@mdpi.com
www.mdpi.com/journal/mathematics

www.ingramcontent.com/pod-product-compliance
Lightning Source LLC
LaVergne TN
LVHW070048120526
838202LV00101B/1845